# Energy Harvesting and Storage Devices

The book discusses the materials, devices, and methodologies that can be used for energy harvesting including advanced materials, devices, and systems. It describes synthesis and fabrication details of energy storage materials. It explains use of high-energy density thin films for future power systems, flexible and biodegradable energy storage devices, fuel cells and supercapacitors, nanogenerators for self-powered systems, and innovative energy harvesting methodologies.

Features:

- Covers all relevant topics in energy harvesting research and focuses on the current state-of-the-art techniques and materials for this application.
- Showcases the true potential of the nature in energy harvesting industry by discussing various harvesting mechanisms based on renewable and sustainable energy sources.
- Explains the recent trends in flexible and wearable energy storage devices that are currently being used in IoT-based smart devices.
- Overviews of the state-of-the-art research performed on design and development of energy harvesting devices.
- Highlights the interdisciplinary research efforts needed in energy harvesting and storage devices to transform conceptual ideas to working prototypes.

This book is aimed at graduate students and researchers in emerging materials, energy engineering, including harvesting and storage.

# Energy Harvesting and Storage Devices

## Sustainable Materials and Methods

Edited by
Laxman Raju Thoutam, J. Ajayan, and D. Nirmal

CRC Press
Taylor & Francis Group
Boca Raton London New York

CRC Press is an imprint of the
Taylor & Francis Group, an **informa** business

Cover design image: shutterstock

First edition published 2024
by CRC Press
2385 NW Executive Center Drive, Suite 320, Boca Raton, FL 33431

and by CRC Press
4 Park Square, Milton Park, Abingdon, Oxon, OX14 4RN

*CRC Press is an imprint of Taylor & Francis Group, LLC*

ISBN: 9781032375083 (hbk)
ISBN: 9781032375090 (pbk)
ISBN: 9781003340539 (ebk)

DOI: 10.1201/9781003340539

Typeset in Times
by Newgen Publishing UK

# Contents

# Preface

The human life strives to scout the unknowns of the universe from different perspectives, ranging from atomic scale to space exploration. These endeavors require tremendous amounts of energy to discover new and novel phenomena that help to uncover the mysteries of the universe. On the other hand, it is estimated that the rise in population and increased use of automation tools will lead to an unprecedented energy-crisis that might topple world economies. The advent of semiconductor technology in the recent past to be on a par with Moore's law has made devices smaller (nano-scale) with high-precision speeds, however, its technology roadmap is limited by power handling issues at such small scales. Global initiatives are put into practice to design and develop myriad sustainable energy harvesting mechanisms and corresponding energy storage devices to hinder the energy-crisis. The fundamental principle that governs the energy harvesting mechanism is "energy is neither created nor destroyed but can be converted from one form to another." Energy harvesting techniques have the capability to generate energy from different sources, including thermal, wind, solar, vibrations, mechanical motions, and electromagnetic radiations. The vast applicability and abundance of these resources has made the energy harvesting industry boom in the last decade and it grew from $125 million in 2010 to $5 billion in 2020. However, energy harvesting is a tedious task and is not trivial.

The book discusses the fundamentals of materials, devices, and methodologies that can be used for energy harvesting. It provides a comprehensive coverage of the advanced materials, devices, and systems that can be used for energy harvesting; synthesis and fabrication details of energy storage materials; recent developments in solar cells. With the advent of the Internet of Things (IoT) in all societal and economic sectors, including agriculture, healthcare, infrastructure, manufacturing, healthcare, and wearable sensors, powering up of different devices to communicate requires sustainable and reliable energy sources. Much of the present IoT-based devices rely on battery-operated devices for its small size, cost, and convenience. However, this requires a major overhaul as we continue to investigate the safer and green technologies for future generations. The book describes the use of high-energy density thin films for future power systems, flexible and biodegradable energy storage devices, fuel cells and supercapacitors, and innovative energy harvesting methodologies that can replace batteries and potentially lead to sustainable and efficient energy cycles.

The book tries to put together different perspectives from scientists and engineers in the design and development of novel and sustainable energy harvesting and storage devices. The book gives a detailed overview in making of sustainable energy harvesting device that include material selection, design, synthesis, and fabrication coupled with efficient storage techniques. The book highlights the interdisciplinary research efforts needed in energy harvesting and storage devices to transform conceptual ideas to working prototypes.

# About the Editors

**Laxman Raju Thoutam** is Assistant Professor at Amrita School of Nanosciences and Molecular Medicine, Amrita Vishwa Vidypeetham, Kochi, India. Dr. Thoutam has more than 10 years of cleanroom fabrication experience in the design and development of novel micron and nanoscale devices and heterostructures for functional applications. Dr. Thoutam did his postdoctoral research on the synthesis and characterization of high-quality complex oxide thin films and heterostructures for functional device applications at University of Minnesota, Minneapolis, USA. He received his Ph.D. in Nanoscience (2016) and Master of Science in Electrical Engineering (2011) from Northern Illinois University, DeKalb, Illinois, USA. He serves as a referee for many international repute journals and has worked as a committee member for international materials science conferences and workshops. His primary research focuses on exploring the structure–property relationships of low-dimensional material systems. His present research area includes implementation of novel electrostatic gating techniques to different material systems; analysis of fundamental electronic transport in semimetals, complex oxide perovskite thin films and heterostructures, wide band gap semiconductors, and low-dimensional materials.

**J. Ajayan** received his B.Tech Degree in Electronics and Communication Engineering from Kerala University in 2009, M.Tech and Ph.D. Degree in Electronics and Communication Engineering from Karunya University, Coimbatore, INDIA, in 2012 and 2017, respectively. He is an Associate Professor in the Department of Electronics and Communication Engineering at SR University, Telangana, India. He has published more than 150 research articles (including 70 SCI articles) in various journals and international conferences. He has published three books, more than 20 book chapters, and 3 patents. He is a reviewer of more than 30 journals in various publishers. He is a Guest Editor with various journals. He has served more than 10 conferences on the technical advisory/reviewer committee. His areas of interest are microelectronics, semiconductor devices, nanotechnology, RF-integrated circuits, and photovoltaics.

**D. Nirmal** is Full Professor and Head of Electronics and Communication Engineering, Karunya Institute of Technology and Sciences. He specialized in VLSI Design after his Bachelor of Engineering and received his Ph.D in Information and Communication Engineering from Anna University. His research interests include nanoelectronics, GaN technology, device and circuit simulation – GSL, sensors, HEMT, beyond 5G, and nanoscale device design and modeling. He is a founding chair of IEEE Electron Device Society Coimbatore chapter and volunteered for several committees in IEEE. He has a funding project tune of 1.2 Cores from various agencies like DRDO (Defence Research and Development Organization), Ministry of Electronics and Information Technology, ISRO (Indian Space Research Organization), and AICTE (All India Council of Technical Education). He is a recipient of various awards namely IEI-Young Engineer award, IETE Smt. Manorama Rathore memorial award 2022 from IETE and Young Scientist Award 2019 from the Academy of Sciences. Professor Nirmal has made more than 150+ peer reviewed research publications and three patents to his credits. He is also a Senior IEEE member. He has delivered many keynote talks and lectures in national and international level conferences/faculty development programs.

# Contributors

**Uday Kumar Adapa** Department of Metallurgical Engineering, University College of Engineering, Science & Technology Hyderabad. JNTUH, Kukatpally, Hyderabad, Telangana, India

**Amala Mithin Minther Singh Amirthaiah** Department of Mechanical Engineering, DMI College of Engineering, Chennai, India

**Manju Arora** CSIR-National Physical Laboratory, Dr. K.S. Krishnan Marg, New Delhi, India

**Arun Banotra** Department of Nanosciences and Materials, Central University of Jammu, J&K, India

**Shilpa Chakra Chidurala** Center for Nano Science and Technology, University College of Engineering, Science & Technology Hyderabad. JNTUH, Kukatpally, Hyderabad, Telangana, India

**Meenakshi Dhiman** Chitkara University Institute of Engineering and Technology, Chitkara University, Punjab, India

**Anushya Ganesan** Department of Physics,St. Joseph College of Engineering, Chennai, India

**Josep M. Guerrero** Villum Investigator Center for Research on Microgrids – CROM Department of Energy Technology – Aalborg University, Denmark

**Mohanraj Jayavelu** Vel Tech Rangarajan Dr.Sagunthala R&D Institute of Science and Technology, Chennai, India

**Srilatha K.** ACE Engineering College, Hyderabad, India

**Kalai Kumar Kamaraj** Department of Computer Science and Engineering, DMI College of Engineering, Chennai, India

**Ramkumar Kannan** Mother Theresa Institute of Engineering and Technology, Palamaner, Andhra Pradesh, India

**Baljinder Kaur** Chitkara University Institute of Engineering and Technology, Chitkara University, Punjab, India

**Balwinder Kaur** Department of Physics, Govt. Degree College, R.S. Pura, Jammu, India

**Saleem Khan** Department of Nanosciences and Materials, Central University of Jammu, J&K, India

**Shirisha Konda** Center for Nano Science and Technology, University College of Engineering, Science & Technology Hyderabad. JNTUH, Kukatpally, Hyderabad, Telangana, India

**Maneesh Kumar** Department of Electrical Engineering, Indian Institute of Technology, Roorkee, India

**Ashok Mahalingam** New Generation Materials Laboratory, Department of Physics, National Institute of Technology, Tiruchirappalli, Tamil Nadu, India

**Vaishali Misra**   Department of Nanosciences and Materials, Central University of Jammu, J&K, India

**Valliammai Muthuraman**   Vel Tech Rangarajan Dr.Sagunthala R&D Institute of Science and Technology, Chennai, India

**Vinodhkumar Nallathambi**   Vel Tech Rangarajan Dr.Sagunthala R&D Institute of Science and Technology, Chennai, India

**Amin Nozariasbmarz**   Department of Materials Science and Engineering, Pennsylvania State University, University Park, PA, USA

**Bed Poudel**   Department of Materials Science and Engineering, Pennsylvania State University, University Park, PA, USA

**Shashank Priya**   Department of Materials Science and Engineering, Pennsylvania State University, University Park, PA, USA

**Sammaiah Pulla** Department of Mechanical Engineering, SR University, Warangal, Telangana, India

**Sarika Raj** Department of Chemistry, Fatima College (Autonomous), Madurai, India

**Madhuri Sakaray**   Center for Nano Science and Technology, University College of Engineering, Science & Technology Hyderabad. JNTUH, Kukatpally, Hyderabad, Telangana, India

**Sachidananda Sen**   Department of Electrical and Electronics Engineering, SR University, Warangal, India

**Archna Sharma**   Department of Physics, University of Jammu, Jammu, J&K (UT), India

**Ajay Singh**   Department of Physics, GGM Science College (Constituent College of cluster University of Jammu) Canal Road, Jammu, India

**Saurabh Singh** Department of Materials Science and Engineering, Pennsylvania State University, University Park, PA, USA

**Vishal Singh** Department of Nanosciences and Materials, Central University of Jammu, J&K, India

**Ananthakumar Soosaimanickam**   R&D Division, Intercomet S.L, Madrid, Spain

**Moorthy Babu Sridharan** Crystal Growth Centre, Anna University, Chennai, India

**Saravanan Krishna Sundaram**   Sri Sairam Institute of Technology, Chennai, India

**Tsunehiro Takeuchi** Research Center for Smart Energy Technology, Toyota Technological Institute, Nagoya 468-8511, Japan; CREST, Japan Science and Technology Agency, Tokyo, Japan

**Rakesh Kumar Thida**   Center for Nano Science and Technology, University College of Engineering, Science & Technology Hyderabad. JNTUH, Kukatpally, Hyderabad, Telangana, India

**Laxman Raju Thoutam** Amrita School of Nanosciences and Molecular Medicine, Amrita Vishwa Vidyapeetham, Kochi, Kerala, India

**Bala Narasaih Tumma** Department of Chemical Engineering, University College of Engineering, Science & Technology Hyderabad. JNTUH, Kukatpally, Hyderabad, Telangana, India

**Shreya Tumma** Department of Electrical Engineering, Indian Institute of Technology, Tirupati, India

**Divya Velpula** Center for Nano Science and Technology, University College of Engineering, Science & Technology Hyderabad. JNTUH, Kukatpally, Hyderabad, Telangana, India

**Shunli Wang** Southwest University of Science and Technology, Aalborg University, China

**Biny R. Wiston** New Generation Materials Laboratory, Department of Physics, National Institute of Technology, Tiruchirappalli, Tamil Nadu, India

**Yanxin Xie** Southwest University of Science and Technology, China

**Manisha Yadav** Department of Nanosciences and Materials, Central University of Jammu, J&K, India

**Xiaoyong Yang** Southwest University of Science and Technology, China

**Chunmei Yu** Southwest University of Science and Technology, China

# 1 Piezoelectric Materials for Energy Harvesting Applications

*Laxman Raju Thoutam and Sammaiah Pulla*

## 1.1 INTRODUCTION

Intelligent or smart materials are the need of the hour for the design of efficient energy harvesting and storage systems to cater energy needs of future generations. The physical and chemical functionality of the smart materials vary in response to external stimuli like temperature, pressure, electric field, magnetic field, light, pH, electrolyte/chemical solutions, mechanical vibrations, and environmental changes. In general, it is expected that smart materials tend to change their intrinsic material properties, including its shape, size, stiffness, damping, and color upon external perturbations. The dynamic responses of the smart materials are then tailored to convert one form of energy to another for useful practical applications. The innate responses of the smart materials to external stimuli are highly reversible in nature that aid in the design and development of high-yield large-scale reliable commercial transducers for different energy harvesting and storage systems. The self-healing and self-adaptability properties of the smart materials excellently fit into healthcare, space, and military applications. The rapid advancement and recent technological breakthroughs in the synthesis, characterization, and fabrication techniques in the fields of materials science and nanotechnology further accentuated the use of smart materials in micro- and nanoelectromechanical systems (MEMS and NEMS) for sophisticated energy harnessing and storage systems. The recent surge in the use of the Internet of Things (IoT) enabled consumer devices into everyday lives, augmenting the extensive use of self-sustainable hand-held portable low-power electronic devices. This demand has triggered the usage of smart materials in power, energy, and consumer electronics industries. It is expected that the smart materials industry will reach a high market value of $73 billion by 2022 indicating its vast potential affecting global economies [1]. The salient features of smart materials, including the immediate response, self-actuation, high selectivity, locally directive response make them an ideal choice for the design of next-generation energy-efficient devices [2]. Piezoelectric materials, are a special class of smart materials that are widely explored for energy storage and conversion applictions. The chapter briefly discusses the role of different piezoelectric materials in energy harvesting and storage mechanisms. The chapter highlights the salient features of piezoelectric materials and outlines different methodologies to increase the piezoelectric properties to achieve higher output energy conversion efficiencies. The chapter extensively discusses the physical mechanisms and applications of single crystal, polycrystal, polymers, and composite piezoelectric materials for myriad energy scavenging applications.

DOI: 10.1201/9781003340539-1

## 1.2 PIEZOELECTRIC MATERIALS

Piezoelectricity refers to the ability of a material to generate electricity upon application of external mechanical stress or tension. In other words, piezoelectric material can convert mechanical energy into electrical energy and vice versa. Piezoelectricity is intrinsic to a material and its origin stems from ionically bonded positive and negative ions in its internal atomic structure. With no external force applied, the positive and negative ions are randomly scattered inside the material yielding net zero charge and thus maintaining charge neutrality. However, application of external mechanical stress on it, will force the ions inside the materials to align in an orderly fashion to create net positive and negative charge resulting in net electric dipole indicating dielectric piezoelectricity. The amount of charge depends on the applied external mechanical stress on the material as depicted in Figure 1.1.

In breif, a neutral material is converted into a charged material and this charge can be used to power up or trigger an external load circuit for useful energy-based sensing or power applications. It is also to be noted that the converse effect, i.e., internal atomic structure changes upon the application of external charge indicating inverse piezoelectricity effect. The shape and size of the material deforms, i.e., compresses or elongates with external applied field as depicted in Figure 1.2. The deformation process inside the material helps to convert applied electrical energy to mechanical energy and is a reversible process, i.e., material retains its original shape after the removal of external applied electrical energy.

Piezoelectric materials have the unique capability of harvesting energy from ambient atmospheric conditions that decreases the dependency of batteries for power generation. Piezoelectric energy harvesting typically harnesses microwatt to milliwatt power that has the potential to energize modern low-power electronics [3]. Piezoelectric energy harvesting offers persistent and continuous energy supply when compared to conventional photovoltaic energy sources. The continuous energy generation of the piezoelectric materials depends on a system's initial mechanical energy of the system, which primarily depends on its nominal working conditions and does not depend on varying external conditions. The piezoelectric energy transducers can be incorporated into natural habitats or manmade structures like buildings, bridges, movable parts, space stations and even in the human body [4–6]. The innovations and advancements in materials science research over the past few decades have largely helped the piezoelectric materials to be available in large-sized single crystals to small-sized thin films or coatings for functional energy scavenging applications. The use of advanced material synthesis and characterization techniques with controlled defect density has

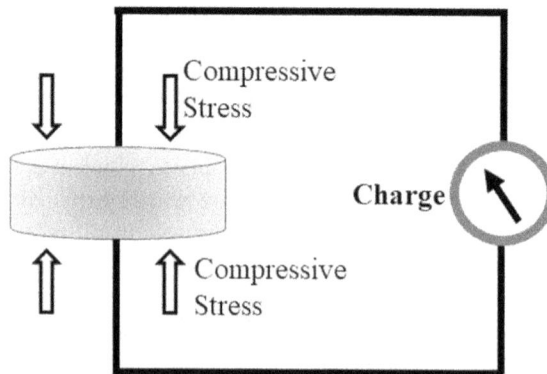

**FIGURE 1.1** Illustration of direct piezoelectric effect. A compressive stress (indicated by vertical arrow marks) is applied on the circular piezoelectric material that produces a corresponding charge in the external circuit.

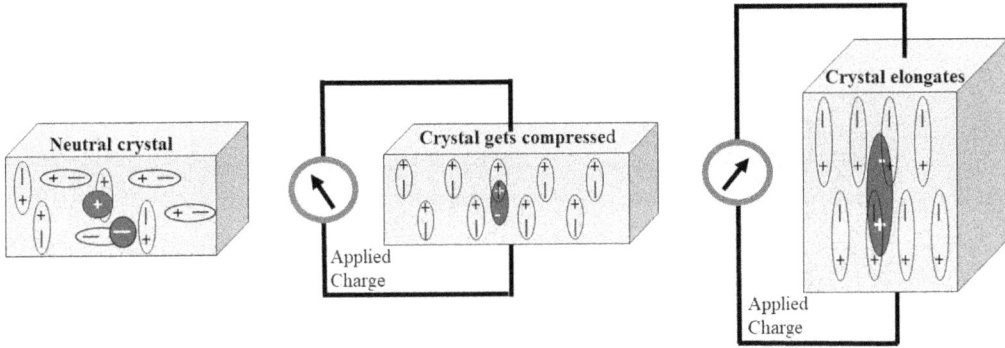

**FIGURE 1.2** Illustration of converse piezoelectric effect. The left figure shows neutral crystal in its pristine state at zero external applied charge. The crystal compresses (center) and elongates (right) upon the application of external applied charge.

created a myriad new and composite piezoelectric materials with enhanced mechanical, thermal, electromechanical, and biocompatible properties to access, extract, and convert the vibrational energy in nature and man-made structures into useful electrical energy [7]. Recent innovations in nanotechnology fabrication tools facilitates creation of complex piezoelectric-based MEMS and NEMS devices that can preserve and sustain higher-energy conversion rates and high-power density for modern electronics [3].

Since the discovery of piezoelectric material in the 1980s, piezoelectric materials have continuously evolved and transpired in the realization of a number of ground-breaking discoveries for various energy harvesting applications [3,7]. For example, the use of an everyday gas lighter uses the piezoelectric material to create an electric spark that lights up the gas. The mechanical movement of the spark wheel against the piezoelectric material creates an electric spark that helps to ignite the gas as depicted in Figure 1.3(a). Recent breakthroughs in technology showed promising applications of piezoelectric materials like the smart placement of piezoelectric material inside the shoe of a walking person as shown in Figure 1.3(b), which helps to convert shoe movement into useful electrical energy that can be used to charge low-power portable and wearable electronics [8,9].

The first piezoelectricity was experimentally demonstrated by applying external pressure on quartz, tourmaline, and rochelle salt materials to generate electrical charge by P. Currie and J. Currie in 1880 [10,11]. Since then, many other natural and artificial materials came into existence that lack a center of symmetry within their internal atomic structure and have the capability to show piezoelectric effect.

## 1.3 TYPES OF PIEZOELECTRIC MATERIALS

The choice of a particular piezoelectric material for a prerequisite application depends on its intrinsic piezoelectric properties, design flexibility, rigidity, size, shape, and the type of environment where it is employed [12,13]. The strength and potential of a particular piezoelectric material are characterized by its piezoelectric strain constant ($d$), piezoelectric voltage ($g$), electromechanical coupling factor ($k$), mechanical quality factor ($Q$), and dielectric constant ($\varepsilon$). The piezoelectric strain constant ($d$) is defined as the amount of induce polarization per unit stress (or unit electrical field) applied. The piezoelectric voltage ($g$) is defined as the amount of induced electric field per unit stress applied. The electromechanical coupling factor ($k$) is the square root of the mechanical-electrical energy conversion efficiency factor. The mechanical quality factor ($Q$) describes the amount of damping in the piezoelectric material [14]. The $k$ factor is very crucial for energy harvesting applications as it describes the amount of input mechanical energy to output electrical energy.

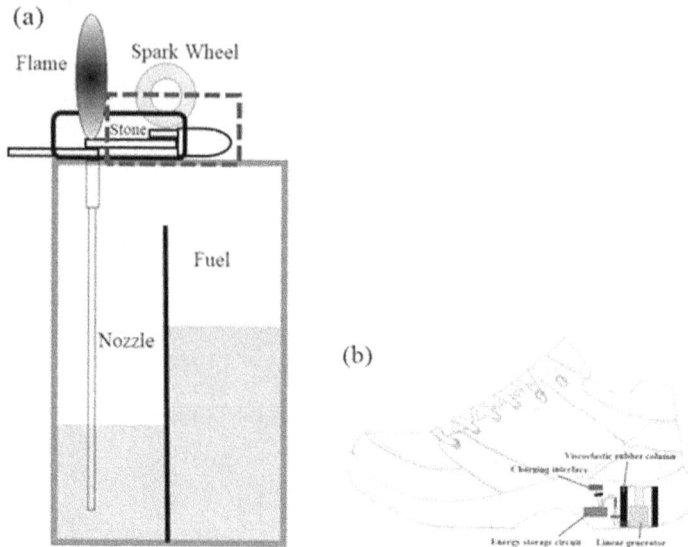

**FIGURE 1.3** (a) Illustration of working principle of gas lighter. The dashed box on the figure shows the usage of piezoelectric material to generate the electric arc that lights up the gas. (b) Illustration of the placement of piezoelectric material inside a shoe to generate electrical energy. Figure 3(b) is reprinted using creative commons attribution license from https://doi.org/10.3389/fmats.2019.00221 [9].

The discovery of ferroelectric materials (a subset of piezoelectric materials) like barium titanate ($BaTiO_3$) and lead zirconate titanate (PZT) had revolutionized the field of piezoelectric materials in making useful commercial products [3,12]. Ferroelectric materials exhibit spontaneous polarization with the formation of electric dipoles in the unit cell of the material and retain its polarization in the absence of the external applied charge. The polarization direction and strength can be altered by application of external charge to suit the energy harvesting applications. The internal atomic structure of the ferroelectric material has a significant impact on its piezoelectric energy harvesting capabilities. The ferroelectric behavior does not persist in all piezoelectrical materials and is absent in naturally occurring piezoelectric materials like quartz. The most studied ceramic materials, like $BaTiO_3$ and PZT, have their domains aligned in the direction of the applied electric field when they pass through their Curie point (Curie point is the temperature above which piezoelectric materials lose their piezoelectricity). PZT dominates the piezoelectric material industry for its spontaneous polarization and the ability to produce output voltage of the order 50–100 V [3]. However, the presence of minute amounts of lead can lead to hazardous environmental and health hazards with detrimental effects. Researchers are pursuing various avenues using the latest artificial intelligence and machine learning (AI and ML) computational tools and experimental synthesis techniques like hybrid molecular beam epitaxy systems in search for the lead-free-based piezoelectric materials with superior properties that can be put into commercial applications. Piezoelectric materials are primarily categorized into four categories, viz., natural, polymers, perovskites, and organic types, as depicted in Figure 1.4 based on structural properties.

### 1.3.1 SINGLE-CRYSTAL PIEZOELECTRIC MATERIALS

The natural single-crystal piezoelectric materials are intrinsically anisotropic dielectric materials with non-centrosymmetric crystal lattice. The natural crystal materials like quartz, rochelle salt, topaz, tourmaline-group minerals, silk, wood, enamel, bone, hair, rubber, dentin are some commonly

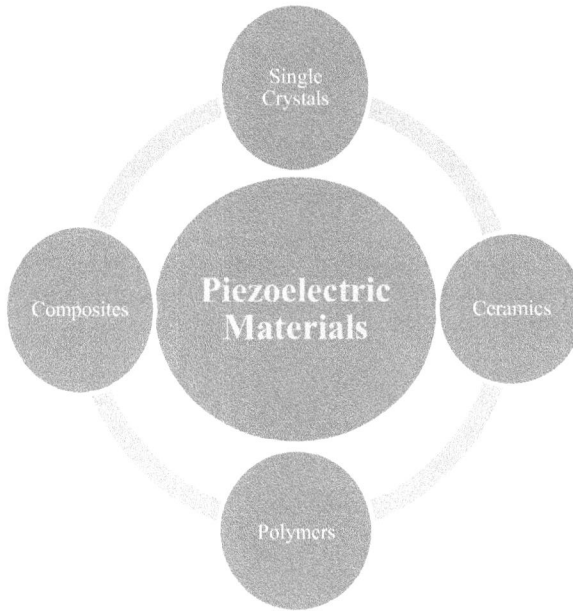

**FIGURE 1.4**   Classification piezoelectric materials.

available single-crystal piezoelectric materials in nature [15]. These natural piezoelectric materials are usually single crystals with high piezoelectric capability. The single-crystal nature of these natural materials give rise to ordered arrangement of dipoles over longer length scales, yielding stronger piezoelectric behavior when compared to other piezoelectric materials. The centimeter-scale long-range order of the single crystal usually tends to have minimal or no defect center that hinder the piezoelectric capability of the crystal. Lithium niobate ($LiNbO_3$) and its doped versions like lead magnesium niobate-lead titanate solution (PMN-PT), and lead zinc niobate-lead titanate (PNZ-PT) are the commonly used single-crystal piezoelectric materials for commercial applications, including energy transducers, accelerometers, actuators, and other biomedical applications [13–16].

   $LiNbO_3$ exists in colorless solid and organic solvent that is colorless, chemically stable, and insoluble in water. $LiNbO_3$ crystals are ferroelectric in nature and have extremely high melting and Curie temperatures. $LiNbO_3$ single crystals as long as 40 mm in diameter and 50–100 m in length can be grown using the standard Czlochlarski technique [17]. Additionally, spherical lithium niobate nanocrystals with a diameter of around 10 nm can be synthesized by coating a mesoporous silica matrix with a mixture of an aqueous solution of $LiNO_3$ and $NH_4NbO(C_2O_4)_2$, then heating for 10 min in an infrared furnace [18]. These substances are made of oxygen octahedron with a high Curie temperature of 1210°C. The availability from centimeter scale to nanoscale of $LiNbO_3$ single crystals coupled with its high electromechanical and electrooptical coupling coefficients makes it a promising candidate for electro-photonic and energy-based applications [19].

   The efficiency of a vibration-based piezoelectric transducer strongly depend on its operational frequency since energy distribution follows a Gaussian curve and maximum energy is generated around the resonant frequency. Thus, the physical size of piezoelectric material plays a pivotal role in designing miniaturized energy harvesters. With the advent in MEMS and NEMS in the last few decades, microscale energy harvesting has seen new avenues and routes to harness energy from the vibrations of the piezoelectrical crystals particularly from $LiNbO_3$ [20,21]. It is important to note that the everyday available vibrational sources like human walking, car engine, refrigerator, and high vacuum vents in the office tend to produce frequencies less than 250 Hz [14]. However, piezoelectric energy harvesting using vibrational or mechanical energy sources typically operate at low frequencies [14]. The resonant frequency of the chosen piezoelectric material should not interfere or overlap with its environment. This necessitates innovative designs of the piezoelectric material with different form factors, sizes, and shapes; advanced energy harvesting conditioning electronic circuitry to tap the energy and additional frequency tuning circuits that aid in eliminating unwarranted frequency components.

The use of thin and flat structures enables the piezoelectric material to have high surface area for spontaneous movements and to maintain less weight with smaller dimensions that are compatible with modern electronic circuits. The cantilever geometry is most commonly used in vibrational-based energy harvesting system due to its simplistic design features, small sizes and yet produce large mechanical strain for high-energy conversions [14,22]. For example, a thin layer of a piezoelectric material ($LiNbO_3$) is glued onto silicon substrate with oxide coating using chromium/gold interface layer, as depicted in Figure 1.5(a). The Cr/Au gold layer is patterned using MEMS technology to form the back contact and a thin layer of aluminum is deposited on the $LiNbO_3$ layer to form the top contact works as a simple cantilever structure to convert mechanical vibrations to useful electrical energy.

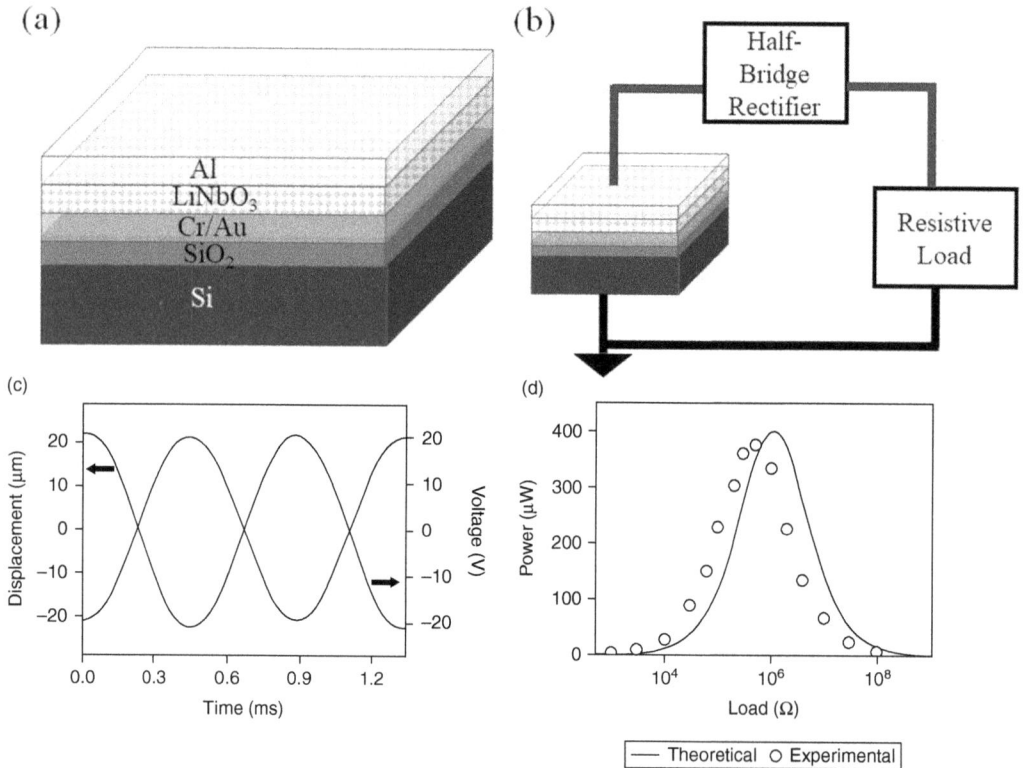

**FIGURE 1.5** (a) Schematic representation of $LiNbO_3$ piezoelectric transducer. The Si wafer is coated with silicon dioxide ($SiO_2$). Chromium/gold (Cr/Au) acts as back electrode and aluminum (Al) acts as top electrode. (b) Illustration of $LiNbO_3$ piezoelectric transducer in a typical electronic circuit for energy harvesting. (c) The displacement (left axis) and the corresponding voltage (right axis) generated on the $LiNbO_3$ transducer. (d) Experimental and theoretical power harvested using the $LiNbO_3$ energy transducer. Figures 1.5(c) and 1.5(d) are reprinted with permission from G. Clementi et al., Mechanical Systems and Signal Processing, 149, 107171, 2021 [20].

The cantilever heterostructure in Figure 1.5(a) is then connected to a half-wave bridge circuit to convert the as-generated electrical energy into DC power to drive the external resistive load circuit as shown in Figure 1.5(b). The external mechanical deformation on a cantilever piezoelectric material causes local strain within, which yields displacement of ionic charges resulting in net electric dipole charges. The harvested power for the simple cantilever geometry shown in Figure 1.5a can be estimated as

$$P = \frac{4\alpha f_0 R_L}{\left(f_0 R_L C_0 + 1\right)^2} u_M^2 \qquad (1.1)$$

where $\alpha$ is the electromechanical force factor, $f_0$ is the resonant frequency, $C_0$ is the clamped capacitance, $u_M$ is the displacement magnitude, and $R_L$ is the equivalent resistive load [20]. The LiNbO$_3$ energy harvester shown in Figure 1.5(a) is perturbed with an external shaker at its resonance frequency under 3.4 g base acceleration. The mechanical displacement of the cantilever and the resulting voltage of the energy transducer is shown in Figure 1.5(c) [20]. A peak voltage around 20 V is observed for a maximum cantilever displacement of 20 μm, which resulted in a root mean square (rms) value of 14V [20]. The as-generated rms value is then rectified using a half-wave bridge rectifier circuit to produce a maximum power of 380 μW in the external load circuit. The experimental measured power closely matches the calculated harvested power (equation 1.1), as shown in Figure 1.5(d), shows the efficacy of the energy harvester circuit, which can be used to power low-power electronics that operate in microwatt regime.

The use of nanogenerators for energy scavenging inside the human body has been the trending research subject for the last few years. Nanoenergy generators have the capability to extract energy from cardiac movements, blood circulation, and muscle contraction/relaxation. However, the realization of such energy generators for biomedical applications mandates the design of flexible, compact, and bio-compatible piezoelectric materials [23]. The piezoelectric $(1 - x)$Pb(Mg$_{1/3}$Nb$_{2/3}$)O$_{3-x}$PbTiO$_3$ (PMN-PT) single-crystal thin film on flexible polyethylene terephthalate (PET) plastic substrate is demonstrated to self-power cardiac peacemaker inside the human body [24]. The flexible PMN-PT-based energy transducer has generated a maximum output voltage of 8.2 V and an output current of 145 μA under a continuous vibrational excitation, which has the potential to power up 50 commercial light emitting diodes [24]. The variations in stoichiometric compositions of the single-crystal piezoelectric material with different dopant elements and concentrations can further improve the output electrical energy for practical applications. However, synthesis of single-crystal piezoelectric material critically depends on the purity of the raw materials, type of single-crystal growth, optimum growth conditions that elevate the cost of the piezoelectric material for commercial applications.

## 1.3.2 Ceramics Piezoelectric Materials

The high cost of single-crystal piezoelectric materials limits their applicability for large-scale mass production in energy-harvesting-based applications. On the other hand, the ceramic piezoelectric materials, which are polycrystalline in nature can be synthesized using conventional low-cost powder mixing, grinding, and sintering processes. Polycrystalline materials lack the long-range crystalline order in comparison to single-crystalline material and are usually made up of granular structures separated by grain boundaries. The short-range order can potentially host a multitude of grain boundary defects (charged and uncharged), trap charges, introduce dislocations (charged and uncharged) that adversely affects the ceramic piezoelectric material intrinsic properties for energy scavenging applications. Ceramic piezoelectric materials typically host perovskite ABO$_3$ structure, such as barium titanate (BaTiO$_3$), lead zirconate titanate (Pb[Zr$_x$Ti$_{1-x}$]O$_3$ ~ PZT), and potassium niobate (KNbO$_3$) [13,14]. The perovskite structure allows high structural stability by the incorporation of different dopants at cation and anion sites with different compositional ratios yielding enhanced properties to yield high-energy conversion efficiencies [25]. The discovery of $BaTiO_3$ ceramic piezoelectric material with a high dielectric constant of 1110 in the year 1947 has triggered researchers to focus on the ceramic-based piezoelectric applications [26]. However, the low Curie temperature around 120°C has hindered its commercialization. PZT is the most commonly used ceramic piezoelectric material due to its simple synthesis technique from the powder mixing technique, high Curie temperature and dielectric constant that depends on its stoichiometric composition in $PbZrO_3$–$PbTiO_3$ solid solution.

For example, the tuning of $PbZrO_3$ chemical composition in $PbZrO_3$–$PbTiO_3$ solid solution results in the change of dielectric constant from 300–1600, and its electromechanical coupling coefficient varies from 0.05 to 0.2 [27,28]. The PZT solid solution phase diagram shown in Figure 1.6 indicates the variation of Curie temperature with respect to the $PbTiO_3$ mol % in PZT solid solution and reaches a maximum value around 420°C [29–32]. Please note the change in crystal structure from symmetrical cubic phase to asymmetrical rhombohedral/tetragonal suggests the formation of electric polarization mechanisms resulting in piezoelectricity within the piezoelectric ceramic PZT. The peak value of the piezoelectric coefficient and the dielectric constant of PZT appear near the morphotropic phase boundary (MPB), where the rhombohedral and tetragonal phases co-exist with equivalent free energies [27,31]. The MPB typically exists at the nominal composition of 48 mol% $PbTiO_3$ in $PbZrO_3$–$PbTiO_3$ solid solution, as shown in Figure 1.6 [27,31]. The co-existence of two crystalline phases allows maximum orientation of microdomains during the poling process (poling is defined as the process of aligning all individual dipole moments of piezoelectric materials into a prerequisite direction). The crystal orientation of a specific compositional PZT also plays a major role in achieving the desired piezoelectric coefficient. The theoretical calculation predicts that for a tetragonal PZT, the piezoelectric coefficient monotonically decreases as the crystal deviates from the spontaneous polarization direction [001], whereas it is larger along the perovskite direction [001] when compared to spontaneous polarization direction [111] in rhombohedral phase [33]. The tunability of the chemical composition in the PZT perovskite structure coupled with its structural stability, ability to synthesize the same in thick film, thin film, and nano forms, has accentuated its increased usage in piezoelectric-based energy harvesting applications.

However, the high toxicity and its environmental issues has been a bottleneck for its widespread usage in consumer electronics. The extensive use of compositional engineering coupled with structural and interface engineering has the potential to synthesize a new family of lead-free piezoelectric ceramics for future energy harvesting and storage applications. The viable and lead-free piezoelectric ceramics that show promising piezoelectric properties include the family of potassium sodium niobate [(K, Na) $NbO_3$–KNN], barium titanate ($BaTiO_3$–BT), bismuth sodium titanate [(Bi, Na) $TiO_3$–BNT], and bismuth ferrite ($BiFeO_3$–BFO) [34]. The compositional variations and the addition of different dopant atoms in the KNN, BT and BNT material systems can be engineered to tailor its unique internal crystal structure to create polymorphic phase transition or morphotropic phase boundaries in their phase diagrams for enhanced piezoelectric properties [31].

### 1.3.2.1 KNN-Based Piezoelectric Ceramics

The KNN-based piezoelectric ceramic materials discovered in the 1960s possess excellent electrical properties, high Curie temperature, and reasonable piezoelectric coefficients [50]. The use of compositional engineering with the addition of external dopants like $Li^+$, $Ba^{2+}$, $Sb^{5+}$, $Ta^{5+}$, and $SrTiO_3$ in the host compound [(K, Na) $NbO_3$] has proven to be a feasible option in achieving high room temperature piezoelectric coefficients and Curie temperature as shown in Figures 1.7(a) and 1.7(b), respectively [41–46]. The compositional engineering also aids to tailor the rhombohedral-orthogonal (R-O), orthogonal-tetragonal (O-T), rhombohedral-tetragonal (R-T), and

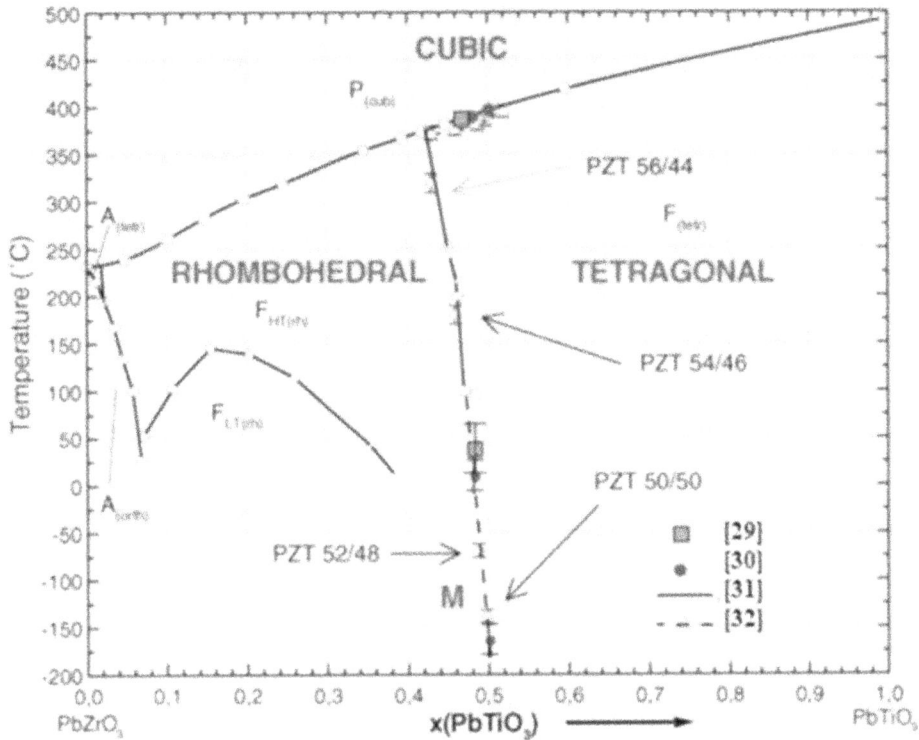

**FIGURE 1.6** The phase diagram of PZT piezoelectric ceramic material. The Curie temperature is dependent on the molar ratio of PbTiO$_3$. The PZT perovskite crystal structure exists in rhombohedral, cubic, and tetragonal phases. The morphotropic phase boundary (*M*) separates the rhombohedral and tetragonal phases. The various legends in the figure stands for the following: $A_{orth}$, antiferroelectric in orthorhombic phase; $F_{LT(rh)}$, low-temperature ferroelectric in rhombohedral phase; $F_{HT(rh)}$, high-temperature ferroelectric in rhombohedral phase; $P_{cub}$, paraelectric in cubic phase; $F_{tetr}$, ferroelectric in tetragonal phase. Figure 1.6 is reprinted with the permission from A. Bouzid et al., Journal of the European Ceramic Society, 25, 3213–3221, 2005 [32].

rhombohedral-orthogonal-tetragonal (R-O-T) phase boundaries in the KNN phase diagram, which shows a direct impact on its piezoelectric properties [51]. The KNN family with the R-T phase boundary possesses the highest piezoelectric coefficient when compared to other phase boundaries [52]. However, the creation of new phase boundaries in KNN-based piezoelectric ceramics adversely affect its thermal stability, which can result in poor reliability of the energy harvesting devices over the long run. This can be eliminated by introduction of multiple dopants Bi, Sb, and Zr in KNN and this technique is successfully employed by Gao et al., and found the highest room temperature dielectric value of 2900, piezoelectric coefficient of 520 pC/N with enhanced thermal stability by introduction of multiple dopants [53]. Gao et al., argued that the enhanced piezoelectric properties are due to the co-existence of multiple ferroelectric phases at MPB or PPT, evolution of average tetragonal phase structure and is largely due to the presence of high-density nanoscale structure heterogeneity. The presence of potassium in KNN-based ceramics pose a serious problem due to its highly reactive nature and strong affinity to absorb moisture altering its intrinsic properties. The optimum control of growth conditions, the choice of synthesis technique and poor thermal stability of KNN-based piezoceramic materials are some of the challenging aspects that need to be addressed for practical realization of energy harvesting and storage applications.

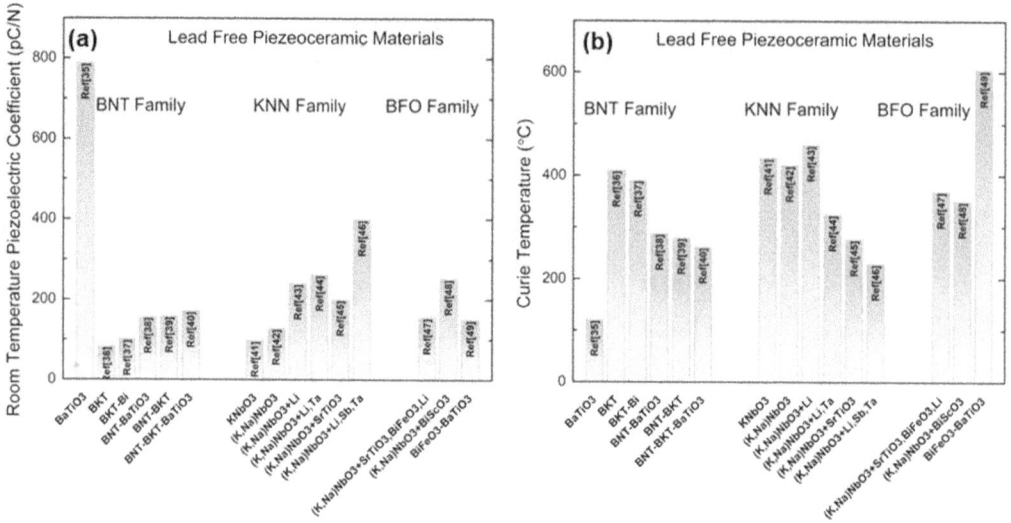

**FIGURE 1.7**  (a) Room temperature piezoelectric coefficients for various lead-free piezoelectric ceramics. (b) Curie temperature for various lead-free piezoelectric ceramics.

### 1.3.2.2  BaTiO$_3$-Based Piezoelectric Ceramics

BaTiO$_3$, a prototype polymorphic phase transition (PPT)-based ceramic first discovered in the 1940s has been conventionally used as a dielectric material largely due to it high dielectric constant. The ceramic BaTiO$_3$ has four polymorphs and undergoes temperature-driven phase change in the order of cubic (C), tetragonal (T), orthorhombic (O), and rhombohedral (R) (with decrease in temperature). The room temperature high dielectric constant around 1000–3600 along the $a$-axis of the perovskite crystal structure enables it to hold high voltages for capacitive-based energy storage applications [26,54]. However, its low Curie temperature of 120°C has hindered its widespread usage in making energy storage devices [see Figure 1.7(b)]. The use of structural engineering methods coupled with a templated grain growth synthesis technique helps to decrease the internal domain and grain sizes within its crystal structure BaTiO$_3$ to achieve a maximum room temperature piezoelectric coefficient of 788 pC/N [36]. This value is four times higher when compared to pure BaTiO$_3$ ceramic with a piezoelectric coefficient of 190 pC/N grown using conventional solid-state reaction method [55]. The compositional engineering and addition of multiple dopants into BaTiO$_3$ has resulted in the synthesis of stable piezoelectric ceramic $(1 - x)$Ba(Ti$_{1-y}$Sn$_y$)O$_{3-x}$(Ba$_{1-z}$Ca$_z$)TiO$_3$ [BTS$_{0.1}$1 $- x$BCT] with a piezoelectric coefficient of 700 pC/N and shows a moderate value of >600 pC/N within a wide compositional range grown using the solid-state method of sintering and calcination [56]. The obtained high values of piezoelectric coefficients are due to the broader coexistence of three ferroelectric phases (T+O+R) at the nanoscale regime, which results in low-energy barriers and smooth polarization changes within coexisting phases. The occurrence of broader R-T phase boundary and the presence of wide co-existence regions also helps to achieve high thermal stability with minimal (5%) variation of piezoelectric coefficient within a temperature window of 10–40°C [56]. However, the use of elevated temperature around 1500°C during sintering methods substantially raises the production cost of the piezoelectric ceramic for its commercial applications. The use of alternative low-temperature growth methods to synthesize stable BaTiO$_3$-based piezoelectric ceramics; and new techniques to raise Curie temperature needs to be explored for its commercialization.

### 1.3.2.3  BNT-Based Piezoelectric Ceramics

Bismuth sodium titanate (Bi$_{0.5}$Na$_{0.5}$TiO$_3$–BNT) materials first discovered in 1960 are relaxor ferroelectric materials and are known for their large remnant polarization with rhombohedral crystal structure at room temperature [57]. The large remnant polarization and high conductivity makes it hard to sufficiently pole a pure BNT for piezoelectric energy harvesting applications. The pure BNT phase diagrams show a first phase change from rhombohedral to tetragonal structure at low temperature (255°C to 425°C) and a second phase change from tetragonal to cubic at high temperature (520°C to 540°C) [58]. It is observed that the piezoelectric coefficients of the BNT-based family is nominally low [see Figure 1.7(a)] when compared to other lead-free piezoelectric ceramics due to the difficulty in polarizations and its inverse relationship with the depolarizing temperature. Similar to KNN and BT piezoelectric ceramics, various approaches like chemical modification, grain size control, addition of metal oxides, and optimization of heat treatments are tested to increase the piezoelectric coefficients of BNT-based piezoelectric ceramics [59]. The presence of $A$-site disorder in the rhombohedral perovskite structure of BNT tends to create compositional fluctuations resulting in the formation of polar nanoregions (PNR), i.e., the formation of nanoscale and mesoscale local structures within the host structure favors the occurrence of relaxor ferroelectric characteristics in BNT-based piezoelectric ceramics.

The experimental evidence of the formation of PNRs were explored vigorously to estimate the effect of PNRs on piezoelectric properties of pure BNT and BNT-based binary systems (BNT-BT and BNT-BKT). The use of advanced characterization techniques like high-energy x-ray diffraction and in situ transmission electron microscope (TEM) on $(1 - x)Na_{0.5}Bi_{0.5}TiO_3 - xK_{0.5}Bi_{0.5}TiO_3$ (BNT-BKT) binary system revealed the PNR's largely rely on the material's initial structure and the phase transitions and domain texturing dictate the polarization processes [60]. The formation of PNRs in morphotropic composition of 0.8BNT-0.2BKT is schematically illustrated in Figure 1.8. The experimental findings show that the 0.8BNT-0.2BKT system in in pseudo-cubic structure and is comprises of nanodomains in its original state [see Figure 1.8(a)]. The virgin state consisted of short-range-ordered nanodomain structure with coexisting in-phase ($a^0a^0c^+$) and antiphase ($a^-a^-a^-$) tilting of the oxygen octahedral tilting. An external compressive stress induces an irreversible transformation to textured rhombohedral domains as seen in Figure 1.8(b). The advanced in situ TEM enabled the poling process to be performed in situ without disturbing the sample dynamics. It is observed that the gradual increase of applied electric field (less than coercive field) favored the merging of initial nanodomains enabling the formation of new separate larger domains with local ordering [see

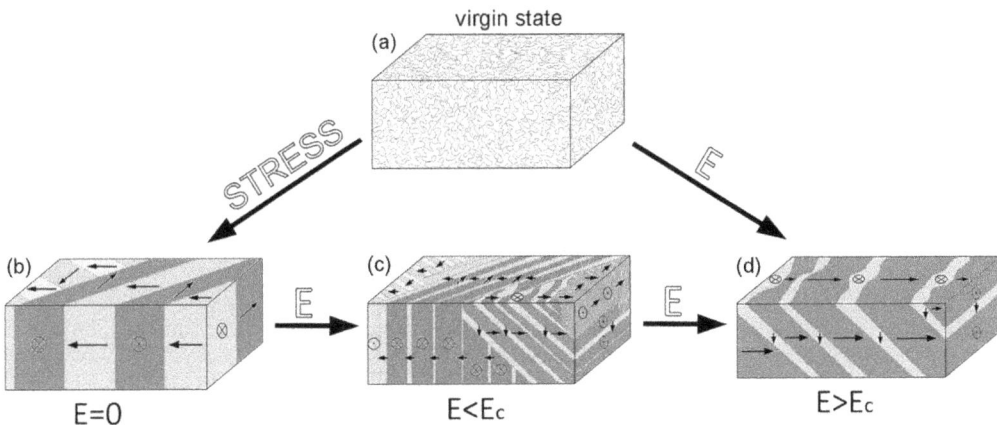

**FIGURE 1.8** Schematic illustration of domain formation and reorientation process in the presence of applied electric field and stress in 0.8BNT-0.2BKT sample (a) intrinsic state depicting the presence of nanodomains; (b) formation of lamellar structure with the application of stress; (c) formation of nanostripe-like intermediate domains during poling process; (d) formation of single domains when applied electric field ($E$) is greater than coercive field ($E_c$). Figure 1.8 is reprinted with permission from M. Otonicar et al., Acta Materialia, 127, 319–331, 2017 [60].

Figure 1.8(c)]. It is to be noted that the application of electric field tends to reduce the tilting of the system that has a direct impact on the local displacements of the cation and anion within the distorted perovskite enabling the creation of long-range ferroelectric ordering [61]. At this stage, a coupling mechanism is seen between the antiphase octahedral tilting in the rhombohedral structure and the induced polar ordering. With further increase in applied electric field, domain reorientation of the lamellar structure allows creation of additional narrow stripe-like domains and is due to pre-induced long-range ferroelectric order. The application of electric field above the coercive field results in the formation of single-domain regions, while preserving some narrow strain-compensating, stripe-like domain regions [60].

Thus, the chemical compositional variations in the parental BNT-based systems' creation of new binary and ternary BNT-based systems enables the remnant polarization ($P_r$) to be decreased and the electric field induced polarization saturation ($P_s$) to be increased, as depicted in Figure 1.9. The decrease in $P_r$ and increase in $P_s$ also shortens the hysteresis window, which has a direct impact on the energy storage capacity of the piezoelectric ceramic. The energy storage capacity of a piezoelectric energy harvester depends on the shape and the window of the $P$-$E$ hysteresis curve. The stored energy can be calculated by the numerical integration of the area between polarization axis and the curves of $P$-$E$ loops. Ideally, the charge retention and storage capacity of a piezoelectric material is a measure of difference between $P_s$ and $P_r$. It is evident from Figure 1.9 that piezoelectric ceramics with polar nanoregions tend to possess a high $P_s$ and small $P_r$ enabling maximum energy storage density. H. Qian et al., increased the compositional variation in BNT-BKT ceramics and experimentally observed the decrease in hysteresis window, consistent with Figure 1.9. The sample 0.96(0.75BNT-0.25BKT)-0.04BA showed a decreased remnant polarization value of 3.8 $\propto$C/cm$^2$ when compared to the high remnant polarization value of 23.5 $\propto$C/cm$^2$ in 0.75BNT-0.25BKT parental compound at room temperature [62]. The BNT-based piezoceramics can be incorporated into flexible and modular designs for energy harvesting applications. N. Raj et al. recently synthesized 0.8BNT-0.2BKT compound incorporated onto a plastic substrate Ecoflex silicone rubber to realize

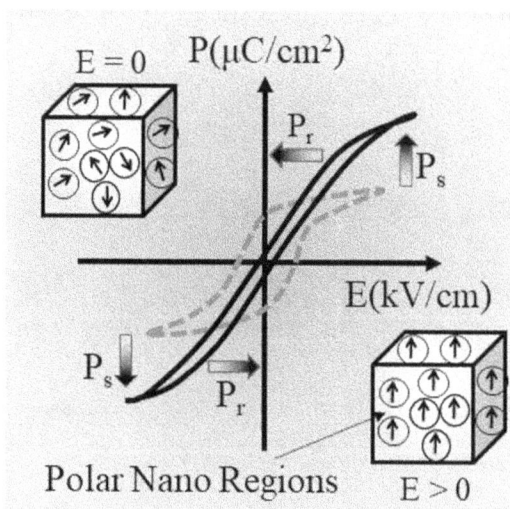

**FIGURE 1.9** Schematic illustration of the increase in saturation polarization ($P_s$) and the decrease of remnant polarization ($P_r$) with the formation of polar nano regions in BNT-based ceramics. The polar nanoregions are randomly oriented at zero applied electric fields (top left inset) and tend to align in the applied field direction (bottom right inset) yielding enhanced piezoelectric properties. The dashed gray line is the typical $P$-$E$ loop hysteresis curve for a conventional piezoelectric ceramic without any polar nanoregions.

human-body-based piezoelectric nanogenerators (PENG) [63]. The modular and flexible PENGs when subjected to bending and compressive modes yielded optimum $(V, I)$ values of (12.5 V, 110 nA) and (22.5 V, 210 nA), respectively, and are practically realized to convert human-body movements to useful electrical energy.

The nanodomain engineering on BNT-based ceramics able to achieve high-energy storage densities for future generation of energy harvesting methods and practices. The advancements in microfabrication and nanofabrication techniques help to tailor micro- or nanodomains of a piezo-electric ceramic material to achieve high piezoelectrical coefficients. However, the intricate impact of nanodomain formation, physical control of orientation with in the polar nanoregions and its relation with macroscopic physical parameters has to be explored further to completely understand its under-lying physical mechanism. The complete theoretical of nucleation of physical polar nanoregions and domains needs to be uncovered to be able to synthesize a new class of nanodomain engineered piezoceramic materials. The intrinsic behavior and motion of the PNRs, and compositional-derived macroscopic disorder need to be taken into account to completely understand the fatigue behavior of BNT-based piezoceramics, as their fatigue behavior deviates from conventional piezoelectric ceramics (fatigue is defined as the decrease in polarization with an increasing number of polariza-tion reversals) [59]. The inverse relationship between the polarization coefficient versus depolariza-tion temperature in BNT-based ceramics limits its widespread usage of energy storage applications [34]. The BNT-based piezoceramics also suffer from the thermal depolarization temperature, which affects the thermal stability of the energy harvester.

### 1.3.2.4 BFO-Based Piezoelectric Ceramics

The presence of low depolarizing temperature is one of the long-standing obstacles for the com-mercialization of BNT-based piezoelectric ceramics. $BiFeO_3$ (BFO) is a single-phase multiferroic material with both ferroelectric and magnetic ordering at room temperature [64]. BFO-based piezo-electric ceramics tend to have low room temperature piezoelectric coefficients when compared to PZT, but their extra ordinary high Curie temperature (>600 °C) renders its feasibility for high tem-perature actuator in the field of automobiles and aerospace applications [48,64].

As with other types of piezoelectric ceramics, introduction of single and multiple dopants, com-positional variations, phase engineering, domain engineering, and creation of binary and ternary systems have been successfully employed on BFO piezoelectric ceramic to improve its piezoelectric applications (see Refs. [51,64] for additional information). H. Tao et al. has studied the system-atic evolution of phase structure of $Bi_{1-x}A_xFeO_3$ when doped with rare earth elements that include A: Eu, Pr, Sm, Dy, and La [66]. The phase structure of the parental compound $Bi_{1-x}A_xFeO_3$ is closely dependent on the amount of the dopant and type of the rare earth dopant element, as shown in Figure 1.10. For example, with an increase in Eu content in $Bi_{1-x}Eu_xFeO_3$, the phase structure changes from rhombohedral→ rhombohedral + orthorhombic →orthorhombic. The piezoelectric properties of the $Bi_{1-x}A_xFeO_3$ are also affected by the type of rare earth dopant element, and the maximum piezoelectric coefficient for A: Eu, Pr, Sm, and Dy tend to appear in rhombohedral phase except for the rare earth dopant element La (see Figure 1.10 top for the maximum value of piezo-electric coefficient).

The binary system $BiFeO_3$-$BiFeO_3$ (BFO-BT) is widely studied due to it high Curie tempera-ture (>580 °C) and is prone to substitutional and compositional defects due to the volatile nature of $Bi_2O_3$ secondary phase, which strongly depends on the sintering temperature [67]. This can poten-tially lead to high leakage current density that shortens the energy storage capacity of BFO-BT-based piezoelectric actuators. The optimum sintering temperature and the dwell time are the two important factors that affect the homogeneity of the ceramic growth. Typically, longer dwell time results in abnormal grain growth causing inhomogeneous grains in the ceramics, and shorter dwell time results in incomplete grain growth, which leads to the formation of defects in the ceramics. L. Zhu et al. studied the effects of sintering temperature and dwell time on $0.7BiFeO_3$-$0.3BaTiO_3$

| | Maximum piezoelectric coefficient | | | | |
|---|---|---|---|---|---|
| | 46 pc/N (x = 0.1) | 42 pc/N (x = 0.1) | 45 pc/N (x = 0.075) | 31 pc/N (x = 0.05) | 41 pc/N (x = 0.1) |
| 0.30 | | | | | |
| 0.25 | Orthorhombic | Triclinic | Orthorhombic | Orthorhombic | Monoclinic |
| 0.20 | | | | | |
| 0.15 | Rhombohedral + Orthorhombic | Rhombohedral + Orthorhombic | Triclinic + Orthorhombic | Rhombohedral + Orthorhombic | Triclinic |
| 0.10 | | | | | |
| 0.05 | Rhombohedral | Rhombohedral | Rhombohedral | Rhombohedral | Rhombohedral |

$Bi_{1-x}Eu_xFeO_3$   $Bi_{1-x}Pr_xFeO_3$   $Bi_{1-x}Sm_xFeO_3$   $Bi_{1-x}Dy_xFeO_3$   $Bi_{1-x}La_xFeO_3$

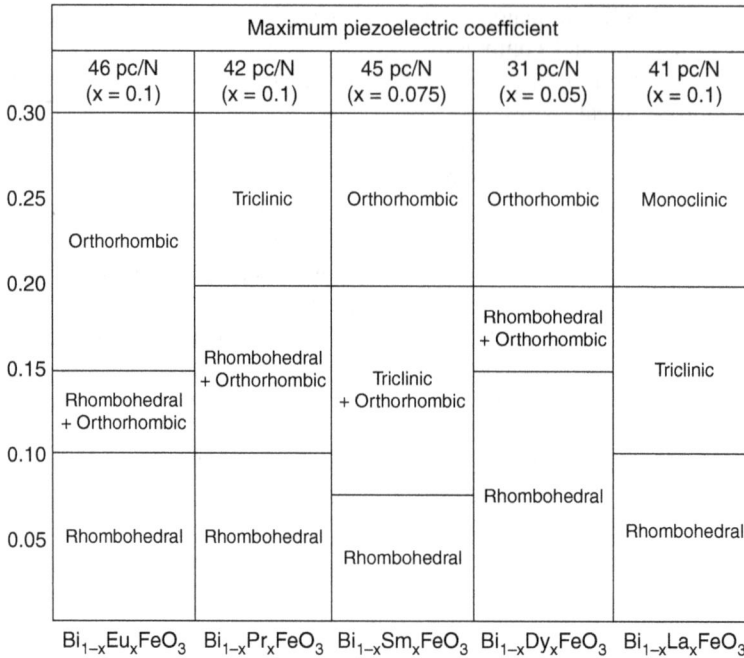

**FIGURE 1.10** Illustration of phase diagrams of $Bi_{1-x}A_xFeO_3$ piezoelectric ceramic with different single dopant elements (A: Eu, Pr, Sm, Dy, and La). The compositional variation in each doped $Bi_{1-x}A_xFeO_3$ compound is represented by the variable $x$ and is its variation can be seen on the vertical axis. The maximum piezoelectric coefficient and its corresponding composition is listed on the top of the figure. The figure is reprinted with the permission from H. Tao et al., Materials & Design, 120, 83–89, 2017 [66].

(BFO-0.3BT) and experimentally achieved a reasonable piezoelectric coefficient of 208pC/N at a sintering temperature of 1000°C and a dwell time of 6 h, which is comparable to conventional PZT-based piezoelectric ceramics [68]. A substantially high piezoelectric coefficient of ~405pC/N with a high Curie temperature of 450°C is recently reported in Ga- modified BFO-BTO piezo-electric ceramics via the formation of R-T phase boundary and optimizing quenching process during the ceramic growth [28]. However, BFO is sensitive to ambient conditions and usually comes with oxygen vacancies and other impurities affecting its room-temperature conductivity. The variation in conductivity is a difficult problem to overcome that invariably affects the reli-ability and the fatigue performance of BFO-based actuator applications [65]. The research is focused on developing new and innovative ceramic growth techniques that aid in realizing high-performance defect-free BFO-based piezoceramics for advanced energy-based harvesting and storage applications.

### 1.3.3 POLYMER PIEZOELECTRIC MATERIALS

The polymers are special class of carbon-based materials with a long chain of polymers. Their unique modular and flexible structure is quite different from rigid single crystals and brittle ceramic piezoelectric materials. The flexible nature of the polymers allows them to be installed and operated in complex architectures and yet withhold a high strain to extract maximum output electrical energy, and are generally used in twisting- or bending-based energy-based harvesting applications. The most commonly studied piezoelectric polymers for energy harvesting and storage applications are polyvinylidene fluoride (PVDF), polyvinylidene fluoride-trifluoro ethylene (PVDF-TrFE),

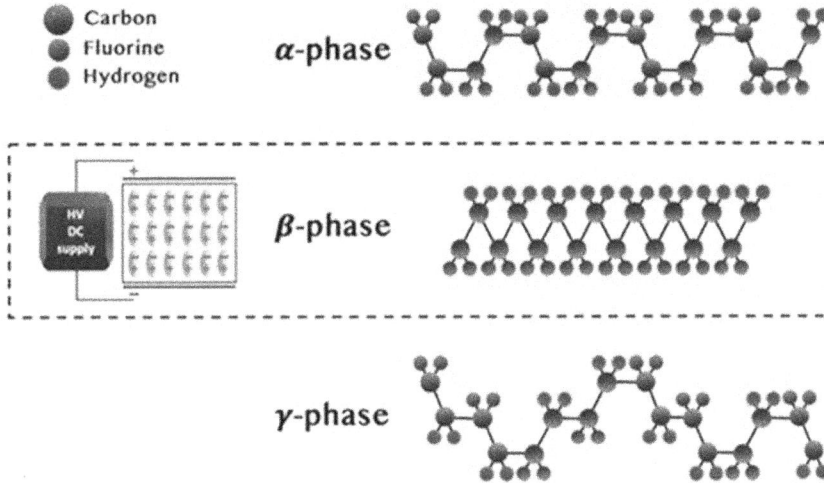

**FIGURE 1.11**  Illustration of different polymorphs in PVDF polymer piezoelectric material. The β-phase is the most widely used for piezoelectric applications. The figure is reprinted with permission from G. Kalumudina et al., Sensors, 2020, 20, 5214 under Creative Commons License 4.0 [69].

polyamides (PA), polylactic acids (PLA), cellulose etc. [13]. PVDF is a semi-crystalline polymer with a repetitive unit of $CH_2$-$CF_2$ that are embedded in an amorphous matrix and is widely used for commercial applications, including healthcare, self-powered, and wearable sensors. The PVDF's piezoelectricity is due to the existence of crystalline $\alpha$ phase, $\beta$ phase, and $\gamma$ phase structures (see Figure 1.11), among which $\beta$ polar phase possesses the highest piezoelectric coefficient due to the alignment of all dipoles in the same direction as that of applied electric field [70,71]. The use of innovative techniques like selective increase of $\beta$ phase, thermal annealing [72], compositional variations via additives/filler incorporation [73] and high electric field polarization [74] have proven to be successful in enhancing PVDS piezoelectric properties.

However, it is be noted that pure PVDF crystal tends to have rigid structure and tends to show nominal piezoelectric properties at low applied electric fields. Alternatively, the use of compositional engineering in PVDF with random co-polymers paved way for the synthesis of high-quality PVDF-TrFE polymer structure with 50:50 composition with MPB behavior, much like single crystal and polycrystalline piezoelectric ceramics that showed a piezoelectric coefficient of 63.5pC/N [75]. The incorporation of additives/fillers in the PVDF aids in increasing the nucleating agents during poling process via addition of conductive fillers might act as bridges between insulating piezoelectric material in the original polymer that enhances the overall piezoelectric properties. A similar value of piezoelectric coefficient of 62pC/N was realized in biaxially oriented poly(vinylidene fluoride) (BOPVDF) that contains pure highly aligned $\beta$ crystals [76].

Polymer piezoelectric materials being organic can be easily synthesized using simple and cost-effective techniques like spin coating, dip coating, and electrospinning. These solvent-based techniques enable the synthesis of piezoelectric polymers in different sizes, shapes, and compositions to meet the complex design challenges of portable and flexible energy harvesting modular architectures. A simple nanogenerator using a single PVDF nanofiber is fabricated using direct-write electrospinning technique on plastic substrate is shown in Figure 1.12(a). Mechanical stress is induced on the PVDF nanofiber by bending the plastic substrate, which results in the creation of polarization charges that create a potential difference between the two ends of the nanogenerator [77]. The time varying measured output voltage and output current upon continuous stretch-release

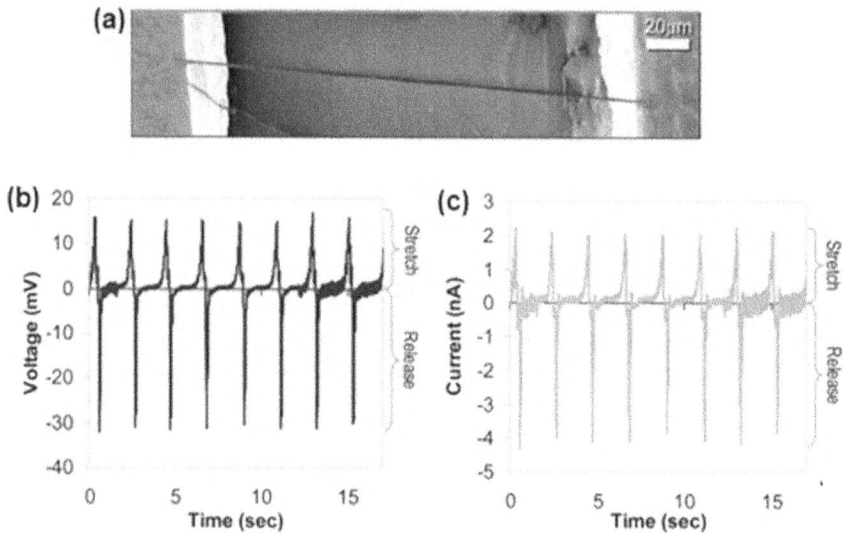

**FIGURE 1.12**   (a) Scanning electron microscope of a single PVDF nanofiber with two contacts on a plastic substrate, representation of a single nanogenerator. (b) Transient output voltage and (c) transient output current measured upon the application of strain at 2 Hz. The figure is reprinted with permission from C. Chang et al., Nano Lett. 10, 726–731, 2010 [77].

cycles is depicted in Figures 1.12(b) and 1.12(c), respectively. The maximum measured output voltage and the current are of the order of ~30 mV and ~5 nA, respectively, which is sufficient to drive low-power nanodevices and sensors. It is to be noted that the as-spun nanofiber polymer PVDF possess a huge internal resistance limiting the amount of leakage current during the short-hold-release cycles for maximum energy conversion efficiency. The energy conversion efficiency of the as-spun PVDF nanofibers depends on the growth parameters like voltage polarity, humidity of the electrospinning technique. The surface chemistry, crystallinity and the amount of $\beta$ phase of the as-spun PVDF nanofibers alters with humidity condition (higher humidity yields high $\beta$ phase), and the fiber network grown in high humidity condition have resulted in a maximum voltage and current of the order ~4 V and ~300 nA [78].

However, the piezoelectric coefficient of polymer materials lack behind the ceramic oxide materials. F. Xia et al. recently added an additional monomer chlorofluoroethylene (CFE) to P(VDF-TrFE) to synthesize a P(VDF-TrFE-CFE) terpolymer. The CFE increases the distance between polymer chains and alters TTTT chain conformations into TG$^+$TG$^-$ and T$_3$G$^+$T$_3$G$^-$, which helps to eliminate the pristine ferroelectric phase and favors the formation of relaxor ferroelectric phase that shows improved piezoelectric performance when compared to pristine P(VDF-TrFE) [79]. X. Chen et al. recently introduced a very small amount of fluorinated alkyne (FA) monomers (< 2%) as the fourth monomer into P(VDF-TrFE-CFE) to synthesize a relaxor ferroelectric tetrapolymer P(VDFTrFE-CFE-FA). The addition of small amounts of FA has resulted in 60% more power generation than state-of-the-art piezoelectric polymers [80, 81].

The low-cost, small-volume, high-density, and flexible piezoelectric polymers seem to be an excellent alternative for bio-compatible self-powered implantable energy harvesters. Future research must be focused on the design of the next generation of eco-friendly polymer piezoelectric materials with high crystallinity with molecular chain oriented in the film plane, and should be easily crystallized into a polar material with its axis perpendicular to the chain axis at low applied electric fields to produce maximum piezoelectric coefficients to match piezoelectric single and polycrystalline materials.

### 1.3.4  COMPOSITE PIEZOELECTRIC MATERIALS

The high degree of mechanical flexibility in piezoelectric polymers and the high value of piezoelectric coefficients in piezoelectric oxide-based ceramics are combined to yield composite piezoelectric materials for optimum performance. The structural composition of composite piezoelectric material consists of oxide ceramics in the form of particles, fibers, rods and embedded in a polymer matrix. The structural geometry for two-phase composites can be divided based on their connectivity in each phase (1, 2, or 3 dimensionally) into 10 structures via 0-0, 0-1, 0-2, 0-3, 1-1, 1-2, 1-3, 2-2, 2-3, and 3-3 composite structures. For example, a two-phase composites connectivity (number of directions through which a material is continuous) is defined as "3" if it self-connected in all three ($x$, $y$, $z$) directions; di-phase composite denoted by "$m$-$n$," where $m$ denotes the connectivity of an active phase (ceramic oxides) and $n$ stands for inactive phase (polymers); "1" if the phase is self-connected in single $z$ direction (e.g., see Figure 1.13) [82]. Among all the composites, the 0-3 composite is widely used because it is simple to make, where in active piezoelectric ceramic particles dispersed in a three-dimensionally connected polymer matrix. The 0-3 composite can take many form factors via thin sheets, molded shape, fibers, nanoparticles, and extruded bars [83]. The early successful attempts on realizing the synthesis of the 0-3 composite were made using the PZT fillers in the polyurethane matrix. The PZT polyurethane composite resulted in an increase of piezoelectric coefficient with increase in PZT filler content in the composite, and a maximum value of 45pC/N at 60% PZT content [84].

The well-studied individual systems, including PZT and PVDF made into a composite has great potential to show enhanced piezoelectric properties and a high degree of mechanical flexibility. Tiwari et al. synthesized a piezoelectric composite film of PZT particles embedded in the PVDT polymer with 0-3 connectivity showed an increase of dielectric constant and aids in improving the amount of $\beta$ phase by 75% when compared to pristine PVDF polymer [84]. The composite piezoelectric materials aid in designing compact and flexible nanogenerators to harness the energy from the ambient environment. The nanogenerator based on PVDF and PZT nanohybrids with 30% weight PZT filler is recently used to convert mechanical energy from human movements like bending, finger tapping, and feet taping to produce a maximum output voltage of 55 V and a power density of 36 μW/cm$^{-2}$ [85].

The computational tools help in analyzing the effect of size, form, shape, volume ratio, and aspect ratio of the fillers in the polymer matrix, which helps the material researchers to optimize the design space and growth condition for future nanogenerators for energy harvesting. The BaTiO$_3$ piezoceramic due to its high dielectric constant can serve as excellent filler material for the design of eco-friendly nanogenerators. P. Bai et al. recently developed a finite-element analysis model

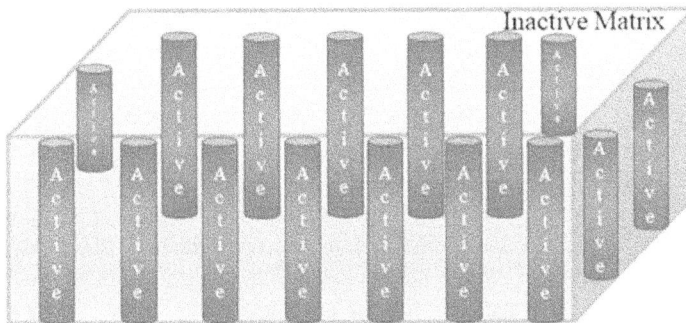

**FIGURE 1.13** Illustration of piezoelectric composite material. The cylindrical rods act as active phase materials (piezoelectric single crystals and ceramics, example: PZT), which are embedded into inactive phase matrix (piezoelectric polymer, example: PVDF).

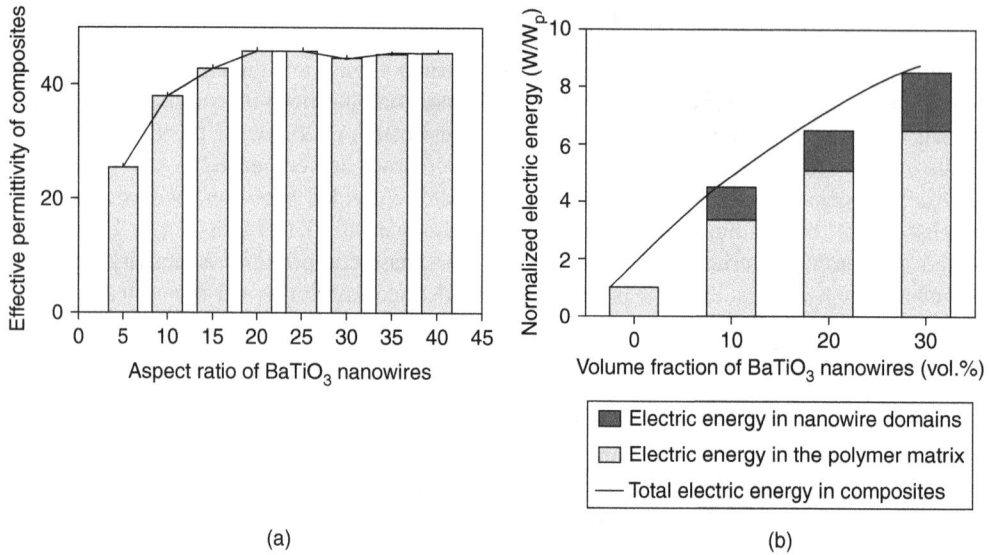

(a)                                                                                          (b)

**FIGURE 1.14**  (a) Variation of effective permittivity of the BT-PVDF composite with respect to aspect ratio of BT nanowires. (b) Variation of normalized energy storage with respect to volume fraction of BT nanowires. The image is reprinted from P. Bai et al., AIP Advances 11, 045018 (2021) under creative commons license BY 4.0 [86].

for BaTiO$_3$ (nanowires): PVDF composite, which allows random distribution of filler material. The model helps to figure out the effect of BaTiO$_3$ nanowires aspect ratio and volume fraction on the effective permittivity and normalized output energy, respectively. The effective permittivity of the composite initially increases with the aspect ratio of BaTiO$_3$ nanowires, followed by saturation at high aspect ratios [see Figure 1.14(a)]. Thus, a maximum value of effective permittivity is achieved for small aspect ratios. The presence of fillers in the polymer matrix distributes electric field strength inside the composite materials and is critically dependent on fillers volume fraction [86]. The electric field locally varies and results in maximum energy accumulation in the polymer matrix when compared to nanofillers. With increase in volume fraction of BaTiO$_3$ nanowire filler, the normalized energy output increases and it is 8.55 times higher than the pristine PVDF, as shown in Figure 1.14(b) [86]. The FEM simulations closely match the experimental results in the piezoelectric composite BaTiO$_3$ nanowires (prepared by double hydro-thermal method) were dispersed in PVDF. The composite material achieved a maximum dielectric constant and piezoelectric constant of 64 and 61pC/N, respectively, and is much higher when compared with BaTiO$_3$ spherical rod fillers [87].

Piezoelectret foam, a cellulose dielectric polymer recently gained interest due to the presence of permanent electric charge in its structure. The application of the polarization process deposits the charge inside the voids in the structure. The trapped charge acts as macroscopic dipoles yielding piezoelectricity. The piezoelectret foam nominally has a high piezoelectric coefficient of 250 pC/N, which is seven times higher than the conventional PVDF [3]. J. Zhong et al. synthesized a new type of flexible piezoelectret generator based on polyethylene terephthalate (PET) electret film and the ethylene vinyl acetate copolymer (EVA) adhesive layer. The flexible PET/EVA electret structure has air bubbles inside its structure, which trap the charges and act as megascopic electric dipoles. The flexible PET/EVA nanogenerator has a piezoelectric coefficient of ~6300 pC/N and showed stable performance with no degradation even after ~90000 cycles, producing a ~0.444mW power under harsh environments (in the presence of high moisture and at a temperature of 70°C) [88].

The use of other fillers like lead nickel niobate-lead zirconate titanate, lead magnesium niobate-lead titanate, zinc oxide, single-walled carbon nanotubes, graphene, graphene oxide, were tested successfully with different aspect ratios, shapes, sizes, and volumes to harness energy from environmental and human-body movements to power nanodevices, sensors, and low-power electronics devices [3,13,14]. The recent surge in the interest of various two-dimensional (2D) materials like layered 2D nanosheets, hexagonal boron-nitride ($h$-BN) and $MX_2$ ($M$ = Mo or W; $X$ = S, Se, or Te) has opened up a new avenue to design and develop next-generation 2D piezoelectric nanomaterials (PNM)-based nanogenerators. The increased surface area, highly flexible mechanical structure, ability to withhold large strain, and deformation facilitates them to be used in complex and high strain environments [89, 90]. The current research is highly focused to design and fabricate of 2D PNM-based nanogenerators for commercial energy harvesting applications.

## 1.4  CHALLENGES AND OPPORTUNITIES

The depletion of fossil fuels and the increased usage of portable electronics demands the design and development of self-contained, sustainable, eco-friendly and non-renewable energy sources. Piezoelectric materials are the "*go-to*" materials for energy harvesting and storage applications. The choice of the type of piezoelectric material for energy harvesting and storage primarily depends on its intended application and ambient environment conditions, where it is used. The comparison of different types of properties for the various types of piezoelectric materials is listed in Table 1.1. It is clear that high piezoelectric coefficients are obtained for single-crystal and ceramic piezoelectric materials at the expense of low mechanical flexibility, which limits their applications in complex environments that include high degree curvatures. On the other hand, high flexibility and low piezoelectric coefficients of polymers and composite piezoceramics limits their usage to microwatt-based power generation applications.

The effect of phase boundaries and physics of MPB in single-crystal piezoelectric; formation and role of PNRs in achieving high piezoelectric coefficients in ceramic piezoelectric; the structural modifications and its intricate effect on the formation of microscope and macroscopic dipoles with the introduction of additional monomers (e.g., copolymer, terpolymer, tetrapolymer and higher) in polymer and composite piezoelectric materials needs a comprehensive theoretical understanding to design and develop future cost-effective, environmentally benign, biocompatible and flexible piezoelectric-based energy harvesters and storage systems.

The growing demands of the IoT technologies mandates the use of self-powered and wearable sensors for various data transfer and data mining application. The portability of the IoT devices and the dependence of power on external battery supplies will become obsolete in the coming years.

## TABLE 1.1
## Comparison of Piezoelectric Materials

|  | Single Crystal | Ceramics | Polymers | Composites |
|---|---|---|---|---|
| *Physical Structure* | Inorganic | Inorganic | Organic | Organic/Inorganic |
| *Mechanical Flexibility* | Low | Low | High | High |
| *Synthesis Technique(s)* | Complex | Complex | Easy | Complex |
| *Density* | High | High | Low | Medium |
| *Piezoelectric Coefficient* | High | High | Low | Medium |
| *Dielectric Constant* | High | High | Low | Medium |
| *Electromechanical Coupling Factor* | High | Medium | Low | Low |
| *Bio-compatibility* | Low | Low | High | Medium |

The shortcomings and the trade-offs in the selection of a particular piezoelectric material for a given flexible high-power application can be realized by the (i) exploration of new piezoelectric material including metamaterials; (ii) emergence of new computational tools to optimize material composition and process parameters; (iii) innovation in fabrication technologies to create stacks of piezoelectric layer; (iv) domain engineering; and (v) use of self-assembly templates to create complex piezoelectric materials.

## REFERENCES

[1]   "Smart materials market by application (transducers, actuators & motors, sensors, structural materials, and coatings) and end-user industry (industrial, defense & aerospace, automotive, consumer electronics, healthcare, and others)," Global Opportunity Analysis and Industry Forecast, 2015–2022. Allied Market Research. Information available at webpage: www.alliedmarketresearch.com/smart-material-market.

[2]   K.T. Alade, A.F. Lawal and D. Akinyele, "Smart materials and technologies for next generation energy-efficient buildings," Special Issue on the Foundational Support Systems, IEEE Smart Grid Resource Center, April 2017.

[3]   M. Safaei, H.A. Sodano and S.R. Anton, "A review of energy harvesting using piezoelectric materials: state-of-the-art a decade later (2008–2018)," *Smart Mater. Struct.* 28, 11300, 2019.

[4]   M. A. Karami and D. J. Inman, "Powering pacemakers from heartbeat vibrations using linear and nonlinear energy harvesters" *Appl. Phys. Lett.* 100, 042901, 2012.

[5]   S.F. Ali, M.I. Friswell and S. Adhikari "Analysis of energy harvesters for highway bridges" *J. Intell. Mater. Syst. Struct.* 22, 1929–38, 2011.

[6]   S.R. Anton, A. Erturk and D.J. Inman, "Multifunctional unmanned aerial vehicle wing spar for low-power generation and storage" *J. Aircr.*, 49, 292–301, 2012.

[7]   S. R. Anton and H.A. Sodano, "A review of power harvesting using piezoelectric materials (2003–2006)," *Smart Mater. Struct.* 16, R1, 2007.

[8]   J. Zhao and Z. You, "A shoe-embedded piezoelectric energy harvester for wearable sensors," *Sensors,* 14, 12497–12510, 2014.

[9]   B. Xu and Y. Li "Force analysis and energy harvesting for innovative multi-functional shoes," *Front. Mater.*, 6, 2019.

[10]   J. Curie and P. Curie, "Development, via compression, of electric polarization in hemihedral crystals with inclined faces," *Bull. Soc. Minerolog. France*, 3, 90–93, 1880.

[11]   "March 1880: The Curie Brothers Discovered Piezoelectricity," APS News, *This Months in Physics History*, 23, 3, March 2014. Information available on APS website at www.aps.org/publications/apsn ews/201403/physicshistory.cfm.

[12]   F. Narita and M. Fox "A review on piezoelectric, magnetostrictive, and magnetoelectric materials and device technologies for energy harvesting applications," *Adv. Eng. Mater.*, 20, 170074, 2017.

[13]   S. Mishra, L. Unnikrishnan, S.K. Nayak, and S. Mohanty, "Advances in piezoelectric polymer composites for energy harvesting applications: A systematic review," *Macromol. Mater. Eng.* 1800463, 2018.

[14]   H. Li, C. Tian and Z. D. Deng, "Energy harvesting from low frequency applications using piezoelectric materials," *Appl. Phys. Rev.*, 1, 041301, 2014.

[15]   A. Manbachi and R.S.C. Cobbold, "Development and application of piezoelectric materials for ultrasound generation and detection." *Ultrasound*, 19(4), 187–196, 2014.

[16]   Q. Zhou, K. H. Lam, H. Zheng, W. Qiu and K. K. Shung, "Piezoelectric single crystals for ultrasonic transducers in biomedical applications," *Prog. Mater. Sci.* 66, 87, 2014.

[17]   K. Kitamura, J.K. Yamamoto, N. Iyi, S. Kirnura and T. Hayashi, "Stoichiometric $LiNbO_3$ single crystal growth by double crucible Czochralski method using automatic powder supply system," *J. Crystal Growth*, 116, 327–332, 1992.

[18]   A. Grigas and S. Kaskel, "Synthesis of $LiNbO_3$ nanoparticles in a mesoporous matrix," *Beilstein J. Nanotechnol.* 2, 28–33, 2011.

[19]   V. Ya. Shur, "Lithium niobate and lithium tantalate-based piezoelectric materials." Advanced Piezoelectric Materials: Science and Technology, Woodhead Publishing Series in Electronic and Optical Materials, Cambridge, 204–238, 2010.

[20] G. Clementi, G. Lombardi, S. Margueron, M. Suarez, A. Miguel, E. Lebrasseur, S. Ballandras, J. Imbaud, F. Lardet-Vieudrin, L. Gauthier-Manuel, B. Dulmet, M. Lallart and A. Bartasyte, "LiNbO$_3$ films - a low-cost alternative lead-free piezoelectric material for vibrational energy harvesters," *Mechan. Sys. Sig. Process.,* 149, 107171, 2021.

[21] V. Zakhar, G. Andrey, K. Daniil, V.I Marina, I.L Oleg, S.K Viktor, N. D. Vladimir, G.K. Boris, H. Zhubing and A.A. Oleg, "Piezoelectric energy harvester based on LiNbO$_3$ thin films," *Materials* 13, 3984, 2020.

[22] G. Clementi, M. Ouhabaz, S. Margueron, M.A. Suarez, F. Bassignot, L. Gauthier-Manuel, D. Belharet, B. Dulmet and A. Bartasyte, "Highly coupled and low frequency vibrational energy harvester using lithium niobate on silicon," *Appl. Phys. Lett.* 119, 013904, 2021.

[23] Z.L. Wang, "Self-powered nanosensors and nanosystem,." *Adv. Mater.* 24: 280–285, 2012.

[24] G. Hwang, H. Park, J.H Lee, S. Oh, K. Park, M. Byun, H. Park, G. Ahn, K. Jeong, K. No, H. Kwon, S. Lee, B. Joung and K. Lee, "Self-powered cardiac pacemaker enabled by flexible single crystalline PMN-PT piezoelectric energy harvester, " *Adv. Mater.,* 26(28), 4880–4887, 2014.

[25] P. Gonnard and M. Troccaz, "Dopant distribution between A and B sites in the PZT ceramics of type ABO$_3$," *J/ Solid-State Chem.* 23, 321–326, 1978.

[26] C.A. Randall, R.E. Newnham and L.E. Cross, "History of the First Ferroelectric Oxide, BaTiO3," The Pennsylvania State University, University Park, USA, 2004.

[27] B. Jaffe, R.S. Roth and S. Marzullo, "Piezoelectric properties of lead zirconate-lead titanate solid-solution ceramics," *J. Appl. Phys.* 25, 809, 1954.

[28] M.H Lee, D.J. Kim, J.S. Park, S.W. Kim, T.K Song, M.H. Kim, W.J. Kim, D. Do and I.K Jeong, "High-Performance lead-free piezoceramics with high Curie temperatures," *Adv. Mater.* 27, 6976–6982, 2015.

[29] R. Ragini, R. Ranjan, S.K. Mishra and D. Pandey, "Room temperature structure of Pb(Zr$_x$Ti$_{1-x}$)O$_3$ around the morphotropic phase boundary region: a Rietveld study," *J. Appl. Phys.,* 92(6), 3266–3274, 2002.

[30] B. Noheda, D.E. Cox, G. Shirane, J.A. Gonzalo, L.E. Cross and S.-E. Park, "A monoclinic ferroelectric phase in the Pb(Zr$_x$Ti$_{1-x}$)O$_3$ solid solution" *Appl. Phys. Lett.* 1999, 74(14), 2059–2061, 1999.

[31] B. Jaffe, W.R. Cook and H. Jaffe, "Piezoelectric ceramics," Academic Press, London, p. 317, 1971.

[32] A. Bouzid, E.M. Bourim, M. Gabbay and G. Fantozzi "PZT phase diagram determination by measurement of elastic moduli," *J. Eur. Ceramic Soc.* 25, 3213–3221, 2005.

[33] X. Du, J. Zheng, U. Belegundu and K. Uchinoa "Crystal orientation dependence of piezoelectric properties of lead zirconate titanate near the morphotropic phase boundary," *Appl. Phys. Lett.* 72, 2421 1998.

[34] J. Wu, "Perovskite lead-free piezoelectric ceramics," *J. Appl. Phys.* 127, 190901, 2020.

[35] S. Wada, K. Takeda, T. Muraishi, H. Kakemoto, T. Tsurumi and T. Kimura, "Domain wall engineering in lead-free piezoelectric grain-oriented ceramics," *Ferroelectrics,* 373, 11, 2008.

[36] Y. Hiruma, K. Marumo, R. Aoyagi, H. Nagata and T. Takenaka, "Ferroelectric and piezoelectric properties of (Bi$_{1/2}$K$_{1/2}$)TiO$_3$ ceramics fabricated by hot-pressing method," *J. Electroceram.* 21, 296, 2008.

[37] Y. Hiruma, H. Nagata and T. Takenaka, "Grain-size effect on electrical properties of (Bi$_{1/2}$K$_{1/2}$)TiO$_3$ ceramics," *Japan. J. Appl. Phys.* 1, 46, 1081, 2007.

[38] C.G. Xu, D.M. Lin and K.W. Kwok, "Structure, electrical properties and depolarization temperature of (Bi0.5Na$_{0.5}$)TiO$_3$–BaTiO$_3$ lead-free piezoelectric ceramics," *Solid State Sci.* 10, 934, 2008.

[39] K. Yoshii, Y. Hiruma, H. Nagata and T. Takenaka, "Electrical Properties and Depolarization Temperature of (Bi$_{1/2}$Na$_{1/2}$)TiO$_3$–(Bi$_{1/2}$K$_{1/2}$)TiO$_3$ Lead-free Piezoelectric Ceramics." *Japan. J. Appl. Phys.* 1, 45, 4493, 2006.

[40] S.J. Zhang, T.R. Shrout, H. Nagata, Y. Hiruma and T. Takenaka, "Piezoelectric properties in (K0.5Bi$_{0.5}$)TiO$_3$-(Na$_{0.5}$Bi$_{0.5}$)TiO$_3$-BaTiO$_3$ lead-free ceramics," *IEEE Trans. Ultrason. Ferroelectr. Freq. Control,* 54, 910, 2007.

[41] H. Nagata, K. Matsumoto, T. Hirosue, Y. Hiruma and T. Takenaka, "Fabrication and electrical properties of potassium niobate ferroelectric ceramics," *Japan. J. Appl. Phys.* 1, 46, 7084, 2007.

[42] R.E. Jaeger and L. Egerton, "Hot pressing of potassium-sodium niobates," *J. Am. Ceram. Soc.* 45, 209, 1962.

[43]  E. Hollenstein, M. Davis, D. Damjanovic and N. Setter, "Piezoelectric properties of Li- and Ta-modified (K0.5Na$_{0.5}$)NbO$_3$ ceramics," *Appl. Phys. Lett.* 87, 182905, 2005.

[44]  J.L. Zhang, X.J. Zong, L. Wu, Y. Gao, P. Zheng and S.F. Shao, "Polymorphic phase transition and excellent piezoelectric performance of (K0.55Na$_{0.45}$)$_{0.965}$Li$_{0.035}$Nb$_{0.80}$Ta$_{0.20}$O$_3$ lead-free ceramics," *Appl. Phys. Lett.* 95, 022909, 2009.

[45]  R. Wang, R.J. Xie, K. Hanada, K. Matsusaki, H. Kawanaka, H. Bando, T. Sekiya and M. Itoh, "Enhanced piezoelectricity around the tetragonal/orthorhombic morphotropic phase boundary in (Na,K)NbO$_3$–ATiO$_3$ solid solutions," *J. Electroceram.* 21, 263, 2008.

[46]  R.Z. Zuo, J. Fu and D.Y. Lv, "Phase transformation and tunable piezoelectric properties of lead-free (Na0.52K$_{0.48-x}$Li$_x$)(Nb$_{1-x-y}$Sb$_y$Ta$_x$)O$_3$ System," *J. Am. Ceram. Soc.* 92, 283, 2009.

[47]  E.Z. Li, R. Suzuki, T. Hoshina, H.Takeda and T. Tsurumi, "Dielectric, piezoelectric, and electro-mechanical phenomena in (K$_{0.5}$Na$_{0.5}$)NbO$_3$–LiNbO$_3$–BiFeO$_3$–SrTiO$_3$ ceramics," *Appl. Phys. Lett.* 94, 132903, 2009.

[48]  H.L. Du, W.C. Zhou, F. Luo, D.M. Zhu, S.B. Qu, Y. Li and Z.B. Pei, "High T$_m$ lead-free relaxor ferroelectrics with broad temperature usage range: 0.04BiScO$_3$–0.96(K$_{0.5}$Na$_{0.5}$)NbO$_3$," *J. Appl. Phys.* 104, 044104, 2008.

[49]  S.O. Leontsev and R.E. Eitel, "Progress in engineering high strain lead-free piezoelectric ceramics," *Sci. Technol. Adv. Mater.* 11, 044302, 2010.

[50]  N. Zhang, T. Zheng and J. Wu "Lead-free (K,Na)NbO$_3$-based materials: Preparation techniques and piezoelectricity," *ACS Omega* 2020, 5(7), 3099–3107, 2020.

[51]  T. Zheng, J. Wu, D. Xiao and J. Zhu, "Recent development in leadfree perovskite piezoelectric bulk materials," *Prog. Mater. Sci.* 98, 552–624, 2018.

[52]  J. Wu, D. Xiao and J. Zhu "Potassium-sodium niobate lead-free piezoelectric materials: past, present, and future of phase boundaries," *Chem. Rev.* 115, 2559–2595, 2015.

[53]  X. Gao, Z. Cheng, Z. Chen, Q. Guo, B. Li, H. Sun, Q. Gu, Y. Liu, X. Meng, X. Zhang, J. Wang, H. Hao, Q. Shen, J. Wu, X. Liao, S.P. Ringer, H. Liu, L. Zhang, W. Chen, F. Li and S. Zhang, "The mechanism for the enhanced piezoelectricity in multi-elements doped (K,Na)NbO$_3$ ceramics," *Nat. Commun.* 12, 881, 2021.

[54]  S.H. Wemple, M. Didomenico Jr. and I. Camlibe, "Dielectric and optical properties of melt-grown BaTiO$_3$," *J. Phys. Chem. Solids* 29, 1797–1803, 1968.

[55]  R. Bechmann, "Elastic, piezoelectric, and dielectric constants of polarized barium titanate ceramics and some applications of the piezoelectric equations," *J. Acoust. Soc. Am.* 28, 347 1956.

[56]  C.L. Zhao, H.J. Wu, F. Li, Y. Cai, Y. Zhang, D. Song, J.G. Wu, X. Lyu, J. Yin, D.Q. Xiao, J.G. Zhu and S.J. Pennycook, "Practical high piezoelectricity in barium titanate ceramics utilizing multiphase convergence with broad structural flexibility" *J. Am. Chem. Soc.* 140, 15252, 2018.

[57]  G.A. Smolenskii, V.A. Isupov, A.I. Agranovskaya and N.N. Krainik "New ferroelectrics of complex composition" *IV. Sov. Phys. Solid State* 2, 2651, 1961.

[58]  J. Haoa, W. Lia, J. Zhaia and H. Chenc, "Progress in high-strain perovskite piezoelectric ceramics," *Mater. Sci. Eng.: R: Reports* 135, 1–57, 2019.

[59]  X. Zhou, G. Xue, H. Luo, C.R. Bowen and D. Zhang, "Phase structure and properties of sodium bismuth titanate lead-free piezoelectric ceramics," *Prog. Mater. Sci.* 122, 100836, 2021.

[60]  M. Otonicar, J. Park, M. Logar, G. Esteves, J.L. Jones and B. Jancar "External-field-induced crystal structure and domain texture in (1-x)Na$_{0.5}$Bi$_{0.5}$TiO$_3$-xK$_{0.5}$Bi$_{0.5}$TiO$_3$ piezoceramics" *Acta Materialia* 127, 319–331, 2017.

[61]  Y. Guo, Y. Liu, R.L. Withers, F. Brink and H. Chen, "Large electric field-induced strain and antiferroelectric behavior in (1-x)(Na0.5Bi$_{0.5}$)TiO$_3$-x BaTiO$_3$ ceramics," *Chem. Mater.* 23, 219–228, 2011.

[62]  H. Qian, Z. Yu, M, Mao, Y. Liu and Y. Lyu, "Nanoscale origins of small hysteresis and remnant strain in Bi0.5Na$_{0.5}$TiO$_3$-based lead-free ceramics," *J. Eur. Ceram. Soc.* 38, 361–369, 2018.

[63]  N.P.M.J. Raj, G. Khandelwal and S.J. Kim "0.8BNT–0.2BKT ferroelectric-based multimode energy harvester for self-powered body motion sensors," *Nano Energy* 83, 105848, 2021.

[64]  J.G. Wu, Z. Fan, D.Q. Xiao, J.G. Zhu and J. Wang, "Multiferroic bismuth ferrite-based materials for multifunctional applications: Ceramic bulks, thin films and nanostructures," *Prog. Mater. Sci.* 84, 335, 2016.

[65] G. Catalan and J.F. Scott, "Physics and applications of bismuth ferrite," *Adv. Mater. 21*, 2463–2485, 2009.

[66] H. Tao, J. Lv, R. Zhang, R. Xiang and J. Wu "Lead-free rare earth-modified BiFeO$_3$ ceramics: Phase structure and electrical properties," *Mater. Des.* 120, 83–89, 2017.

[67] C. Zhou, H. Yang, Q. Zhou, G. Chen, W. Li, H. Wang, "Effects of Bi excess on the structure and electrical properties of high-temperature BiFeO$_3$-BaTiO$_3$ piezoelectric ceramics," *J. Mater. Sci. Mater. Electron.* 24, 1685–1689, 2013.

[68] L.F. Zhu, B.P. Zhang, J.Q. Duan, B.W. Xun, N. Wang, Y.C. Tang and G.L. Zhao, "Enhanced piezoelectric and ferroelectric properties of BiFeO$_3$-BaTiO$_3$ lead-free ceramics by optimizing the sintering temperature and dwell time," *J. Eur. Ceram. Soc.* 38, 3463–3471, 2018.

[69] G. Kalimuldina, N. Turdakyn, I. Abay, A. Medeubayev, A. Nurpeissova, D. Adair, Z. Bakenov, "A review of piezoelectric PVDF film by electrospinning and its applications," *Sensors* 20, 5214, 2020.

[70] G.D. Zhu, Z.G. Zeng, L. Zhang and X.J. Yan, "Piezoelectricity in beta-phase PVDF crystals: A molecular simulation study," *Comput. Mater. Sci.* 44, 224–229, 2008.

[71] X.M. Wang, F.Z. Sun, G.C. Yin, Y.T. Wang, B. Liu and M.D. Dong, "Tactile-sensing based on flexible PVDF nanofibers via electrospinning: A review," *Sensors* 18, 16, 2018.

[72] H.Y. Pan, B. Na, R.H. Lv, C. Li, J. Zhu and Z.W. Yu, "Polar phase formation in poly(vinylidenefluoride) induced by melt annealing," *J. Polym. Sci. Pt. B-Polym. Phys.* 50, 1433–1437, 2012.

[73] A. Lund, C. Gustafsson, H. Bertilsson and R.W. Rychwalski, "Enhancement of beta phase crystals formation with the use of nanofillers in PVDF films and fibres," *Compos. Sci. Technol.* 71, 222–229, 2011.

[74] Z.H. Liu, C.T. Pan, L.W. Lin, J.C. Huang and Z.Y. Ou, "Direct-write PVDF nonwoven fiber fabric energy harvesters via the hollow cylindrical near-field electrospinning process," *Smart Mater. Struct.* 23, 11, 2014.

[75] Y. Liu, H. Aziguli, B. Zhang, W. Xu, W. Lu, J. Bernholc and Q.Wang. "Ferroelectric polymers exhibiting behaviour reminiscent of a morphotropic phase boundary," *Nature* 562, 96–100, 2018.

[76] Y. Huang, G. Rui, Q. Li, E. Allahyarov, R. Li, M. Fukuto, G. Zhong, J. Xu, Z. Li, P. Taylor and L. Zhu "Enhanced piezoelectricity from highly polarizable oriented amorphous fractions in biaxially oriented poly(vinylidene fluoride) with pure β crystals," *Nat. Commun.* 12, 675 2021.

[77] C. Chang, V.H. Tran, J. Wang, Y.K. Fuh and L. Lin "Direct-write piezoelectric polymeric nanogenerator with high energy conversion efficiency," *Nano Lett.* 10, 726–731, 2010.

[78] P.K. Szewczyk, A. Gradys, S. Kim, L. Persano, M.M. Marzec, A.P. Kryshtal, T. Busolo, A. Toncelli, D. Pisignano, A. Bernasik, S. Narayan, P. Sajkiewicz and U. Stachewicz "Enhanced piezoelectricity of electrospun polyvinylidene fluoride fibers for energy harvesting," *ACS Appl. Mater. Interfaces* 12, 13575–13583, 2020.

[79] F. Xia, Z.-Y. Cheng, H.S. Xu, H.F. Li, Q.M. Zhang , G.J. Kavarnos, R.Y. Ting, G. Abdul-Sadek and K.D. Belfield, "High electromechanical responses in a poly (vinylidene fluoride–trifluoroethylene–chlorofluoroethylene) terpolymer," *Adv. Mater.* 14, 1574–1577, 2002.

[80] X. Chen, H. Qin, X. Qian, W. Zhu, B. Li, B. Zhang, W. Lu, R. Li, S. Zhang, L. Zhu, F.D.D. Santos, H. Bernholc and Q.M. Zhang, "Relaxor ferroelectric polymer exhibits ultrahigh electromechanical coupling at low electric field," *Science* 375, 1418, 2022.

[81] Penn State. "Enhancing the electromechanical behaviour of a flexible polymer: Researchers improved the material's electricity generation efficiency by 60%." ScienceDaily, 24 March 2022. www.sciencedaily.com/releases/2022/03/220324143755.htm

[82] R.E. Newnham, D.P. Skinner and L.E. Cross, "Connectivity and piezoelectric-pyroelectric composites," *Mater. Res. Bull.* 13, 525–536, 1978.

[83] K.A. Hannert, A. Safari, R.E. Newnham and J. Runt, "Thin film 0-3 polymer/piezoelectric ceramic composites: Piezoelectric paints," *Ferroelectrics* 100, 255–260, 1989.

[84] V. Tiwari and G. Srivastava "Structural, dielectric and piezoelectric properties of 0-3 PZT/PVDF composites," *Ceram. Int.* 41, 8008, 2015.

[85] S. Wankhade , S. Tiwari, A. Gaur and P. Maiti, "PVDF–PZT nanohybrid based nanogenerator for energy harvesting applications," *Energy Rep.* 6, 358–364, 2020.

[86] P. Bai, S. Wang1, J. Jia, H. Wang and W. Yang, "Effect of BaTiO$_3$ nanowire on effective permittivity of the PVDF composites," *AIP Adv.* 11, 045018, 2021.

[87] W. Choi, K. Choi, G. Yang, J. C. Kim and C. Yu "Improving piezoelectric performance of lead-free polymer composites with high aspect ratio BaTiO$_3$ nanowires," *Polymer Test.* 53, 143–148, 2016.

[88] J. Zhonga, Q. Zhong, X. Zang, N. Wu, W. Li, Y. Chua and L. Lin "Flexible PET/EVA-based piezoelectret generator for energy harvesting in harsh environments," *Nano Energy*, 37, 268–274, 2017.

[89] Y. Nan, D. Tan, J. Shao, M. Willatzen and Z.L. Wang "2D materials as effective cantilever piezoelectric nano energy harvesters," *ACS Energy Lett.* 6, 2313–2319, 2021.

[90] M.H. Lee and W. Wu "2D materials for wearable energy harvesting," *Adv. Mater. Technol.* 2101623, 2022.

# 2 An Overview of Fuel Cells and Supercapacitors

*Shreya Tumma, Divya Velpula, Uday Kumar Adapa,
Shirisha Konda, Rakesh Kumar Thida, Madhuri Sakaray,
Shilpa Chakra Chidurala, Bala Narasaih Tumma, and
Srilatha K.*

## 2.1 INTRODUCTION

Even if the demand for energy changes throughout the day, the amount of electricity that can be created is practically constant during short periods. Developing technology that can store electrical energy and deploy it to meet demand whenever it is needed would be a big step forward in power distribution. Storage devices for electricity can help fulfil this aim by managing the amount of power necessary to provide consumers at peak load. Renewable energy, whose power production cannot be regulated by grid operators, can also benefit from these devices.

They can also balance microgrids to ensure that generation and load are well matched. Storage devices can offer frequency management to keep the demand on the network balanced with the power generated, resulting in a consistent power supply for high-tech industrial facilities. As a result, power electronics and energy storage have a lot of potential for revolutionizing the electric power business. In several applications, like stationary power generation, battery replacement, and transportation, fuel cells may be able to replace the already existing power equipment. They are already commercially accessible for uses such as portable power sources, transportation, and small-scale power generation.

Fuel cells can be explained as electrochemical cells that create electricity by an electrochemical reaction of hydrogen as well as oxygen where the products obtained are heat, water [1]. There are two electrodes, the cathode and anode, which are divided by an F that make up a fuel cell. It is this electrolyte making the motion of the protons possible. At the anode, anion undergoes oxidation, whereas, at the cathode, cation undergoes reduction. William Grove, a lawyer and scientist, demonstrated the first rudimentary basic electrochemical cell in 1839. In his experiment, water was electrolyzed into oxygen and hydrogen by running electricity via a battery, which was then replaced by a multimeter, which produced a modest current.

The burning or combustion of hydrogen fuel occurs in a simple reaction, which is depicted as follows:

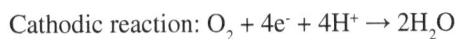

$$2H_2 + O_2 \rightarrow 2H_2O \text{ ----------} \tag{1}$$

$$\text{Anodic reaction: } 2H_2 \rightarrow 4H^+ + 4e^-$$

$$\text{Cathodic reaction: } O_2 + 4e^- + 4H^+ \rightarrow 2H_2O$$

So, at the end of both equations, we get water and energy.

Supercapacitors (SC), sometimes known as ultracapacitors, have a high capacitance value, significantly more than conventional capacitors but fewer voltage limitations than typical capacitors.

DOI: 10.1201/9781003340539-2

It can store ten to one hundred times the energy of electrolytic capacitors per unit volume or mass, takes and distributes charge considerably faster than batteries, and can endure a lot more discharge and charge cycles than rechargeable batteries. In buses, autos, cranes, elevators, and trains these are used for storing energy for short periods, regenerative braking, or burst-mode power delivery.

The component that stores energy electrostatically is called a capacitor. Two conducting electrodes are divided by an insulating dielectric material in prevalent electrochemical capacitors.

The ratio of electric charge on each electrode ($Q$) to the potential difference ($V$) between them is simply the capacitance ($C$) of an electrochemical capacitor so that

$$C = \frac{Q}{V}$$

The energy stored in a capacitor is related to the charge of the electrode, at each interface and the potential difference ($V$) is directly proportional to its capacitance ($C$):

$$E = \frac{1}{2} * CV^2$$

Supercapacitors are in the middle of the energy-power continuum, having higher energy densities than dielectric capacitors and higher power densities than batteries. The typical power density and energy density of commercially available supercapacitors are 10–20 kW/kg and 4–5 Wh/kg, respectively, however, the latest improvements indicate that these numbers might have been surpassed recently. Despite having lower energy densities than batteries, supercapacitors offer some benefits that batteries do not, including a practically longest lifespan. They also have the capacity to charge quickly (as opposed to battery materials), have a larger range of temperature, are environmentally friendly, have improved safety, and dependability, and are maintenance-free.

Supercapacitors, unlike batteries, are primarily thought of as pulse power sources, with discharge times going from some seconds to less than one second. Supercapacitors are now in high demand in the transportation industry, particularly passenger transport, like tramways, buses, as a primary and secondary energy source. Supercapacitors are great for starting engines and brakes using regenerative energy because they can store charge. They are also utilized in renewable energy (to change the pitch of a wind turbine's rotor blades, for example), industrial machinery (forklifts, cranes), electronics, and consumer products.

## 2.2  FUEL CELLS

### 2.2.1  Concept and History of Fuel Cells

The breakthrough of these fuel cells dates back to 1838. In the "*London and Edinburgh Philosophical Magazine and Journal of Science*," published in December 1838, a Welsh physicist and barrister, Sir William Grove, discussed his first rudimentary electrochemical cells with the use of copper sulphate and dilute acid solution, as well as copper, porcelain plates, and sheet iron [1]. In a similar publication published in 1839, Christian Friedrich Schonbein wrote about the earliest primitive fuel cell that he had built. Francis Thomas Bacon, an engineer, constructed a stationary fuel cell, which gave a power output of 5 kW in 1950 [2]. Since the mid-1960s, NASA has been using this fuel cell, alkaline fuel cell. This cell is familiar as the Bacon cell. In honor of the fuel cell industry and America's contribution to fuel cell research, the US Senate established October 8, 2015, National Hydrogen and Fuel Cell Day. The stack power densities of the second-generation MIRAI were 4.4 kWl$^{-1}$ and 5.4 kWl$^{-1}$, respectively, up 42 percent and 54 percent from the previous version [3].

### 2.2.2 Components of Fuel Cells

A fuel cell mainly has three components [4].

#### 2.2.2.1 Anode

The anode, the negative part of the cell transports the liberated electrons from $H_2$ molecules to an external circuit. The anion undergoes oxidation here. Also, it is here where the hydrogen molecule splits into protons and electrons.

#### 2.2.2.2 Cathode

The cathode, the positive part of the cell carries electrons from the outer circuit to the catalyst's surface, and they merge with oxygen and ions of hydrogen to form $H_2O$.

#### 2.2.2.3 Electrolyte

Electrolyte is responsible for the movement of protons in the fuel cell. It maintains the proper proportion of ions to travel through the anode and cathode. The fuel type is determined by the electrolyte utilized.

### 2.2.3 Working Principle of Fuel Cells

Operation of these cells is straightforward. It involves electricity-generating reaction between hydrogen and oxygen. One of these fuel cells was employed as a potable water and fuel source during the Apollo space mission [5].

Hydrogen and oxygen are passed into the electrolyte (here, concentrated sodium hydroxide) via carbon electrodes in this fuel cell as shown in the Figure 2.1.

The cell reaction are as follows [4]:

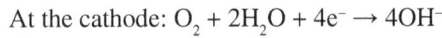

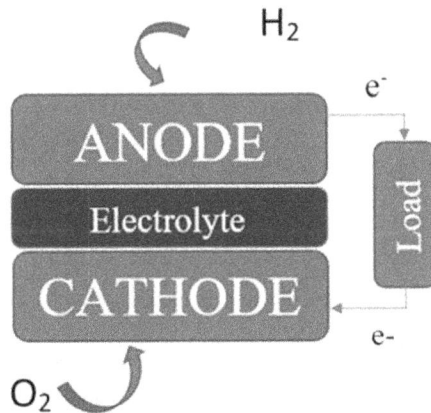

$$\text{At the cathode: } O_2 + 2H_2O + 4e^- \rightarrow 4OH^-$$

$$\text{At the anode: } 2H_2 + 4OH^- \rightarrow 4H_2O + 4e^-$$

$$\text{Net reaction: } 2H_2 + O_2 \rightarrow 2H_2O\text{----------} \tag{2}$$

**FIGURE 2.1**  Block diagram of fuel cell.

Anyway, the reaction is having a slow rate. We can speed up the reaction using a catalyst like platinum or palladium. Depending on the form of energy, they produce heat, water, and trace quantities of nitrogen dioxide and other pollutants. Fuel cells have an energy efficiency of 40 to 60% [6].

To ensure continuous isothermal operation for optimal electric power generation, water and heat by-products must be frequently removed. As a result, water and temperature control are important aspects of fuel cell design and operation [4].

### 2.2.4   CLASSIFICATION OF FUEL CELLS

Depending on the type of electrolyte, fuel cells are divided into various types. Each fuel cell has its specialities, advantages, and drawbacks, and is designed for a particular usage. Power outputs, operating temperatures, electrical efficiency, and usual applications are all different. Some of the types include proton exchange membrane fuel cell, Alkaline fuel cell, solid oxide fuel cell, direct methanol fuel cell, molten carbonate fuel cell [7].

#### 2.2.4.1   Proton Exchange Membrane Fuel Cell

Early in the 1960s, General Electric developed the polymer electrolyte membrane fuel cell, known to be a proton exchange membrane fuel cell (PEMFC), for the Gemini space missions. This fuel cell contains electrodes, a polymer membrane, bipolar plates, and catalyst. In PEMFC, at the anode, hydrogen undergoes oxidation whereas at the cathode, oxygen undergoes reduction. Platinum-based electrodes are used as electrodes in PEMFC, whereas a water-based acidic polymer membrane is used as electrolyte [8]. Because of the low operating temperatures (50–100°C), these cells run on pure hydrogen, platinum electrodes and can achieve an efficiency of 60% [9]. Polymer membrane cells are used in portable communication, stationary application, transportation, etc., [8].

#### 2.2.4.2   Solid Oxide Fuel Cell

SFCs were first built by Baur and Preis, which used magnetite as the oxidant and coke as the fuel in 1937. Later in the1960s, there was more research on SFCs with more emphasis on submarine, space, and military applications, in particular. To conduct negative oxygen ions from the cathode to the anode, these cells employ an oxide substance as the electrolyte [1]. In recent times, proton-conducting solid oxide fuel cells are being developed where protons are transported instead of negative oxygen ions through the electrolyte, which enables the usage of the cell at low temperatures also. Traditional SFC's work at very high temperatures (500–1000 °C). Though these cells are resistant to carbon monoxide poisoning, they are vulnerable to sulfur poisoning [10]. The working of this cell is as shown in Figure 2.2.

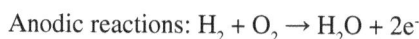

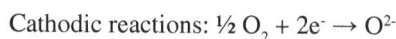

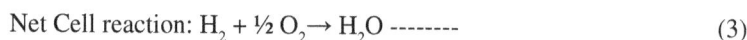

$$\text{Anodic reactions: } H_2 + O_2 \rightarrow H_2O + 2e^-$$

$$\text{Cathodic reactions: } \tfrac{1}{2} O_2 + 2e^- \rightarrow O^{2-}$$

$$\text{Net Cell reaction: } H_2 + \tfrac{1}{2} O_2 \rightarrow H_2O \text{ --------} \tag{3}$$

These cells achieve up to 60% efficiency and are used in power generation [11].

#### 2.2.4.3   Alkaline Fuel Cell (AFC)

AFCs were the earliest fuel cell types, which received substantial development and employed by NASA for some space shuttle missions like Gemini and Apollo. This fuel cell was first developed by the British inventor, Francis Thomas Bacon in 1939. After its inventor, the alkaline fuel cell is also

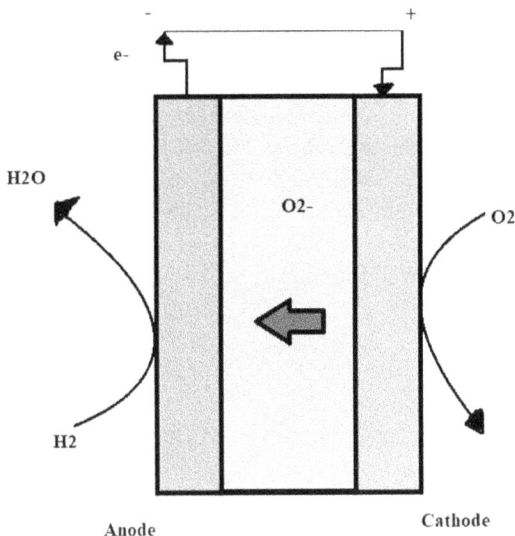

**FIGURE 2.2** Solid oxide fuel cell.

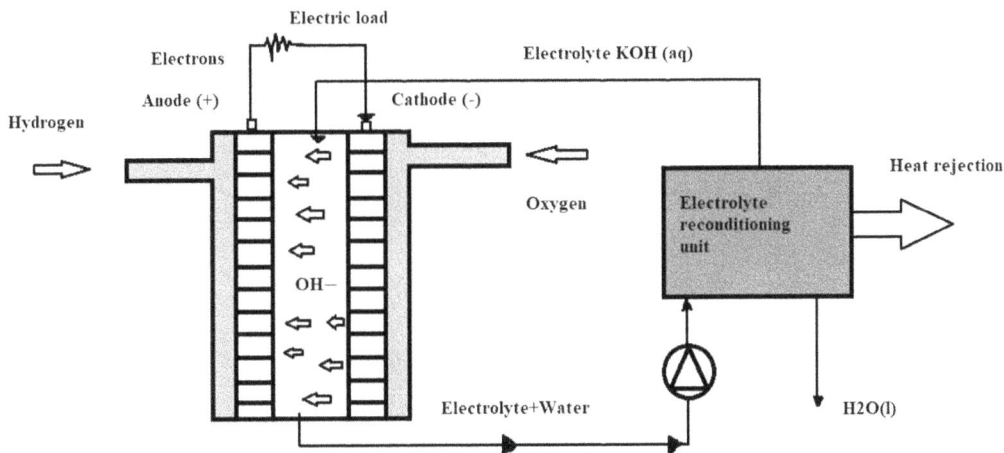

**FIGURE 2.3** Alkaline fuel cell.

called the Bacon cell. Thomas Bacon used potassium hydroxide as the electrolyte, and porous gas-diffusion as the electrode. The working temperature of this cell is around 250°C, and can achieve an efficiency of up to 70% [8].

The working of alkaline fuel cells is shown in Figure 2.3.

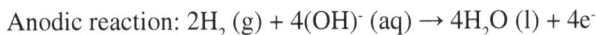

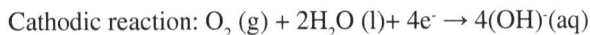

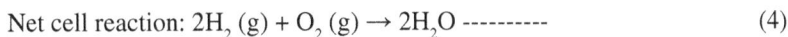

$$\text{Anodic reaction: } 2H_2 \text{ (g)} + 4(OH)^- \text{ (aq)} \rightarrow 4H_2O \text{ (l)} + 4e^-$$

$$\text{Cathodic reaction: } O_2 \text{ (g)} + 2H_2O \text{ (l)} + 4e^- \rightarrow 4(OH)^- \text{(aq)}$$

$$\text{Net cell reaction: } 2H_2 \text{ (g)} + O_2 \text{ (g)} \rightarrow 2H_2O \text{ ----------} \tag{4}$$

#### 2.2.4.4 Molten Carbonate Fuel Cell

MCFCs were first discovered in the 1950s by two Dutch researchers, J.A.A. Ketelaarand G.H.J. Broers, at the University of Amsterdam [12]. A molten carbonate salt mixture in a beta-alumina solid electrolyte (BASE) matrix is used as the electrolyte in these cells [13]. These work at high temperatures around 600°C. Unlike the others, MCFCs do not need an additional reformer to transform more energy-dense fuels to hydrogen. They are transformed to hydrogen by themselves due to their high operating temperatures through a method called internal reforming, which also decreases cost.

Working of a MCFC is as shown in Figure 2.4.

Cell reactions [14]:

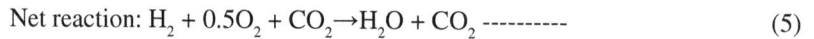

$$\text{Anodic reaction: } H_2 + CO_3^{2-} \rightarrow H_2O + CO_2 + 2e^-$$

$$\text{Cathodic reaction: } 0.5O_2 + CO_2 + 2e^- \rightarrow CO_3^{2-}$$

$$\text{Net reaction: } H_2 + 0.5O_2 + CO_2 \rightarrow H_2O + CO_2 \text{ ----------} \tag{5}$$

#### 2.2.4.5 Direct Methanol Fuel Cell (DMFC)

DMFCs were built in the 1990s by scientists from different universities in the US. This is a low-temperature cell where the operating temperatures are around 50 to 120°C. Due to the thermal signatures, low noise, and no effluent characteristics of DMFCs, they are used in the military. These are also used in power necessities like portable power packs and chargers or mobile electronic devices. The catalysts used here are Pt-Ru/C as the anode and Pt/C as the cathode [14]. The efficiency of a direct methanol cell is relatively low around 30% because of its low rate of reaction [15] and working is as shown in Figure 2.5.

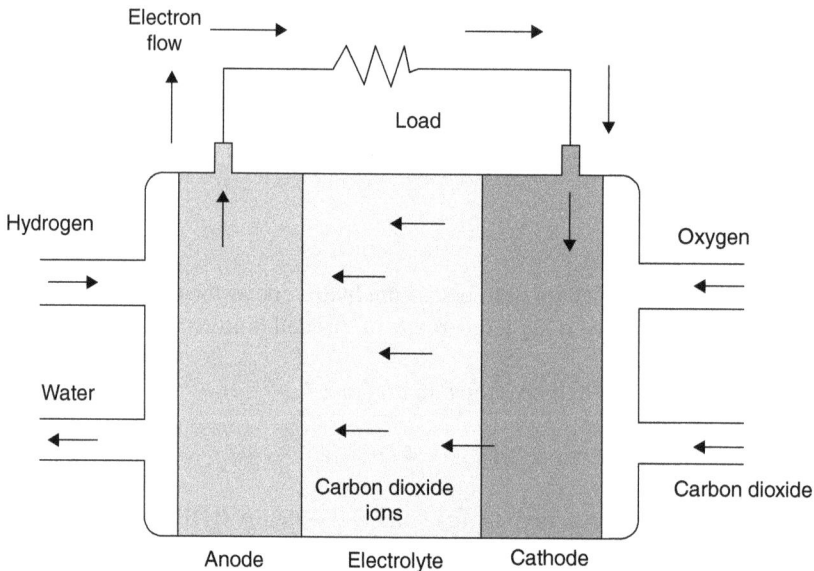

**FIGURE 2.4** Molten carbonate fuel cell.

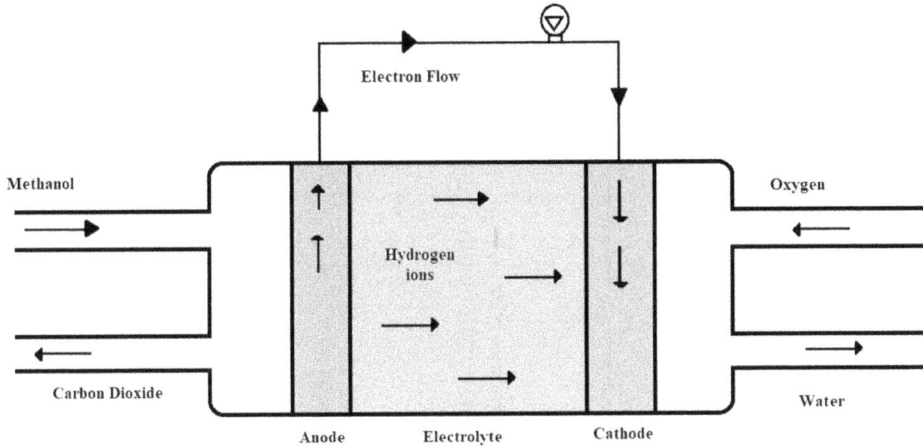

**FIGURE 2.5** Direct methanol fuel cell.

At the cell [14].

$$\text{Anodic reaction: } CH_3OH + H_2O \rightarrow CO_2 + 6H^+ + 6e^-$$

$$\text{Cathodic reaction: } 3\backslash2\ O_2 + 6H^+ + 6e^- \rightarrow 3H_2O$$

$$\text{Net reaction: } CH_3OH + 3/2O_2 \rightarrow CO_2 + 2H_2O \text{ --------} \tag{6}$$

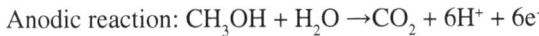

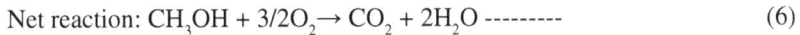

## 2.2.5 Fabrication of Electrode Materials

The materials with which the electrodes are made have a high impact on the efficiency of the machine. Recent researches have enabled electrochemistry to explore emerging potentials. Some of the examples are given below in Table 2.1.

## 2.2.6 Types of Electrode Materials and Electrolytes

There are numerous types of electrolytes and electrode materials depending on the fuel cell type [23].

### 2.2.6.1 Polymer Electrolyte Membrane Fuel Cells

In PEMFCs, a proton-conducting polymer membrane like Perfluorosulfonic acid acts as the electrolyte, a platinum anode and nickel cathode are used.

### 2.2.6.2 Direct-Methanol Fuel Cells

Similar to PEMFCs, DMFCs also employ a proton-conducting polymer membrane as an electrolyte.

### 2.2.6.3 Alkaline Fuel Cells

In AFCs, an alkaline electrolyte like KOH or any alkaline membrane that is capable of conducting hydroxide ions instead of protons is used. KOH is employed as an electrolyte in the fuel cell, and an activated carbon cathode and nickel anode are used.

**TABLE 2.1**
**Fabrication of Electrode Materials**

| Precursor | Reducing Agent/ Oxidizing/ Potential | Synthesis Method | Material | Morphology | Size/ Thickness (nm) | Ref. |
|---|---|---|---|---|---|---|
| $K_2PtCl_4$, $CoCl_2$, Te NWs | KOH | Dual-template method | PtCo MNTs | Mesoporous | 20 nm | [16] |
| $HClO_4$, $CuSO_4$, $PtCl_6^{2-}$ | -0.50 V vs Ag/AgCl | Electrodeposition | Pt-Cu, Pt-Ni, and Pt-Co | Nodular | 125 nm | [17] |
| $PtCl_4$, polyaniline, rGO | Dimethylammonium bromide | Chemical coprecipitation | Pt@rGO-Polyaniline | Monodisperse distribution of Pt nanoparticles on the rGO- Polyaniline hybrid support | 3.96 nm | [18] |
| Potassium tetrachloropalladate (II) | 0.300 V | Electrodeposition | Pd | Thorn flower Each Thorn | 1 μm 10 nm | [19] |
| $H_2PtCl_6.6H_2O$, $PdCl_2$ | -4 V | Electrodeposition | PtPd | 3D Pd foam film | 5–10 μm | [20] |
| Aniline, potassium tetrachloroplatinate, walled carbon nanotubes | -0.2 to 1.0 V | Electrodeposition | Platinum-polyaniline-multiwalled carbon nanotubes | Sugarcoated haws structure | 260 nm | [21] |
| Indole monomer, silver nitrate | Ammonium persulfate | Liquid/liquid interfacial method | Ag-polyindole | Hallow-sphere, pentagons | 21 nm | [22] |

#### 2.2.6.4 Molten Carbonate Fuel Cells (MCFCs)

The electrolyte in MCFCs is molten carbonate salt immobilized in a porous matrix that is capable of conducting carbonate ions. Lithium-, sodium-, and potassium-based carbonates are employed as an electrolyte in the fuel cell, and a nickel, chromium, and aluminium alloy anode, and porous nickel cathode are used.

#### 2.2.6.5 Solid Oxide Fuel Cells

In SOFCs, a coating of ceramic serves as the electrolyte, conducting oxide ions. SDC/GDC [samaria-doped ceria (SDC), gadolinium-doped ceria (GDC)] is employed as an electrolyte in the fuel cell, and a nickle anode and perovskite cathode are used.

### 2.2.7 CHARACTERIZATION STUDIES WITH A TABLE

Given below, Table 2.2 provides information of the key role of metal and their hybridization in enhancing catalytic activity.

The decrease of onset potential and peak potential, increase of peak current reflects high efficiency of catalyst.

## 2.3 SUPERCAPACITORS

### 2.3.1 CONCEPT AND HISTORY OF SUPERCAPACITOR

During the 1950s, supercapacitors were invented. Initially, experiments were done by General Electric and Standard Oil of Ohio (SOHIO). The initial electrochemical supercapacitors had a capacity of about 1 F. SOHIO was unable to discover purpose for the platform and instead licensed the design to the Nippon Electric Company (NEC) in Japan. SOHIO patented this form of supercapacitor in 1971.

In 1975, NEC released the "Supercapacitor," the first commercially functional electrochemical capacitor (EC). Though ECs are now often referred to as supercapacitors or ultracapacitors, NEC's label of ECs with a similar name is the sole "true" supercapacitor. Following NEC's commercialization, several other firms began designing their own ECs. Panasonic introduced the "Gold Cap" supercapacitor to the commercial market in 1982, with a high equivalent series resistance (ESR) [33].

**TABLE 2.2**
**Characterization Studies of Fuel Cell Material**

| Name of Catalyst | Onset Potential (V) | Peak Potential (V) | Peak Current | Ref. |
|---|---|---|---|---|
| Pd/graphene | 0.600 | 0.880 | 7 mA cm$^{-2}$ | [24] |
| Pd/NS-graphene | 0.472 | 0.880 | 12.5 mA cm$^{-2}$ | [25] |
| Pd/C$_3$N$_4$-RGO | 0.450 | 0.800 | 1550 mA mg$^{-1}$ | [26] |
| Pd$_{65}$Ag$_{35}$/C | 0.435 | 0.865 | 629 mA mg$^{-1}$ | [27] |
| Pd-Ag(1:1)/CNTs | 0.436 | 0.886 | 0.950 mA cm$^{-2}$ | [28] |
| Pt/C | 0.525 | - | 18 mA cm$^{-2}$ | [29] |
| Pd-Ni(1:1)/C | 0.421 | 0.941 | 1.50 mA cm$^{-2}$ | [30] |
| Pd/C 1:1 | 0.600 | - | 0.310 mA cm$^{-2}$ | [31] |
| Polyindole | -0.1 | 0.4 | 1.4 mA cm$^{-2}$ | [32] |
| Ag-polyindole | 0.01 | 0.3 | 3.4 mA cm$^{-2}$ | [22] |
| Pd-chrysanthemum flowers | −0.504 | -0.1 | 1.33 mA cm$^{-2}$ | [19] |
| Pd-thorns | −0.543 | 0.2 | 15.50 mA cm$^{-2}$ | [19] |

The Pinnacle Research Institute created the first electric double-layer capacitor (EDLC) supercapacitor for military use in 1982. The ESR of the first EDLC was anyway less. Then, 10 years later, Maxwell Laboratories launched an extensive range of EDLC supercapacitors with low ESR called "Boost Cap," which has a normal capacity of 1 kF to the commercial market in 1992. Since 2007, researchers have been working on new hybrid supercapacitors. Supercapacitors of this type should have a higher nominal voltage as well as a larger gravimetric and volumetric energy density than ordinary EDLC supercapacitors. The majority of current supercapacitors possess capacities of 1000 F and charge-discharge currents in the tenths to hundredths of amperes range. The key benefit in terms of application is their ability to run at very high currents. It is due to this feature that supercapacitors fill space between conventional capacitors and batteries (accumulators) in energy storage systems [34].

Today, most electronic firms, including Maxwell, Tecate, and Murata Group, produce ECs. The technology is mostly applied to energy and transportation problems. The automobile sector, hybrid transportation systems across the world, utility vehicles, grid stability, and rail-system power models are all examples of current usage. Tecate Group's HC Series ultracapacitors have a voltage of 2.7 V, capacitance of up to 150 F, and a current of up to 65 A. Murata's high-performance supercapacitor (EDLC) DMF Series has the greatest output power with a release of 50 W per piece. Murata also praises its ability to balance high peak loads for energy storage, consumer electronics, and rapid charge/discharge cycles [33].

Combining supercapacitors and fuel cells for maximum energy storage and quick refueling is one of the most innovative uses of already-commercialized technology. ABB's quick charging station, for example, lets electric buses be fully charged in less than 10 min. We are quite close to having self-contained supercapacitor batteries. In comparison to lithium-ion cells, a prototype supercapacitor battery developed by researchers at the University of Central Florida charges more quickly and can be recharged many more times without losing any performance [35].

Another idea that is likely to transform the capacitor industry and is currently under research is developing ECs with graphene to build supercapacitors, which are light in weight with the capability of storing energy between 150 F/g and 550 F/g at a fourth of the expense of existing EC designs. When tying their energy systems to things like supercapacitor-based engine starters and charging stations, several cities employing hybrid technology for public transportation have also experienced improvements in total energy storage and charge cycles [33].

Supercapacitors' most likely future applications are rapid charging and energy storage. Most of the applications developed, have revolutionised our understanding of energy storage. It could be a long time before a commercially effective supercapacitor battery is developed. Nonetheless, the discovered supercapacitor applications are an excellent realization of a component of an old technique that are just being developed.

Strengths of supercapacitors are listed as follows [36]:

- Internal resistance or effective series resistance is extremely low, resulting in excellent efficiency.
- High specific power.
- Rapid charging and fast transient response.
- Wide operating temperature range.
- Simple charging method.
- Deep discharge and charge.
- Very high rates of discharge and charge.
- Almost zero maintenance and long life.
- Environment friendly.
- As they can tolerate short circuits and reverse polarity, they provide increased safety. There is also no risk of fire or explosion. Rugged, since they have an epoxy resin sealed case, which is non-corrosive.

Limitations of supercapacitors are listed as follows [36]:

- In comparison to an electrochemical battery, the quantity of energy stored per unit weight is significantly smaller. It also has a volumetric energy density of around 1/10,000 that of gasoline.
- To successfully recover and store energy, switching equipment and advanced electronic control is required.
- Among all capacitor types, it has the largest dielectric absorption.
- The normal voltage of EDLC cells is 2.7 V. Because most applications need a greater voltage, the cells should be connected in series.

## 2.3.2 STRUCTURE OF SUPERCAPACITOR

A supercapacitor cell consists of two electrodes as shown in Figure 2.6, each of which is attached to a current collector that connects to a circuit, as well as a separator between the electrolyte and electrodes. The electrodes might be the same (symmetric cells) or different (asymmetric cells). The separator, which prevents a short circuit between the electrodes while allowing ionic charge transfer via the electrolyte ions, allows a supercapacitor cell to be as small as possible. As every electrode can be thought of as an individual capacitor, the capacitance values of two electrodes can be thought of as series-connected capacitors. The electrodes are classified based on how much their electric potential changes where they are already charged.

The anode's potential falls while the potential of the positive electrode grows until both electrodes accumulate a charge. Supercapacitors are more often called negative electrodes and positive electrodes than cathodes and anodes.

Hence, the total capacitance can be determined as

$$C_{total} = \frac{C_+ C_-}{C_+ + C_-}$$

Here, $C_-$, $C_+$ are the capacitance of anions and cations in Farads (F).

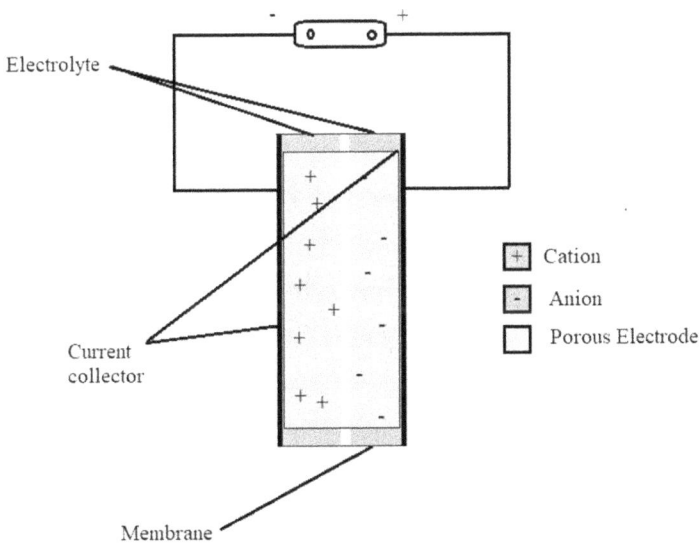

FIGURE 2.6   Structure of supercapacitor.

### 2.3.3  WORKING PRINCIPLE OF THE SUPERCAPACITOR

The supercapacitor, like a regular capacitor, has two parallel plates with a larger surface area (Figure 2.7) but small distance between the plates. Metal plates with electrolytes are used. The plates are separated by a thin layer known as an insulator. An electric double layer is generated when opposing charges build up on either sides of the plates and the insulators are charged. As a result, the supercapacitor is charged, and its capacitance is increased. These capacitors allow for big load currents while delivering great power and low resistance.

#### 2.3.3.1  Working of the Supercapacitor

Energy is stored in capacitors via static electricity, often known as electrostatics. An electrolyte solution with both positively and negatively charged ions is present between the two plates. When a voltage is supplied across the plates, one plate will be positive, while the other will be negative. Anions from the electrolyte solution are attracted to the positively charged plate, while cations are moved to the anode.

On both sides of the plates, a thin coating of ions is deposited. As a result, an electrostatic double layer is formed, which is similar to two capacitors connected in series. As the distance between the charge layers of both the resulting capacitors is so small, they each have a large capacitance value. $(C1 \times C2)/(C1 + C2)$ may be used to compute the total capacitance of the supercapacitor.

#### 2.3.3.2  Parameters Involved in the Supercapacitor

##### 2.3.3.2.1  Columbic Efficiency

The Columbic efficiency can be evaluated by the parameters deduced from the charge and discharge curve. It is the ratio of discharging time to charging time.

$$\text{Coulombic efficiency} = \frac{\text{Discharging time}}{\text{Charge time}} \times 100$$

It is calibrated by comparing the first cycle with the final cycle by the cyclic stability method.

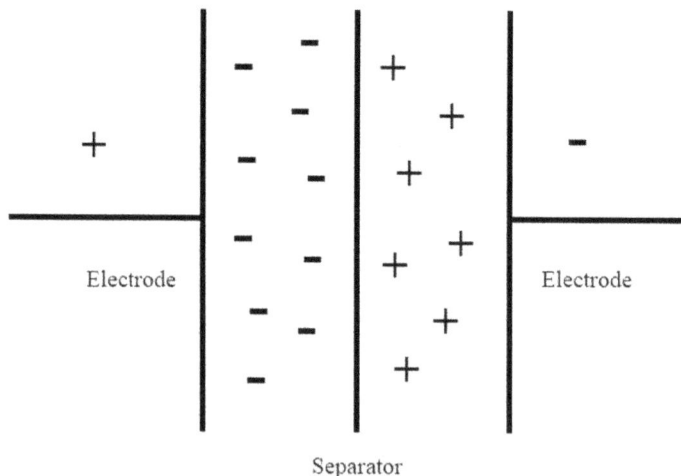

FIGURE 2.7  Working principle of SC.

*2.3.3.2.2    Energy Density (E)*
The most significant criterion to consider, when assessing a supercapacitor is its energy density. The electrode and electrolyte materials utilized have a significant impact on the specific energy density. Different electrode and electrolyte interactions can influence the energy density of a supercapacitor with different specific capacitance.

Specific energy densities can be expressed as

$$E = \frac{1}{2} * C * (dV)^2 \quad Wh/kg$$

where $C$ is the specific capacitance (F/g).

$\quad$ $dV$ is voltage range of the material of the electrode (V).

*2.3.3.2.3    Power Density (P)*
Power density is another important performance indicator of supercapacitors because it depicts the energy stored in the device, it talks about how quickly the power is delivered to an external load. The quantity of power per unit volume is known as power density.

$$P = \frac{E}{t} W/kg$$

Here $E$ = specific energy density (W/kg).

$\quad$ $t$ = discharging time (hours or seconds).

### 2.3.4    CLASSIFICATION OF THE SUPERCAPACITOR

These are classified into three types, viz., EDLCs, pseudocapacitors and hybrid capacitors.

### 2.3.4.1    EDLC

EDLCs are electrochemical capacitors, where the non-faradic process of charge accumulation is responsible for the capacitance in EDLC. It reversibly adsorbs electrolyte ions onto the large surface area of electrochemically stable active carbon materials to electrostatically store charged particles.

This is the most prevalent form of supercapacitor, accounting for the bulk of the commercial market. A liquid electrolyte is used in EDLC supercapacitors. Most of these electrolytes contain dissolved salts such as tetra ethyl ammonium tetra fluoroborate (TEABF4) or lithium hexa fluoroarsenate, which are dissolved in aprotic solvents such as propylene carbonate (PC), dimethyl carbonate (DME), or ethylene carbonate (EC). Ionic liquids such as tri ethyl sulfonium bis (tri fluoromethyl sulfonyl) imide or 1-ethyl-3-methylimdazolium thiocyanate, as well as an extensive group of ionic liquids, are employed as electrolytes (experimentally).

On the phase interface between the electrodes and the electrolyte, electrostatic interaction stores energy in Helmholtz double layers. The potential dependency of the surface energy accumulated electrostatically at the contact of capacitor electrodes causes double-layer capacitance. For supercapacitors like this, the exchange of electrons or redox reaction will not happen, the energy is held non-faradaically. The electrodes' massive surface area and the Helmholtz layer's thickness are important for reaching extremely high capacity. In millions of cycles, EDLC supercapacitors have proven to be durable and cyclable. Activated carbon (AC) is extensively employed as an electrode material in EDLC supercapacitors because of its huge specific surface area (SSA) [37,38].

### 2.3.4.2 Pseudocapacitor

Enhancement of capacitance values can be obtained by occurrence of quick faradic reactions. Pseudocapacitance is defined as the charge transported during these faradic events and is proportional to voltage. Distinctive to EDLCs, pseudocapacitors working under the principle of the faradic process to charge storage with accomplished redox reactions, is electrosorption. Pseudocapacitors can have larger capacitances and energy densities than EDLCs, thanks to the faradic approach.

Only a few companies sell this form of supercapacitor commercially, which is significantly less popular than EDLC supercapacitors. In terms of operation, they resemble batteries. At the time of discharge and charge, the electrolyte and the electrode undergo a reduction-oxidation reaction and transfer of energy. The energy of molecular bonds, not the "dielectric" layer, is used to indicate energy. The disadvantage of these systems is that the electrodes are strained and degrade more quickly during charging and discharging than in the electrostatic storage theory. It has to do with raising supercapacitors' internal resistance. Both electrodes of pseudosupercapacitors are constructed of pseudocapacitive materials such as manganese dioxide ($MnO_2$) or ruthenium oxide ($RuO_2$). In comparison to EDLC supercapacitors, there is less stability and cyclability, as well as a longer time response (lower discharge rate) and poorer charging efficiency.

### 2.3.4.3 Hybrid Supercapacitor

A hybrid supercapacitor is indeed aiming to achieve better properties than EDLCs and pseudocapacitors, with the principle of making use of non-faradic and faradic methods to store charge. Compared to EDLCs, hybrid capacitors have higher energy, higher power densities, and superior cycle stability. It is the most recent generation of supercapacitors. This most sophisticated supercapacitor incorporates the EDLC and pseudosupercapacitors, as well as preceding supercapacitor kinds. It has a larger gravimetric and volumetric energy density as well as the ability to produce large currents. Hybrid supercapacitors have a greater energy density because to the faradaic reaction that happens on the anode, which is often made of pseudocapacitor. The cathode commonly consists of activated carbon, which has a double layer that stores electrostatic energy. Due to the electrostatic interaction among the electrode surface on the cathode side and charge carriers, hybrid supercapacitors may provide enormous currents. In terms of design and operation, these are comparable to lithium-ion batteries.

### 2.3.5 Electrode Materials

The electrodes for supercapacitors were made by impregnation of some nanostructures. Below, Table 2.3 discusses the synthesis of these electrode materials for SC.

### 2.3.6 Types of Electrode Materials

### 2.3.6.1 Carbon Based

Due to their outstanding characteristics, like large area, less expense, environmental friendliness, and simplicity of synthesis, carbonaceous materials are often employed as electrode materials for EDLCs. Pore size of carbon-based materials, which is smaller than 1 nm, determines their physicochemical qualities.

Additionally, their chemical, thermal, and electrochemical stability (in a variety of solutions ranging from strongly acidic to mildly basic media), strong electrical conductivity, and nice rectangular form of CV curves all point to carbon-based materials being useful capacitive materials. The specific surface area has a huge impact on electrode capacitance. As a result of the high porosity and huge specific surface area of the carbon-based materials, they have shown increased capacitance.

The creation of the Helmholtz layer on the electrode material's surface is necessary to store charge on a carbon-based electrode. Secondly, the presence of solvated ions pushed by high electrostatic

**TABLE 2.3**
**Synthesis of Electrode Materials for SC**

| Precursor | Reducing Agent | Synthesis Method | Material | Morphology | Size (nm) | Ref. |
|---|---|---|---|---|---|---|
| Nickel nitrate hexahydrate, reduced grapheneoxide | Sodium hydroxide | Microwave irradiation method | NiO@rGO nanocomposite | Nanocircular plate | 60 nm | [35] |
| Ammonia meta vanadate | Moringaoleifera leaves | Microwave irradiation method | $V_2O_5$ nanoparticles | - | 25nm | [34] |
| Zincnitrate, manganese nitrate, graphite flakes | Sucrose, $NaNO_3$, $KMnO_4,H_2SO_4,H_2O_2,HCl$ | Auto-combustion method, Hummer's method | $ZnMn_2O_4/rGO$ nanocomposite | Porous structure | 35nm | [36] |
| Nickel nitrate, magnesium nitrate, reduced graphene oxide | $NH_4OH$ | Co-precipitation method | NiMgOH-rGO nanocomposites | - | 40 nm | [39] |
| Nickel nitrate, magnesium nitrate, reduced graphene oxide | $NH_4OH$ | Microwave-assisted method | NiMgOH-rGO nanocomposites | flakes | 29 nm | [40] |
| Graphite flakes, Tin chloride dehydrate | $NH_4OH$, $N_2H_4H_2O$ | Ultrasonication | $Graphene/SnO_2$ Nanocomposite | Tetragonal and Spherical | 33 nm | [41] |
| Tetrabutylammoniumhexamolybdate, Sodium molybdate | $C_4H_6O_3$, $C_3H_7NO$, HCl, $C_{16}H_{36}BrN$ | Thermal decomposition | $MoO_2$ nanorods | Nanorods | 80 nm | [42] |
| MAX phase ($Ti_3AlC_2$), ferric nitrate, cobalt nitrate | HF, $C_2H_6OS$, $NH_2CONH_2$,NaOH, $NH_4OH$, $C_{18}H_{34}O_2$ | Co-precipitation | $CoFe_2O_4$ nanoparticles-decorated 2D MXene | Accordion | 86 nm | [43] |

forces causes the creation of a thicker outer Helmholtz layer, which causes charge storage. From solvated ions and thermal motion, the indicated layer can be extended to a thicker layer, the Gouy–Chapman diffuse layer. The zeta potential ($\xi$-potential) is a potential difference between the thin inner layer and the thicker outside layer that can be used to describe the degree of charge storage in an EDLC device. Furthermore, to get greater $E_d$, it is critical to increase the capacitance of carbon-based materials [44].

Materials based on carbon could be employed in a SC device in a variety of forms, including powder, fiber, monoliths, and foils. Additional parameters such as the material's structure, electrical conductivity, and pore size distribution have a substantial impact on the electrochemical performance of a supercapacitor device. Carbon aerogels, graphene, activated carbon materials (ACs), carbon nanotubes (CNT), and carbon cloth (CC), and numerous carbon-based composites are examples of carbon-based materials that can be utilized as electrode materials in SC devices [45].

### 2.3.6.2   Metal Oxide

The frequent redox-active materials utilized are transition-metal-based compounds, specifically in oxide and hydroxide form. Transition-metal oxides (TMOs) and transition-metal hydroxides [TM(OHs)] contain a variety of transition metals, including Fe, Ni, Ti, Co, Mo, Nb, and V. As a result of their low resistance and high specific capacitance compared to carbon-based materials, and their large conductivity, TMOs are an alternative for high-energy SC electrodes. The most common redox pseudocapacitive materials are TMOs and conducting polymers (CPs), which have fast-reversible reduction-oxidation reactions and hence large specific capacitances and long working periods. As specified by B.E. Conway, there are several TMOs with charge storage mechanisms based on faradic behavior, such as $RuO_2$ [46], $Fe_3O_4$ [47], $MnO_2$ [48], NiO [35], $Co_3O_4$ [49], $V_2O_5$ [34], and CuO [50].

Furthermore, layered transition-metal hydroxides [TM(OH)s] with a large gap between their layers, like $Ni(OH)_2$[51], FeOOH[52], and $Co(OH)_2$[53], enable adequate redox processes, increasing the capacitance of the electrodes in SCs. Anyway, as the electrical conductivity of their equivalent oxides is greater, the TM(OH)s may be able to store more charge than their oxide counterparts [54]. When compared to their oxide counterparts, TM(OH)s have reduced structural stability and, as a result, inferior rate capability or cycle performance. However, there are several ways for overcoming this constraint and improving the material's cycle stability, including layered engineering of the TM(OH)s via the anion and cation doping process [40].

Based on their electrochemical reaction in the hybrid supercapacitor (HSC) device, the battery-type TM may be separated into two primary classes: conversion-type and intercalation-type. Although TM-based materials are often associated with batteries, they can also exhibit capacitive properties. The researchers discovered that creating a highly porous nanostructure and increasing the concentration of hydrous states like HOH and MnOH in the thick oxide layer facilitated liquid-phase and solid-phase diffusion.

There are several examples of various mixed-MOs such as CuO@NiO, $CuO/Cu_2O$, $Cu/Cu_2O$@ TiO, $Fe_2O_3$@TiO, ZnO@$Co_3O_4$, $Fe_2O_3$/$Co_3O_4$, $Fe_2O_3$/$NiCo_2O$, NiO@ZnO, and ZnO@$ZnCo_2O_4$; they might be manufactured as electrodes for SCs to deliver a high performance by generating a large surface area and involving several functions in the metal-organic framework (MOF) [55].

### 2.3.6.3   Polymers

Another pseudocapacitive electrode material that has gained a lot of interest is different CP interest due to their high $E_d$, high capacitance, adjustable redox activity, high voltage windows, and their ease of manufacture, less cost, low environmental impact. As no structural changes like phase transitions happen during the charge/discharge operation, CPs may hold the charge in its entirety. Due to the bigger redox storage capacities of CPs and surface areas, they can provide a higher capacitance.

CPs exhibit greater conductivity, capacitance, and low equivalent series resistance than other carbon-based electrode materials [32]. Capacitances in CPs are formed via fast-reversible redox processes involving conjugated polymer chains, in which ions travel to the polymer backbone during doping and are released back into the electrolyte solution during de-doping. Despite all of the benefits, redox in CPs creates stress, which limits stability, deterioration of the material during many charge-discharge cycles. Low-power densities, which result from sluggish ion diffusion rates in bulk, are another issue of CPs that hinders their performance.

The most extensively used method for the manufacture of CPs is the chemical or electrochemical oxidation of monomers. The polymers are produced in one of the common states. Polymers are more likely to be oxidised, or "$p$-doped," than reduced, or "$n$-doped," when they are synthesised. Polymer backbone is positively charged in the $p$-doped state and has high electrical conductivity (between 1 and 100 $Scm^{-1}$). Due to their steady performances, $p$-dopable polymers (polypyrrole and PANI) are currently getting greater interest from researchers.

## 2.3.7 ELECTROLYTES

The electrolyte provides ionic conductivity, permitting charge compensation on both electrodes, one of the most critical components SCs. The electrolyte not only impacts the performance of SC but also electric double layer (EDL) relies heavily on creation and the reversible redox process for charge storage [40,56].

Organic electrolytes with a cell voltage regime of 2.5–2.8 V are now used in most corporate SCs. Acetonitrile (ACN) is used as a solvent in most organic electrolyte-based SCs, while propylene carbonate is used in others. The nature of the electrolyte, which includes (1) the size and kind of ion; (2) the concentration of ions; (3) interaction between ions, electrolyte and electrode materials; and (5) EDL capacitance and pseudocapacitance, as well as the energy/power densities and cycle life of the SC, are all affected by the potential window. SC life span and self-discharge may be affected by the interaction of the ion with the solvent, as well as the electrolyte with the electrode material.

Solid/quasi-solid-state electrolytes and liquid electrolytes and are two types of electrolytes. Liquid electrolytes are separated into organic electrolytes, aqueous electrolytes, in general, whereas solid or quasi-solid-state electrolytes are classified into inorganic electrolytes and organic electrolytes (Figure 2.8). Yet, no electrolyte has been discovered that possesses the required qualities. Every electrolyte has its set of benefits and drawbacks.

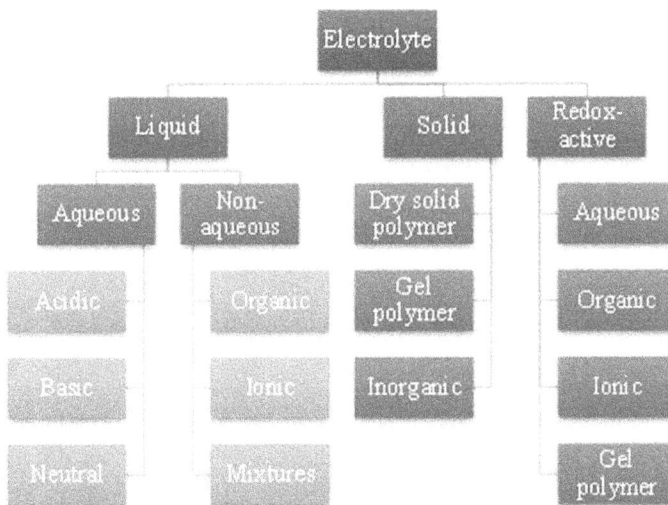

**FIGURE 2.8** Classification of electrolyte.

**TABLE 2.4**
**Characterization Studies for SC Electrodes**

| Electrode | Electrolyte | $C_s$ (Fg$^{-1}$) | $E_d$ (WhKg$^{-1}$) | $P_d$(kWkg$^{-1}$) | Retention (Cycles) | Ref. |
|---|---|---|---|---|---|---|
| Graphitic carbon nitride (g-C3N4) | 6M KOH | 265 | 7.47 | - | 94% (5000) | [60] |
| vanadium nitride/nitrogen-doped graphene (VN/NG) | 1M H$_2$SO$_4$ | 445 | 51.2 | 28.8 | 98% (10000) | [61] |
| Al-doped NiO@MWCNTs | 1M KOH | 192 | 15 | 21.6 | 100% (5000) | [62] |
| NiO-CNT | 1M Na$_2$SO$_4$ | 878 | 85.7 | 11.2 | - | [63] |
| Few-layer graphene (FLG) | 1M H$_2$SO$_4$ | 186 | 58.12 | 37.5 | - | [64] |
| silver doped graphene | Ion diffusion electrolyte | 437 | 1.13 | 88 | 89% (5000) | [65] |

SCs with aqueous electrolytes, for example, have high capacitance and conductivity, but the narrow decomposition voltage of the aqueous electrolytes limits their operating voltage. Aqueous electrolytes, on the other hand, have been widely studied, over 80% of the literature for SCs used them [34]. This is mostly because they are affordable and can be used in the laboratory without specific equipment, considerably simplifying the production and assembly procedures. In general, aqueous electrolytes are divided as acid, alkaline, or neutral solutions, and are commonly used electrolytes [57–59].

Although organic electrolytes can function at higher voltages, their ionic conductivity is often lower. Solid-state electrolytes avoid the leakage problems associated with liquid electrolytes, but they have a low conductivity. To circumvent the limits of targeted electrolytes, much research into alternative electrolyte molecules has been undertaken in order to improve the SC's overall performance.

### 2.3.8 CHARACTERIZATION STUDIES WITH A TABLE

Below, Table 2.4 discusses the characterizations studies for SC electrodes.

## 2.4 TECHNOLOGY CHALLENGES OF FUEL CELLS AND SUPERCAPACITORS

Energy storage and conversion are critical links between energy generation and use. To solve energy and environmental problems, fuel consumption must be clean and efficient. An electrochemical cell's objective is to transform the oxidant, chemical energy held in the fuel directly into DC electric energy, with heat and H$_2$O as leftovers. Because of the demand for clean energy, the scarcity of fossil fuel supplies, and the potential of a fuel cell to create electricity without engaging any moving mechanical parts, fuel cells and electricity generating technologies have gotten a lot of attention.

A fuel cell system's design may be viewed as a decision-making process that entails identifying several design options and picking the optimal one. If it fits the design standards and makes a trade-off between the many designs aims, it may be considered an excellent design. Performance, dimension (weight and size), emissions, output power, quick start-up, and reaction to changes in load, longevity, and operability in harsh conditions are among the parameters and aims of a fuel cell system. Noise, which is significant in some applications, is also included. The use of computer-based and modeling optimization in the design of fuel cell systems is being given a lot of attention.

Supercapacitors are the most versatile and widely utilized for delivering electrical energy in a short amount of time and in applications requiring a long shelf life. As a result, there is a huge market for supercapacitors, and long-term growth is necessary for effective commercialization and innovation. Due to their excellent qualities, supercapacitors are commonly utilized in transportation, military, consumer electronics, manufacturing, and other industries. Anyway, these devices have some flaws too [66].

## 2.5 CONCLUSIONS

Energy storage and conversion are critical links between energy generation and its usage in the field of supercapacitors and fuel cells. The coupling of fuel cells and supercapacitors may be used for quick charging time and increased energy storage. Fuel cells can operate at higher temperatures with effective efficiency depending upon the electrolyte used whereas supercapacitors have the largest dielectric absorption compared to other types. The benchmark considered while assessing a supercapacitor are energy density, power density, columbic efficiency. In EDLC, the energy is held non-faradaically, in pseudocapacitors it is held through quick faradic reaction, and hybrid supercapacitors use non-faradic and faradic methods to store charge larger gravimetric and volumetric energy density. The researchers discovered that by innovating a highly porous nanostructure and increasing the concentration of hydrous states in the thick oxide layer facilitated liquid-phase and solid-phase diffusion. These technologies have been playing a pivotal role in building a bridge between energy storage and conversion leading to a clean energy without affecting the environmental concerns.

## ACKNOWLEDGEMENT

The authors gratefully acknowledge F.No. CRG/2019/007040, DST SERB; File No. SP/YO/2019/1599, DST SYST and Sanction No. SR/WOS-A/CS-13/2019, DST Women Scientist (WOS-A), New Delhi.

## REFERENCES

[1] Kale S. A review paper on electrical system consisting of fuel cell. *International Engineering Research Journal.* 2016; 109–116.

[2] Yilmaz AE, Ispirli MM. An investigation on the parameters that affect the performance of hydrogen fuel cell. *Procedia – Social Behav. Sci.* 2015;195:2363–2369

[3] Jiao K, Xuan J, Du Q, et al. Designing the next generation of proton-exchange membrane fuel cells. *Nature.* 2021;595:361–369.

[4] Maheshwari K, Sharma S, Sharma A, et al. Fuel cell and its applications: A eeview. *Int. J. Eng. Res. Technol.* 2018;7:6–9.

[5] Sharaf OZ, Orhan MF. An overview of fuel cell technology: Fundamentals and applications. *Renew. Sustain, Energy Rev.* 2014;32:810–853.

[6] Haragirimana A. *Fuel Cells.* 2006. New York: Springer.

[7] Holbrook JA, Arthurs D, Cassidy E. Understanding the Vancouver hydrogen and fuel cells cluster: A case study of public laboratories and private research. *Eur. Plann. Stud.* 2010;18:317–328..

[8] Dincer I, Bicer Y. 4.22 electrochemical energy conversion. *Comp. Energy Syst.* 2018;4:856–894.

[9] Moseley PT. *Fuel Cell Systems Explained.* 2001.

[10] Viswanathan B, Fuel cells. Energ. Sources. 2017;329–356.

[11] Hussain S, Yangping L. Review of solid oxide fuel cell materials: cathode, anode, and electrolyte. Energy Trans. 2020;4:113–126.

[12] Jiang SP, Yan Y. Molten Carbonate fuel cells. *Mater. High-Temp. Fuel Cells.* 2016; 4:341–371.

[13] Steilen M, Jo L. Hydrogen conversion into electr;icity and thermal energy by fuel cells: Use of H2-systems and batteries. *Electrochem. Energ. Stor. Renew. Sources Grid Balanc.* 2015;10:143–158.

[14] Colpan CO, Nalbant Y, Ercelik M. 4.28 fundamentals of fuel cell technologies organic. *Comp. Energy Syst.* 2018;4:1107–1130.

[15]    Glusen A, Muller M ,Stolten D. 45% cell efficiency in DMFCs via process engineering. 2020;20:1–8.

[16]    Sossina MH. Materials for fuel cells. *Mater. Today.* 2003;6:24–29.

[17]    Blaabjerg F, Sangwongwanich A. A review on superrcapacitor material and developments. *Turk. J. Mater..* 2020; 5:10–24.

[18]    Eris S, Das Z, Yıldız Y, et al. Nanostructured polyaniline-rGO decorated platinum catalyst with enhanced activity and durability for methanol oxidation. 2017;43:1337–1343.

[19]    Divya V, Mondal S, Sangaranarayanan M V. Shape-controlled synthesis of palladium nanostructures from flowers to thorns: Electrocatalytic oxidation of ethanol. *J. Nanosci. Nanotechnol.* 2018;19:758–769.

[20]    Liu J, Cao L, Huang W, et al. Direct electrodeposition of PtPd alloy foams comprised of nanodendrites with high electrocatalytic activity for the oxidation of methanol and ethanol. *J. Electroanalyt. Chem.* 2012;686:38–45.

[21]    Eswaran M, Dhanusuraman R, Tsai P. One-step preparation of graphitic carbon nitride / Polyaniline / Palladium nanoparticles based nanohybrid composite modi fi ed electrode for e ffi cient methanol electro-oxidation. *Fuel.* 2019;251:91–97.

[22]    Chakra Ch S, Velpula D, Shireesha K, Impact of synthetic strategies for the preparation of polymers and metal-polymer hybrid composites in electrocatalysis applications. *Synth. Met.* 2021;282.

[23]    Haile M. Materials for fuel cells. *Mater. Today.* 2003;6:24–29.

[24]    Zhang X, Zhu J, Tiwary CS, et al. Palladium nanoparticles supported on nitrogen and sulfur dual-doped graphene as highly active electrocatalysts for formic acid and methanol oxidation. *ACS Appl. Mater. Interfaces.* 2016;8:10858–10865.

[25]    Gómez JCC, Moliner R, Lázaro MJ. Palladium-based catalysts as electrodes for direct methanol fuel cells: A last ten years review. *Catalysts.* 2016;6:130.

[26]    Zhang W, Yao Q, Wu X, et al. Intimately coupled hybrid of graphitic carbon nitride nanoflakelets with reduced graphene oxide for supporting Pd nanoparticles: A stable nanocatalyst with high catalytic activity towards formic acid and methanol electrooxidation. *Electrochim. Acta.* 2016;200:131–141.

[27]    Yin Z, Zhang Y, Chen K, et al. Monodispersed bimetallic PdAg nanoparticles with twinned structures: Formation and enhancement for the methanol oxidation. *Sci. Rep.* 2014;4:4288.

[28]    Wang Y, Sheng ZM, Yang H, et al. Electrocatalysis of carbon black- or activated carbon nanotubes-supported Pd-Ag towards methanol oxidation in alkaline media. *Int. J. Hydrogen. Energy.* 2010;35:10087–10093.

[29]    Shen PK, Xu C, Zeng R, et al. Electro-oxidation of methanol on NiO-promoted Pt/C and Pd/C catalysts. *Electrochem. Solid-State Lett.* 2006;9:39–42.

[30]    Liu Z, Zhang X, Hong L. Physical and electrochemical characterizations of nanostructured Pd/C and PdNi/C catalysts for methanol oxidation. *Electrochem Commun.* 2009;11:925–928.

[31]    Calderón JC, Nieto-Monge MJ, Pérez-Rodríguez S, et al. Palladium–nickel catalysts supported on different chemically-treated carbon blacks for methanol oxidation in alkaline media. *Int. J. Hydrogen. Energy.* 2016;41:19556–19569.

[32]    Velpula D, Shireesha K, Chakra Ch S. Impact of synthetic strategies for the preparation of polymers and metal-polymer hybrid composites in electrocatalysis applications. *Synth. Met.* 2021;282:116956.

[33]    Blaabjerg F, Sangwongwanich A. A review on supercapacitor materials and developments. *Turk. J. Mater.* 2020;5:10–24.

[34]    Velpula D, Konda S, Vasukula S, et al. Microwave radiated comparative growths of vanadium pent-oxide nanostructures by green and chemical routes for energy storage applications. *Mater. Today Proc.* 2021;47:1760–1766.

[35]    Rakesh Kumar T, Shilpa Chakra CH, Madhuri S, et al. Microwave-irradiated novel mesoporous nickel oxide carbon nanocomposite electrodes for supercapacitor application. *J. Mater. Sci. Mater. Electron.* 2021;32:20374–20383.

[36]    Kommu P, Singh GP, Chakra CS, et al. Preparation of ZnMn2O4 and ZnMn2O4/graphene nano composites by combustion synthesis for their electrochemical properties. *Mater. Sci. Eng. B Solid-State Mater. Adv. Technol.* 2020;261:114647.

[37]    Lämmel C, Schneider M, Weiser M, et al. Investigations of electrochemical double layer capacitor (EDLC) materials – A comparison of test methods. *Materwiss Werksttech.* 2013;44:641–649.

[38]    Lei C, Markoulidis F, Ashitaka Z, et al. Reduction of porous carbon/Al contact resistance for an elec-tric double-layer capacitor (EDLC). *Electrochim. Acta.* 2013;92:183–187.

[39]  Shireesha K, Kumar TR, Rajani T, et al. Novel NiMgOH-rGO-based nanostructured hybrids for electrochemical energy storage supercapacitor applications: Effect of reducing agents. *Crystals.* 2021;11:1144.

[40]  Shireesha K, Divya V, Pranitha G, et al. A systematic investigation on the effect of reducing agents towards specific capacitance of NiMg@OH/ reduced graphene oxide nanocomposites. *Mater. Technol..* 2021;37:1–13.

[41]  Eedulakanti SR, Gampala AK, Venkateswara Rao K, et al. Ultrasonication assisted thermal exfoliation of graphene-tin oxide nanocomposite material for supercapacitor. *Mater. Sci. Energy Technol.* 2019;2:372–376.

[42]  Rajeswari J, Kishore PS, Viswanathan B. One-dimensional $MoO_2$ nanorods for supercapacitor applications. *Electrochem. Commun.* 2009;11:2–6.

[43]  Ayman I, Rasheed A, Ajmal S, et al. Batteries and Energy Storage CoFe2O4 Nanoparticles-Decorated 2D MXene: A Novel Hybrid Material for Supercapacitors Applications Sustainable Energy Technologies Center, 2020. College of Engineering, King Saud University, PO.

[44]  Sun J, Huang Y, Sze Sea YN, et al. Recent progress of fiber-shaped asymmetric supercapacitors. *Mater. Today Energy.* 2017;5:1–14.

[45]  Samantara AK, Das JK, Behera JN. Supercapacitors based on graphene and its hybrids [Internet]. *Fundamentals and Supercapacitor Applications of 2D Materials.* 2021. Amsterdam: Elsevier.

[46]  Majumdar D. An overview on ruthenium oxide composites – challenging material for energy storage applications. *Mater. Sci. Res. India.* 2018;15:30–40.

[47]  Wang X, Huang H, Li G, et al. Hydrothermal synthesis of 3D hollow porous $Fe_3O_4$ microspheres towards catalytic removal of organic pollutants. *Nanoscale Res. Lett.* 2014;9:3–7.

[48]  Chou S, Wang J, Chew S, et al. Electrochemistry communications electrodeposition of $MnO_2$ nanowires on carbon nanotube paper as free-standing, flexible electrode for supercapacitors. *Electrochem. Commun.* 2008;10:1724–1727.

[49]  Jang G, Ameen S, Akhtar MS, et al. Cobalt oxide nanocubes as electrode material for the performance evaluation of electrochemical supercapacitor. *Ceram. Int.* 2018;44:588–595.

[50]  Dubal DP, Gund GS, Holze R, et al. Surfactant-assisted morphological tuning of hierarchical CuO thin films for electrochemical supercapacitors. *Dalt. Trans.* 2013;42:6459–6467.

[51]  Cui H, Xue J, Ren W, et al. Ultra-high specific capacitance of β-$Ni(OH)_2$ monolayer nanosheets synthesized by an exfoliation-free sol-gel route. *J. Nanoparticle Res.* 2014;16:2601.

[52]  Du K, Wei G, Zhao F, et al. Urchin-like FeOOH hollow microspheres decorated with $MnO_2$ for enhanced supercapacitor performance. *Sci. China Mater.* 2018;61:48–56.

[53]  Zhang L, Cai P, Wei Z, et al. Synthesis of reduced graphene oxide supported nickel-cobalt-layered double hydroxide nanosheets for supercapacitors. *J. Colloid Interface Sci.* 2021;588:637–645.

[54]  Shireesha K, Kumar TR, Rajani T, et al. Novel NiMgOH-rGO-based nanostructured hybrids for electrochemical energy storage supercapacitor applications. *Crystals.* 2021;11:1144.

[55]  Liang R, Du Y, Xiao P, et al. Transition metal oxide electrode materials for supercapacitors: A review of recent developments. *Nanomaterials.* 2021;11:1248.

[56]  Velpula D, Konda S, Vasukula S, et al. Microwave radiated comparative growths of vanadium pentoxide nanostructures by green and chemical routes for energy storage applications. *Mater. Today Proc.* 2021;47:1760–1766.

[57]  Jiménez-Cordero D, Heras F, Gilarranz MA, et al. Grape seed carbons for studying the influence of texture on supercapacitor behaviour in aqueous electrolytes. *Carbon NY.* 2014;71:127–138.

[58]  Galiński M, Lewandowski A, Stepniak I. Ionic liquids as electrolytes. *Electrochim. Acta.* 2006;51:5567–5580.

[59]  Brandt A, Ramirez-Castro C, Anouti M, et al. An investigation about the use of mixtures of sulfonium-based ionic liquids and propylene carbonate as electrolytes for supercapacitors. *J. Mater. Chem. A.* 2013;1:12669–12678.

[60]  Lin R, Li Z, Abou I, et al. A general method for boosting the supercapacitor performance of graphitic carbon nitride / graphene hybrids. *J. Mater. Chem. A Mater. Energy Sustain.* 2017;5:25545–25554.

[61]   Balamurugan J, Karthikeyan G, Duy T, et al. Facile synthesis of vanadium nitride / nitrogen-doped graphene composite as stable high performance anode materials for supercapacitors. *J. Power Sources*. 2016;308:149–157.

[62]   Chen J, Peng X, Song L, et al. Facile synthesis of Al-doped NiO nanosheet arrays for supercapacitors. *Roy. Soc. Open Sci*. 2018;5:1–10.

[63]   Roy A, Ray A, Saha S, et al. NiO-CNT composite for high performance supercapacitor electrode and oxygen evolution reaction. *Electrochim. Acta*. 2018;283:327–337.

[64]   Purkait T, Singh G, Singh M, et al. Large area few-layer graphene with scalable preparation from waste biomass for high-performance supercapacitor. *Sci. Rep*. 2017;7:1–14.

[65]   Jiao S, Li T, Xiong C, et al. A facile method to prepare silver doped graphene combined with polyaniline for high performances of filter paper based flexible electrode. *Nanomaterials*. 2019;9:1–12.

# 3 Promising Electrode Materials for Hybrid Supercapacitors

*Biny R. Wiston and Ashok Mahalingam*

## 3.1 INTRODUCTION

Ever since the ancient Greeks rubbed amber with animal fur to generate electrical charges on the surface, the history of energy devices had begun. Over time, the knowledge about energy and electricity has been broadening. In the last three centuries, the field of energy devices has witnessed ground-breaking revolutions. Leyden jar, a 'vintage condenser' consisting of a glass vial holding an aqueous electrolyte with a metal rod dipped, stored static electricity. A metal foil was coated inside the jar and another metal foil was outside the jar forming the electrodes, whereas the glass jar acted as a dielectric [1]. In the mid-18th century, Benjamin Franklin conducted an experiment with a key, storm-cloud, and kite to demonstrate the transfer of electrical charges from (storm) clouds to his hand. His observations and insights, later lead to a better understanding about electricity and charge transport. In the late-18th century and early-19th century, Galvani invented 'animal electricity' and Alessandro Volta invented the first 'electric battery' that stored direct current (DC). Nevertheless, more awareness about electricity aroused after discovery of sub-atomic particles like proton, electrons, and neutrons by Sir J. J. Thomson and Ernest Rutherford in the late 19th century [2].

Modern portable and wearable electronic gadgets like mobile phones, activity trackers (smartwatches and sensors), medical implants (pacemakers, insulin pumps), and many more have made our life convenient. The need to devise alternative, renewable energy solutions has become the need of the hour due to a shortage of conventional fuel systems. While the rising environmental concerns like global warming, carbon-dioxide ($CO_2$) emission, and greenhouse effect have fastened the development of green and renewable energy resources. Renewable energy systems, including solar cell, wind energy, have been extensively explored by the research community. But there are few practical difficulties while replacing conventional energy systems with renewable energy resources. The electricity generated through renewable resources is relatively lower when the demand is highest. To troubleshoot this issue, energy-storage devices are developed to manage the power supply, when demand is high, and resource is lower. A wide range of green energy-storage systems with zero (or) negligible $CO_2$ emission, including capacitors, supercapacitors, batteries, and fuel cells, have been developed and are in usage.

Lithium-ion batteries (LIB), first proposed by M. S. Whittingham in the 1970s, are the most familiar energy-storage technology. The first commercial LIB was developed at Oxford University and commercialized by Sony in the 1990s [3,4]. In 2019, J. B. Goodenough, M. S. Whittingham, and Yoshino Akira were awarded the Nobel Prize for their revolutionary works on lithium-ion battery technology. Lithium, being the most important component of the conventional lithium-ion and other emerging lithium-based battery chemistries like lithium-sulfur, lithium-air, and lithium-bromine

DOI: 10.1201/9781003340539-3

batteries, is extensively mined in certain countries like Bolivia, Argentina, Chile, the United States, Australia, and China. Whereas the other world nations rely on imported lithium from these countries. The desire for self-reliant energy-storage technologies engendered the development of lithium-free energy-storage systems. Supercapacitors (SCs) are emerging energy-storage systems, which have fastest charged and longer cycling (about 10,000 cycles) accompanied with high safety, and are expected to replace conventional batteries, providing a lithium-free economy. While capacitors have high specific power with poor specific energy, Lithium-ion batteries have high specific energy with inadequate specific power. Whereas supercapacitors have note-worthy specific power and improvable specific energy. Though the first patent on SCs was filed by H. I. Becker of General Electric in 1957 [5], they became more familiar in 1990s following the thrive for electric vehicles. At present, supercapacitor technology has been commercially available in a variety of applications. Maxwell Technologies was amongst the forerunners to introduce supercapacitors to the world market. Automobile, maglev trains, aircraft, and braking retrieval systems are amongst the top clients of supercapacitors [6,7]. Supercapacitors have also penetrated the field of defense. Due to the widening scope of utilization of SCs, both governmental and private organizations are investing heavily on research and development related to SCs. In the present chapter, innovative electrode materials for hybrid supercapacitors have been discussed in detail.

## 3.2 FUNDAMENTAL COMPONENTS OF SUPERCAPACITOR DEVICE

Supercapacitors are electrochemical devices that typically possess two electrodes electrically separated by a separator and connected ionically through an electrolyte as shown in Figure 3.1. Each component of the supercapacitor device plays a role in their device performance and efficiency. Supercapacitors resemble batteries in their prototype and engineering. Concepts like the Helmholtz model, the Gouy-Chapman-Stern model, etc., have been proposed to study the mechanism of charge storage in a supercapacitor. The highly recognized mechanism of charge storage in supercapacitors is presented in Figure 3.2. During the process of charging, the anions and cations get accumulated on the electrode surface. The ions present in the electrolyte dissociate into positive ions and negative ions, and travel towards positive and negative electrodes, respectively. They form a double layer of ions on the electrode-electrolyte interface. Whereas, during discharge process, the charges accumulated on one electrode go back to the other electrode through the external circuit and the electrolyte ions recombine [9]. After complete discharge, the supercapacitor goes back to the state before charging. In supercapacitors, the process of charge storage is a reversible reaction.

### 3.2.1 CURRENT COLLECTOR

The device performance in a supercapacitor relies on both the electrode-electrolyte and electrode-current collector interface. While most works emphasis on improving the properties of electrode material, it is equally important that the choice of current collector used is looked upon. There are quite a few preferences of the nature of current collectors like foam-type, foil-type, mesh-type, flexible, transparent, fiber-type, etc. At time, the choice of current collector depends on the electrolyte utilized and is categorized in Figure 3.3. Metal-based current collectors have good ion/electron transportation properties. Still, they have a few disadvantages like, they at times react with acidic (or) alkaline electrolytes to produce intermediate reactions that affect the process of charge storage. Readily available metallic current collectors are stainless steel, aluminum foil, nickel foil, copper foam, nickel foam, etc. Nickel foam (NF) is most preferred amongst the researchers owing to its porosity, high intrinsic strength, and excellent conductivity. However, in an alkaline KOH electrolyte, nickel foam may be corrosive and produce capacitance due to the pre-existence of nickel hydroxide/oxide on the surface of NF that may produce potential errors during the electrochemical

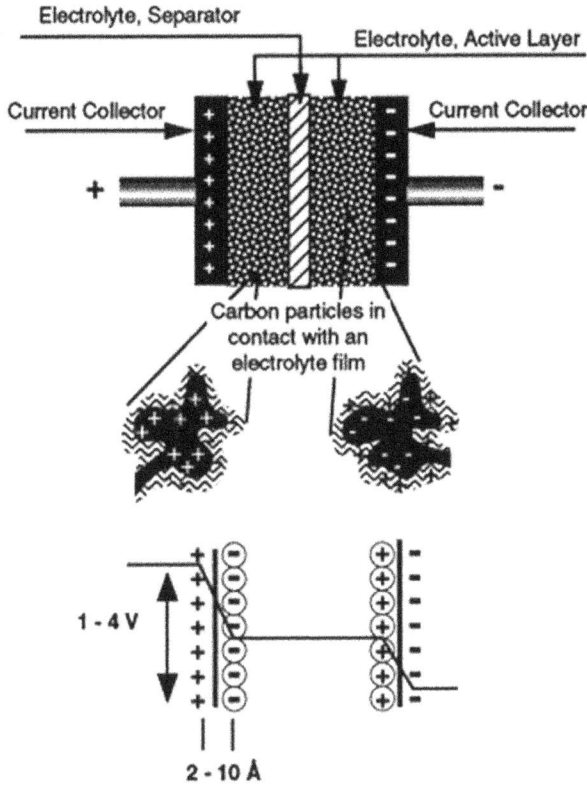

**FIGURE 3.1** Illustration of supercapacitor. Reprinted from [8], Copyright (2000), with permission from Elsevier.

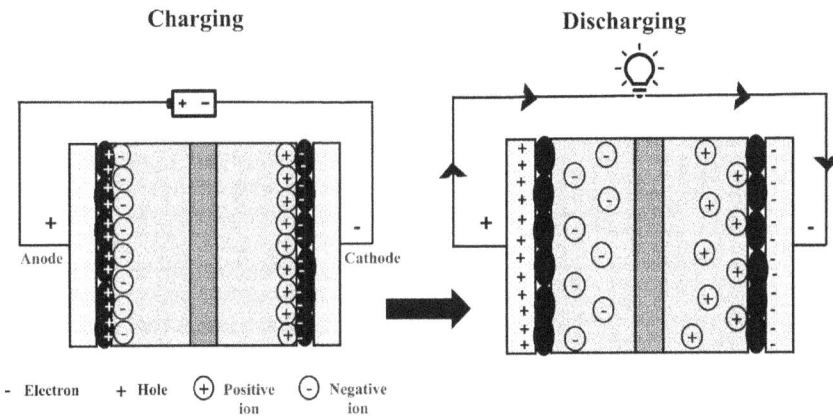

**FIGURE 3.2** Schematics of charging and discharging mechanism in supercapacitors.

measurement [10]. Hence cleaning is crucial in metallic current collectors. An ideal current collector is expected to be light-weight, tough, and stable in any corrosive environment. In that case carbon-based current collectors like carbon felt, graphite foil, graphene sponge, etc., gratify the commercial requirements. They are cheaper than metal-based current collectors, eco-friendly, and are highly stable in most electrolytes.

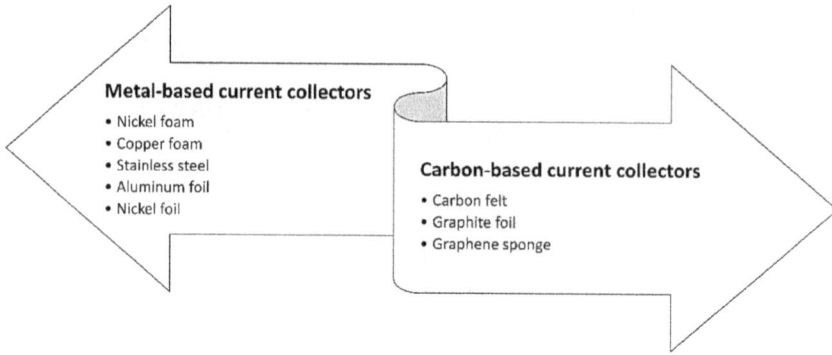

**FIGURE 3.3** Types of current collectors.

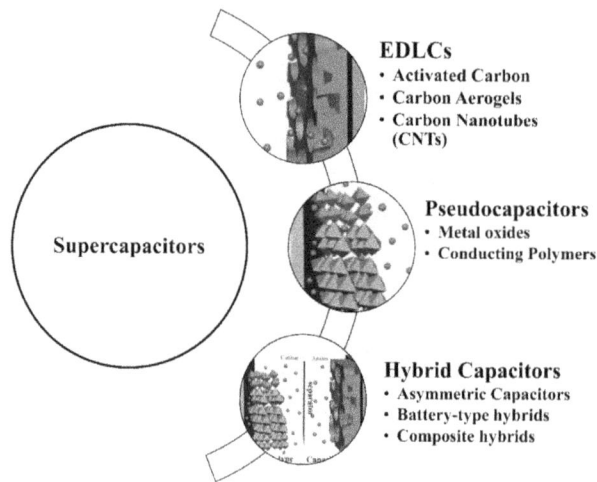

**FIGURE 3.4** Classification of supercapacitors. Reprinted from [11], Copyright (2020), with permission from Elsevier.

### 3.2.2 Electrode Material

It is popularly known that electrodes play a pivotal role in supercapacitor device performance. Conventionally supercapacitors employ electrode materials of high porosity and high specific surface area (SSA). Electrode materials generally used for supercapacitors include carbon materials, transition metal oxides, and conducting polymers. The mechanism behind storage of electrical charges in an electrode material may differ. With vast improvements in energy-storage materials, numerous novel materials have been tested for supercapacitor application. Depending on the mechanism of charge storage in an electrode material, supercapacitors may be classified into three types namely, electric double layer capacitors (EDLCs), pseudocapacitors, and hybrid supercapacitors as illustrated in Figure 3.4. When carbon-based materials are utilized as a supercapacitor electrode, the charge storage occurs by electrostatic adsorption of the electrolyte ions over the electrode surface. The supercapacitors employing such a charge-storage mechanism are named as EDLC. Whereas, in the other type of electrode materials called pseudocapacitive materials, the charge storage is mainly through fast faradaic oxidation and reduction reactions occurring at the electrode-electrolyte interface. Materials like transition metal oxide and conducting polymers having multiple oxidation states and good conductivity are employed as pseudocapacitive electrodes. EDLCs are highly stable and possess high cycling stability but have lower specific capacitance. Whereas pseudocapacitors have several times more specific capacitance than EDLCs but are poorer in stability during repetitive charging-discharging, due to the plausible structural changes that may degrade their cyclability. In hybrid capacitors, one electrode engages EDLC-type material, and another electrode employ pseudocapacitive material. This configuration holds potential to achieve high specific capacitance and high cycling stability in the same device.

### 3.2.3 ELECTROLYTE

Electrolytes act as the electrochemical contact between the two electrodes. The electrolyte-electrode interaction plays a major role in the performance of SCs, also called electrochemical supercapacitors (ESs) as shown in Figure 3.5. Electrolytes can be broadly classified as aqueous, organic, ionic liquids, solid-state, or quasi-solid-state, as well as redox-active electrolytes [12]. The working potential window, operating temperature, specific power, internal resistance, equivalent series resistance (ESR), and cycling stability of the supercapacitors are greatly influenced by the thermodynamic stability, concentration, viscosity, boiling point, freezing point, ionic conductivity, etc., of the electrolyte [13]. Moreover, the porosity of the electrode and the electrolyte ion size must match to achieve high proficiencies. Aqueous electrolytes may be acidic, basic or neutral in nature. Though aqueous electrolytes have higher ionic conductivity, and mobility, their operating potential is limited, as the water-splitting voltage is about 1.23 V theoretically and around 1.8 V practically [14]. Meanwhile, organic, ionic, redox-active,

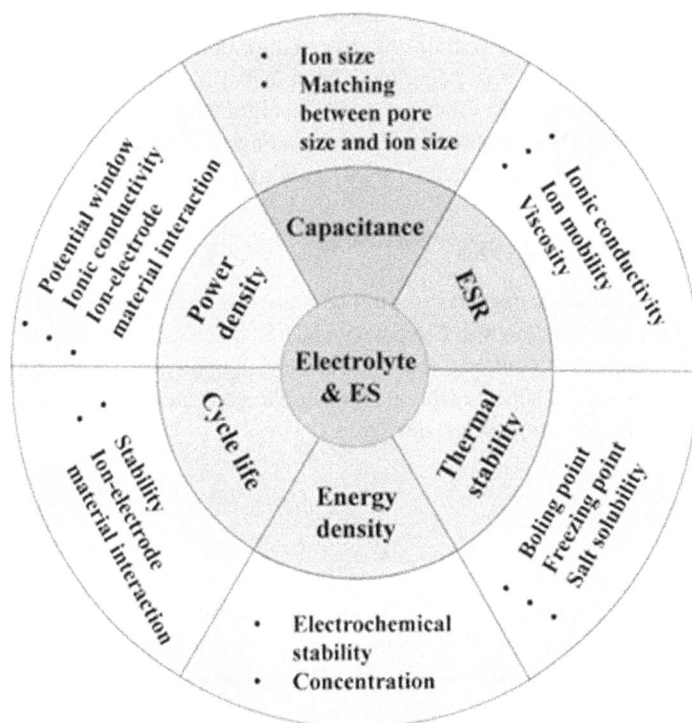

**FIGURE 3.5** Influence of electrolyte. Reproduced from [12] with permission from the Royal Society of Chemistry.

and solid-state electrolytes can provide a working window of about 4.0 V, respectively. Solid-state electrolytes like solid polymer electrolyte, gel polymer electrolyte, and polyelectrolyte are gaining more attention owing to the growing future with flexible, all-solid-state, wearable, and micro-supercapacitors. An ideal electrolyte must have good ionic conductivity, sufficient chemical stability to the electrode material, low cost, be eco-friendly, have low flammability, and wider operating temperature. Each electrolyte has their own advantages and disadvantages, it is preferential to choose an electrolyte that serves best for the targeted SC performance.

### 3.2.4 Separator

Separator is a thin electrical barrier placed between the two electrodes to provide minimal resistance to electron transfer inhibiting electrical short-circuit and chemical leakage [15]. They have good mechanical stability, support ion mobility between the two electrodes. Separators are inert, porous film filled with electrolyte (or) microporous membranes. Most used separators include cellulose paper, glass fibers, and polymer-based membranes [16]. The chemical composition, thickness, surface morphology, porosity, and pore-size distribution, play a substantial role in the internal resistance, specific capacitance, specific energy, and specific power of the supercapacitor device [12,17].

### 3.2.5 Binders

Binders are cohesive agents that bind the active material and current collectors together. They possess suitable pore dimensions alongside good mechanical strength and are expected to withstand for wider temperature ranges. Polymers like polytetrafluoroethylene (PTFE), polyvinyl acetate (PVA), poly (vinylidene difluoride) (PVDF), polyvinylpyrrolidone (PVP), etc., are the widely used binders [18]. PVDF is the most common binder owing to their sensible electrochemical activity but are costly. In the search for cost-effective, eco-friendly binders, nature-derived polymers like carboxymethyl cellulose, chitosan, guar gum, agarose, etc., have been tested as binders for electrode fabrication [19,20]. However, the usage of binders may act as dead weight that reduces the electrode-electrolyte contact and affect the porosity and pore distribution leading to ineffective usage of surface-active sites. Hence, binder-free supercapacitors are rapidly gaining interest amongst the researchers.

## 3.3  ASSESSMENT PARAMETERS

The performance of a supercapacitor device could be evaluated by determining parameters like specific capacitance, potential window, specific power, specific energy, cycling stability, rate retention, and charge transfer resistance. Potential window is the working potential unique for an electrochemical cell. In macroscopic scale, supercapacitors behave like capacitors. Capacitance ($C$) is the ratio of charge and the potential developed in a device.

$$C = \frac{Q}{\Delta V} = \frac{\varepsilon_0 \varepsilon_e A}{d} \tag{3.1}$$

where '$Q$' is the charge stored in the electrode material (Coulomb, C), '$\Delta V$' is the potential window (Volt, V), '$\varepsilon_o$' is the dielectric constant of vacuum and '$\varepsilon_e$' is the dielectric constant of electrolyte, '$A$' is the area of electrode, '$d$' is the distance of charge separation. The electrochemical techniques like cyclic voltammetry, galvanostatic charge discharge, electrochemical impedance spectroscopy, etc., is being utilized to analyze the origin of charge storage in an electrode material. The electrochemical testing is usually performed in two different configurations namely half-cell testing (three-electrode system) and full-cell testing (two-electrode system). Three-electrode system configuration involves the usage of counter (usually inert) and reference (with standard potential) electrodes in addition to the working electrode (test electrode). In two-electrode configuration, two-equivalent electrodes are utilized.

### 3.3.1 CYCLIC VOLTAMMETRY

Cyclic voltammetry (CV) is a graph of current response obtained on applying the potential linearly across the electrodes. In three-electrode system configuration, the potential is applied between the working and reference electrode whereas the current is measured between the working and counter electrode respectively. The specific capacitance ($C_s$) can be calculated from the CV curves using the formula,

$$C_s = \frac{\int I \times dV}{v \times \Delta V} \tag{3.2}$$

where '$\int I \times dV$' is the integral area under the CV curve, '$v$' is the scan rate, '$\Delta V$' is the potential window. Moreover, the information regarding the electrochemical kinetics involved during charge storage can be derived from CV curves by relating the scan rate and current response. The total current response may be considered as the sum of the capacitive current generated at the double layer formed at the electrode-electrolyte interface and the faradaic diffusion-controlled current developed on the electrode surface.

$$\text{Total current response } (I_{total}) = \text{capacitive current } (I_{cap}) + \text{diffusive current } (I_{diff}) \tag{3.3}$$

$$I = a\, v^b \tag{3.4}$$

$$\log I = \log a + b \log v \tag{3.5}$$

where '$a$', '$b$' are variable parameters, '$I$' current, and '$v$' is the scan rate. Equation (3.5) is called the power law that contains details on charge transfer kinetics. The value of '$b$' is the slope of the graph between log (peak current) and log (scan rate). If the value of '$b = 1$', the capacitive charge storage mechanism predominates, whereas when '$b = 0.5$' and below, diffusion-controlled reactions dominate. As the value of '$b$' varies between 0.5 and 1, the electrode materials are said to be in the 'transitional' region. The graphical representation is given in Figure 3.6.

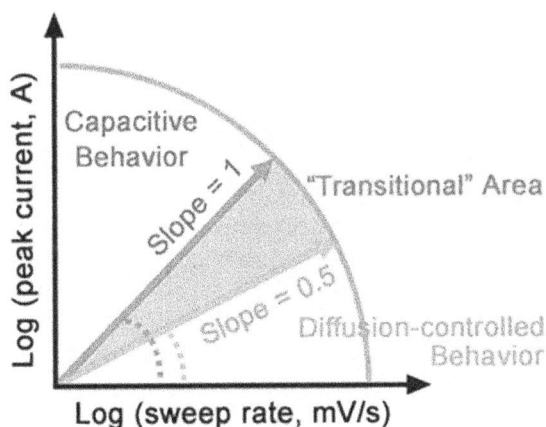

**FIGURE 3.6** Power law dependence of peak current on sweep rate. Reproduced with permission from [21] Copyright 2018 Wiley-VCH.

### 3.3.2    GALVANOSTATIC CHARGE DISCHARGE

The galvanostatic charge-discharge (GCD) studies are performed at constant current. In a three-electrode system, the current is applied between the counter and working electrode. Whereas, in two-electrode system configuration, the constant current is maintained between the positive and negative electrodes. The variation in potential developed as a function of time is measured. Specific capacitance defined as the charge stored per gram of the electrode material with respect to the potential window can be calculated from GCD graphs. The unit of specific capacitance is farad per gram $(F\ g^{-1})$. In three-electrode system configuration, the specific capacitance $\left(C_s\right)$ can be calculated from the formula,

$$C_s = \frac{I}{m} \cdot \frac{t}{\Delta V} \tag{3.6}$$

where '$I$' is the current applied, '$t$' is the discharge time, '$m$' is the electrode material loading and '$\Delta V$' is the potential window. In the two-electrode system, the specific capacitance $(C_{sp})$ can be evaluated using the formula,

$$C_{sp} = 2 \times \frac{I}{m} \cdot \frac{t}{\Delta V} \tag{3.7}$$

The Coulombic efficiency $(\eta)$ is the ratio of time taken to discharge $(t_d)$ and the charging time $(t_c)$.

$$\eta\left(\%\right) = \frac{t_d}{t_c} \times 100 \tag{3.8}$$

Ragone plot estimating the charge-storage efficiency of an energy-storage device could be plotted using specific energy and specific power. In two-electrode system configuration, the specific power and specific energy could be evaluated. Specific energy (E) is the amount of energy stored in a supercapacitor device and has the unit Watt hour per kilogram (W h kg$^{-1}$).

$$E = \frac{1}{2 \times 3.6} \cdot C_{sp} \Delta V^2 \tag{3.9}$$

Specific power (P) is the amount of charge delivered per unit time and has a unit Watt per kilogram (W kg$^{-1}$).

$$P = \frac{E \times 3600}{t} \tag{3.10}$$

Cycling stability is the number of charge-discharge cycles the electrode material withstands, until its capacitance drops a minimal percentage (%) of its initial capacitance.

### 3.3.3    ELECTROCHEMICAL IMPEDANCE SPECTROSCOPY

The electrical properties and charge transfer characteristics of an electrode material could be estimated using Electrochemical Impedance Spectroscopy (EIS). Nyquist plot (also called the complex plane plot in case of a three-electrode system) is a graph between the real part of impedance versus the negative of its imaginary part, which must be plotted with equal $X$-axis and $Y$-axis. The

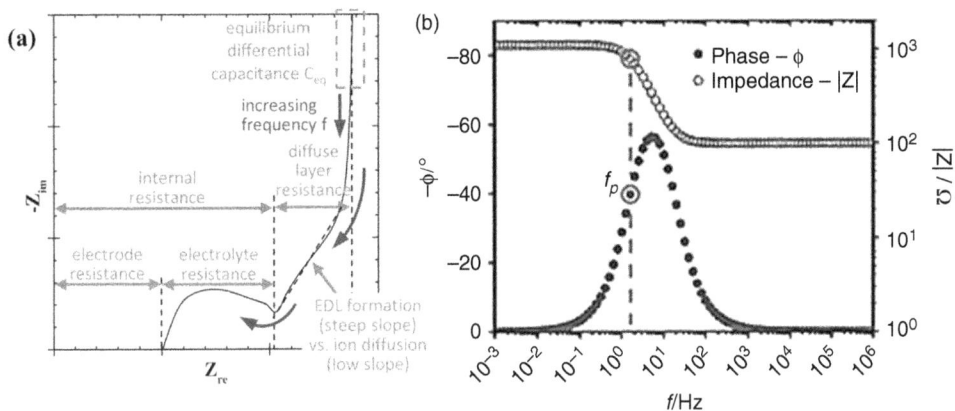

**FIGURE 3.7** Simulated model for (a) Nyquist and (b) Bode plots. Reprinted (adapted) with permission from [22] and [23]. Copyright 2018, 2020 American Chemical Society.

electrode resistance, contact resistance, electrolyte solution resistance, diffusion resistance, total internal resistance, overall resistance, and charge transportation kinetics occurring during the process of charge storage could be distinguished and interpreted from the Nyquist plot. Moreover, the electrical circuit (known as Randle circuit), equivalent to the obtainable Nyquist plot could be simulated numerically thereby approximating the electrical components involved in the electrochemical cell. A Bode plot is another graphically represented EIS information, relating the frequency $f$, with the phase angle $\varphi$, and the magnitude of impedance $|Z|$. They may be named as Bode-phase and Bode-Z plots, respectively. The frequency dependance of impedance could be assessed through the Bode plot. The model representation of the simulated Nyquist plot and Bode plot are presented in Figure 3.7.

## 3.4 TYPES OF CHARGE STORAGE

Based on the current response from the electrode materials, the mechanism of charge storage could be categorized into three types broadly, namely capacitive, pseudocapacitive, and faradaic type of charge storage. The mechanism of charge storage in an electrode material can be evaluated using their current response to applied potential. Based on the current response, electrode materials could be classified as capacitive, pseudocapacitive, and faradaic materials, as demonstrated in Figure 3.8. In capacitive materials, the charges are stored through non-faradaic charge accumulation on the electrode-electrolyte interface, with no electron transfer reaction [24]. Most carbonaceous materials without any additional elements, exhibit pure capacitive type of charge storage. Whereas, subject to the existence of functional groups on the surface of the electrode material, they may exhibit surface redox activity leading to pseudocapacitive charge storage. On the other hand, pseudocapacitive materials like metal oxides/hydroxides, conducting polymers, etc., store charge on the surface or, near the surface through fast faradaic redox reactions. In faradaic materials, the redox reactions dominate during the charge-storage process. Intercalation pseudocapacitance may arise in electrode materials, as electrolyte ions intercalate in the layers of the electrode material and partial faradaic redox reactions take place. In pseudocapacitive materials, there is no phase change occurring in the electrode material. Whereas, in typical battery electrodes (faradaic type), the redox peaks occurring at the bulk are more prominent and has more probability for change in the crystallographic phase of the material.

In EIS spectra shown in Figure 3.9, the capacitive, pseudocapacitive, and faradaic nature of the electrode materials reflect in the shape of the Nyquist plot. The 90° (or) 45° line at the

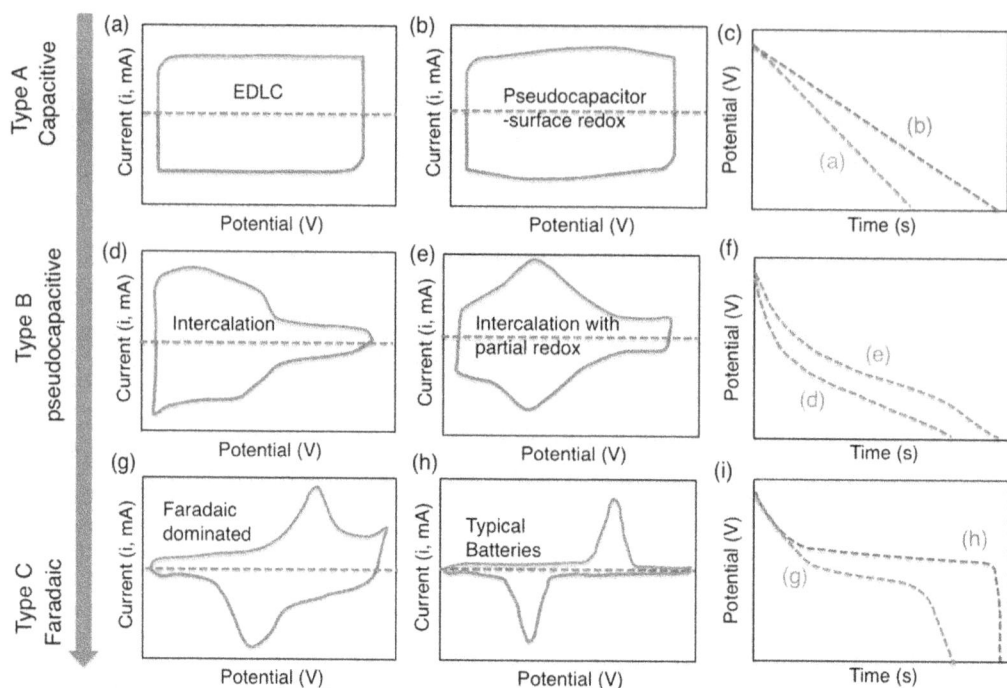

**FIGURE 3.8** Types of charge storage. Reprinted (adapted) with permission from [25]. Copyright 2018 American Chemical Society.

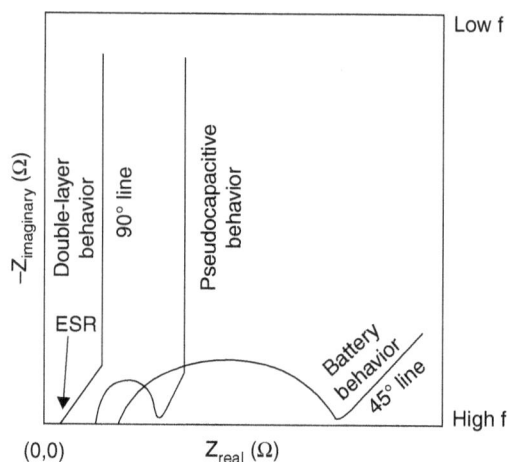

**FIGURE 3.9** Typical Nyquist plots for a capacitive, pseudocapacitive and faradaic material. Reprinted (adapted) with permission from [26]. Copyright 2019 American Chemical Society.

lower frequency region typify the capacitive, and faradaic nature of charge storage, respectively. In capacitive-type materials, there is an intercept at $X$-axis instead of the semi-circle at the high-frequency region. Whereas, in pseudocapacitive and battery-type materials at high frequencies, the semi-circle arises. Typically, the intercept at $X$-axis of the Nyquist plot (or) the semi-circle denotes the equivalent series resistance.

## 3.5 CONCEPT OF HYBRID SUPERCAPACITOR DEVICE

With the fast growth of energy-storage devices, the concept of a hybrid supercapacitor device has also evolved. To the authors' knowledge, the concept of hybrid energy device configuration was initially proposed by B. E. Conway et al., in 2004 [27]. The usage of electrode materials with dissimilar charge-storing behavior in a single device may enhance the overall charge-storage efficacy of the device together with potential window, specific energy, specific power, etc. Through this integration, the merits of diverse charge-storage processes may lead to disproportionate electroactive sites occurring at each electrode surface, providing chances for a wider range of working voltage window, improved specific power, and high specific energy competencies. There are three types of hybrid supercapacitor device configurations, namely, asymmetric capacitors, battery-type hybrids, and composite hybrids.

### 3.5.1 Asymmetric Capacitors

Asymmetric capacitors (ASCs) are a type of hybrid supercapacitor, involving electrode materials with different charge-storage mechanisms. In other words, electrodes with capacitive and faradaic redox charge-storage mechanisms are assigned as anode (positive electrode) and cathode (negative electrode) materials for fabrication of ASC. The synergistic effect of the distinctive processes broadens the specific energy and potential window of the device. Generally, one electrode material utilizing the faradaic charge-storage mechanism and another capacitive electrode material are combined in an asymmetric capacitor. To attain mass balance in asymmetric supercapacitor, the following equation must be satisfied.

$$\frac{M_+}{M_-} = \frac{C_{s-} \times \Delta V_-}{C_{s+} \times \Delta V_+} \tag{3.11}$$

where '$M_+$' is the mass of positive electrode, '$M_-$' is the mass of negative electrode, '$C_{s+}$' is the specific capacitance of positive electrode, '$C_{s-}$' is the specific capacitance of negative electrode, '$\Delta V_+$' and '$\Delta V_-$' are the potential windows of positive and negative electrodes, respectively. The asymmetric capacitor device nickel oxide‖carbon (negative‖positive electrode), containing metal oxide and carbon as electrodes, had specific capacitance of 38 F g$^{-1}$ at 1.5 V potential window in 6 M KOH aqueous electrolyte [28]. ASC assembled with exfoliated $Ti_3C_2T_x$ and transition metal chalcogenide ($MoS_{3-x}$) coated 3D-printed nanocarbon framework as electrodes employing polyvinyl alcohol (PVA)/$H_2SO_4$ gel electrolyte, had good charge-storage performance [29]. Meanwhile, unusual combination of electrode materials like vanadium nitride@activated carbon‖α-manganese oxide ASC had a specific energy of 597.5 μW h cm$^{-2}$ with a wider operating potential of 2.4 V [30]. $Mn(OH)_2$‖$Fe_2O_3$ ASC exhibited a superior specific energy of 5.125 mW h cm$^{-3}$ at specific power 14.239 mW cm$^{-3}$ and superior cycling stability of 97.1% even after 8000 cycles [31]. Coupling redox-active materials with well-regulated porosity and capacitive electrode material with accessible surface sites, open the opportunity to fabricate optimal asymmetric capacitors with extended operating voltage and superior device performance.

### 3.5.2 Battery-Type Hybrids

Materials that exhibit battery-like charge storage have been tested for supercapacitor electrodes so that they may be used to build highly efficient hybrid energy-storage devices. Most battery-grade materials demonstrate a highly faradaic charge-storage mechanism. Electrode materials like metal sulfides/hydroxides/phosphides, having typically faradic behavior, are examined as supercapacitor electrodes resulting in extraordinary charge-storage competencies. Development of devices with inherent battery-type and capacitive-type electrode materials may largely influence the overall device performance. The poor cyclability of the battery-type materials may be balanced by the highly stable carbonaceous electrode with robust surface architecture. Meanwhile, the synergistic effect of capacitive electrode with high power efficacy and battery-electrode with high energy efficacy are expected to be accomplished in the hybrid device. Such devices tend to deliver the longer life span, superior capacities, superior power, and energy proficiencies. Hydrothermally fabricated $Ni_3S_4$ microflower had battery-like electrochemical activity with a capacitance of 1703 F g$^{-1}$ at 1 A g$^{-1}$ [32]. Supercapacitor device fabricated with β-Ni(OH)$_2$@Ni foam‖activated carbon (AC) had high specific energy of 74.2 W h kg$^{-1}$ at specific power 776.9 W kg$^{-1}$, with 89.9% capacitance retention even after 10,000 cycles [33]. NiCo-layered double hydroxide (LDH)@NiOOH‖AC device exhibited specific energy of 51.7 W h kg$^{-1}$ at specific power of 599 W kg$^{-1}$ [34]. FeMoO$_4$ nanosheets chemically deposited over Ni foam had dual charge-storage nature (pseudocapacitive, battery-type) with a specific capacity of 158 mA h g$^{-1}$ at 2 A g$^{-1}$, when used as positive electrode [35].

### 3.5.3 Composite Hybrids

Electronic conductivity of metal oxides could be considerably improved through incorporation of conductive materials like carbonaceous materials (activated carbon, graphene oxide, reduced graphene oxide, carbon dots, etc.), polymers, etc. Through this strategy the specific surface area of the electrode material could be enhanced. By combining different electroactive materials in a composite hybrid, enhanced charge-storage performance could be attained. Such composite electrode materials could be employed as both anode and cathode material. Rational designing of the electrode surface may lead to swifter charge mobility, better conductivity, rapid diffusion of charges, improved surface accessible sites for charge transportation and storage. Carbon/metal oxide composite, CNT/MnO$_2$/graphene-embedded carbon cloth had superior specific energy of 10 mWh cm$^{-3}$ [36]. Solvothermally prepared metal sulfide/polymer/carbon based Cu$_2$FeSnS$_4$/PVP/rGO-decorated nanocomposite delivered 328 F g$^{-1}$ (45.55 mA h g$^{-1}$) at 0.5 A g$^{-1}$ [37]. MnO$_2$ nanowire@Co-Ni LDH prepared two-stage processing involving chemical etching and coprecipitation method as shown in Figure 3.10., had specific capacitance of 1436 F g$^{-1}$ at 1 A g$^{-1}$ [38]. Chemically precipitated Fe$_3$O$_4$/MXene/rGO exhibited a specific capacitance of 46 F g$^{-1}$ at 0.2 A g$^{-1}$ in the 5 M LiCl electrolyte [39].

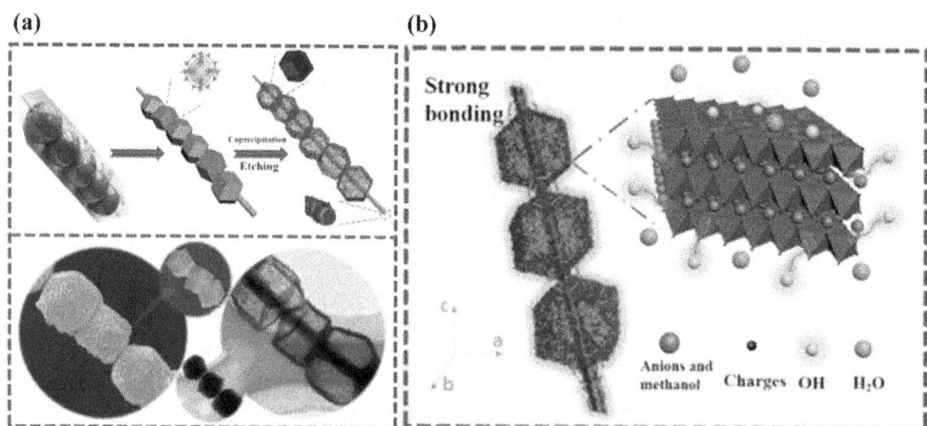

**FIGURE 3.10** Schematic representation of MnO$_2$@Co-Ni LDH electrode. Reprinted from [38], Copyright 2019, with permission from Elsevier.

## 3.6 PERSPECTIVES ON ELECTRODE MATERIALS FOR HYBRID SUPERCAPACITORS

### 3.6.1 CARBON MATERIALS

Carbon has a variety of allotropes ranging from zero-dimensional (0D) to three-dimensional (3D) structures, with different microstructures ranging from carbon nanofibers, carbon nanotubes (CNTs), graphene sheets, fullerenes, etc. They are highly stable in acidic, basic, and neutral electrolytes in a range of temperature with tunable physical properties based on the chemical/physical treatments, degree of graphitization, microstructure, etc. Carbonaceous materials without surface functionalization, mostly exhibit a capacitive type of charge storage.

#### 3.6.1.1 Graphene

Carbon-based two-dimensional (2D) planar sheets with honeycomb-like structure called graphene, can be derived from natural graphite, and have good mechanical strength, high electrical conductivity, with prospects for facile property modification. Moreover, they are abundant and cost-effective electrode materials. They have high specific surface area (SSA) about 2600 $m^2$ $g^{-1}$, which may lead to enhanced charge-storage sites on the surface of the electrode material [40]. Graphene may be prepared via physical (chemical) exfoliation, physical (chemical) reduction, Hummer's method, chemical vapor deposition (CVD), flame-induced reduction, arc discharge, electrochemical synthesis, unzipping of carbon nanotube (CNT), etc. [41]. Graphene prepared via chemical modification had a specific capacitance of 205 F $g^{-1}$ and electrochemical performance more than CNT-based supercapacitors [42]. J. Ma et al., prepared fullerenes ($C_{60}$)/graphene composite via solution method to attain a specific capacitance of 135.36 F $g^{-1}$ at 1 A $g^{-1}$ [43]. Holey graphene (HG) film with free-standing nature had capacitance of 303 F $g^{-1}$ (406 F $cm^{-3}$) at 0.5 A $g^{-1}$ [44]. Symmetric supercapacitor fabricated using graphene hydrogel had a wider potential window and higher specific energy in 1 M $Li_2SO_4$ electrolyte than that in 1 M $H_2SO_4$ electrolyte [45]. Graphene oxide (GO) prepared by a modified Hummer method, and thermal reduction had a specific capacitance of 731 F $g^{-1}$ at 10 mV $s^{-1}$ [46]. GO prepared via chemical reduction, hydrothermal method, and thermal reduction, deposited over porous current collectors, had attractive properties like feasible charge transportation and decent charge-storage performances [47].

#### 3.6.1.2 Activated Carbon

Activated carbon (AC) are highly porous carbon structures having minimal pore volume, and high SSA available for reaction. These surface properties of AC may favor the charge-storage performance of supercapacitor device. AC generally are low cost, highly stable, easy-to-prepare, and efficient electrode material, mostly utilized in commercial supercapacitors. They can be prepared from carbon-rich precursors like organic compounds (or) biomass, through the process of carbonization at inert atmosphere [48–50]. Through controlled activation and graphitization, the surface properties, area, and pore-size distribution can be engineered. Activation maybe through physical methods (or) chemical methods. Physical activation includes $CO_2$ gas purging at high temperature [51], steam treatment [52,53], etc. Whereas, chemical activation can be performed using the activating agents like $ZnCl_2$ [54], $H_3PO_4$ [55], $K_2CO_3$ [56], KOH [57], etc. Pores of different dimensions (micropores, mesopores, macropore) can be generated through optimizing the concentration of activating agent utilized during the activation process. In most cases, the activation process is done pre-carbonization and the residuals are removed after carbonization. Biomass including cocoa pods [58], areca midrib [59], rotten potato [60], cornhusk [61], lotus stalk [62], rice husk [63], human hair [64], etc., have been used as precursors for activated carbon and are reported to exhibit fair electrochemical performances. The only disadvantage of activated carbon electrodes is that specific capacitance of such materials could not be enhanced beyond a certain level (~150 F $g^{-1}$) owing to the limited SSA available for electrochemical performance.

### 3.6.1.3 Functionalized Carbon

To enhance the charge-storage properties of carbon, heteroatoms like nitrogen (N), sulfur (S), phosphorous (P), boron (B), oxygen (O) can be self-doped or simulated functionalized over the carbon prepared. The presence of these heteroatoms (possessing polarization effect) may enhance the wettability, electronic conductivity, and thereby upgrade the capacitive properties of the carbonaceous electrode material. In certain cases, these heteroatoms maybe present in the biomass and can act as self-dopants [65]. E. Taer et al. reported formation of electroactive self-oxygen doped carbon derived from dried banana leaves [66]. S. Yaglikci et al. utilized sodium thiosulfate to dope sulfur over the surface of AC prepared from waste tea [67]. Y. Feng et al. utilized $K_2S$ powder as the sulfur source to be doped in microporous carbon [68]. However, the poor stability and limited faradaic redox reactions owing to the presence of dissimilar functional groups are the drawbacks of functionalized carbon-based supercapacitors.

### 3.6.2 TRANSITION METAL OXIDES

Transition metal oxides (TMOs) are generally faradaic redox materials with numerous surface storage sites, larger SSA, through efficient pore engineering. They possess multiple oxidation states that may facilitate improved charge-storage efficiencies [69]. However, they are prone to poor electrical conductivity, poor chemical stability, and sluggish ion diffusion [70]. Redox-active electrode materials like ruthenium oxide ($RuO_2$), manganese oxide ($MnO_2$), nickel oxide (NiO), cobalt oxide ($Co_3O_4$), iron oxide ($Fe_3O_4$), vanadium oxide ($V_2O_5$), etc., have been studied for their superior charge-storage performance and remarkable reversibility. Moreover, they can be prepared in different morphologies through various physical and chemical synthesis strategies. Based on the surface morphology, the surface-active sites may vary and relative variations in their electrochemical activities could be observed.

### 3.6.2.1 Ruthenium Oxide

Ruthenium oxide ($RuO_2$) is a transition metal oxide with variable oxidation states (+4, +3, and + 2), high conductivity ($10^5$ S cm$^{-1}$), and good reversibility [71]. Trasatti proposed faradaic charge-storage mechanism of $RuO_2$ in aqueous acidic electrolyte [72,73].

$$RuO_2 + nH^+ + ne^- \leftrightarrow RuO_{2-n}(OH)_n \ (0 \leq n \leq 2)$$

$RuO_2$ has rutile structure with high theoretical capacitance of the order 1400~2000 Fg$^{-1}$ [74,75]. However, the high cost, less abundance, weak stability, and severe agglomeration of $RuO_2$ hinder the practicability of this electrode material. Different substrates were utilized by different groups to achieve the best performance out of $RuO_2$. Electrodeposited $RuO_2$ over CNT substrate had a specific capacitance of 1170 F g$^{-1}$ at 10 mV s$^{-1}$ [76]. Electrooxidized $RuO_2$ deposited over CNT substrate via atomic layer deposition method had 170 times specific capacitance than CNT electrodes [77]. $RuO_2$ coated over chemically activated carbon cloth had more than thrice higher specific capacitance than coated over bare carbon cloth [78]. Various $RuO_2$-based composites have been proven to demonstrate enhanced charge-storage efficacies. Electrospun poly(3-methylthiophene)/poly(ethylene oxide)/$RuO_2$ nanofibers exhibited a specific capacitance three times that of pristine $RuO_2$ [79]. $RuO_2$@$Co_3O_4$/N-doped GO composite had 97.1% capacitance retention even after 5000 cycles [80]. However, they are less abundant and are a costly electrode material choice in commercial scale.

### 3.6.2.2 Manganese Oxide

Manganese oxide ($MnO_2$) are naturally abundant, relatively cheaper, eco-friendly electrode materials with high theoretical capacitance of about ~1360 F g$^{-1}$ as manganese (Mn) ion undergoes

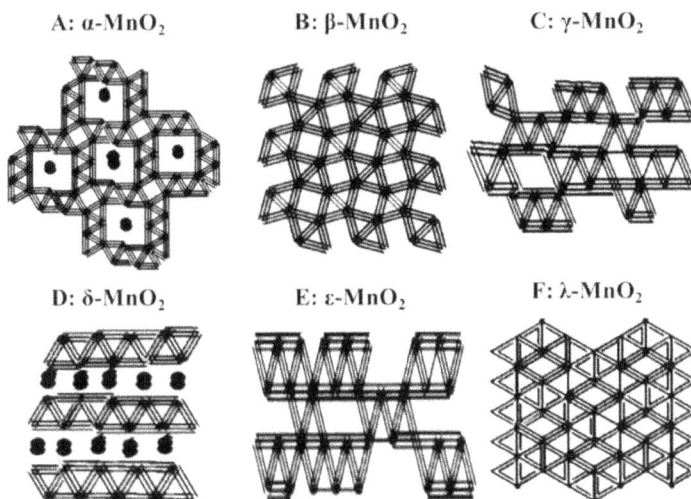

**FIGURE 3.11** Crystal structures of polymorphs of $MnO_2$. Reprinted (adapted) with permission from [95]. Copyright 2013 American Chemical Society. Reprinted from [96], [97] Copyright (1995), (2017) with permission from Elsevier.

reversible transformations $Mn^{4+} \leftrightarrow Mn^{3+}$ in the potential window 0.8 V [81,82]. $MnO_2$ was tested for the supercapacitor activity for the first time in 1999 [83]. The $MnO_6$ octahedral of $MnO_2$ could be arranged in various ways to form crystallographic forms like $\alpha$, $\beta$, $\gamma$, $\delta$, $\varepsilon$, $\lambda$, etc. as demonstrated in Figure 3.11 [84]. In $\alpha$-phase of $MnO_2$, rutile structure with [2 × 2] tunnel form is accomplished. Whereas spinel structure with [1 × 1] tunnel denoted $\beta$-$MnO_2$. In $\gamma$-$MnO_2$, pyrolusite (with [1 × 1] tunnels) and ramsdellite structures (with [1 × 2] tunnels) are intergrown. Meanwhile, $\delta$-$MnO_2$ has layered structure and $\varepsilon$-$MnO_2$ has structure similar to $\gamma$-$MnO_2$. Three-dimensional tunnel networks were observed in $\lambda$-$MnO_2$ [85–88]. Manganese oxide could be formed using facile hydrothermal, coprecipitation, microwave method, etc. However, poor electrical conductivity ($10^{-5}$ to $10^{-6}$ S cm$^{-1}$), poor stability, lower diffusion constant (~$10^{-13}$ cm$^2$ V$^{-1}$ s$^{-1}$) restrict the electrochemical performance of $MnO_2$ [89]. Hence, manganese-based composites like $MnO_2$@MWCNT fibers [90], graphene/$MnO_2$ [91], ZnO/$MnO_2$ [92], $\delta$-$MnO_2$@Fe-CNF [93], PANi/$MoS_2$-$MnO_2$ [94], have been investigated.

### 3.6.2.3 Nickel Oxide

Nickel oxide (NiO) are inexpensive, abundant, layered $p$-type TMOs with an attractive theoretical capacitance of 2584 F g$^{-1}$ in the potential window 0.5 V [98,99]. Nickel oxide could be formed using electrodeposition, hydrothermal, chemical liquid precipitation, electrodeposition, sol-gel technique, and template synthesis method. Hydrothermally synthesized NiO nanoflakes over Ni foam had 91.6% capacitance retention even after 1000 cycles [100]. Electrodeposited NiO film over stainless-steel substrate had a specific capacitance of 458 F g$^{-1}$ [101]. Electrospun NiO nanotubes had specific capacitance of 670 Fg$^{-1}$ at 1 Ag$^{-1}$, which is 30% of the theoretical capacitance [102]. Sol-gel prepared NiO with different particle sizes were obtained by controlling the calcination temperature. The influence of particle size on the charge-storage performance was studied (Figure 3.12) and it was found that NiO with an average particle size 8 nm had maximum specific capacitance of 549 F g$^{-1}$ at 1 mV s$^{-1}$ [103]. Meanwhile, Jayachandran et al. 2021 studied the influence of electrolyte on the electrochemical behavior of NiO [104]. Composites like NiO/rGO prepared via hydrothermal method demonstrated 1016.6 F g$^{-1}$ specific capacitance with 94.9% retention even after 5000 cycles [105]. Microwave synthesized NiO@rGO had 395 F g$^{-1}$ in aqueous 6 M KOH electrolyte [106].

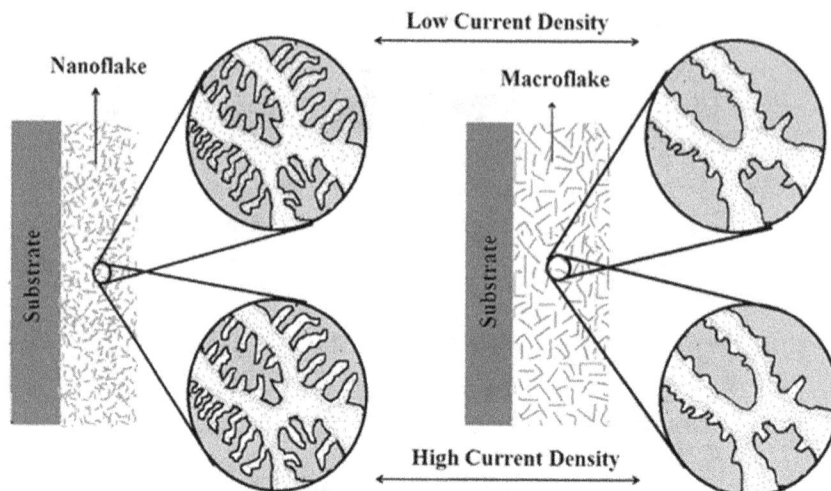

**FIGURE 3.12** Effect of particle size on electrolyte ion diffusion in NiO electrode. Reproduced from [107] with permission from the Royal Society of Chemistry.

Still, the swift capacity deterioration, inherently weak electrical conductivity ($\sim 10^{-4}$ S cm$^{-1}$), and superior resistivity restrict their practical utility [98,107].

### 3.6.2.4 Cobalt Oxide

Cobalt oxide ($Co_3O_4$, $CoO_x$) are attractive spinel TMOs with $Co^{2+}$ and $Co^{3+}$ ions in tetrahedral and octahedral sites, respectively [108]. $Co_3O_4$ are relatively less abundant and can store charge via intercalative pseudocapactiance when used as electrodes in supercapacitors. Theoretically, cobalt oxide possesses high specific capacitance (CoO-4292 F g$^{-1}$, $Co_3O_4$-3560 F g$^{-1}$) [109] but poor electrical conductivity ($10^{-4}$ to $10^{-2}$ S cm$^{-1}$) [110,111]. Cobalt oxide ($Co_3O_4$) film deposited on copper substrate through SILAR (successive ionic layer adsorption and reaction) method [112], and chemical deposition method [113] were tested for supercapacitor application. MOF-derived $Co_3O_4$ had specific energy of 54.6 W h kg$^{-1}$ at specific power 360.6 W kg$^{-1}$ [114]. Recently, research on cobalt-free (or) low-cobalt electrode materials for energy-storage devices has been trending [115–118]. The high toxicity, high cost, and fast deterioration of cobalt availability, hinder the usage cobalt-based electrode materials [119,120].

### 3.6.2.5 Iron Oxide

Iron-based oxides ($Fe_xO_y$) like wüstite (FeO), hematite ($Fe_2O_3$), magnetite ($Fe_3O_4$) are attractive electrode materials owing to their non-toxic nature, low cost, and natural abundance [121]. They also possess large theoretical capacitance (3625 F g$^{-1}$) and rich reversible redox activity ($Fe^{2+}/Fe^{3+}$) [122]. Self-assembly $Fe_3O_4$ nanostructures prepared through ethylene glycol mediated method delivered a specific capacitance of 183 F g$^{-1}$ at 1 A g$^{-1}$ [123]. Mono-dispersed $Fe_3O_4$ nanocrystals prepared via sol-gel method had a specific capacitance of 185 F g$^{-1}$ at 1 mA [124]. Hydrothermally deposited $\alpha$-$Fe_2O_3$ nanoparticles on activated carbon cloth had 96.6% capacitance retention even after 10,000 cycles [125]. Laser-induced hydrothermally grown iron oxide nanoparticles over Au–Ti substrate [Figure 3.13 (a–d)] had capacitance of 9.37 mF cm$^{-2}$ [126]. $Fe_2O_3$–rGO electrodes fabricated via unique photonic technique over PET substrate had 70% capacitance retention even after 5000 cycles [127]. $Fe_3O_4$@ 2D-crumpled porous carbon had outstanding specific power and specific energies 8000 W kg$^{-1}$ and 115.5 W h kg$^{-1}$, respectively [128]. $Fe_2O_3$ decorated graphite foam–CNT hierarchical structure processed by CVD (carbon) and ALD ($Fe_2O_3$) techniques had capacitance of $\sim$470.5

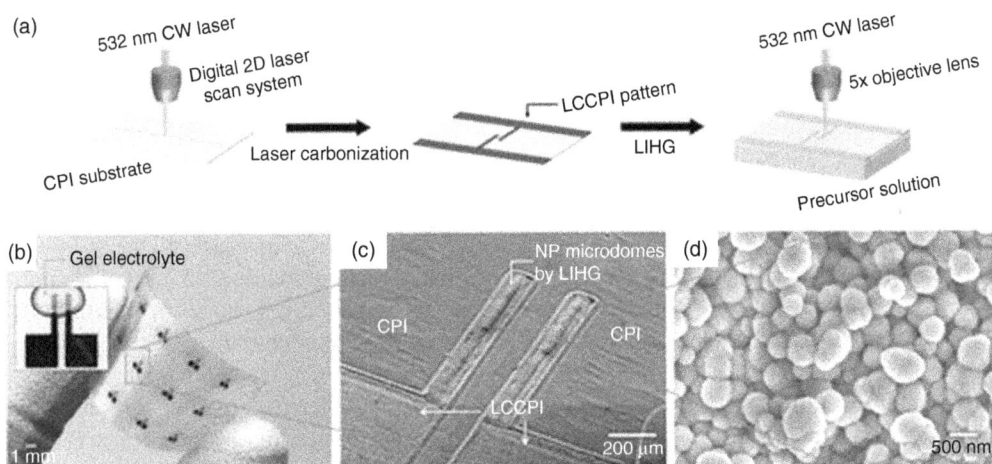

**FIGURE 3.13** (a) Representation of flexible micro-supercapacitor fabricated via laser-assisted fabrication method. (b) Picture (c-d) SEM images. Reprinted (adapted) with permission from [131] Copyright (2022) American Chemical Society.

mF cm$^{-2}$ [129]. However, their inadequate electrical conductivity ($\sim 10^{-14}$ S cm$^{-1}$) [130] and deficient electrochemical activity (specific capacitance $\sim$100 to 500 F g$^{-1}$) hamper their practical application.

### 3.6.2.6 Vanadium Oxide

Vanadium oxide are a less toxic, relatively cheap electrode material possessing unique layered structure, note-worthy theoretical capacitance (2120 F g$^{-1}$), and excellent electrochemical activity that depend on the multiple oxidation states V$^{3+}$, V$^{4+}$, and V$^{5+}$ of vanadium [132–134]. VO$_x$ grown over carbon cloth had specific capacitance of 637 F g$^{-1}$ at 0.5 mA cm$^{-2}$ in 5 M LiCl without any capacitance loss even after 1,00,000 cycles [134]. V$_2$O$_5$-embedded carbon fibers prepared via electrospinning method had specific capacitance 606 F g$^{-1}$ at 0.5 A g$^{-1}$ [133]. V$_3$O$_7$ decorated rGO-PANi composite fabricated by thermal explosive method recorded specific capacitance of 579 F g$^{-1}$ at 0.2 A g$^{-1}$ [135]. However, vanadium oxides have poor conductivity and are easily soluble in aqueous electrolytes, which limit their practical usage.

### 3.6.3 BIMETALLIC OXIDES

Bimetallic oxides (BMOs) are typical mixed metal oxides (MMOs) with a minimum of one transition metal and are intended to possess improved electronic conductivity, more surface-active sites, enhanced stability, with enriched electrochemical redox activity than pristine metal oxides owing to the synergistic effect [70,136,137]. Nickel cobalt oxide (Ni$_x$Co$_{3-x}$O$_4$, 0<$x \leq$1) spinel had higher electrical conductivity (10$^{-1}$–10 S cm$^{-1}$) than pristine constituent metal oxides, leading to higher electrochemical activity than the pristine metal oxides [138]. Ni$_x$Co$_{3-x}$O$_4$ prepared from bimetallic MOF-74–NiCo structure exhibited a capacitance of 797 F g$^{-1}$ [139]. In a report on nickel manganese mixed oxides (Ni$_y$Mn$_{1-y}$O$_x$; 0 $\leq y \leq$ 0.4) prepared through *in situ* inclusion of Ni, it was revealed that Ni$_{0.2}$Mn$_{0.8}$O$_x$ had low resistivity (2 × 10$^4$ Ohm cm$^{-1}$) with a specific capacitance of 380 F g$^{-1}$ [136]. 3D-Mn/Zn BMO prepared via microwave assisted method had a capacitance of 251 F g$^{-1}$ at 0.5 A g$^{-1}$ [140]. Solvothermal synthesized iron cobaltite (FeCo$_2$O$_4$) had specific capacitance of 854 F g$^{-1}$ at 5 A g$^{-1}$ [141]. ZnCo$_2$O$_4$ (ZCO) formed via combustion method had specific capacitance of 843 F g$^{-1}$ at 1 A g$^{-1}$ in aqueous electrolyte [142]. Hydrothermally fabricated S-doped zinc-cobalt-nickel oxide (S-ZNCO) nanorods over Ni foam had an improved capacitance of 2919 F g$^{-1}$ at 1 A g$^{-1}$ owing the oxygen vacancies and defects developed through anion doping [143]. Hierarchical composite structure $ZnMn_2O_4@ZnFe_2O_4$ microspheres had superior capacitance of 1025 F $g^{-1}$ at 10 mA $cm^{-2}$ with 95.8% capacitance retention after 3000 cycles and marginal internal resistance [144].

### 3.6.4 CONDUCTIVE POLYMERS

Conducting polymers (CPs) are pseudocapacitive electrode materials with good flexibility, comparatively low-cost, high theoretical capacitance, and remarkable conductivity ascribed to conjugated carbon bonds of the polymeric entities but limited stability [145,146]. Polyaniline (PANi), polythiophene (PTh), polypyrrole (PPy), and their derivatives shown in Figure 3.14(a) are the most used supercapacitor electrodes [147]. On insertion of counter ions, the CP undergoes redox reactions as shown in Figure 3.14(b). The incorporation of $p$-type charge carriers occurs during the process of oxidation (de-doped during reduction). Whereas $n$-type charges are doped during reduction (de-doped during oxidation). The mechanism of charge storage in CP can be described as follows [145]:

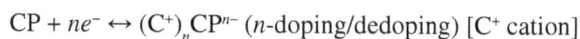

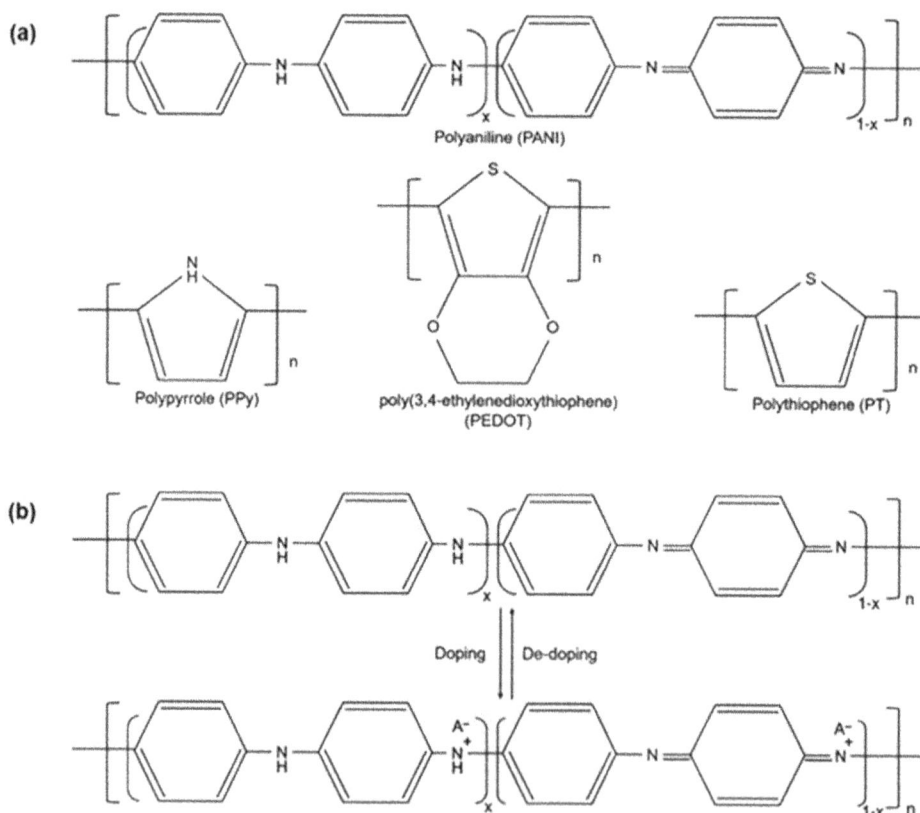

$$CP \leftrightarrow CP^{n+}(A^-)_n + ne^- \ (p\text{-doping/dedoping}) \ [A^- \text{ anion}]$$

$$CP + ne^- \leftrightarrow (C^+)_n CP^{n-} \ (n\text{-doping/dedoping}) \ [C^+ \text{ cation}]$$

**FIGURE 3.14**   (a) Chemical representation of few CP. (b) Doping/de-doping of PANI. Reproduced from [153] with permission from the Royal Society of Chemistry.

Secondary supraparticles of PANi prepared via emulsion supramolecular strategy had 93.9% capacitance retention after 10,000 cycles [148]. Benzene tetracarboxylic acids doped polyaniline had specific energy 7.42 W h kg$^{-1}$ with specific power 200 W kg$^{-1}$ [149]. Copolymerization of poly(aniline-co-pyrrole)/polyvinyl alcohol (PACP/PVA) prepared through *in situ* polymerization had capacitance of 1267 mF cm$^{-2}$ at 1 mA cm$^{-2}$ [150]. ZIF-67@PANi hybrids had 92.3% capacitance retention after 9000 cycles [151]. Highly flexible Ti@PEDOT-PANi electrode prepared via template-assisted electrochemical deposition method had capacitance of 2876 mF cm$^{-2}$ at 5 mA cm$^{-2}$ [152]. On fine tuning the degree of dopants added to CPs, the flexibility, metallic nature, electronic, optical, and electrochemical properties can be varied.

### 3.6.5 TRANSITION METAL DICHALCOGENIDES

Transition metal dichalcogenides (TMDs) having chemical formula $MX_2$ [$M$-transition metal, $X$-chalcogens like sulfide (S), selenide (Se), telluride (Te)] are graphene-analogues with high SSA, multiple oxidation states, good electrical conductivity, mechanical strength, and stability. Mostly TMDs exist in three phases, namely, semiconductor-phases 2H, 3R, and metallic phase 1T having hexagonal, rhombohedral, and trigonal crystal structures, respectively. Few-layered (or) monolayer TMDs exhibit better charge-storage performances compared to the bulk state. Interlayer electron hopping is the major drawback in TMDs. They can store charges through both surface-controlled capacitive mechanism and diffusion-controlled faradaic charge-storage mechanism.

#### 3.6.5.1 Molybdenum-Based TMDs

Molybdenum disulfide ($MoS_2$), molybdenum diselenide ($MoSe_2$), molybdenum ditelluride ($MoTe_2$) are few molybdenum based two-dimensional TMD nanosheets. Molybdenum disulfide ($MoS_2$) hold high SSA, good conductivity, and excellent mechanical strength (Young's modulus, $E = 0.33$ TPa) [154]. They exhibit remarkable charge-storage performances when exfoliated into few-layers/monolayers (or) chemically grown through deposition methods. $MoS_2$ nanostructures can be formed via versatile synthesis strategies like exfoliations (physical and chemical), chemical methods like hydrothermal, chemical vapor deposition, etc. [155,156]. The first report on the electrochemical activity of $MoS_2$ demonstrated the dominance of capacitive-type charge storage [157]. L. Jiang et al. 2015, developed 1T-2H phase hybridized $MoS_2$ monolayers with specific capacitance of 366.9 F g$^{-1}$ and 92.2% retention even after 1000 cycles [158]. Raman et al. 2021, spin-coated exfoliated $MoS_2$ dispersion over silver mesh-like network and devised flexible micro-supercapacitor with areal capacitance of 207 mF cm$^{-2}$ [159], specific energy 28.78 µWh cm$^{-2}$ and specific power 1.03 mW cm$^{-2}$ at 6 mA cm$^{-2}$. Vattikuti et al. 2018, hydrothermally developed $MoS_2$/$MoO_3$ heterostructure with specific capacitance 287.7 F g$^{-1}$ at 1 A g$^{-1}$ [160].

Molybdenum diselenide ($MoSe_2$) and molybdenum ditelluride ($MoTe_2$) are layered structures having more inter-layer distance and lower Gibbs free energy than $MoS_2$, leading to more metallic nature [161]. $MoSe_2$ can be prepared with varied morphologies. They have two prominent phases namely 1T and 2H, of which 2H is the most stable structure. Hydrothermally grown few-layered $MoSe_2$ exhibited a specific capacitance of 49.7 F g$^{-1}$ at 2 mV s$^{-1}$ in symmetric device configuration [162]. $MoSe_2$ nanosheets fabricated through *in situ* selenization technique had a specific capacity of 46 mA h g$^{-1}$ at 2 A g$^{-1}$ [163]. Hydrothermal prepared $MoSe_2$–rGO composite showcased a specific capacitance of 814 F g$^{-1}$ at 1 A g$^{-1}$ [164]. Ultrathin 1T′-$MoTe_2$ nanosheets prepared through chemical strategy had a specific capacitance of 1393 F g$^{-1}$ at 1 A g$^{-1}$ [165]. Few layer $MoTe_2$ prepared through liquid-phase exfoliation had three times specific capacitance than that of bulk [166].

#### 3.6.5.2 Tungsten-Based TMDs

Tungsten-based TMDs are impressive materials with high intrinsic conductivity, faster ion diffusion with potential to act as efficient electrode material [167]. Interwoven tungsten disulphide ($WS_2$) nanoplates supported on carbon fiber cloth had a specific capacitance of 399 F g$^{-1}$ at 1.0 A g$^{-1}$ [168].

Hydrothermally prepared $WS_2$ nanoflowers had a specific capacitance of 119 F g$^{-1}$ at 1 A g$^{-1}$ in the symmetric capacitor configuration [169]. $WS_2$ encapsulated CNTs had a specific capacitance of 536 F g$^{-1}$ at 1 A g$^{-1}$ [170]. Tungsten disulphide ($WS_2$) and tungsten diselenide ($WSe_2$) prepared via chemical vapor transport technique had superior cycling stability of 80 and 99%, respectively, even after 20,000 cycles [171]. Chakravarty and Late 2015, reported $WSe_2$ micro/nanorods synthesized via microwave and hydrothermal methods for supercapacitor application [172]. But their charge-storage efficiencies were very low. Xia et al. 2021, developed symmetric supercapacitor using 1T-$WSe_2$/graphene to achieve specific energy of 48.2 W h kg$^{-1}$ at 250 W kg$^{-1}$ [173].

### 3.6.6 TRANSITION METAL CARBIDES AND/OR NITRIDES

Transition metal carbides and/or nitrides (MXenes) are 2D metal carbides and carbonitrides, which are intrinsically conductivity with good charge transfer properties [174]. MXenes are generally synthesized through etching of MAX phase precursors. There are more than 25 unique compositions of MXenes known so far. Titanium carbide $Ti_3C_2T_x$ is the most-studied phase owing to their excellent conductivity, tunable surface morphology, stability, etc. MXenes generally store charge through a faradaic charge-storage mechanism in the $H_2SO_4$ electrolyte [175]. $Ti_3C_2T_x$ prepared via ball-milling method exhibited specific capacitance of 481 F g$^{-1}$ at 5 mV s$^{-1}$ [176]. Symmetric supercapacitor developed with $Ti_3C_2T_x$ had 100% capacitance retention even after 3000 cycles [177]. MXene/metal oxide hierarchical structures like hydrothermally grown NiO sheets over $Ti_3C_2T_x$ had enhanced charge-storage efficiency, by more than six times that of pristine [178]. MXenes being versatile in nature, have been cast as inks by various research groups as shown in Figure 3.15(a, b). All MXene micro-supercapacitor merging 3D-printed stamps and 2D titanium carbide or carbonitride inks had a capacitance of 61 mF cm$^{-2}$ at 25 µA cm$^{-2}$ [179]. Highly stable ligand-capped water-based MXene ink had capacitance of 108 mF cm$^{-2}$ (721 F cm$^{-3}$) without any current collector [180]. Various novel MXenes like $V_2NT_x$ [181], $Nb_2CT_x$ [182], etc., have also been synthesized and studied as supercapacitor electrodes.

### 3.6.7 TRANSITION METAL HYDROXIDE

Transition metal hydroxides like layered double hydroxide (LDH) are two-dimensional layered structures with cationic clay materials intercalated with charge-equivalent anions leading to charge neutrality. The unique structure with easily modifiable host layer cations and anions, tunable interlayer spacing, larger SSA, facile synthesis methods, etc., make LDH preferential electrode material. The general formula for LDH involve $[M^{2+}_{1-x}M^{3+}_x(OH)_2]A^{m-}_{x/m}.nH_2O$, which includes divalent cation $M^{2+}$, trivalent cation $M^{3+}$, interlayer anion $A^{m-}$ with valency '$m$', and '$n$' water molecules, respectively [185]. Nickel- and cobalt-based LDHs are the most studied energy-storage material. Nickel hydroxide $Ni(OH)_2$ being a traditional battery-grade material acts as a high-performance supercapacitor electrode owing to its high stability, well-defined electrochemical signature, low cost, and high theoretical capacity [186]. Hierarchical $NiO@Ni(OH)_2$ nanoarrays electrodeposited over Ni foam had capacitance of 2509 mA cm$^{-2}$ at 1 mA cm$^{-2}$ [187]. Multi-active metals in the cationic layers are proven to have improved charge-storage performances than the single-metal hydroxides. NiCo-LDH coated over surface-modified halloysite nanotubes had specific capacity of 1401 C g$^{-1}$ at 1 A g$^{-1}$ [188]. CoAl LDH prepared via hydrothermal method shown to have capacitance of 983 F g$^{-1}$ at 1 A g$^{-1}$ [189]. Hydrothermally prepared NiCoAl-LDH microspheres had specific capacitance of 451 F g$^{-1}$ at 5 A g$^{-1}$ [190]. Cu-doped cobalt copper carbonate hydroxide @NiCo LDH fibre electrodes had higher effective charge transportation with capacitance of 1.97 F cm$^{-2}$ (1237 F g$^{-1}$/193.3 mAh g$^{-1}$) and 90.8% retention even after 30,000 cycles [191]. The weaker mechanical strength and possible phase transition hamper the usage of LDH in marketable supercapacitors.

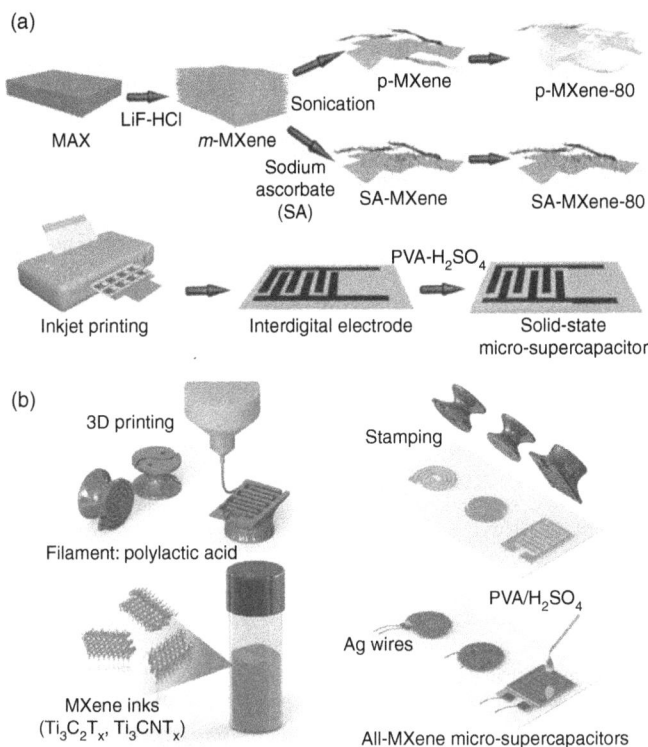

**FIGURE 3.15** Micro-supercapacitor fabricated using MXenes. (a) Inkjet printing, Reprinted from [183] Copyright (2020), with permission from Elsevier. (b) 3D printing and stamping, Reproduced with permission from [184] Copyright 2018 Wiley-VCH.

### 3.6.8 Transition Metal Phosphide

Transition metal phosphides (TMPs) are materials with significant thermal stability and high electronic conductivity attracting more energy researchers recently. Moreover, phosphorus (P) has low electronegativity than oxygen (O), sulfur (S), and selenium (Se), inducing exceptional redox signatures and facile charge transportation [192]. Biphasic nickel phosphide $Ni_5P_4$–$Ni_2P$ ($Ni_xP_y$) nanosheets had an ultrahigh capacity of 1272 C g$^{-1}$ at 2 A g$^{-1}$ [193]. Hierarchical cobalt phosphide, CoP hollow microspheres prepared via hydrothermal synthesis had a specific capacitance of 449 F g$^{-1}$ at 1 A g$^{-1}$ [194]. Copper phosphide ($Cu_3P$) nanorod coated over Cu foam by solution-free dry approach had specific capacity of 664 mA h g$^{-1}$ [195]. Multi-metal TMP present the possibility to develop more surface-active sites and enhanced electrical conductivity than single TMP owing to delocalization of electrons and synergistic effect [196]. Heterogeneous Ni–Co–P micro-flowers had excellent rate retention of 72% even at 30 A g$^{-1}$ [197]. However, they have poor initial coulombic efficiency that are to be improved.

### 3.7 CONCLUSIONS AND OUTLOOK

With the advent of next-generation electrical vehicles, smart gadgets, power electronics, the need for sustainable, highly competent energy solutions is expanding. Hybrid supercapacitors being highly competent with conventional batteries and conventional supercapacitors, open ventures for advanced energy-storage systems with superior charge-storage competencies. With immense progress of advanced energy-storage materials, and the widening need for efficient energy-storage devices, the specification of electrode materials has been upgraded enormously. Numerous electrode materials ranging from carbonaceous materials, transition metal oxides, conducting polymers, have been reported for supercapacitor application. Recently, novel transition metal dichalcogenides/hydroxide/ carbides/nitrides/phosphide, etc., have been designed, fabricated, and tested for supercapacitor application. This chapter elaborated the different mechanisms possible for charge storage in an electrode material. The distinctions in charge-storage behaviors of electrode materials were traceable through the prominent differences in their electrochemical signatures. The surface properties like morphology, pore-size distribution, surface-active sites, electrical properties, conductivity could be engineered by modifying the synthesis method, elemental compositions, oxygen vacancies, defects, etc. The enriched surface parameters may lead to enhanced energy storage in the supercapacitor device developed. A well-designed and engineered, hybrid supercapacitors with highly efficiency have the potential to be the next-generation energy-storage devices.

## REFERENCES

[1] A. Yu, V. Chabot, J. Zhang, *Electrochemical Supercapacitors for Energy Storage and Delivery: Fundamentals and Applications*, 1st ed., Boca Raton: CRC Press, 2017. https://doi.org/10.1201/ b14671

[2] B.E. Conway, *Introduction and Historical Perspective*, in: Electrochemical Supercapacitors, Boston: Springer, 1999: pp. 1–9. https://doi.org/10.1007/978-1-4757-3058-6_1

[3] M.S. Whittingham, Electrical energy storage and intercalation chemistry, *Science*. 192 (1976) 1126–1127. https://doi.org/10.1126/science.192.4244.1126

[4] H. Qiao, Q. Wei, Functional Nanofibers in Lithium-Ion Batteries, in: *Functional Nanofibers and Their Applications, Amsterdam: Elsevier*, 2012: pp. 197–208. https://doi.org/10.1533/9780857095 640.2.197

[5] H. Becker, Low voltage electrolytic capacitor, US2800616A, 1957. https://worldwide.espacenet. com/patent/search/family/023677461/publication/US2800616A?q=pn%3DUS2800616

[6] J. Shandle, Supercapacitors Find Applications in Hybrid Vehicles, Smartphones, and Energy Harvesting (n.d.). www.mouser.in/applications/new-supercapacitor-applications/ (accessed April 3, 2022).

[7] K.A. Shah, Supercapacitors: Fundamentals and Applications, 11/1/2018. (n.d.). www.electronicsf oru.com/technology-trends/learn-electronics/supercapacitors-fundamentals-applications (accessed April 3, 2022).

[8] R. Kötz, M. Carlen, Principles and applications of electrochemical capacitors, *Electrochimica Acta*. 45 (2000) 2483–2498. https://doi.org/10.1016/S0013-4686(00)00354-6

[9] G. Jiang, R.A. Senthil, Y. Sun, T.R. Kumar, J. Pan, Recent progress on porous carbon and its derivatives from plants as advanced electrode materials for supercapacitors, *Journal of Power Sources*. 520 (2022) 230886. https://doi.org/10.1016/j.jpowsour.2021.230886

[10] W. Xing, S. Qiao, X. Wu, X. Gao, J. Zhou, S. Zhuo, S.B. Hartono, D. Hulicova-Jurcakova, Exaggerated capacitance using electrochemically active nickel foam as current collector in electrochemical measurement, *Journal of Power Sources*. 196 (2011) 4123–4127. https://doi.org/10.1016/ j.jpowsour.2010.12.003

[11] H. Liu, X. Liu, S. Wang, H.-K. Liu, L. Li, Transition metal based battery-type electrodes in hybrid supercapacitors: A review, *Energy Storage Materials*. 28 (2020) 122–145.https://doi.org/10.1016/ j.ensm.2020.03.003

[12] Zhong, Y. Deng, W. Hu, J. Qiao, L. Zhang, J. Zhang, A review of electrolyte materials and compositions for electrochemical supercapacitors, *Chemical Society Re*views. 44 (2015) 7484–7539. https://doi.org/10.1039/C5CS00303B

[13] G. Wang, L. Zhang, J. Zhang, A review of electrode materials for electrochemical supercapacitors, *Chemical Society Reviews*. 41 (2012) 797–828. https://doi.org/10.1039/C1CS15060J

[14] X. Li, L. Zhao, J. Yu, X. Liu, X. Zhang, H. Liu, W. Zhou, Water Splitting: From Electrode to green energy system, *Nano-Micro Letters*. 12 (2020) 131. https://doi.org/10.1007/s40820-020-00469-3

[15] K.D. Verma, P. Sinha, S. Banerjee, K.K. Kar, M.K. Ghorai, Characteristics of Separator Materials for Supercapacitors, K.K. Kar (Ed.), *Handbook of Nanocomposite Supercapacitor Materials I*, Cham: Springer International Publishing, 2020: pp. 315–326. https://doi.org/10.1007/978-3-030-43009-2_11

[16] S. Ahankari, D. Lasrado, R. Subramaniam, Advances in materials and fabrication of separators in supercapacitors, *Materials Adv*ances 3 (2022) 1472–1496. https://doi.org/10.1039/D1MA00599E

[17] K. Tõnurist, T. Thomberg, A. Jänes, I. Kink, E. Lust, Specific performance of electrical double layer capacitors based on different separator materials in room temperature ionic liquid, *Electrochemistry Communications*. 22 (2012) 77–80. https://doi.org/10.1016/j.elecom.2012.05.029

[18] Z. Zhu, T. Shuihua, J. Yuan, X. Qin, Y. Deng, R. Qu, G.M. Haarberg, Effects of various binders on supercapacitor performances, *International Journal of Electrochemical Sci*ence (2016) 8270–8279. https://doi.org/10.20964/2016.10.04

[19] G. Landi, L. La Notte, A.L. Palma, A. Sorrentino, M.G. Maglione, G. Puglisi, A comparative evaluation of sustainable binders for environmentally friendly carbon-based supercapacitors, *Nanomaterials*. 12 (2021) 46. https://doi.org/10.3390/nano12010046

[20] P. Ruschhaupt, A. Varzi, S. Passerini, natural polymers as green binders for high-loading supercapacitor electrodes, *ChemSusChem*. 13 (2020) 763–770. https://doi.org/10.1002/cssc.201902863

[21] J. Liu, J. Wang, C. Xu, H. Jiang, C. Li, L. Zhang, J. Lin, Z.X. Shen, Advanced energy storage devices: Basic principles, *Analytical Methods, and Rational Materials Design, Advanced Sci*ence 5 (2018) 1700322. https://doi.org/10.1002/advs.201700322

[22] A.R.C. Bredar, A.L. Chown, A.R. Burton, B.H. Farnum, Electrochemical impedance spectroscopy of metal oxide electrodes for energy applications, *ACS Applied Energy Mater*ials 3 (2020) 66–98. https://doi.org/10.1021/acsaem.9b01965

[23] B.-A. Mei, O. Munteshari, J. Lau, B. Dunn, L. Pilon, Physical interpretations of Nyquist plots for EDLC electrodes and devices, *Journal of Physical Chemistry C*. 122 (2018) 194–206. https://doi.org/10.1021/acs.jpcc.7b10582

[24] B. Vidhyadharan, N.K.M. Zain, I.I. Misnon, R.A. Aziz, J. Ismail, M.M. Yusoff, R. Jose, High performance supercapacitor electrodes from electrospun nickel oxide nanowires, *Journal of Alloys and Compounds*. 610 (2014) 143–150. https://doi.org/10.1016/j.jallcom.2014.04.211

[25] Y. Gogotsi, R.M. Penner, Energy storage in nanomaterials–capacitive, pseudocapacitive, or battery-like? *ACS Nano*. 12 (2018) 2081–2083. https://doi.org/10.1021/acsnano.8b01914

[26] T.S. Mathis, N. Kurra, X. Wang, D. Pinto, P. Simon, Y. Gogotsi, Energy storage data reporting in perspective – guidelines for interpreting the performance of electrochemical energy storage systems, Advances in *Energy Mater*ials 9 (2019) 1902007. https://doi.org/10.1002/aenm.201902007

[27] W.G. Pell, B.E. Conway, Peculiarities and requirements of asymmetric capacitor devices based on combination of capacitor and battery-type electrodes, *Journal of Power Sources*. 136 (2004) 334–345. https://doi.org/10.1016/j.jpowsour.2004.03.021

[28] D.-W. Wang, F. Li, H.-M. Cheng, Hierarchical porous nickel oxide and carbon as electrode materials for asymmetric supercapacitor, *Journal of Power Sources*. 185 (2008) 1563–1568. https://doi.org/10.1016/j.jpowsour.2008.08.032

[29] K. Ghosh, M. Pumera, MXene and $MoS_{3-x}$ coated 3D-printed hybrid electrode for solid-state asymmetric supercapacitor, *Small Methods*. 5 (2021) 2100451. https://doi.org/10.1002/smtd.202100451

[30] G. Qu, Z. Wang, X. Zhang, S. Zhao, C. Wang, G. Zhao, P. Hou, X. Xu, Designing flexible asymmetric supercapacitor with high energy density by electrode engineering and charge matching mechanism, *Chemical Engineering Journal*. 429 (2022) 132406. https://doi.org/10.1016/j.cej.2021.132406

[31] J. Li, S. Luo, B. Zhang, J. Lu, W. Liu, Q. Zeng, J. Wan, X. Han, C. Hu, High-performance asymmetric $Mn(OH)_2$//$Fe_2O_3$ supercapacitor achieved by enhancing and matching respective properties of cathode and anode materials, *Nano Energy*. 79 (2021) 105410. https://doi.org/10.1016/j.nanoen.2020.105410

[32] H. Wang, M. Liang, D. Duan, W. Shi, Y. Song, Z. Sun, Rose-like $Ni_3S_4$ as battery-type electrode for hybrid supercapacitor with excellent charge storage performance, *Chemical Engineering Journal*. 350 (2018) 523–533. https://doi.org/10.1016/j.cej.2018.05.004

[33] J. Li, Y. Liu, W. Cao, N. Chen, Rapid *in situ* growth of β-$Ni(OH)_2$ nanosheet arrays on nickel foam as an integrated electrode for supercapacitors exhibiting high energy density, *Dalton Trans*actions 49 (2020) 4956–4966. https://doi.org/10.1039/D0DT00687D

[34]  H. Liang, J. Lin, H. Jia, S. Chen, J. Qi, J. Cao, T. Lin, W. Fei, J. Feng, Hierarchical NiCo-LDH@ NiOOH core-shell heterostructure on carbon fiber cloth as battery-like electrode for supercapacitor, *Journal of Power Sources*. 378 (2018) 248–254. https://doi.org/10.1016/j.jpowsour.2017.12.046

[35]  H.W. Nam, C.V.V.M. Gopi, S. Sambasivam, R. Vinodh, K.V.G. Raghavendra, H.-J. Kim, I.M. Obaidat, S. Kim, Binder-free honeycomb-like $FeMoO_4$ nanosheet arrays with dual properties of both battery-type and pseudocapacitive-type performances for supercapacitor applications, *Journal of Energy Storage*. 27 (2020) 101055. https://doi.org/10.1016/j.est.2019.101055

[36]  L. Lyu, K. Seong, J.M. Kim, W. Zhang, X. Jin, D.K. Kim, Y. Jeon, J. Kang, Y. Piao, CNT/high mass loading $MnO_2$/graphene-grafted carbon cloth electrodes for high-energy asymmetric supercapacitors, *Nano-Micro Letters*. 11 (2019) 88. https://doi.org/10.1007/s40820-019-0316-7

[37]  M. Isacfranklin, R. Yuvakkumar, G. Ravi, B. Saravanakumar, M. Pannipara, A.G. Al-Sehemi, D. Velauthapillai, Quaternary $Cu_2FeSnS_4$ /PVP/rGO composite for supercapacitor applications, *ACS Omega*. 6 (2021) 9471–9481. https://doi.org/10.1021/acsomega.0c06167

[38]  H. Luo, B. Wang, T. Liu, F. Jin, R. Liu, C. Xu, C. Wang, K. Ji, Y. Zhou, D. Wang, S. Dou, Hierarchical design of hollow Co-Ni LDH nanocages strung by $MnO_2$ nanowire with enhanced pseudocapacitive properties, *Energy Storage Materials*. 19 (2019) 370–378. https://doi.org/10.1016/j.ensm.2018.10.016

[39]  T. Arun, A. Mohanty, A. Rosenkranz, B. Wang, J. Yu, M.J. Morel, R. Udayabhaskar, S.A. Hevia, A. Akbari-Fakhrabadi, R.V. Mangalaraja, A. Ramadoss, Role of electrolytes on the electrochemical characteristics of $Fe_3O_4$/MXene/RGO composites for supercapacitor applications, *Electrochimica Acta*. 367 (2021) 137473. https://doi.org/10.1016/j.electacta.2020.137473

[40]  B. Tale, K.R. Nemade, P.V. Tekade, Graphene based nano-composites for efficient energy conversion and storage in solar cells and supercapacitors: *A Review, Polymer-Plastics Technology and Materials*. 60 (2021) 784–797. https://doi.org/10.1080/25740881.2020.1851378

[41]  Y.B. Tan, J.-M. Lee, Graphene for supercapacitor applications, *Journal of Materials Chemistry A*. 1 (2013) 14814. https://doi.org/10.1039/c3ta12193c

[42]  Y. Wang, Z. Shi, Y. Huang, Y. Ma, C. Wang, M. Chen, Y. Chen, Supercapacitor devices based on graphene materials, *Journal of Physical Chemistry C*. 113 (2009) 13103–13107. https://doi.org/10.1021/jp902214f

[43]  J. Ma, Q. Guo, H.-L. Gao, X. Qin, Synthesis of $C_{60}$/graphene composite as electrode in supercapacitors, fullerenes, *Nanotubes and Carbon Nanostructures*. 23 (2015) 477–482. https://doi.org/10.1080/1536383X.2013.865604

[44]  B. Zhu, H. Li, Y. Chen, H. Liu, Facile synthesis and high volumetric capacitance of holey graphene film for supercapacitor electrodes with optimizing preparation conditions, *Soft Materials*. 20 (2022) 137–148. https://doi.org/10.1080/1539445X.2021.1928703

[45]  B. Yang, X. Yu, T. Wang, T. Sun, Effective and high-voltage supercapacitors with graphene hydrogel in neutral aqueous solution, *Integrated Ferroelectrics*. 191 (2018) 126–132. https://doi.org/10.1080/10584587.2018.1457382

[46]  A.N. Fouda, M.K.A. Assy, G. El Enany, N. Yousf, Enhanced capacitance of thermally reduced hexagonal graphene oxide for high performance supercapacitor, fullerenes, *Nanotubes and Carbon Nanostructures*. 23 (2015) 618–622. https://doi.org/10.1080/1536383X.2014.943889

[47]  Y. Xie, Overview of supercapacitance performance of graphene supported on porous substrates, *Materials Technology*. 32 (2017) 355–366. https://doi.org/10.1080/10667857.2016.1242198

[48]  D. Puthusseri, V. Aravindan, S. Madhavi, S. Ogale, 3D micro-porous conducting carbon beehive by single step polymer carbonization for high performance supercapacitors: the magic of in situ porogen formation, *Energy and Environmental Science* 7 (2014) 728–735. https://doi.org/10.1039/C3EE42551G

[49]  S. Sundriyal, V. Shrivastav, H.D. Pham, S. Mishra, A. Deep, D.P. Dubal, Advances in bio-waste derived activated carbon for supercapacitors: Trends, challenges and prospective, *Resources, Conservation and Recycling*. 169 (2021) 105548. https://doi.org/10.1016/j.resconrec.2021.105548

[50]  Y. Liu, H. Wang, C. Li, S. Wang, L. Li, C. Song, T. Wang, Hierarchical flaky porous carbon derived from waste polyimide film for high-performance aqueous supercapacitor electrodes, *International Journal of Energy Research*. 46 (2022) 370–382. https://doi.org/10.1002/er.7106

[51]  R. Farzana, R. Rajarao, B.R. Bhat, V. Sahajwalla, Performance of an activated carbon supercapacitor electrode synthesised from waste Compact Discs (CDs), *Journal of Industrial and Engineering Chemistry*. 65 (2018) 387–396. https://doi.org/10.1016/j.jiec.2018.05.011

[52]  C.M. Ashraf, K.M. Anilkumar, B. Jinisha, M. Manoj, V.S. Pradeep, S. Jayalekshmi, Acid washed, steam activated, coconut shell derived carbon for high power supercapacitor applications, *Journal of the Electrochemical Soc*iety 165 (2018) A900–A909. https://doi.org/10.1149/2.0491805jes

[53]  L. Qin, Z. Hou, S. Lu, S. Liu, Z. Liu, E. Jiang, Porous carbon derived from pine nut shell prepared by steam activation for supercapacitor electrode material, *International Journal of Electrochemical Science* (2019) 8907–8918. https://doi.org/10.20964/2019.09.20

[54]  S.S. Gunasekaran, S. Badhulika, Effect of pH and activation on macroporous carbon derived from cocoa-pods for high performance aqueous supercapacitor application, *Materials Chemistry and Physics.* 276 (2022) 125399. https://doi.org/10.1016/j.matchemphys.2021.125399

[55]  Z. Zapata-Benabithe, C.D. Castro, G. Quintana, Kraft lignin as a raw material of activated carbon for supercapacitor electrodes, *Journal of Material Science: Materials in Electronics.* 33 (2022) 7031–7047. https://doi.org/10.1007/s10854-022-07884-9

[56]  J. Mu, Q. Li, X. Kong, X. Wu, J. Sunarso, Y. Zhao, J. Zhou, S. Zhuo, Characterization of hierarchical porous carbons made from bean curd via $K_2CO_3$ activation as a supercapacitor electrode, *ChemElectroChem.* 6 (2019) 4022–4030. https://doi.org/10.1002/celc.201900962

[57]  A. Adan-Mas, L. Alcaraz, P. Arévalo-Cid, Félix.A. López-Gómez, F. Montemor, Coffee-derived activated carbon from second biowaste for supercapacitor applications, *Waste Management.* 120 (2021) 280–289. https://doi.org/10.1016/j.wasman.2020.11.043

[58]  Y. Yetri, A.T. Hoang, Mursida, D. Dahlan, Muldarisnur, E. Taer, M.Q. Chau, Synthesis of activated carbon monolith derived from cocoa pods for supercapacitor electrodes application, *Energy Sources, Part A: Recovery, Utilization, and Environmental Effects.* (2020) 1–15. https://doi.org/10.1080/15567036.2020.1811433

[59]  R. Farma, M. Kusumasari, I. Apriyani, A. Awitdrus, The production of carbon electrodes from lignocellulosic biomass of areca midrib through a chemical activation process for supercapacitor cells application, *Energy Sources, Part A: Recovery, Utilization, and Environmental Effects.* (2021) 1–11. https://doi.org/10.1080/15567036.2021.2000068

[60]  A. Wang, K. Sun, R. Xu, Y. Sun, J. Jiang, Cleanly synthesizing rotten potato-based activated carbon for supercapacitor by self-catalytic activation, *Journal of Cleaner Production.* 283 (2021) 125385. https://doi.org/10.1016/j.jclepro.2020.125385

[61]  C.J. Raj, R. Manikandan, M. Rajesh, P. Sivakumar, H. Jung, S.J. Das, B.C. Kim, Cornhusk mesoporous activated carbon electrodes and seawater electrolyte: The sustainable sources for assembling retainable supercapacitor module, *Journal of Power Sources.* 490 (2021) 229518. https://doi.org/10.1016/j.jpowsour.2021.229518

[62]  L. Zhang, J. Yuan, S. Su, Y. Cui, W. Shi, X. Zhu, Porous active carbon derived from lotus stalk as electrode material for high-performance supercapacitors, *Journal of Wood Chemistry and Technology.* 41 (2021) 46–57. https://doi.org/10.1080/02773813.2020.1861020

[63]  M. Yu, Z. Song, C. Zhang, X. He, One-step synthesis of mesoporous carbons from mixed resources by microwave-assisted phosphoric acid activation for supercapacitors, *Materials Technology.* 32 (2017) 701–705. https://doi.org/10.1080/10667857.2017.1344370

[64]  D. Bal Altuntaş, S. Aslan, Y. Akyol, V. Nevruzoğlu, Synthesis of new carbon material produced from human hair and its evaluation as electrochemical supercapacitor, Energy Sources, Part A: Recovery, *Utilization, and Environmental Effects.* 42 (2020) 2346–2356. https://doi.org/10.1080/15567036.2020.1782536

[65]  D. Chen, L. Yang, J. Li, Q. Wu, Effect of self-doped heteroatoms in biomass-derived activated carbon for supercapacitor applications, *ChemistrySelect.* 4 (2019) 1586–1595. https://doi.org/10.1002/slct.201803413

[66]  E. Taer, R. Taslim, A. Apriwandi, Ultrahigh capacitive supercapacitor derived from self-oxygen doped biomass-based 3D porous carbon sources, *ChemNanoMat.* 8 (2022) e202100388. https://doi.org/10.1002/cnma.202100388

[67]  S. Yaglikci, Y. Gokce, E. Yagmur, Z. Aktas, The performance of sulphur doped activated carbon supercapacitors prepared from waste tea, *Environmental Technology.* 41 (2020) 36–48. https://doi.org/10.1080/09593330.2019.1575480

[68]  Y. Feng, H. Huang, W. Yang, W. Huang, Y. Xia, Y. Yi, X. Zhang, S. Zhang, Sulfur-doped microporous carbons developed from coal for enhanced capacitive performances of supercapacitor electrodes, *Integrated Ferroelectrics.* 188 (2018) 44–56. https://doi.org/10.1080/10584587.2018.1454763

[69] Z. Wu, Y. Zhu, X. Ji, C.E. Banks, Transition Metal Oxides as Supercapacitor Materials, in: K.I. Ozoemena, S. Chen (Eds.), *Nanomaterials in Advanced Batteries and Supercapacitors, Cham: Springer International Publishing*, 2016: pp. 317–344. https://doi.org/10.1007/978-3-319-26082-2_9

[70] C. An, Y. Zhang, H. Guo, Y. Wang, Metal oxide-based supercapacitors: progress and prospectives, *Nanoscale Advances* 1 (2019) 4644–4658. https://doi.org/10.1039/C9NA00543A

[71] D. Majumdar, T. Maiyalagan, Z. Jiang, Recent progress in ruthenium oxide-based composites for supercapacitor applications, *ChemElectroChem*. 6 (2019) 4343–4372. https://doi.org/10.1002/celc.201900668

[72] C. Zhan, D. Jiang, Understanding the pseudocapacitance of $RuO_2$ from joint density functional theory, *Journal of Physics: Condensed Matter*. 28 (2016) 464004. https://doi.org/10.1088/0953-8984/28/46/464004

[73] S. Trasatti, Physical electrochemistry of ceramic oxides, *Electrochimica Acta*. 36 (1991) 225–241. https://doi.org/10.1016/0013-4686(91)85244-2

[74] D.A. McKeown, P.L. Hagans, L.P.L. Carette, A.E. Russell, K.E. Swider, D.R. Rolison, Structure of hydrous ruthenium oxides: implications for charge storage, *Journal of Physical Chemistry B*. 103 (1999) 4825–4832. https://doi.org/10.1021/jp990096n

[75] L.Y. Chen, Y. Hou, J.L. Kang, A. Hirata, T. Fujita, M.W. Chen, Toward the theoretical capacitance of $RuO_2$ reinforced by highly conductive nanoporous gold, Advanced *Energy Materials*. 3 (2013) 851–856. https://doi.org/10.1002/aenm.201300024

[76] I.-H. Kim, J.-H. Kim, Y.-H. Lee, K.-B. Kim, Synthesis and characterization of electrochemically prepared ruthenium oxide on carbon nanotube film substrate for supercapacitor applications, *Journal of the Electrochemical Society*. 152 (2005) A2170. https://doi.org/10.1149/1.2041147

[77] R. Warren, F. Sammoura, F. Tounsi, M. Sanghadasa, L. Lin, Highly active ruthenium oxide coating via ALD and electrochemical activation in supercapacitor applications, *Journal of Material Chemistry A*. 3 (2015) 15568–15575. https://doi.org/10.1039/C5TA03742E

[78] C. Mevada, M. Mukhopadhyay, Enhancement of electrochemical properties of hydrous ruthenium oxide nanoparticles coated on chemically activated carbon cloth for solid-state symmetrical supercapacitor application, *Materials Chemistry and Physics*. 245 (2020) 122784. https://doi.org/10.1016/j.matchemphys.2020.122784

[79] S. Subramani, S. Rajiv, Fabrication of poly(3-methylthiophene)/poly(ethylene oxide)/ruthenium oxide composite electrospun nanofibers for supercapacitor application, Journal of Material Science: Materials in Electronics. 33 (2022) 9558–9569. https://doi.org/10.1007/s10854-021-07549-z

[80] S. Ramesh, K. Karuppasamy, A. Sivasamy, H.-S. Kim, H.M. Yadav, H.S. Kim, Core shell nanostructured of $Co_3O_4@RuO_2$ assembled on nitrogen-doped graphene sheets electrode for an efficient supercapacitor application, *Journal of Alloys and Compounds*. 877 (2021) 160297. https://doi.org/10.1016/j.jallcom.2021.160297

[81] P. Wang, Y.-J. Zhao, L.-X. Wen, J.-F. Chen, Z.-G. Lei, Ultrasound–microwave-assisted synthesis of $MnO_2$ supercapacitor electrode materials, Industrial and *Engineering Chemistry Research*. 53 (2014) 20116–20123. https://doi.org/10.1021/ie5025485

[82] N. Jabeen, Q. Xia, S.V. Savilov, S.M. Aldoshin, Y. Yu, H. Xia, Enhanced pseudocapacitive performance of $\alpha$-$MnO_2$ by cation preinsertion, *ACS Applied Material Interfaces*. 8 (2016) 33732–33740. https://doi.org/10.1021/acsami.6b12518

[83] H.Y. Lee, J.B. Goodenough, Supercapacitor behavior with KCl electrolyte, *Journal of Solid State Chemistry*. 144 (1999) 220–223. https://doi.org/10.1006/jssc.1998.8128

[84] F. Nawaz, H. Cao, Y. Xie, J. Xiao, Y. Chen, Z.A. Ghazi, Selection of active phase of MnO2 for catalytic ozonation of 4-nitrophenol, *Chemosphere*. 168 (2017) 1457–1466. https://doi.org/10.1016/j.chemosphere.2016.11.138

[85] D. Portehault, S. Cassaignon, E. Baudrin, J.-P. Jolivet, Structural and morphological control of manganese oxide nanoparticles upon soft aqueous precipitation through $MnO_4^{-}$/$Mn^{2+}$ reaction, *Journal of Material Chemistry*. 19 (2009) 2407. https://doi.org/10.1039/b816348k

[86] K.A. Stoerzinger, M. Risch, B. Han, Y. Shao-Horn, Recent Insights into Manganese Oxides in Catalyzing Oxygen Reduction Kinetics, *ACS Catalysis*. 5 (2015) 6021–6031. https://doi.org/10.1021/acscatal.5b01444

[87] C. Julien, M. Massot, S. Rangan, M. Lemal, D. Guyomard, Study of structural defects in α-$MnO_2$ by Raman spectroscopy, *Journal of Raman Spectroscopy*. 33 (2002) 223–228. https://doi.org/10.1002/jrs.838

[88] D.A. Kitchaev, H. Peng, Y. Liu, J. Sun, J.P. Perdew, G. Ceder, Energetics of $MnO_2$ polymorphs in density functional theory, *Physical Reviews B*. 93 (2016) 045132.https://doi.org/10.1103/PhysRevB.93.045132

[89] Jian-Gan Wang ED1–Zoran Stevic, Engineering Nanostructured $MnO_2$ for High Performance Supercapacitors, in: *Supercapacitor Design and Applications*, Rijeka: IntechOpen, 2016: p. Ch. 3. https://doi.org/10.5772/65008

[90] P. Shi, L. Li, L. Hua, Q. Qian, P. Wang, J. Zhou, G. Sun, W. Huang, Design of amorphous manganese oxide@multiwalled carbon nanotube fiber for robust solid-state supercapacitor, *ACS Nano*. 11 (2017) 444–452. https://doi.org/10.1021/acsnano.6b06357

[91] G. Yu, L. Hu, N. Liu, H. Wang, M. Vosgueritchian, Y. Yang, Y. Cui, Z. Bao, Enhancing the supercapacitor performance of graphene/$MnO_2$ nanostructured electrodes by conductive wrapping, *Nano Letters*. 11 (2011) 4438–4442. https://doi.org/10.1021/nl2026635

[92] C.J. Raj, M. Rajesh, R. Manikandan, J.Y. Sim, K.H. Yu, S.Y. Park, J.H. Song, B.C. Kim, Two-dimensional planar supercapacitor based on zinc oxide/manganese oxide core/shell nano-architecture, *Electrochimica Acta*. 247 (2017) 949–957. https://doi.org/10.1016/j.electacta.2017.07.009

[93] J. Li, B. Hu, P. Nie, X. Shang, W. Jiang, K. Xu, J. Yang, J. Liu, Fe-regulated δ-$MnO_2$ nanosheet assembly on carbon nanofiber under acidic condition for high performance supercapacitor and capacitive deionization, *Applied Surface Science*. 542 (2021) 148715. https://doi.org/10.1016/j.apsusc.2020.148715

[94] H. Heydari, M. Abdouss, S. Mazinani, A.M. Bazargan, F. Fatemi, Electrochemical study of ternary polyaniline/$MoS_2$–$MnO_2$ for supercapacitor applications, *Journal of Energy Storage*. 40 (2021) 102738. https://doi.org/10.1016/j.est.2021.102738

[95] D.M. Robinson, Y.B. Go, M. Mui, G. Gardner, Z. Zhang, D. Mastrogiovanni, E. Garfunkel, J. Li, M. Greenblatt, G.C. Dismukes, Photochemical water oxidation by crystalline polymorphs of manganese oxides: structural requirements for catalysis, *Journal of the American Chemical Soci*ety. 135 (2013) 3494–3501. https://doi.org/10.1021/ja310286h

[96] F. Nawaz, H. Cao, Y. Xie, J. Xiao, Y. Chen, Z.A. Ghazi, Selection of active phase of $MnO_2$ for catalytic ozonation of 4-nitrophenol, *Chemosphere*. 168 (2017) 1457–1466. https://doi.org/10.1016/j.chemosphere.2016.11.138

[97] G. Pistoia, A. Antonini, D. Zane, M. Pasquali, Synthesis of Mn spinels from different polymorphs of $MnO_2$, *Journal of Power Sources*. 56 (1995) 37–43. https://doi.org/10.1016/0378-7753(95)80006-3

[98] X. Gao, H. Zhang, E. Guo, F. Yao, Z. Wang, H. Yue, Hybrid two-dimensional nickel oxide-reduced graphene oxide nanosheets for supercapacitor electrodes, *Microchemical Journal*. 164 (2021) 105979. https://doi.org/10.1016/j.microc.2021.105979

[99] P. Lamba, P. Singh, P. Singh, A. Kumar, P. Singh, Bharti, Y. Kumar, M. Gupta, Bioinspired synthesis of nickel oxide nanoparticles as electrode material for supercapacitor applications, *Ionics*. 27 (2021) 5263–5276. https://doi.org/10.1007/s11581-021-04245-0

[100] Y. Zheng, H. Ding, M. Zhang, Preparation and electrochemical properties of nickel oxide as a supercapacitor electrode material, *Materials Research Bulletin*. 44 (2009) 403–407. https://doi.org/10.1016/j.materresbull.2008.05.002

[101] S.T. Navale, V.V. Mali, S.A. Pawar, R.S. Mane, M. Naushad, F.J. Stadler, V.B. Patil, Electrochemical supercapacitor development based on electrodeposited nickel oxide film, *RSC Advances*. 5 (2015) 51961–51965. https://doi.org/10.1039/C5RA07953E

[102] B. Vidhyadharan, N.K.M. Zain, I.I. Misnon, R.A. Aziz, J. Ismail, M.M. Yusoff, R. Jose, High performance supercapacitor electrodes from electrospun nickel oxide nanowires, *Journal of Alloys and Compounds*. 610 (2014) 143–150. https://doi.org/10.1016/j.jallcom.2014.04.211

[103] S. Pilban Jahromi, A. Pandikumar, B.T. Goh, Y.S. Lim, W.J. Basirun, H.N. Lim, N.M. Huang, Influence of particle size on performance of a nickel oxide nanoparticle-based supercapacitor, *RSC Advances*. 5 (2015) 14010–14019. https://doi.org/10.1039/C4RA16776G

[104] M. Jayachandran, S. Kishore babu, T. Maiyalagan, M.R. Kannan, R. Goutham kumar, Y. Sheeba Sherlin, T. Vijayakumar, Effect of various aqueous electrolytes on the electrochemical performance

of porous NiO nanocrystals as electrode material for supercapacitor applications, *Materials Letters.* 302 (2021) 130415. https://doi.org/10.1016/j.matlet.2021.130415

[105] P. Cao, L. Wang, Y. Xu, Y. Fu, X. Ma, Facile hydrothermal synthesis of mesoporous nickel oxide/ reduced graphene oxide composites for high performance electrochemical supercapacitor, *Electrochimica Acta.* 157 (2015) 359–368. https://doi.org/10.1016/j.electacta.2014.12.107

[106] T. Rakesh Kumar, C.H. Shilpa Chakra, S. Madhuri, E. Sai Ram, K. Ravi, Microwave-irradiated novel mesoporous nickel oxide carbon nanocomposite electrodes for supercapacitor application, *Journal of Materials Science: Materials in Electronics.* 32 (2021) 20374–20383. https://doi.org/10.1007/s10 854-021-06547-5

[107] G. Lee, Y. Cheng, C.V. Varanasi, J. Liu, Influence of the nickel oxide nanostructure morphology on the effectiveness of reduced graphene oxide coating in supercapacitor electrodes, *Journal of Physical Chem*istry C. 118 (2014) 2281–2286. https://doi.org/10.1021/jp4094904

[108] S.A. Makhlouf, Z.H. Bakr, K.I. Aly, M.S. Moustafa, Structural, electrical, and optical properties of $Co_3O_4$ nanoparticles, *Superlattices and Microstructures.* 64 (2013) 107–117. https://doi.org/10.1016/ j.spmi.2013.09.023

[109] S.J. Uke, V.P. Akhare, D.R. Bambole, A.B. Bodade, G.N. Chaudhari, Recent advancements in the cobalt oxides, manganese oxides, and their composite as an electrode material for supercapacitor: *A Review, Frontiers in Materials.* 4 (2017) 21. https://doi.org/10.3389/fmats.2017.00021

[110] V. Patil, P. Joshi, M. Chougule, S. Sen, Synthesis and characterization of $Co_3O_4$ thin film, *Soft Nanoscience Letters.* 02 (2012) 1–7. https://doi.org/10.4236/snl.2012.21001

[111] M. Masalovich, O. Zagrebelnyy, A. Nikolaev, A. Baranchikov, O. Shilova, A. Ivanova, Development of pseudocapacitive materials based on cobalt and iron oxide compounds for an asymmetric energy storage device, *Electrochimica Acta.* 410 (2022) 139999. https://doi.org/10.1016/j.electa cta.2022.139999

[112] S.G. Kandalkar, J.L. Gunjakar, C.D. Lokhande, Preparation of cobalt oxide thin films and its use in supercapacitor application, *Applied Surface Science.* 254 (2008) 5540–5544. https://doi.org/10.1016/ j.apsusc.2008.02.163

[113] S.G. Kandalkar, D.S. Dhawale, C.-K. Kim, C.D. Lokhande, Chemical synthesis of cobalt oxide thin film electrode for supercapacitor application, *Synthetic Metals.* 160 (2010) 1299–1302. https://doi. org/10.1016/j.synthmet.2010.04.003

[114] Y. Lu, Y. Liu, J. Mo, B. Deng, J. Wang, Y. Zhu, X. Xiao, G. Xu, Construction of hierarchical struc- ture of $Co_3O_4$ electrode based on electrospinning technique for supercapacitor, *Journal of Alloys and Compounds.* 853 (2021) 157271. https://doi.org/10.1016/j.jallcom.2020.157271

[115] Arumugam Manthiram, L . Wangda, S. Lee, Low-cobalt and cobalt-free, high-energy cathode materials for lithium batteries,US Patent, US11233239, 2022.

[116] Z. Cui, Q. Xie, A. Manthiram, a cobalt- and manganese-free high-nickel layered oxide cathode for long-life, safer lithium-ion batteries, Advances in *Energy Materials.* 11 (2021) 2102421. https://doi. org/10.1002/aenm.202102421

[117] S. Lee, W. Li, A. Dolocan, H. Celio, H. Park, J.H. Warner, A. Manthiram, In-depth analysis of the deg- radation mechanisms of high---nickel, low/no-cobalt layered oxide cathodes for lithium-ion batteries, Advances in *Energy Materials.* 11 (2021) 2100858. https://doi.org/10.1002/aenm.202100858

[118] N. Zhang, N. Zaker, H. Li, A. Liu, J. Inglis, L. Jing, J. Li, Y. Li, G.A. Botton, J.R. Dahn, Cobalt-free nickel-rich positive electrode materials with a core–shell structure, *Chemical Materials.* 31 (2019) 10150–10160. https://doi.org/10.1021/acs.chemmater.9b03515

[119] Y. Kim, W.M. Seong, A. Manthiram, Cobalt-free, high-nickel layered oxide cathodes for lithium- ion batteries: Progress, challenges, and perspectives, *Energy Storage Materials.* 34 (2021) 250–259. https://doi.org/10.1016/j.ensm.2020.09.020

[120] L. Noerochim, S. Suwarno, N.H. Idris, H.K. Dipojono, Recent development of nickel-rich and cobalt- free cathode materials for lithium-ion batteries, *Batteries.* 7 (2021) 84. https://doi.org/10.3390/batte ries7040084

[121] B. Xu, M. Zheng, H. Tang, Z. Chen, Y. Chi, L. Wang, L. Zhang, Y. Chen, H. Pang, Iron oxide-based nanomaterials for supercapacitors, *Nanotechnology.* 30 (2019) 204002. https://doi.org/10.1088/ 1361-6528/ab009f

[122] X. Chen, K. Chen, H. Wang, D. Xue, Composition design upon iron element toward supercapacitor electrode materials, *Materials Focus.* 4 (2015) 78–80. https://doi.org/10.1166/mat.2015.1213

[123] N. Manikandan, B. Lakshmi, S. Shivakumara, Preparation of self-assembled porous flower-like nanostructured magnetite ($Fe_3O_4$) electrode material for supercapacitor application, *Journal of Solid State Electrochemistry*. 26 (2022) 887–895. https://doi.org/10.1007/s10008-021-05097-4

[124] E. Mitchell, R.K. Gupta, K. Mensah-Darkwa, D. Kumar, K. Ramasamy, B.K. Gupta, P. Kahol, Facile synthesis and morphogenesis of superparamagnetic iron oxide nanoparticles for high-performance supercapacitor applications, *New Journal of Chemistry*. 38 (2014) 4344–4350. https://doi.org/10.1039/C4NJ00741G

[125] K. Orisekeh, B. Singh, Y. Olanrewaju, M. Kigozi, G. Ihekweme, S. Umar, V. Anye, A. Bello, S. Parida, W.O. Soboyejo, Processing of α-$Fe_2O_3$ nanoparticles on activated carbon cloth as binder-free electrode material for supercapacitor energy storage, *Journal of Energy Storage*. 33 (2021) 102042. https://doi.org/10.1016/j.est.2020.102042

[126] H. Kong, H. Kim, S. Hwang, J. Mun, J. Yeo, Laser-induced hydrothermal growth of iron oxide nanoparticles on diverse substrates for flexible micro-supercapacitors, *ACS Applied Nano Materials*. (2022) acsanm.2c00049. https://doi.org/10.1021/acsanm.2c00049

[127] M. Gaire, N. Khatoon, B. Subedi, D. Chrisey, Flexible iron oxide supercapacitor electrodes by photonic processing, *Journal of Materials Research*. 36 (2021) 4536–4546. https://doi.org/10.1557/s43578-021-00346-8

[128] S. Venkateswarlu, H. Mahajan, A. Panda, J. Lee, S. Govindaraju, K. Yun, M. Yoon, $Fe_3O_4$ nano assembly embedded in 2D-crumpled porous carbon sheets for high energy density supercapacitor, *Chemical Engineering Journal*. 420 (2021) 127584. https://doi.org/10.1016/j.cej.2020.127584

[129] C. Guan, J. Liu, Y. Wang, L. Mao, Z. Fan, Z. Shen, H. Zhang, J. Wang, Iron oxide-decorated carbon for supercapacitor anodes with ultrahigh energy density and outstanding cycling stability, *ACS Nano*. 9 (2015) 5198–5207. https://doi.org/10.1021/acsnano.5b00582

[130] Y. Ding, S. Tang, R. Han, S. Zhang, G. Pan, X. Meng, Iron oxides nanobelt arrays rooted in nanoporous surface of carbon tube textile as stretchable and robust electrodes for flexible supercapacitors with ultrahigh areal energy density and remarkable cycling-stability, *Scientific Reports*. 10 (2020) 11023. https://doi.org/10.1038/s41598-020-68032-z

[131] H. Kong, H. Kim, S. Hwang, J. Mun, J. Yeo, Laser-induced hydrothermal growth of iron oxide nanoparticles on diverse substrates for flexible micro-supercapacitors, *ACS Applied Nano Materials*. 5 (2022) 4102–4111. https://doi.org/10.1021/acsanm.2c00049

[132] H. Qin, S. Liang, L. Chen, Y. Li, Z. Luo, S. Chen, Recent advances in vanadium-based nanomaterials and their composites for supercapacitors, *Sustainable Energy Fuels*. 4 (2020) 4902–4933. https://doi.org/10.1039/D0SE00897D

[133] G. Huang, C. Li, X. Sun, J. Bai, Fabrication of vanadium oxide, with different valences of vanadium,–embedded carbon fibers and their electrochemical performance for supercapacitor, *New Journal of Chemistry*. 41 (2017) 8977–8984. https://doi.org/10.1039/C7NJ01482A

[134] M. Yu, Y. Zeng, Y. Han, X. Cheng, W. Zhao, C. Liang, Y. Tong, H. Tang, X. Lu, Valence-optimized vanadium oxide supercapacitor electrodes exhibit ultrahigh capacitance and super-long cyclic durability of 100 000 cycles, Advances in *Functional Materials*. 25 (2015) 3534–3540. https://doi.org/10.1002/adfm.201501342

[135] K.Y. Yasoda, A.A. Mikhaylov, A.G. Medvedev, M.S. Kumar, O. Lev, P.V. Prikhodchenko, S.K. Batabyal, Brush like polyaniline on vanadium oxide decorated reduced graphene oxide: Efficient electrode materials for supercapacitor, *Journal of Energy Storage*. 22 (2019) 188–193. https://doi.org/10.1016/j.est.2019.02.010

[136] P. Ahuja, S.K. Ujjain, R.K. Sharma, G. Singh, Enhanced supercapacitor performance by incorporating nickel in manganese oxide, *RSC Advances*. 4 (2014) 57192–57199. https://doi.org/10.1039/C4RA09027F

[137] X. Tang, B. Zhang, Y.H. Lui, S. Hu, Ni-Mn bimetallic oxide nanosheets as high-performance electrode materials for asymmetric supercapacitors, *Journal of Energy Storage*. 25 (2019) 100897. https://doi.org/10.1016/j.est.2019.100897

[138] X. Wang, C. Yan, A. Sumboja, P.S. Lee, High performance porous nickel cobalt oxide nanowires for asymmetric supercapacitor, *Nano Energy*. 3 (2014) 119–126. https://doi.org/10.1016/j.nanoen.2013.11.001

[139] S. Chen, M. Xue, Y. Li, Y. Pan, L. Zhu, S. Qiu, Rational design, and synthesis of $Ni_xCo_{3-x}O_4$ nanoparticles derived from multivariate MOF-74 for supercapacitors, *Journal of Material Chemistry A*. 3 (2015) 20145–20152. https://doi.org/10.1039/C5TA02557E

[140] T. Prasankumar, V.S. Irthaza Aazem, P. Raghavan, K. Prem Ananth, S. Biradar, R. Ilangovan, S. Jose, Microwave assisted synthesis of 3D network of Mn/Zn bimetallic oxide-high performance electrodes for supercapacitors, *Journal of Alloys and Compounds*. 695 (2017) 2835–2843. https://doi.org/10.1016/j.jallcom.2016.11.410

[141] S.G. Mohamed, S.Y. Attia, H.H. Hassan, Spinel-structured $FeCo_2O_4$ mesoporous nanosheets as efficient electrode for supercapacitor applications, *Microporous and Mesoporous Materials*. 251 (2017) 26–33. https://doi.org/10.1016/j.micromeso.2017.05.035

[142] J. Bhagwan, Sk. Khaja Hussain, J.S. Yu, Aqueous asymmetric supercapacitors based on $ZnCo_2O_4$ nanoparticles via facile combustion method, *Journal of Alloys and Compounds*. 815 (2020) 152456. https://doi.org/10.1016/j.jallcom.2019.152456

[143] Y. Guo, Y. Wang, Y. Zhang, Y. Zhai, W. Cai, Functional sulfur-doped zinc-nickel-cobalt oxide nanorods materials with high energy density for asymmetric supercapacitors, *Journal of Alloys and Compounds*. 896 (2022) 163053. https://doi.org/10.1016/j.jallcom.2021.163053

[144] A.E. Reddy, T. Anitha, C.V.V. Muralee Gopi, I.K. Durga, H.-J. Kim, Facile synthesis of hierarchical $ZnMn_2O_4$ @$ZnFe_2O_4$ microspheres on nickel foam for high-performance supercapacitor applications, *New Journal of Chemistry*. 42 (2018) 2964–2969. https://doi.org/10.1039/C7NJ04269H

[145] P. Naskar, A. Maiti, P. Chakraborty, D. Kundu, B. Biswas, A. Banerjee, Chemical supercapacitors: a review focusing on metallic compounds and conducting polymers, *Journal of Materials Chemistry A*. 9 (2021) 1970–2017. https://doi.org/10.1039/D0TA09655E

[146] L. Li, J. Meng, M. Zhang, T. Liu, C. Zhang, Recent advances in conductive polymer hydrogel composites and nanocomposites for flexible electrochemical supercapacitors, *Chemical Communications*. 58 (2022) 185–207. https://doi.org/10.1039/D1CC05526G

[147] Q. Meng, K. Cai, Y. Chen, L. Chen, Research progress on conducting polymer-based supercapacitor electrode materials, *Nano Energy*. 36 (2017) 268–285. https://doi.org/10.1016/j.nanoen.2017.04.040

[148] Y. Wang, X. Chu, Z. Zhu, D. Xiong, H. Zhang, W. Yang, Dynamically evolving 2D supramolecular polyaniline nanosheets for long-stability flexible supercapacitors, *Chemical Engineering Journal* 423 (2021) 130203. https://doi.org/10.1016/j.cej.2021.130203

[149] P. Das, S. Mondal, S. Malik, Fully organic polyaniline nanotubes as electrode material for durable supercapacitor, *Journal of Energy Storage*. 39 (2021) 102662. https://doi.org/10.1016/j.est.2021.102662

[150] X.-Y. Tao, Y. Wang, W. Ma, S.-F. Ye, K.-H. Zhu, L.-T. Guo, H.-L. Fan, Z.-S. Liu, Y.-B. Zhu, X.-Y. Wei, Copolymer hydrogel as self-standing electrode for high performance all-hydrogel-state supercapacitor, *Journal of Material Science*. 56 (2021) 16028–16043. https://doi.org/10.1007/s10853-021-06304-3

[151] P.-Y. Liu, J.-J. Zhao, Z.-P. Dong, Z.-L. Liu, Y.-Q. Wang, Interwoving polyaniline and a metal-organic framework grown in situ for enhanced supercapacitor behavior, *Journal of Alloys and Compounds*. 854 (2021) 157181. https://doi.org/10.1016/j.jallcom.2020.157181

[152] F. Niu, X. Han, H. Sun, Q. Li, X. He, Z. Liu, J. Sun, Z. Lei, Connecting PEDOT nanotube arrays by polyaniline coating toward a flexible and high-rate supercapacitor, *ACS Sustainable Chemical Engineering*. 9 (2021) 4146–4156. https://doi.org/10.1021/acssuschemeng.0c09365

[153] Z. Zhao, K. Xia, Y. Hou, Q. Zhang, Z. Ye, J. Lu, Designing flexible, smart and self-sustainable supercapacitors for portable/wearable electronics: from conductive polymers, *Chemical Society Reviews*. 50 (2021) 12702–12743. https://doi.org/10.1039/D1CS00800E

[154] A. Castellanos-Gomez, M. Poot, G.A. Steele, H.S.J. van der Zant, N. Agraït, G. Rubio-Bollinger, Elastic Properties of Freely Suspended $MoS_2$ Nanosheets, *Advances in Materials*. 24 (2012) 772–775. https://doi.org/10.1002/adma.201103965

[155] X. Wang, W. Xing, X. Feng, L. Song, Y. Hu, $MoS_2$/Polymer nanocomposites: Preparation, properties, and applications, *Polymer Reviews*. 57 (2017) 440–466. https://doi.org/10.1080/15583724.2017.1309662

[156] W.-J. Zhang, K.-J. Huang, A review of recent progress in molybdenum disulfide-based supercapacitors and batteries, Inorganic *Chemical Frontiers*. 4 (2017) 1602–1620. https://doi.org/10.1039/C7QI00515F

[157] J.M. Soon, K.P. Loh, Electrochemical double-layer capacitance of $MoS_2$ nanowall films, *Electrochemical Solid-State Letters.* 10 (2007) A250. https://doi.org/10.1149/1.2778851

[158] L. Jiang, S. Zhang, S.A. Kulinich, X. Song, J. Zhu, X. Wang, H. Zeng, Optimizing hybridization of 1T and 2H phases in $MoS_2$ monolayers to improve capacitances of supercapacitors, *Materials Research Letters.* 3 (2015) 177–183. https://doi.org/10.1080/21663831.2015.1057654

[159] V. Raman, D. Rhee, A.R. Selvaraj, J. Kim, K. Prabakar, J. Kang, H.-K. Kim, High-performance flexible transparent micro-supercapacitors from nanocomposite electrodes encapsulated with solution processed $MoS_2$ nanosheets, *Science and Technology of Advanced Materials.* 22 (2021) 875–884. https://doi.org/10.1080/14686996.2021.1978274

[160] S.V.P. Vattikuti, P.C. Nagajyothi, P. Anil Kumar Reddy, M. Kotesh Kumar, J. Shim, C. Byon, Tiny $MoO_3$ nanocrystals self-assembled on folded molybdenum disulfide nanosheets via a hydrothermal method for supercapacitor, *Materials Research Letters.* 6 (2018) 432–441. https://doi.org/10.1080/21663831.2018.1477848

[161] S. Upadhyay, O.P. Pandey, Studies on 2D-molybdenum diselenide ($MoSe_2$) based electrode materials for supercapacitor and batteries: A critical analysis, *Journal of Energy Storage.* 40 (2021) 102809. https://doi.org/10.1016/j.est.2021.102809

[162] S.K. Balasingam, J.S. Lee, Y. Jun, Few-layered $MoSe_2$ nanosheets as an advanced electrode material for supercapacitors, *Dalton Transactions.* 44 (2015) 15491–15498. https://doi.org/10.1039/C5DT01985K

[163] S. Upadhyay, O.P. Pandey, Synthesis of layered 2H–$MoSe_2$ nanosheets for the high-performance supercapacitor electrode material, *Journal of Alloys and Compounds.* 857 (2021) 157522. https://doi.org/10.1016/j.jallcom.2020.157522

[164] Z. Wang, H.Y. Yue, Z.M. Yu, F. Yao, X. Gao, E.H. Guan, H.J. Zhang, W.Q. Wang, S.S. Song, One-pot hydrothermal synthesis of $MoSe_2$ nanosheets spheres-reduced graphene oxide composites and application for high-performance supercapacitor, *Journal of Material Science: Materials in Electronics.* 30 (2019) 8537–8545. https://doi.org/10.1007/s10854-019-01174-7

[165] M. Liu, Z. Wang, J. Liu, G. Wei, J. Du, Y. Li, C. An, J. Zhang, Synthesis of few-layer 1T′-$MoTe_2$ ultrathin nanosheets for high-performance pseudocapacitors, *Journal of Materials Chemistry A.* 5 (2017) 1035–1042. https://doi.org/10.1039/C6TA08206H

[166] R. Hu, H. Qiao, Y. Shu, J. Li, Z. Huang, J. Tao, X. Qi, Liquid-exfoliated molybdenum telluride nanosheets for high-performance supercapacitors, *Journal of Electronic Materials.* 50 (2021) 2277–2286. https://doi.org/10.1007/s11664-021-08742-w

[167] A. Eftekhari, Tungsten dichalcogenides ($WS_2$, $WSe_2$, and $WTe_2$): materials chemistry and applications, *Journal of Materials Chemistry A.* 5 (2017) 18299–18325. https://doi.org/10.1039/C7TA04268J

[168] X. Shang, J.-Q. Chi, S.-S. Lu, J.-X. Gou, B. Dong, X. Li, Y.-R. Liu, K.-L. Yan, Y.-M. Chai, C.-G. Liu, Carbon fiber cloth supported interwoven $WS_2$ nanosplates with highly enhanced performances for supercapacitors, *Applied Surface Science.* 392 (2017) 708–714. https://doi.org/10.1016/j.apsusc.2016.09.058

[169] V.V. Mohan, M. Manuraj, P.M. Anjana, R.B. Rakhi, $WS_2$ nanoflowers as efficient electrode materials for supercapacitors, *Energy Technology.* 10 (2022) 2100976. https://doi.org/10.1002/ente.202100976

[170] B. Hu, X. Qin, A.M. Asiri, K.A. Alamry, A.O. Al-Youbi, X. Sun, $WS_2$ nanoparticles–encapsulated amorphous carbon tubes: A novel electrode material for supercapacitors with a high rate capability, *Electrochemistry Communications.* 28 (2013) 75–78. https://doi.org/10.1016/j.elecom.2012.11.035

[171] M. Habib, A. Khalil, Z. Muhammad, R. Khan, C. Wang, Z. ur Rehman, H.T. Masood, W. Xu, H. Liu, W. Gan, C. Wu, H. Chen, L. Song, $WX_2$(X=S, Se) single crystals: A highly stable material for supercapacitor applications, *Electrochimica Acta.* 258 (2017) 71–79. https://doi.org/10.1016/j.electacta.2017.10.083

[172] D. Chakravarty, D.J. Late, Microwave and hydrothermal syntheses of $WSe_2$ micro/nanorods and their application in supercapacitors, *RSC Advances.* 5 (2015) 21700–21709. https://doi.org/10.1039/C4RA12599A

[173] M. Xia, J. Ning, D. Wang, X. Feng, B. Wang, H. Guo, J. Zhang, Y. Hao, Ammonia-assisted synthesis of gypsophila-like 1T-$WSe_2$/graphene with enhanced potassium storage for all-solid-state supercapacitor, *Chemical Engineering Journal.* 405 (2021) 126611. https://doi.org/10.1016/j.cej.2020.126611

[174] V. Bayram, M. Ghidiu, J.J. Byun, S.D. Rawson, P. Yang, S.A. Mcdonald, M. Lindley, S. Fairclough, S.J. Haigh, P.J. Withers, M.W. Barsoum, I.A. Kinloch, S. Barg, MXene tunable lamellae architectures for supercapacitor electrodes, *ACS Applied Energy Materials*. 3 (2020) 411–422. https://doi.org/10.1021/acsaem.9b01654

[175] M. Hu, H. Zhang, T. Hu, B. Fan, X. Wang, Z. Li, Emerging 2D MXenes for supercapacitors: status, challenges and prospects, *Chemical Society Reviews*. 49 (2020) 6666–6693. https://doi.org/10.1039/D0CS00175A

[176] R. Syamsai, A.N. Grace, $Ta_4C_3$ MXene as supercapacitor electrodes, *Journal of Alloys and Compounds*. 792 (2019) 1230–1238. https://doi.org/10.1016/j.jallcom.2019.04.096

[177] K. Zhu, Y. Jin, F. Du, S. Gao, Z. Gao, X. Meng, G. Chen, Y. Wei, Y. Gao, Synthesis of $Ti_2CT_x$ MXene as electrode materials for symmetric supercapacitor with capable volumetric capacitance, *Journal of Energy Chemistry*. 31 (2019) 11–18. https://doi.org/10.1016/j.jechem.2018.03.010

[178] Q.X. Xia, J. Fu, J.M. Yun, R.S. Mane, K.H. Kim, High volumetric energy density annealed-MXene-nickel oxide/MXene asymmetric supercapacitor, *RSC Advances*. 7 (2017) 11000–11011. https://doi.org/10.1039/C6RA27880A

[179] C.J. Zhang, M.P. Kremer, A. Seral-Ascaso, S.-H. Park, N. McEvoy, B. Anasori, Y. Gogotsi, V. Nicolosi, Stamping of flexible, coplanar micro-supercapacitors using MXene inks, Advances in *Functional Materials*. 28 (2018) 1705506. https://doi.org/10.1002/adfm.201705506

[180] C.-W. Wu, B. Unnikrishnan, I.-W.P. Chen, S.G. Harroun, H.-T. Chang, C.-C. Huang, Excellent oxidation resistive MXene aqueous ink for micro-supercapacitor application, *Energy Storage Materials*. 25 (2020) 563–571. https://doi.org/10.1016/j.ensm.2019.09.026

[181] S. Venkateshalu, J. Cherusseri, M. Karnan, K.S. Kumar, P. Kollu, M. Sathish, J. Thomas, S.K. Jeong, A.N. Grace, New method for the synthesis of 2D vanadium nitride (MXene) and its application as a supercapacitor electrode, *ACS Omega*. 5 (2020) 17983–17992. https://doi.org/10.1021/acsomega.0c01215

[182] J. Xiao, J. Wen, J. Zhao, X. Ma, H. Gao, X. Zhang, A safe etching route to synthesize highly crystalline $Nb_2CT_x$ MXene for high performance asymmetric supercapacitor applications, *Electrochimica Acta*. 337 (2020) 135803. https://doi.org/10.1016/j.electacta.2020.135803

[183] C.-W. Wu, B. Unnikrishnan, I.-W.P. Chen, S.G. Harroun, H.-T. Chang, C.-C. Huang, Excellent oxidation resistive MXene aqueous ink for micro-supercapacitor application, *Energy Storage Materials*. 25 (2020) 563–571. https://doi.org/10.1016/j.ensm.2019.09.026

[184] C.J. Zhang, M.P. Kremer, A. Seral-Ascaso, S.-H. Park, N. McEvoy, B. Anasori, Y. Gogotsi, V. Nicolosi, Stamping of flexible, coplanar micro-supercapacitors using MXene inks, Advances in *Functional Materials*. 28 (2018) 1705506. https://doi.org/10.1002/adfm.201705506

[185] Q. Wang, D. O'Hare, Recent advances in the synthesis and application of layered double hydroxide (LDH) nanosheets, *Chemical Reviews*. 112 (2012) 4124–4155. https://doi.org/10.1021/cr200434v

[186] S. Natarajan, M. Ulaganathan, V. Aravindan, Building next-generation supercapacitors with battery type $Ni(OH)_2$, *Journal of Materials Chemistry A*. 9 (2021) 15542–15585. https://doi.org/10.1039/D1TA03262C

[187] S.A. Mozaffari, S.H. Mahmoudi Najafi, Z. Norouzi, Hierarchical $NiO@Ni(OH)_2$ nanoarrays as high-performance supercapacitor electrode material, *Electrochimica Acta*. 368 (2021) 137633. https://doi.org/10.1016/j.electacta.2020.137633

[188] Y. Wang, F. Zheng, Q. Pan, D. Deng, L. Liu, B. Chen, A three-dimensional NiCo-LDH array modified halloysite nanotube composite for high-performance battery-type supercapacitor, *Journal of Alloys and Compounds*. 884 (2021) 161162. https://doi.org/10.1016/j.jallcom.2021.161162

[189] C. Jing, X. Liu, H. Yao, P. Yan, G. Zhao, X. Bai, B. Dong, F. Dong, S. Li, Y. Zhang, Phase and morphology evolution of CoAl LDH nanosheets towards advanced supercapacitor applications, *CrystEngComm*. 21 (2019) 4934–4942. https://doi.org/10.1039/C9CE00905A

[190] X. Gao, R. Zhang, X. Huang, Y. Shi, C. Wang, Y. Gao, Z. Han, One-step growth of NiCoAl layered double hydroxides microspheres toward high energy density supercapacitors, *Journal of Alloys and Compounds*. 859 (2021) 157879. https://doi.org/10.1016/j.jallcom.2020.157879

[191] Y. Guo, X. Hong, Y. Wang, Q. Li, J. Meng, R. Dai, X. Liu, L. He, L. Mai, Multicomponent hierarchical Cu-doped NiCo-LDH/CuO double arrays for ultralong-life hybrid fiber supercapacitor, Advances in *Functional Materials*. 29 (2019) 1809004. https://doi.org/10.1002/adfm.201809004

[192] L. Xie, S. Chen, Y. Hu, Y. Lan, X. Li, Q. Deng, J. Wang, Z. Zeng, S. Deng, Construction of phosphatized cobalt nickel-LDH nanosheet arrays as binder-free electrode for high-performance battery-like supercapacitor device, *Journal of Alloys and Compounds*. 858 (2021) 157652. https://doi.org/10.1016/j.jallcom.2020.157652

[193] S. Liu, K.V. Sankar, A. Kundu, M. Ma, J.-Y. Kwon, S.C. Jun, Honeycomb-like interconnected network of nickel phosphide heteronanoparticles with superior electrochemical performance for supercapacitors, *ACS Applied Materials and Interfaces*. 9 (2017) 21829–21838. https://doi.org/10.1021/acsami.7b05384

[194] L. Ding, K. Zhang, L. Chen, Z. Yu, Y. Zhao, G. Zhu, G. Chen, D. Yan, H. Xu, A. Yu, Formation of three-dimensional hierarchical pompon-like cobalt phosphide hollow microspheres for asymmetric supercapacitor with improved energy density, *Electrochimica Acta*. 299 (2019) 62–71. https://doi.org/10.1016/j.electacta.2018.12.180

[195] N.R. Chodankar, P.A. Shinde, S.J. Patil, S.-K. Hwang, G.S.R. Raju, K.S. Ranjith, D.P. Dubal, Y.S. Huh, Y.-K. Han, Solution-free self-assembled growth of ordered tricopper phosphide for efficient and stable hybrid supercapacitor, *Energy Storage Materials*. 39 (2021) 194–202. https://doi.org/10.1016/j.ensm.2021.04.023

[196] M. Afshan, S. Kumar, S.T. Aziz, R. Ghosh, M. Pahuja, S.A. Siddiqui, K. Alam, S. Rani, D. Rani, T. Maruyama, S. Riyajuddin, K. Ghosh, Boosting the supercapacitive performance *via* incorporation of vanadium in nickel phosphide nanoflakes: A high-performance flexible renewable energy storage device, *Energy Fuels*. (2022) acs.energyfuels.2c00315. https://doi.org/10.1021/acs.energyfuels.2c00315

[197] S. He, Z. Li, H. Mi, C. Ji, F. Guo, X. Zhang, Z. Li, Q. Du, J. Qiu, 3D nickel-cobalt phosphide heterostructure for high-performance solid-state hybrid supercapacitors, *Journal of Power Sources*. 467 (2020) 228324. https://doi.org/10.1016/j.jpowsour.2020.228324

# 4 Constructive Approach Towards Design of High-Performance Thermoelectrics, Thermal Diodes, and Thermomagnetic Devices for Energy Generation Applications

*Saurabh Singh, Bed Poudel, Amin Nozariasbmarz,*
*Tsunehiro Takeuchi, and Shashank Priya*

## 4.1 INTRODUCTION

Electrical energy plays a vital role in driving daily life activities in our society. Fossil fuel-based electricity consumption has resulted in an increased environmental impact driving climate change [1–6]. Alternative energy harvesting approaches are needed that can provide electricity generation without any harmful impact on the environment. Typically, in an energy conversion process, 40–60% of energy is wasted in the form of heat. Thus, heat represents a major source of wasted energy that can be converted into electricity. One of the promising methods for heat to electricity conversion is based on the thermoelectric effect, which will be discussed at the beginning of this chapter [7–12]. Thermoelectric electricity generation requires no moving component, and it can be designed using earth-abundant materials. Waste heat in the environment is available in a wide temperature ranging from near room temperature to a few hundred degrees Celsius. Based on the temperature range, heat content, and intended thermal management outcomes, approaches based on thermoelectricity, thermal diodes, thermal rectifiers, and thermomagnetic can be utilized. The details of each of these mechanisms are presented in the subsequent sections.

## 4.2 ENERGY HARVESTING USING THERMOELECTRICITY

The schematic of the thermoelectric effect is demonstrated in Figure 4.1. It consists of a thermoelectric generator (TEG) that comprises a thermoelectric module made from several $n$- and $p$-type thermoelectric materials (discussed later); and an electric load (light bulb) in series to the TEG. An electric voltage ($\Delta V$) is generated from the TEG when temperature gradient $\Delta T$ is applied across it.

In order to create the $\Delta T$, one end of the TEG is kept at a fixed temperature via coolant (cryogenic fluid or ice), while the other end is exposed to the available waste heat. Based on the material properties, and chemical and thermal stability of TEG, the hot-side temperature can be fixed.

DOI: 10.1201/9781003340539-4

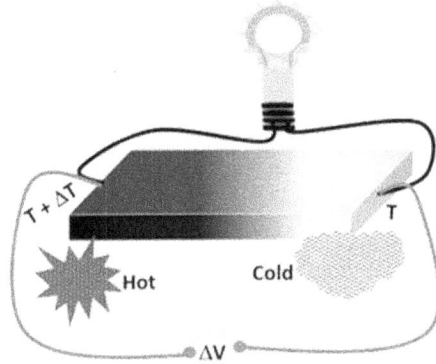

**FIGURE 4.1** Schematic description of electricity generation using thermoelectricity.

Thermoelectric generators utilize the Seebeck effect, discovered in 1821 by Thomas Johann Seebeck [13,14]. The Seebeck effect implies that a material depending upon its metallic, semi-metallic, and semiconductor nature generates the measurable electric voltage, $\Delta V$, when a finite temperature gradient, $\Delta T$, is applied across it. The generated electric voltage is known as thermo-emf voltage [15–18]. The ratio of $\Delta V/\Delta T$ is called the Seebeck coefficient ($S$) of the material. The magnitude of $S$ depends upon the chemical composition and electronic structure of the given material and is independent of the dimension. Materials with dominant electron carrier contribution to the $S$ show a negative sign, whereas, for the case where the hole is the dominant character, it shows a positive sign. The $n$- and $p$-type thermoelectric materials are classified based on the sign of the Seebeck coefficient they possess.

Metals consist of a large number of free electrons, and therefore the majority of them show negative $S$; however, some exceptional metals like silver, copper, and gold exhibit positive $S$ [4]. The positive sign of $S$ in these metals is attributed to the unusual shape of the electronic structure and scattering processes of electrons in the vicinity of the Fermi level [15]. In the case of semiconductors, depending on the type of dopants (electrons or holes), the shift in the position of chemical potentials (conduction band or valence band) occurs, and this results in both negative or positive sign of $S$. Considering the material property at the fundamental level, the magnitude of the $S$ is related to the effective mass of the charge carriers, and an interrelationship between them under the free electron approximation is given as [4–6, 15,19]

$$S = \left( \frac{8\pi^2 k_B^2}{3eh^2} \right) m^* T \left( \frac{\pi}{3n} \right)^{2/3} \tag{4.1}$$

where, $h$, $k_B$, $m^*$, $T$, and $n$ are Planck's constant, Boltzmann's constant, effective mass of charge carriers, absolute temperature, and carrier's density, respectively. In equation (4.1), the magnitude of fundamental physical constants $h$ and $k_B$ are fixed and have positive values. Also considering the absolute value of temperature, $T$, it is always a positive number, i.e., $T > 0$ K. For a fixed value of absolute temperature, the value of carrier density/concentrations, i.e., number of charge carriers per unit volume, is also a positive quantity. Therefore, the only physical quantity in the expression of $S$, which determine the sign is $m^*$. In the case of semiconductor materials having a small band gap between the top of the valence band and the bottom of the conduction band, the values of $m^*$ for holes and electrons are estimated from the electronic band structure plot, i.e., *energy* vs $k$ [15,19].

The values of effective masses for both electrons and holes depend upon the band curvature. Within the parabolic band approximation, the expression for $m^*$ is given as [15]

$$m^* = \left( \frac{\hbar^2}{\dfrac{d^2 E}{dk^2}} \right) \qquad (4.2)$$

It is evident from equation (4.2) that the value of $m^*$ will be larger for the energy band having broad or flat curvature, whereas for the narrow band, the value of $m^*$ will be smaller. For a given temperature, the most probable energy states, which contribute to the transport properties are in the range of $k_B T$, from the top of the valence band to lower energy and the bottom of the conduction band to above energy states. Therefore, with precise calculations of energy band structure and fitting of equation (4.2), the values of effective masses for both holes (VB) and electrons (CB) can be estimated. In a case where the effective mass of electrons is larger, it will have a large contribution to the $S$ and material will possess a negative sign of $S$. In another case where holes have a larger effective mass, the material will show a positive sign of $S$.

By calculating the effective masses of charge carriers, one can understand the magnitude as well as the sign of the Seebeck coefficient for semiconductor materials. The validation of this approach has been successfully demonstrated in different materials including oxides, heusler alloys, nitrides, chalcogenides, and various other thermoelectric materials. Recently, Snyder *et al.* provided a pathway to determine the values of $m^*$ for doped semiconductor materials in a wide temperature range by using the experimental values of Seebeck coefficients and carrier concentrations obtained from the Hall measurement [227]. This helps understand the transport behavior of semiconductor materials. Another tunable parameter in equation (4.1) is carrier concentration. An optimization of $n$ through external dopants can enhance the maximum power factor (*PF*), defined as $PF = S^2\sigma$, where $\sigma$ is the electrical conductivity [19]. The details of the strategic approach for the search of materials with optimized values of power factor, thus the thermoelectric performance, are discussed in a later part. In the next section, a brief discussion on thermoelectric generators is presented.

## 4.3   THERMOELECTRIC GENERATOR (TEG)

In a thermoelectric device, a temperature gradient is created by setting hot-side and cold-side temperatures. The cold-side temperature is maintained around room temperature to create a large gradient. There have been demonstrations at the laboratory level with TE modules exhibiting thermoelectric efficiency of 14% [157], however, the practical efficiency of the commercially available thermoelectric generator is limited to <10% [4–6, 19–24, 144, 157]. A typical thermoelectric generator is shown in Figure 4.2. It consists of several $n$ and $p$ legs that are connected electrically in series and thermally in parallel. The electrical contact between the legs is made using electrically conducting materials such as silver, gold, or eutectic alloys. A thermoelectric generator for large-scale power generation will comprise modules with hundreds of $n$ and $p$ leg pairs. A temperature gradient is applied to generate the output power for the load (electric bulbs, mobile chargers, electronic sensors, etc.) connected between positive and negative electrodes of the TEG as shown in Figure 4.2.

The heat to electricity conversion efficiency of a TEG is directly related to the thermoelectric properties of the materials quantified by a dimensionless parameter known as the *figure of merit* (*ZT*). The expression for thermoelectric efficiency (TE) in terms of *ZT* can be written as [4,6,19]

$$\eta = \frac{T_H - T_C}{T_H} \frac{\sqrt{1 + ZT_{avg}} - 1}{\sqrt{1 + ZT_{avg}} + T_C / T_H} \qquad (4.3)$$

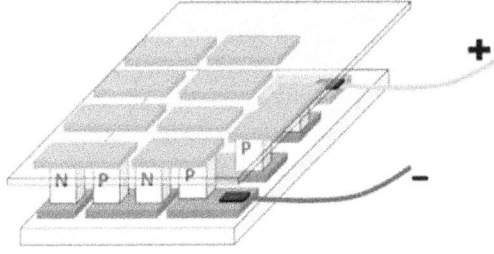

**FIGURE 4.2**    Schematic of thermoelectric generator (TEG) comprising of several *n* and *p* thermoelectric legs.

where, $T_H$, $T_C$, and $ZT_{avg}$ are hot-end temperature, cold-end temperature, and average values of the *figure of merit*, respectively. The first term on the right-hand side of equation (4.3) is known as Carnot efficiency, $\eta_{Carnot} = \left( \dfrac{T_H - T_C}{T_H} \right)$, and it depends on the temperature gradient across TEG. The second term is a reduction factor as a function of the material's figure of merit determined by the $ZT_{avg.}$ The average value of the *figure of merit* for both *n*- and *p*-type materials should be estimated carefully for a realistic estimation of the conversion efficiency of TEGs [12].

The *figure of merit* is the important material parameter for higher thermoelectric efficiency. The value of $ZT$ for a given material can be estimated by knowing Seebeck coefficients ($S$), electrical conductivity ($\sigma$), and thermal conductivity ($\kappa$). In the mathematical form, the expression of $ZT$ at absolute temperature ($T$) can be written as [4,6,19]

$$ZT = \frac{S^2 \sigma}{\kappa} T \qquad (4.4)$$

Equation (4.4) suggests that for large $ZT$ values, a material should have large $S$ and $\sigma$, whereas a small $\kappa$ is required. In the total measured value of $\kappa$, it has contributions from two different modes, i.e.: (i) the motion of electronic charge carriers carries heat and (ii) lattice vibrations help to propagate the heat through the phonon wave packet. It can be written as $\kappa = \kappa_e + \kappa_l$, where $\kappa_e$ and $\kappa_l$ are the electronic and lattice thermal conductivity, respectively. In the case of metal, the first term $\kappa_e$ is related to electrical conductivity via a relation known as Wiedemann–Franz law, i.e., $\kappa_e = L\sigma T$, where $L$ is the Lorentz number. The value of $L$ for most of the metals is $2.44 \times 10^{-8}$ $V^2K^{-2}$ however, it varies in some cases including degenerate semiconductors. Thus, one needs to be careful to adopt the constant value of $L$ while using the Wiedemann–Franz law for the estimation of $\kappa_e$ from the measured value of $\sigma$ [15,19].

From equation (4.4), it is clear that the materials should have large $S$ and $\sigma$, and low $\kappa$ in order to have large $ZT$. However, due to the interdependency of $S$, $\sigma$, and $\kappa_e$ on each other, and their interrelationship via a common physical quantity called carrier concentrations or carrier densities, it is difficult to enhance the $ZT$. Progress has been made via optimization of the carrier concentration, tuning the carrier mobility, and modifying the electronic and phonon structure to find the best condition where both electrons and phonons transport in the materials can be optimum. In the search for a good thermoelectric material, a brief selection criterion is discussed in the next section [4–6, 20–27].

### 4.3.1    SELECTION CRITERIA AND STRATEGY FOR SCREENING THE HIGH-PERFORMANCE TE MATERIALS

To understand the transport properties governed by the electronic motions in solids, a set of theoretical information has been developed within the framework of the quantum theory of solids, which is known as linear response theory. The mathematical representation of this linear response theory is implemented both in the Kubo–Greenwood formula and Boltzmann transport equations [28–31]. These equations are found to be relevant for calculating the temperature-dependent transport coefficients. The temperature-dependent aspects of these transport coefficients are well addressed by the spectral conductivity, Fermi–Dirac distribution function, and temperature-dependent chemical potential. The electronic transport coefficients ($\sigma$, $S$, $\kappa_{el}$) as a function of temperature, which express the thermoelectric properties of the solid material are given as

$$\sigma(T) = \int_{-\infty}^{\infty} \sigma(\varepsilon,T)\left(-\frac{\partial f_{FD}(\varepsilon,T)}{\partial \varepsilon}\right) d\varepsilon \tag{4.5}$$

$$S(T) = -\frac{1}{|e|T} \frac{\int_{-\infty}^{\infty} (\varepsilon-\mu)\sigma(\varepsilon,T)\left(-\frac{\partial f_{FD}(\varepsilon,T)}{\partial \varepsilon}\right) d\varepsilon}{\int_{-\infty}^{\infty} \sigma(\varepsilon,T)\left(-\frac{\partial f_{FD}(\varepsilon,T)}{\partial \varepsilon}\right) d\varepsilon} \tag{4.6}$$

$$_{el}(T) = \frac{1}{e^2 T}\int_{-\infty}^{\infty}(\varepsilon-\mu)^2 \sigma(\varepsilon,T)\left(-\frac{\partial f_{FD}(\varepsilon,T)}{\partial \varepsilon}\right) d\varepsilon$$
$$-\frac{1}{e^2 T}\frac{\left\{\int_{-\infty}^{\infty}(\varepsilon-\mu)\sigma(\varepsilon,T)\left(-\frac{\partial f_{FD}(\varepsilon,T)}{\partial \varepsilon}\right) d\varepsilon\right\}^2}{\int_{-\infty}^{\infty}\sigma(\varepsilon,T)\left(-\frac{\partial f_{FD}(\varepsilon,T)}{\partial \varepsilon}\right) d\varepsilon} \tag{4.7}$$

where, $\sigma(\varepsilon,T)$, $f_{FD}(\varepsilon,T)$, and $\mu$ are known as spectral conductivity, Fermi–Dirac distribution function, and chemical potential (Fermi level $\varepsilon_F$ at 0 K), respectively. At finite temperature, $T \neq 0$ K, for the case of metals; and with external dopants such as electrons and holes, the energy level $\varepsilon_F$ gets changed and is known by the term chemical potential. The expression $f_{FD}(\varepsilon,T) = \frac{1}{e^{(\varepsilon-\mu)/k_B T}+1}$, determines the probability of the energy states occupied by the excited electrons above and below the Fermi level. From equations (4.5), (4.6), and (4.7), one can understand that the transport coefficients $S$, $\sigma$, and $\kappa_e$ at finite temperature are mainly determined by the two functional $f_{FD}(\varepsilon,T)$ and $\sigma(\varepsilon,T)$, and chemical potential $\mu$. The expression for spectral conductivity $\sigma(\varepsilon,T)$ [28] is given as

$$\sigma(\varepsilon,T) = \frac{e^2}{3}N(\varepsilon)v^2\ \tau(\varepsilon,T) \tag{4.8}$$

where, $N(\varepsilon)$, $v$, $\tau$ are the density of states, group velocity, and relaxation time, respectively. Within the constant relaxation time approximations, the value of spectral conductivity mainly depends upon the $N(\varepsilon)$ and $v$. Therefore, for precise information on transport coefficients, these two parameters have to be estimated accurately. The information of $N(\varepsilon)$ and $v$ are obtainable from the electronic structure calculations using density functional theory based computational tools implemented in several DFT codes.

Several thermoelectric materials have been developed by considering the above relationships correlating the fundamental electronic structure and the material transport properties. For the tuning of the electronic contributions in the transport coefficients, it is required to make constructive modifications in electronic structure near chemical potential [28–36]. The sharp change in the density of states near the band edge of a narrow band gap (at least greater than $10\,k_BT$) semiconductor provides large spectral conductivity, and thus a remarkable enhancement in the transport coefficients. Additionally, the controlled tuning of carrier concentration, carrier mobility, and density of states effective mass ($m_d^*$) via suitable doping helps in achieving large Seebeck coefficients and electrical conductivity through band convergence. These strategies have been successfully demonstrated in the case of chalcogenides, silicides, and other champion thermoelectric materials. An important factor in deciding the magnitude of $ZT$ is lattice thermal conductivity. The magnitude of $\kappa_l$ is governed by the phonon structure of the material which is controllable via phonon engineering. The formula for the lattice thermal conductivity is given as [33]

$$\kappa_l = \frac{1}{3}C_v V_g\, l = \frac{1}{3}C_v V_g^{\,2}\tau \tag{4.9}$$

where, $C_v$, $V_g$, $l$, and $\tau$ are the constant volume heat capacity, the group velocity of the phonon vibration mode, mean free path, and phonon relaxation time, respectively. The values of $\kappa_l$ is mainly determined by the modifications of phonon structure. Therefore, by point defect scattering, dislocation scattering, interface scattering, resonance scattering via heavy mass elements substitutions and nano-structuring, low thermal conductivity can be obtained [15, 33].

In summary, a constructive modification in the electronic structure of the materials showing low thermal conductivity is a good approach for finding high performance thermoelectric materials. An improvement in the value of thermoelectric materials' figure of merit has been made in the last several decades, which will be discussed in the next section.

### 4.3.1.1 Thermoelectric Materials

A systematic increase in the value of $ZT$ has been attained in a different class of materials suitable for thermoelectric applications [37–39]. The trend of improvement in $ZT$ with time is shown in Figure 4.3. Several new materials with high $ZT$ values were introduced in the last two decades. In Figure 4.3, the data for chalcogenides, Si-Ge, Heusler alloys, and oxides materials have been compiled

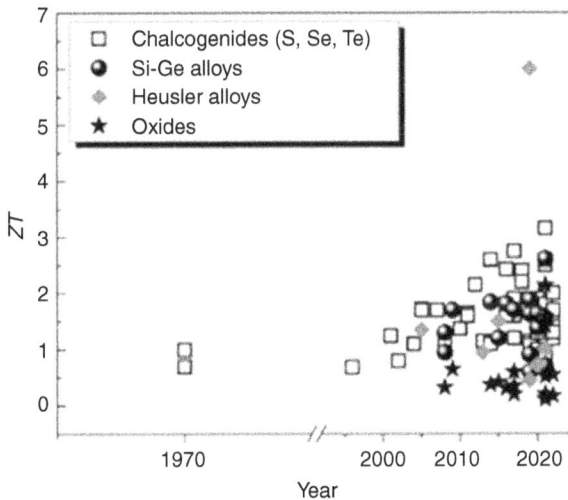

**FIGURE 4.3** Progress in the values of figure-of-merit ($ZT$) of the *state-of-art* thermoelectric (TE) materials in past several decades.

as these materials have shown the potential to make the device for thermoelectric applications [40–138]. There are several other categories of materials such as skutterudites, halides, etc., for which *ZT* has been improved with time; however, they need more refinement to get implemented at the device level. Among the materials shown in Figure 4.3, chalcogenide-based materials have been explored widely due to their high thermoelectric performance near room temperature [89–95]. Some of the chalcogenide materials are found to be stable and they preserve the thermoelectric performance in the mid temperature range of 300–600 K; yet, the performance gets suppressed due to the degradation at high temperatures.

For the high-temperature region, Si-Ge based alloys, silicide (Mg-Si,Sn), Heusler alloys, and oxides are the better choice as their thermoelectric performance with time does not change much [40–138]. Additionally, at high temperature the chemical, mechanical, and thermal stability of these materials fulfills the desired criteria for making thermoelectric devices.

### 4.3.1.2 Thermoelectric Efficiency

The heat to electricity conversion efficiency of a thermoelectric device is characterized by thermoelectric efficiency ($\eta$). Several thermoelectric devices have been demonstrated at the laboratory level [84, 147, 150]. Such devices kept under a temperature gradient of a few hundred degrees Celsius generate reasonable output power as shown in Figure 4.2. One end of the device is maintained at base temperature, mostly at room temperature, and another end at high temperature at a safe point below the chemical and thermal stability limit of *n* and *p* legs, and electrical contact materials. The calculated percentage efficiency is required to be estimated carefully by taking the average *ZT* of the material rather than the peak value of *ZT*. Figure 4.4 shows the thermoelectric efficiency for

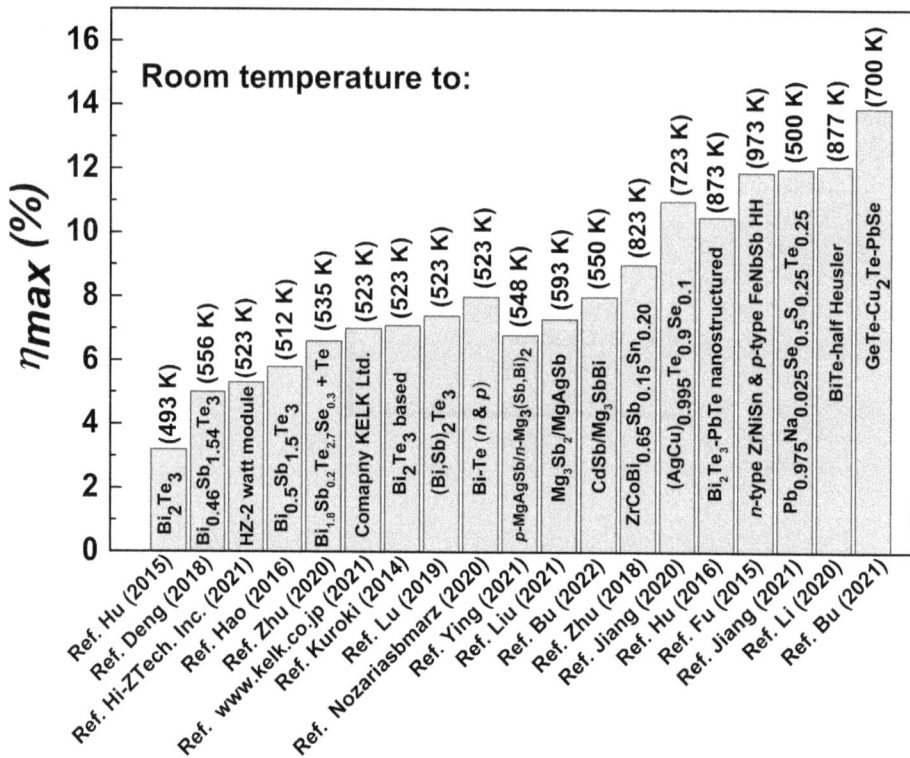

**FIGURE 4.4** Maximum thermoelectric efficiency ($\eta_{max}$) of high-performance thermoelectric materials in recent years.

thermoelectric devices made of chalcogenides, (Mg,Sb)-based alloys [139–157], chalcogenides and Heusler alloys. The best thermoelectric efficiency is achieved up to 14% in GeTe-Cu$_2$Te-PbSe based material [157].

The magnitude of 14% thermoelectric efficiency obtained within a decade is a remarkable achievement in the field of energy harvesting using thermoelectric materials [157]. In addition to thermoelectric devices for energy conversion, heat flow management is also one of the important aspects of thermal management. Control of heat flow direction opens new directions of applications with new designs of devices such as thermal diodes, thermal rectifiers, and thermal actuators. Thermal diodes are beneficial due to asymmetric heat transfer. For making thermal diodes, materials need to have a large change in thermal conductivity with temperature. This can be found in materials that go through the structural and/or thermal phase transition. The application domain is enhanced by finding the materials where an unusual change in thermal conductivity occurs within hundreds of degrees Celsius above room temperature. In the next section, this new field of heat management is discussed.

## 4.3.2 Thermal Diodes

An effective way of heat flow management is possible by designing the device like a diode, known as a thermal diode, which allows the asymmetrical flow of heat. A schematic diagram of a thermal diode made of two different materials **A** and **B**, having different trends of thermal conductivity and connected in series is shown in Figure 4.5.

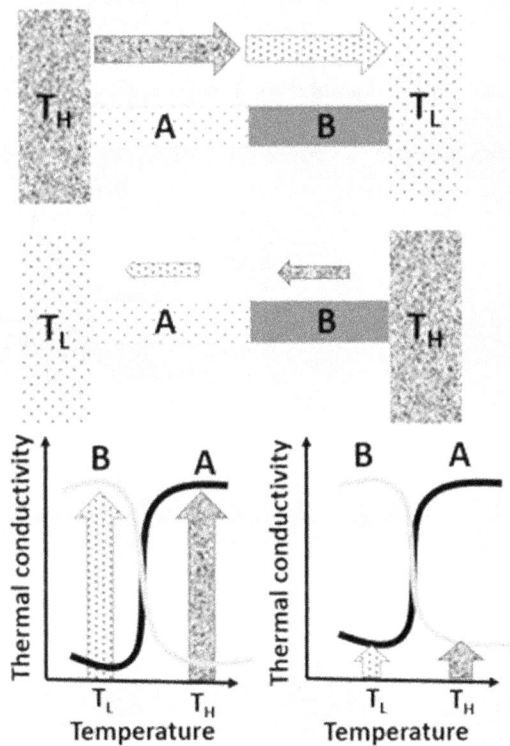

**FIGURE 4.5** Schematic representation of thermal diodes consists of two different materials A and B having different thermal conductivity behavior.

For material **A,** thermal conductivity shows the step increment change with temperature, whereas for material **B** it shows the step decrement in magnitude. Such behavior in thermal conductivity as a function of temperature is found in the materials such as $VO_2$, $Ag_2$(S, Se, Te), manganites, etc. [158–160]. The thermal resistance at the interface of materials **A** and **B** is minimized by applying the heat conducting thermal paste. When one end of material **A** is kept at a high temperature $T_H$ and one end of material **B** is kept at a low temperature, the opposite trend in thermal conductivity between them allows the net heat flow from **A** to **B**. In another situation, when the same configuration of **A** and **B** is attached, but, hot and cold ends get exchanged, i.e., end **A** at a low temperature and end **B** at high temperature, heat flow gets suppressed. This unique characteristic of thermal conductivity possessed by the materials provides an effective heat flow control with unique opportunities required for solar thermal energy harvesting, thermoelectric module, caloric refrigeration, etc.

The control gauge of heat flow management for a given thermal diode is determined by the thermal rectification ratio (TRR) defined as

$$TRR = \frac{\left|q_{fwd}\right| - \left|q_{rev}\right|}{\left|q_{rev}\right|} \cdot 100\% \tag{4.10}$$

where $q_{fwd}$ and $q_{rev}$ are the heat flux in forward and reverse direction when the same temperature gradient is maintained across the device in both directions [158–170].

Figure 4.6 shows the thermal rectification ratio (TRR) percentage achieved from the thermal diodes developed by using the $VO_2$, $MnV_2O_4$ (MVO), $La_{1.98}Nd_{0.02}CuO_4$ (LNCO), nitinol graphite, silver chalcogenides, manganites, and phase change materials (PCMs). Among these, a promising TRR ratio has been demonstrated by Hirata et al. [173, 174] by developing the silver chalcogenide-based material. The TRR of a thermal diode made from PCM-PCM material, $Ag_2S_{0.6}Se_{0.4}$-$Ag_2S_{0.1}Te_{0.9}$ shows about 170%, which can be considered a good material for heat management [160–176].

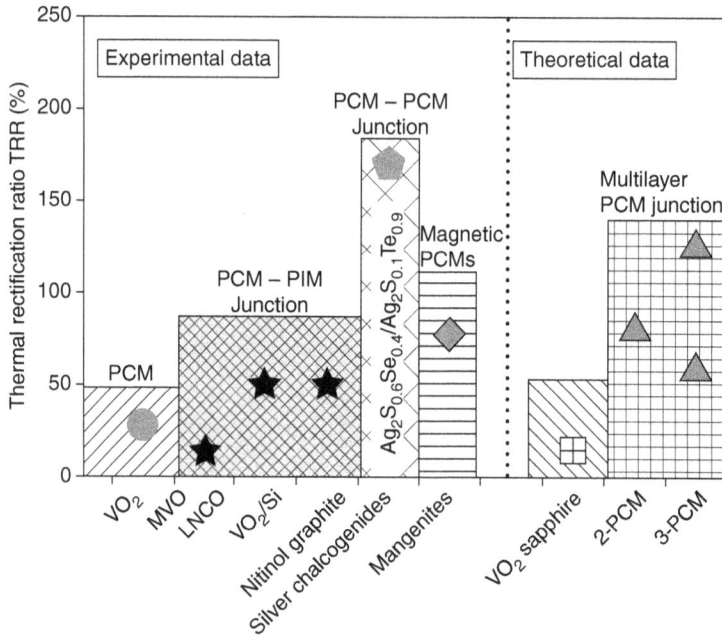

**FIGURE 4.6** Thermal rectification ratio, TRR (%) of the *state-of-the-art* thermal diodes tested using experimental tools and computational modeling. Figure reproduced from Swoboda et al. [176].

So far, we have discussed energy harvesting and thermal management by utilizing the electronic and heat transport mechanism in demonstrated materials. The high-performance thermoelectric materials are limited, and it requires large temperature gradients to develop practical applications. There is a promising energy harvesting approach for capturing ultra-low waste heat. This approach is based upon the thermomagnetic materials that have magnetic phase transition near room temperature. Thermomagnetic devices allow the conversion of heat energy into useful electricity from small temperature gradients (<10 K) [177]. In the next section, we will discuss the thermomagnetic materials, working principle, and design criterion for devices.

## 4.4 ENERGY HARVESTING NEAR ROOM TEMPERATURE USING A THERMOMAGNETIC GENERATOR

Thermomagnetic materials exhibit magnetic phase transition when subjected to thermal energy. By varying the temperature of the material, a reversible order (Ferro) to disorder (Para) magnetic phase of the material is obtained. The sharp magnetic phase transition from the low-temperature phase having ferromagnetic properties to the high-temperature phase with paramagnetic behavior is effective in the design of thermomagnetic (TM) devices. This phase transition behavior of the material when implemented with suitable magnetic and thermal properties into proper geometrical architecture, can provide an efficient power generator [177–179].

Thermomagnetic (*TM*) materials exhibit a large change in magnetization ($\Delta M$) near the magnetic phase transition located around room temperature. Gadolinium (Gd) exhibits favorable magnetic properties and is considered as promising *TM* material but has several limitations. Gadolinium exhibits a moderate change in magnetization ($\Delta M = 77$ emu/g), moderate thermal conductivity ($\kappa$), and ferromagnetic to paramagnetic phase transition, i.e., $T_C = 293$ K, below the room temperature, and brittle characteristics [180]. These factors limit deployment in applications with exposure to cyclic stresses. The output performance of a *TM* device exposed to input heat $Q_{in}$ is determined by the power density, defined as, $P_D = E_M \cdot f$, where $E_M$ is the magnetic energy stored during the thermomagnetic cycle, and $f$ is the number of cycles. High $P_D$ requires high energy per cycle $E_M (= \mu_0 \cdot \Delta M \cdot H)$, as well as high cycle frequency $f$. The large $E_m$ demands large $\Delta M$, whereas, for higher $f$, materials should have large $\kappa$. Therefore, for the best conversion efficiency, $\eta = \dfrac{E_m}{Q_{in}}$ of *TM* devices, the constructive modifications of existing materials at the atomic level via charge, spin, and lattice degree of freedoms are required to impart large improvement in the magnetic moment, thermal conductivity; structural, chemical and thermal stability; and robust mechanical properties [181, 182].

The materials possessing magnetic phase transition near room temperature have the unique advantage of transitioning into the high-temperature phase without any extra thermal energy controlling the structural arrangement. There is plenty of waste heat available with hot-side temperatures just above the room temperature, for example, heat generated in a data mining center. A basic design of a thermomagnetic generator is shown in Figure 4.7.

A ferromagnetic material is attached to the permanent magnet (or an electromagnet). On applying the heat, $q_{in}$, to the FM material, it transforms to the paramagnetic phase which lies just above its Curie temperature. A winding around the FM material generates the emf voltage, i.e., output voltage, by applying alternative heating and cooling cycle that changes the magnetic flux.

Consider that $q_{in}$ and $q_{out}$ are the heat-in and heat-out at the time of heating, and cooling fractions of the thermal cycle and the corresponding generated electrical energy are represented as $w_h$ and $w_c$, respectively. The efficiency of TM generator can be written as:

$$\eta_{TM} = \frac{w_h + w_c}{q_{in}} \tag{4.11}$$

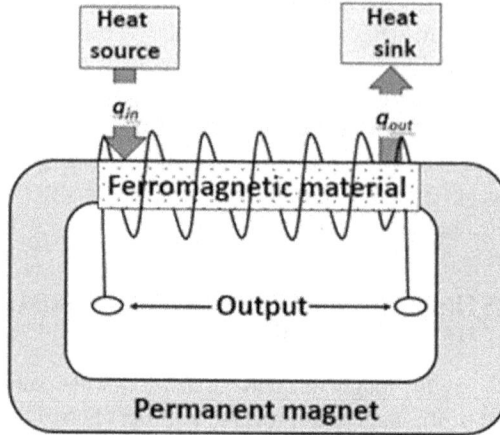

**FIGURE 4.7** Schematic diagram of the thermomagnetic generator. This schematic is generated from Kishore et al. [178].

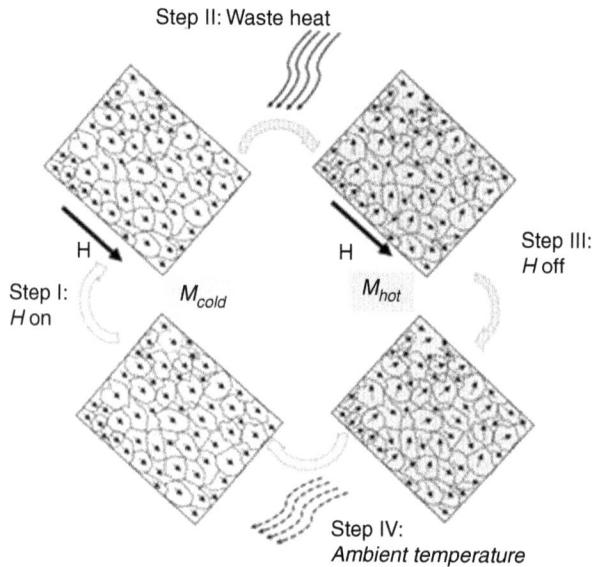

**FIGURE 4.8** Schematic diagram of working principle of thermomagnetic (TM) materials for low waste heat energy harvesting.

### 4.4.1 Working Principle

The schematic representation of the working principle for low grade waste heat conversion using thermomagnetic material is shown in Figure 4.8. A magnetic material goes through four steps of the thermomagnetic cycle in order to convert the waste heat.

As shown in Figure 4.8, FM material kept below Curie temperature ($T_C$) shows spontaneous magnetization on the application of the applied magnetic field. The moments within each domain get aligned (step I) in the direction of the externally applied magnetic field and have the net magnetic moment $M_{cold}$. In this condition, exposure to external waste heat randomizes the moment within each domain and the material's magnetization decreases once the temperature increases above $T_C$, step II, and possesses a small magnetic moment $M_{hot.}$ At this step, if the magnetic field is switched off (step III), the moments remain very small. A further decrease in temperature below, $T_C$, leads to the recovery of magnetic moments in the presence of the magnetic moment. In this whole thermal cyclic process, magnetic flux across the materials changes by the amount of $\Delta M = M_{cold} - M_{hot}$, which increases magnetic energy stored in the material. Utilizing this magnetic energy, a rotational motion can be created with different geometrical designs of the device. An efficient device can be designed by finding the magnetic material with a large change in magnetization, large thermal conductivity, and high mechanical and thermal stability. In the next section, we describe thermomagnetic materials and their use in the design of thermomagnetic devices.

### 4.4.2 STATE-OF-THE-ART THERMOMAGNETIC MATERIALS

Gadolinium (Gd) is one of the prominent materials for making thermomagnetic devices. This is due to its high magnetic moment in low temperature phase, sharp and reversible magnetic phase transition at ~21 degree Celsius (294 K), and large thermal conductivity when it is in the paramagnetic phase (~15 W m$^{-1}$ K$^{-2}$ at 310 K) compared to its ferromagnetic phase. Besides the favorable thermal and magnetic properties, the brittle nature of gadolinium, when exposed to moisture conditions, poses a challenge in designing the device.

Figure 4.9 shows the magnetization as a function of temperature (left figure) at a fixed applied field of 0.1 Tesla for Gd. As the temperature reaches the Curie temperature, a drop in $M$ occurs, and the material transitions into the paramagnetic phase. This change in $\Delta M$ for a given $\Delta T$ across the phase transition is a very important factor in deciding the selection of thermomagnetic material. A sharp change in $M$ with $T$ in the vicinity of $T_C$ is favorable to storing the magnetic energy without any significant loss if the material possesses the second-order phase transition. The right side of Figure 4.9 shows the magnetization as a function of the external field measured at room temperature. In this material, the $M$-$H$ loop is not saturated with the applied 2T field, and the shape of the M-H loop is also broad. For a good TM material, both $M$-$T$ and $M$-$H$ transition in magnetic phase and magnetic saturation, respectively, should be sharp. The sharp thermal switching of the magnetic phase also demands the large thermal conductivity of the material for absorbing and releasing the heat during the heating and cooling cycle.

Thermal conductivity of Gd versus temperature is shown in Figure 4.10. Near room temperature the value of thermal conductivity is 14 Wm$^{-1}$ K$^{-1}$. In the case of Gd, magnetic measurement shows $T_C$

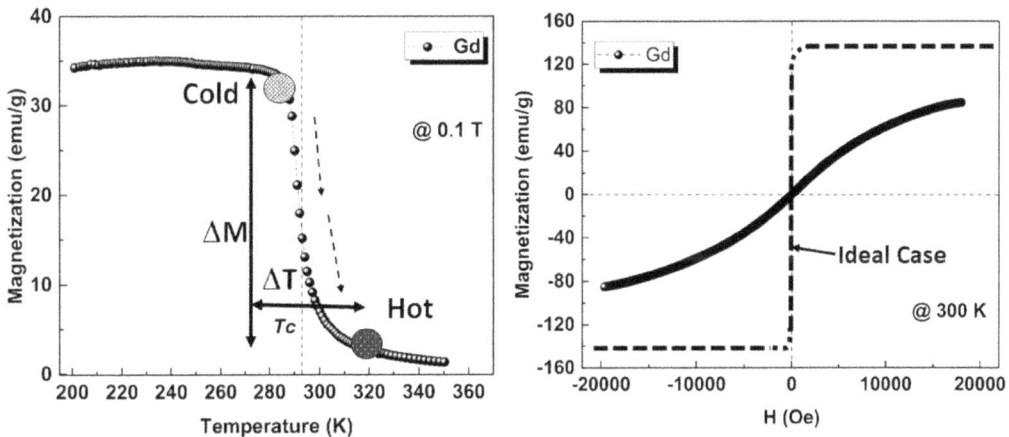

**FIGURE 4.9** Magnetization behavior as a function of temperature (left) at 0.1 Tesla, isothermal magnetization vs. applied field (right) at 300 K of a typical thermomagnetic material (gadolinium).

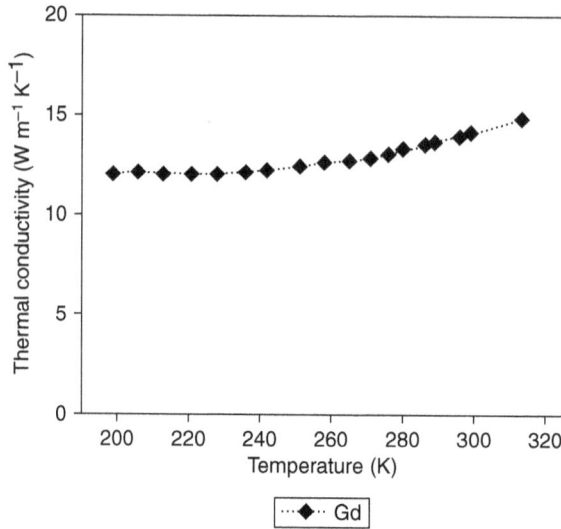

**FIGURE 4.10**   Thermal conductivity of gadolinium. Figure is reproduced from [182].

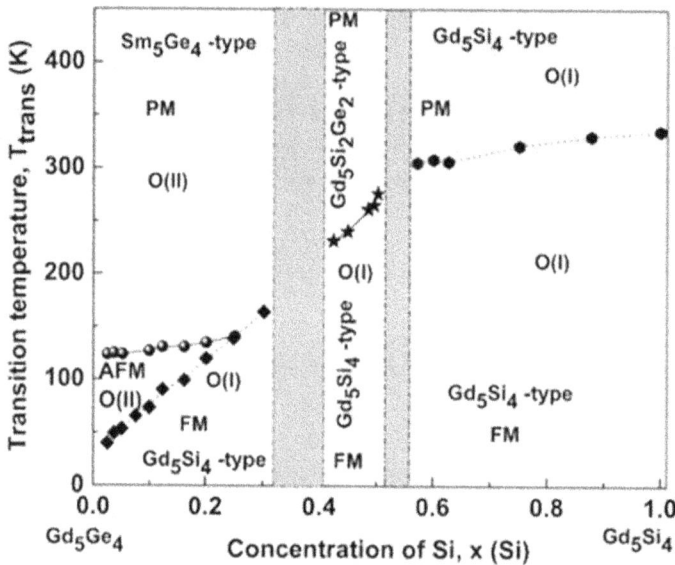

**FIGURE 4.11**   The magnetic phase diagram of the $Gd_5Ge_4$-$Gd_5Si_4$. The data is obtained for this pseudo-binary system at zero field. FM: ferromagnetic, PM: paramagnetic, AFM: antiferromagnetic, O(I) orthorhombic.

~292 K. Around this temperature, thermal conductivity remains the same. This helps in increasing thermal cycling during operation. Since Gd transition temperature, $T_C$, is below room temperature it is difficult to achieve cycling without an external cooling arrangement for the completion of the TM cycle [184]. Therefore, several elemental substitutions have been attempted in Gd as well as alloying with different $p$, $d$, and $f$-block elements to shift the Curie temperature to higher values. Si and Ge are two alloying elements to tune the Curie temperature in Gd-based alloys. The phase diagram of the GdSi-GdGe alloy is shown in Figure 4.11 [181]. An effective change in transition temperature has been observed when Gd-Si and Gd-Ge alloys are prepared. $Gd_5Si_4$ is one of the stable compositions which shows the FM to PM transition at 330 K [180]. This is a remarkable increment in the value of $T_C$. Interestingly, with further addition of Ge at the Si site of $Gd_5Si_4$, the crystal structure remains the same with a favorable change in the value of $T_C$. This result shows that the position of $T_C$ can be tuned for the Gd-Si-Ge alloy.

Here, we can see that with optimization of the Si-Ge ratio of Gd-Si-Ge alloy, the desired transition temperature can be set for energy harvesting using a thermomagnetic device. Although $T_C$ is found to be tunable, thermal conductivity gets suppressed ($\kappa < 10$ $Wm^{-1}K^{-1}$) compared to the Gd metal. The exchange of heat from the material can be made easily during the thermal cycle if the material has large thermal conductivity. For the material having low thermal conductivity, geometrical optimization is required to complete the transfer of heat faster by increasing the contact surface area on which heat flux gets exposed. Considering several important points, various types of thermomagnetic generators have been developed by researchers in past years. Some of these designs are discussed briefly in the next section.

### 4.4.3 Thermomagnetic Materials and Type of Designs for TM Generators

Figure 4.12 shows the plot of change in the magnetic moment per unit mass $\Delta M$, i.e., magnetization, for various materials developed in recent years [184–225]. These materials are classified into different categories, namely, LaFe-based, MnFe-based, SOMT (second-order magnetic transition), GdSi-based, metallic, anti-perovskites, oxides, and amorphous glasses. Among these LaFe-based materials show the best magnetic properties $\Delta M \sim 96$ emu/g near 300 K, followed by a large $\Delta M \sim 109$ emu/g at 320 K. Thermomagnetic devices have been developed using Gd and LaFe-based materials exhibiting good performance. The design of a few thermomagnetic devices is discussed briefly in the next section.

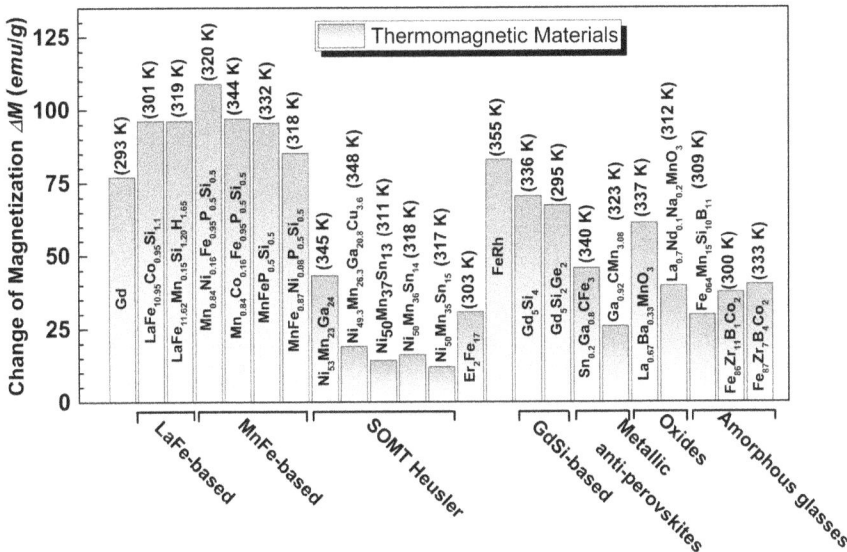

**FIGURE 4.12**  Plot of the change in magnetization for different class of thermomagnetic materials developed in past several years.

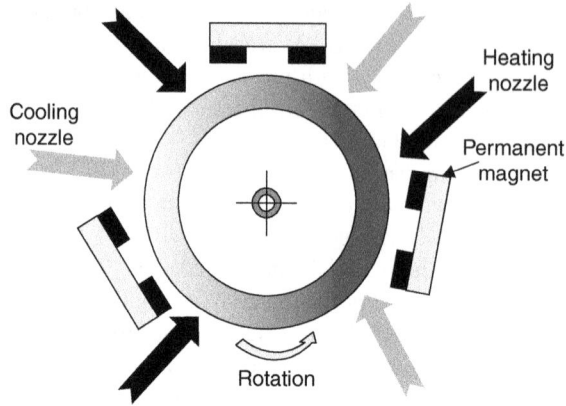

**FIGURE 4.13**   Schematic representation of rotary cylindrical and disk-type thermomagnetic engine developed by Takahashi et al. [183].

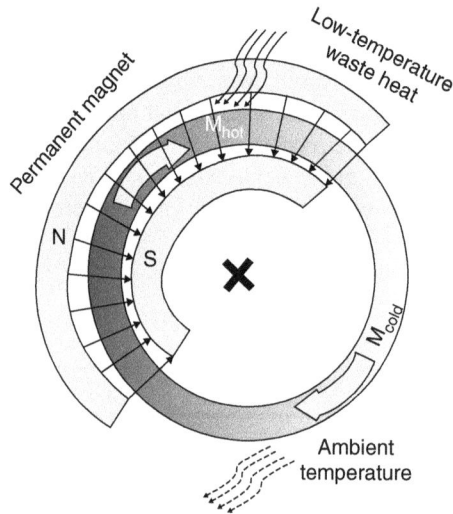

**FIGURE 4.14**   Schematic representation of a rotating ring of thermomagnetic motor. Figure recreated from Dzken et al. [180].

### 4.4.3.1   Wheel-Based Design (Thermomagnetic Motor)

Typically, in the ring-shaped TM device mounted on a movable bearing, rotational motion is created due to the torque generated as a result of the ferromagnetic phase to the paramagnetic phase transition. This process continues as another portion of the ring at a lower temperature in the ferromagnetic phase continues to be available for driving the wheel. This rotational motion of the wheel helps in creating electrical energy by using electromagnetic motors.

A ring-type TM device was introduced by Takahashi et al. [183] as shown in Figure 4.13. TM material is ring-shaped and surrounded by a permanent magnet, cooling nozzle, and heating nozzle. An alternate cooling (11 °C) and heating nozzles (95 °C) maintained at a fixed temperature allow it to be magnetized and demagnetized. This process helps in creating the torque in the ring due to magnetic energy stored in a portion of the ring. This design has lower eddy current losses in comparison to the solid cylindrical shape and exhibits faster rotation. The rotational motion of the ring is utilized in generating the electric power by connecting it to the magnetic coil.

Another ring-like structure has been developed as shown in Figure 4.14. In this design, the permanent magnet is in an arc shape covering the small portion of the ring, and heat is applied to this small portion. Another big portion of the ring is kept at ambient temperature. This creates a change in the magnetization between two different portions of the ring and a rotational motion gets created in the ring. This type of design does not need multiple heating, and cooling nozzles, and also do not require multiple permanent magnets. This type of device structure is low-cost and easy to implement where waste heat is available in a small portion of space. In the ring-like structure, cyclic frequency is mainly determined by the material's thermal conductivity and the change in magnetization. By optimizing the inner and outer diameter of the ring and outer surface thickness the frequency can be enhanced.

### 4.4.3.2 Cantilever Based Design (Thermomagnetic Oscillator)

A cantilever-based design has been developed to achieve small-scale devices as required for sensing-type applications. In this design, as shown in Figure 4.15, TM material is attached at the tip of the flexible cantilever, mostly made from piezoelectric materials, with one end of the cantilever fixed. A permanent magnet is fixed at some distance from the equilibrium horizontal position of TM material. At a temperature less than $T_C$, TM material gets attracted toward the permanent magnet, which results in the magnetization of TM material. With the exposure of heat to the TM, it gets demagnetized at a temperature above $T_C$ and is displaced away from the magnet due to its weight and gravity effect. The tip of the cantilever moves away from the permanent magnet and transitions into the FM phase as it is exposed to ambient temperature which is typically below the $T_C$. The displaced cantilever tip below the mean horizontal position turns upward with a combined effect of restoring force in the cantilever as well as the attraction force between TM and the permanent magnet. This back-and-forth motion continues with magnetization and demagnetization of the TM material. In this process, a coil attached below the TM materials feels a change in the magnetic flux across it. As a result, it generates the induced emf voltage. This is a fast process and has the benefit to generate electricity from a small amount of TM material. Such a design is easy to implement as it requires a vertical up and down motion in an axial direction.

In the above designs, a mechanical motion is required to induce change in the magnetic flux across the TM materials or coil fixed close to it. Further, in this design output power is generated during the thermal cycle but there is some heat loss due to the conduction, convection, and radiation

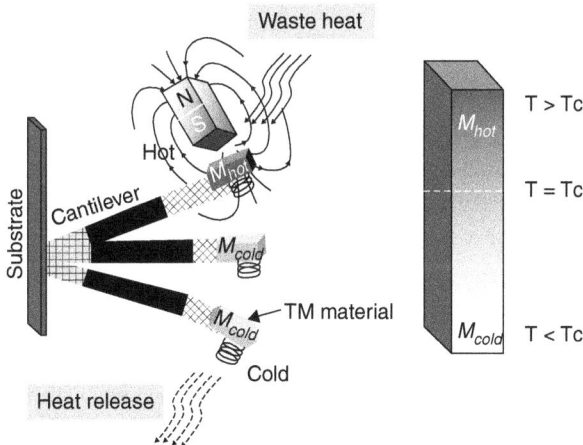

**FIGURE 4.15** Schematic representation of a cantilever based of thermomagnetic motor. Figure reproduced from Dzken et al. [180].

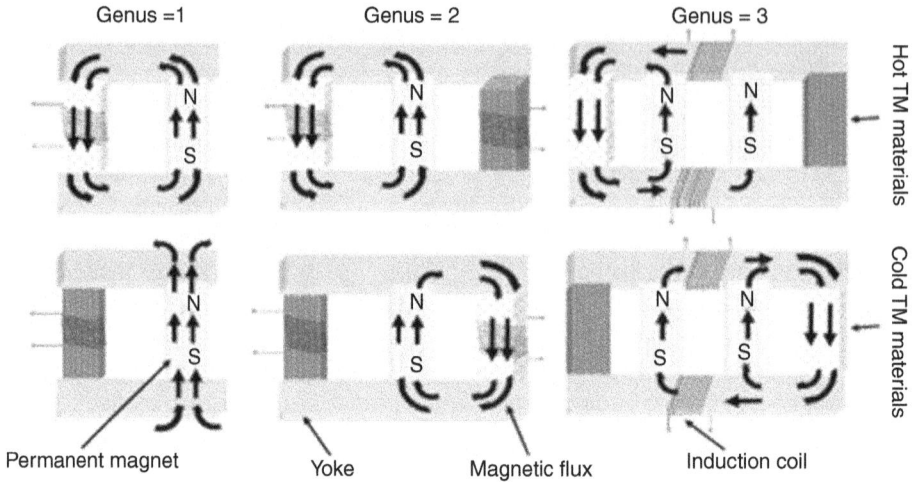

**FIGURE 4.16**    Schematic representation of a pertz-like magnetic flux topology design of thermomagnetic generator. Figure reproduced from Waske et al. [226].

based on the type of heat source used. Also, a mechanical motion during power generation creates challenges in maintaining the strength of the device as some of the materials can break due to the degradation effect which happens with reacting with moisture. Waske et al. [226] proposed a motion-less thermomagnetic generator with a pretzel-like magnetic flux topology. The details of this design are discussed in the next section.

### 4.4.3.3    Magnetic-Flux-Topology-Based Design (Thermomagnetic Generator)

A pretzel-like magnetic-flux-topology-based thermomagnetic generator is schematically shown in Figure 4.16 [226]. This design is very similar to an electric circuit in which La-Fe-Co-Si-based materials are used to make the magnetic circuit in an arrangement such that a sign reversal of magnetic flux can be achieved. In this design, no mechanical movement of the material is involved. The topology of the magnetic circuit is optimized with different sections as shown in Figure 4.16. In an optimized condition with section numbers equal to 3, magnetic flux reversal can be obtained within the magnetic circuit. Here, a hard magnet is used as a source of magnetic flux whereas to conduct the magnetic flux soft, a magnetic yoke is used. With heating and cooling of the magnetic material used as TM material here, a magnetic flux change is obtained, which acts as a switch for the flux passing through the yoke. At a temperature above $T_C$, thermomagnetic materials become hot and become paramagnetic and the flux switch is opened as shown in the bottom left figure of Figure 4.16. The heating and cooling by fluid in this magnetic circuit can generate electrical energy, which is an order higher than that produced by conventional designs like ring and cantilever type. By improving the materials' thermal and magnetic properties one can get even higher performance.

### 4.5    SUMMARY AND FUTURE PROSPECT

In this chapter, we discussed thermoelectric and thermomagnetic materials and their implementation in various energy generation and thermal management devices. A brief introduction was also presented on thermal management using thermal diodes made of phase change materials that exhibit large temperature-dependent thermal conductivity. The basic working principles of thermoelectrics, thermal diodes, and thermomagnetic materials followed by the state-of-the-art materials developed for making high-performance devices are discussed. Further development requires tuning of materials properties such as a large Seebeck coefficient, a large electrical conductivity by electronic structure modifications, and low thermal conductivity by structural modification and phonon engineering. A large step change in thermal conductivity of phase change materials with similar crystal structures would provide thermal diodes for heat flow management. For the waste heat recovery near room temperature, magnetic phase change materials with a large magnetic moment, large thermal conductivity, a sharp change in magnetic phase transition with temperature, and mechanically robust nature are required. Thermomagnetic generators based on these magnetic materials will continue to gain importance for low-grade waste heat recovery applications.

**Acknowledgement:** The authors acknowledge the financial support through the DARPA MATRIX program (TE3 and NETS).

## REFERENCES

[1]     N. Lenssen *et al.* Improvements in the GISTEMP uncertainty mode, *J. Geophys. Res. Atmos.,* 124(12), 6307–6326 (2019).

[2]     A. M. G. K. Tank, F. W. Zwiers, and X. Zhang, Climate data and monitoring, *WCDMP* 72, WMO-TD 1500 (2009).

[3]     A. Riering, *Automobilwoche*, 20200116/BCONLINE/200119942/1465 (2020).

[4]     A. D. Pollock, General principles and theoretical considerations. In *CRC Handbook of Thermoelectrics* Chap. 2 (ed. D. M. Rowe) 7–17 (CRC, Boca Raton, FL, 1995).

[5]     H. J. Goldsmid, Conversion efficiency and figure-of-merit. In *CRC Handbook of Thermoelectrics* Chap. 3 (ed. D. M. Rowe) 19–25 (CRC, Boca Raton, FL, 1995).

[6]     H. J. Goldsmid, Conversion efficiency and figure-of-merit. In *CRC Handbook of Thermoelectrics* Chap. 3 (ed. D. M. Rowe) 19–25 (CRC, Boca Raton, FL, 1995).

[7]     L. E. Bell, Cooling, heating, generating power, and recovering waste heat with thermoelectric systems, *Science* 321, 1457 (2008).

[8]     F. J. DiSalvo, Thermoelectric cooling and power generation, *Science* 285, 703 (1999).

[9]     G. Chen et al., Phonon transport in low dimensional, *Semiconductors Semimet.* 71, 203 (2001).

[10]    J. Yang and T. Caillat, Thermoelectric materials for space and automotive power generation, *MRS Bull.* 31, 224 (2006).

[11]    T. M. Tritt, H. Böttner, and L. Chen, Direct solar thermal energy conversion, *MRS Bull.* 33, 366 (2008).

[12]    D. M. Rowe, *Thermoelectrics Handbook: Macro to Nano* (CRC, Boca Raton, FL, 2005).

[13]    T. J. Seebeck, Magnetische Polarisation der Metalle und Erze durch Temperatur–Differenz, *Abh. Deutsch. Akad. Wiss. Berlin*, 265 (1822).

[14]    T. J. Seebeck, Ueber die magnetische Polarisation der Metalle und Erze durch Temperatur-Differenz, *Ann. Phys.* 82, 253 (1826).

[15]    N. W. Ashcroft, and N. D. Mermin. *Solid State Physics.* (Saunders College Publishing, New York, 1976).

[16]    J. Peltier, A.: Nouvelles expériences sur la caloricité des courans électriques, *Ann. Chim. Phys.* 56, 371 (1834).

[17]    S. B. Riffat and X. Ma, Thermoelectrics: a review of present and potential applications, *Appl. Therm. Eng.* 23, 913 (2003).

[18]    M. Zebarjadi, K. Esfarjani, M. Dresselhaus, Z. Ren, and G. Chen, Perspectives on thermoelectrics: from fundamentals to device applications, *Energy Environ. Sci.* 5, 5147 (2012).

[19]    A. F. Ioffe, L. S. Stil'Bans, E. K. Iordanishvili, T. S. Stavitskaya, A. Gelbtuch and G. Vineyard. Semiconductor thermoelements and thermoelectric cooling. *Phys. Today*, 12(5), p. 42 (1959).

[20]    L. D. Hicks and M. S. Dresselhaus. Effect of quantum-well structures on the thermoelectric figure of merit. *Phys. Rev. B*, 47(19), p.12727 (1993).

[21]    R. Venkatasubramanian, E. Siivol, T. Colpitts and B. O'quinn. Thin-film thermoelectric devices with high room-temperature figures of merit. *Nature*, 413(6856), pp.597–602 (2001).

[22]    I. Terasaki, Y. Sasago and K. Uchinokura. Large thermoelectric power in $NaCo_2 O_4$ single crystals. *Phys. Rev. B*, 56(20), p.R12685 (1997).

[23] R. Funahashi, I. Matsubara, H. Ikuta, T. Takeuchi, U. Mizutani and S. Sodeoka. An oxide single crystal with high thermoelectric performance in air. *Japanese J. Appl. Phys.*, 39(11B), p.L1127 (2000).

[24] J. T. Okada, T. Hamamatsu, S. Hosoi, T. Nagata, K. Kimura and K. Kirihara. Improvement of thermo-electric properties of icosahedral AlPdRe quasicrystals by Fe substitution for Re. *J. Appl. Phys.*, 101(10), p.103702 (2007).

[25] G. A. Slack. *CRC Handbook of Thermoelectrics*, ed. by D. M. Rowe (CRC press, Boca Raton, 1995) pp. 407–440.

[26] J. W. Sharp, E. C. Jones, R. K. Williams, P. M. Martin and S. C. Sales. Thermoelectric properties of $CoSb_3$ and related alloys, *J. Appl. Phys.* 78 (1995) 1013–1018.

[27] T. Takabatake, T. Sasakawa, J. Kitagawa, T. Suemitsu, Y. Echizen, K. Umeo, M. Sera and Y. Bando. Thermoelectric properties of Ce-based Kondo semimetals and semiconductors. *Physica B: Condens. Matter*, 328(1–2), pp.53–57 (2003).

[28] T. Takeuchi, New thermoelectric materials with precisely determined electronic structure and phonon dispersion. In *CRC Handbook of Thermoelectrics and Its Energy Harvesting*, Chap. 7 (ed. D. M. Rowe) (CRC, Boca Raton, FL, 2012).

[29] T. Takeuchi, Conditions of electronic structure to obtain large dimensionless figure of merit for developing practical thermoelectric materials, *Mater. Trans.* 50, 2359–2365 (2009).

[30] T. Takeuchi, Y. Toyama and A. Yamamoto. Role of temperature dependent chemical potential on thermoelectric power. *Mater. Trans.*, 51, pp.1002010989 (2010).

[31] J. M. Ziman. *Principles of the Theory of Solids*. (Cambridge University Press, Cambridge, 1972).

[32] G. J. Snyder and E. S. Toberer. Complex thermoelectric materials. *Materials for Sustainable Energy: A Collection of Peer-Reviewed Research and Review* (Nature Publishing Group, London, 2011), pp.101–110.

[33] E. S. Toberer, A. Zevalkink and G. J. Snyder. Phonon engineering through crystal chemistry. *J. Mater. Chem.*, 21(40), pp.15843–15852 (2011).

[34] C. Nielsen. Thermoelectric materials: Current status and future challenges. *Front. Electron. Mater.*, 1, 1 (2021).

[35] D. M. Rowe, Miniature semiconductor thermoelectric devices. In *CRC Handbook of Thermoelectrics* Chap. 35 (ed. D. M. Rowe) 441–458 (CRC, Boca Raton, FL, 1995).

[36] A. G. McNaughton, Commercially available generators. In *CRC Handbook of Thermoelectrics* Chap. 36 (ed. D. M. Rowe) 459–478 (CRC, Boca Raton, FL, 1995).

[37] J. Yang et al. On the tuning of electrical and thermal transport in thermoelectrics: an integrated theory–experiment perspective, *Npj Comput. Mater.* 2, 15015 (2016).

[38] G. T. Alekseeva et al. Nonisovalent substitution of atoms in the cation sublattice of bismuth telluride, *Phys. Solid State* 38, 1639 (1996).

[39] I. T. Witting et al. The thermoelectric properties of bismuth telluride, *Adv. Electron. Mater.* 5, 1800904 (2019).

[40] B. Hinterleitner, I. Knapp, M. Poneder, Y. Shi, H. Müller, G. Eguchi, C. Eisenmenger Sittner, M. Stöger-Pollach, Y. Kakefuda, N. Kawamoto and Q. Guo. Thermoelectric performance of a metastable thin-film Heusler alloy. *Nature*, 576(7785), pp.85–90 (2019).

[41] X. L. Shi, J. Zou and Z. G. Chen. Advanced thermoelectric design: from materials and structures to devices. *Chem. Rev.*, 120(15), pp.7399–7515 (2020).

[42] M. N. Hasan, H. Wahid, N. Nayan and M. S. Mohamed Ali. Inorganic thermoelectric materials: A review. *Int. J.Energy Res.*, 44(8), pp.6170–6222 (2020).

[43] X. Tong, Z. Liu, J. Zhu, T. Yang, Y. Wang and A. Xia. Research progress of p-type Fe-based skut-terudite thermoelectric materials. *Front. Mater. Sci.*, 15, 1–17 (2021).

[44] J. Wei, L. Yang, Z. Ma, P. Song, M. Zhang, J. Ma, F. Yang and X. Wang. Review of current high-ZT thermoelectric materials. *J. Mater. Sci.*, 1–63 (2020).

[45] K. Xia, C. Hu, C. Fu, X. Zhao and T. Zhu2021. Half-Heusler thermoelectric materials. *Appl. Phys. Lett.*, 118(14), p.140503 (2021).

[46] M. Imada, A. Fujimori and Y. Tokura, Metal-insulator transitions, *Rev. Mod. Phys.* 70, 1039 (1998).

[47] T. Tritt and M. Subramanian, Harvesting energy through thermoelectrics: power generation and cooling, *MRS Bull.* 31, 113 (2006).

[48]  B. Sales, D. Mandrus, and R. K. Williams, Filled skutterudite antimonides: a new class of thermo-electric materials, *Science* 272, 1325 (1996).

[49]  G. Nolas, J. Cohn, G. Slack, and S. Schujman, Semiconducting Ge clathrates: promising candidates for thermoelectric applications, *Appl. Phys. Lett.* 73, 178 (1998).

[50]  X. Shi *et al.*, Multiple-filled skutterudites: high thermoelectric figure of merit through separately optimizing electrical and thermal transports, *J. Am. Chem. Soc.* 133, 7837 (2011).

[51]  C. Fu *et al.*, Thermoelectric properties of FeVSb half-Heusler compounds by levitation melting and spark plasma sintering, *Intermetallics* 32, 39 (2013).

[52]  X. Yan *et al.*, Thermoelectric properties of Ba-Cu-Si clathrates, *Phys. Rev. B* 85, 165127 (2012).

[53]  B. Poudel *et al.*, High-thermoelectric performance of nanostructured bismuth antimony telluride bulk alloys, *Science* 320, 634 (2008).

[54]  J. P. Heremans *et al.*, Enhancement of thermoelectric efficiency in PbTe by distortion of the electronic density of states, *Science* 321, 554 (2008).

[55]  L.-D. Zhao *et al.*, Ultralow thermal conductivity and high thermoelectric figure of merit in SnSe crystals, *Nature* 508, 373 (2014).

[56]  R. Venkatasubramanian, E. Silvola, T. Colpitts, and B. O'quinn, Thin-film thermoelectric devices with high room-temperature figures of merit, in *Materials for Sustainable Energy: A Collection of Peer-Reviewed Research and Review Articles from Nature Publishing Group*, pages 120–125 (Singapore, World Scientific, 2011).

[57]  F. Hsu *et al.*, Cubic AgPb m SbTe$^{2+}$m: bulk thermoelectric materials with high figure of merit, *Science* 303, 818 (2004).

[58]  S. Dresselhaus *et al.*, New directions for low-dimensional thermoelectric materials, *Adv. Mater.* 19, 1043 (2007).

[59]  W. Xie *et al.*, Recent advances in nanostructured thermoelectric half-Heusler compounds, *Nanomaterials* 2, 379 (2012).

[60]  K. Biswas *et al.*, High-performance bulk thermoelectrics with all-scale hierarchical architectures, *Nature* 489, 414 (2012).

[61]  D. Srivastava, F. Azough, R. Freer, E. Combe, R. Funahashi, D. M. Kepaptsoglou, Q. M. Ramasse, M. Molinari, S. R. Yeandel, J. D. Baran and S. C. Parker. Crystal structure and thermoelectric properties of Sr–Mo substituted CaMnO$_3$: A combined experimental and computational study. *J. Mater. Chem. C*, 3(47), pp.12245–12259 (2015).

[62]  S. Ghodke, A. Yamamoto, M. Omprakash, H. Ikuta and T. Takeuchi, Thermoelectric properties of p-type Cr doped MnSiγ prepared by liquid quenching technique, *Mater. Trans.* 58, 160 (2017).

[63]  S. Hiroi, M. Mikami, and T. Takeuchi, Thermoelectric properties of Fe2Val-based thin-films deposited at high temperatures, *Mater. Trans.* 57, 1628 (2016).

[64]  M. Sabarinathan *et al.*, Synthesis of ZnO/SrO nanocomposites for enhanced photocatalytic activity under visible light irradiation, *Appl. Surf. Sci.* 418, 147 (2017).

[65]  A. Yamamoto and T. Takeuchi, The potential of FeVSb half-Heusler phase for practical thermoelectric material, *J. Electron. Mater.* 46, 3200 (2017).

[66]  T. Takeuchi, Special issue on recent progress in thermoelectrics-new analyses and new materials-(thermoelectric conversion materials IX), *Mater. Trans.* 57, 1017 (2016).

[67]  T. Takeuchi, A. Yamamoto and S. Ghodke, Development of thermoelectric materials consisting solely of environmental friendly elements, *Mater. Trans.* 57, 1029 (2016).

[68]  A. Minnich, M. Dresselhaus, Z. Ren, and G. Chen, Bulk nanostructured thermoelectric materials: current research and future prospects, *Energy Environ. Sci.* 2, 466 (2009).

[69]  H. Dong, B. Wen and R. Melnik, Relative importance of grain boundaries and size effects in thermal conductivity of nanocrystalline materials, *Sci. Rep.* 24(1), 7037 (2014).

[70]  C.-W. Nan and R. Birringer, Determining the Kapitza resistance and the thermal conductivity of polycrystals: a simple model, *Phys. Rev. B* 57, 8264 (1998).

[71]  K. Koumoto, I. Terasaki and R. Funahashi, Complex oxide materials for potential thermoelectric applications, *MRS Bull.* 31, 206 (2006).

[72]  K. Koumoto, Y. Wang, R. Zhang, A. Kosuga, and R. Funahashi, Oxide thermoelectric materials: a nanostructuring approach, *Ann. Rev. Mater. Res.* 40, 363 (2010).

[73]  B. Jiang, Y. Yu, H. Chen, J. Cui, X. Liu, L. Xie and J. He. Entropy engineering promotes thermoelectric performance in p-type chalcogenides. *Nature Comm.*, 12(1), 1–8 (2021).

[74] Z. Z. Luo, S. Cai, S. Hao, T. P. Bailey, I. Spanopoulos, Y. Luo, J. Xu, C. Uher, C. Wolverton, V. P. Dravid and Q. Yan. Strong valence band convergence to enhance thermoelectric performance in PbSe with two chemically independent controls. *Ang. Chem.*, 133(1), 272–277 (2021).

[75] Y. Qin, T. Hong, B. Qin, D. Wang, W. He, X. Gao, Y. Xiao and L. D. Zhao. Contrasting Cu roles lead to high ranged thermoelectric performance of PbS. *Adv. Funct. Mater.*, 31(34), p. 2102185 (2021).

[76] W. Lin, J. He, X. Su, X. Zhang, Y. Xia, T. P. Bailey, C. C. Stoumpos, G. Tan, A. J. Rettie, D. Y. Chung and V. P. Dravid. Ultralow thermal conductivity, multiband electronic structure and high thermoelectric figure of merit in TlCuSe. *Adv. Mater.*, 33(34), p. 2104908 (2021).

[77] P. Jood, J. P. Male, S. Anand, Y. Matsushita, Y. Takagiwa, M. G. Kanatzidis, G. J. Snyder and M. Ohta. Na doping in PbTe: solubility, band convergence, phase boundary mapping, and thermoelectric properties. *J. Am. Chem. Soc.*, 142(36), 15464–15475 (2020).

[78] B. Jiang, Y. Yu, J. Cui, X. Liu, L. Xie, J. Liao, Q. Zhang, Y. Huang, S. Ning, B. Jia and B. Zhu. High-entropy-stabilized chalcogenides with high thermoelectric performance. *Science*, 371(6531), 830–834 (2021).

[79] D. Yang, X. Su, F. Meng, S. Wang, Y. Yan, J. Yang, J. He, Q. Zhang, C. Uher, M. G. Kanatzidis and X. Tang. Facile room temperature solventless synthesis of high thermoelectric performance Ag$_2$Se via a dissociative adsorption reaction. *J. Mater. Chem. A*, 5(44), 23243–23251 (2017).

[80] S. Singh, K. Hirata, D. Byeon, T. Matsunaga, O. Muthusamy, S. Ghodke, M. Adachi, Y. Yamamoto, M. Matsunami and T. Takeuchi. Investigation of Thermoelectric Properties of Ag2S$_x$Se$_{1-x}$ ($x$= 0.0, 0.2 and 0.4). *J. Electron. Mater.*, 49(5), 2846–2854 (2020).

[81] P. Jood, R. Chetty and M. Ohta. Structural stability enables high thermoelectric performance in room temperature Ag$_2$Se. *J. Mater. Chem. A*, 8(26), 13024–13037 (2020).

[82] T. Zhu, H. Bai, J. Zhang, G. Tan, Y. Yan, W. Liu, X. Su, J. Wu, Q. Zhang and X. Tang. Realizing high thermoelectric performance in Sb-doped Ag$_2$Te compounds with a low-temperature monoclinic structure. *ACS Appl. Mater. Interfaces*, 12(35), pp.39425–39433 (2020).

[83] S. Yang, Z. Gao, P. Qiu, J. Liang, T. R. Wei, T. Deng, J. Xiao, X. Shi and L. Chen, L. Ductile Ag$_{20}$S$_7$Te$_3$ with excellent shape-conformability and high thermoelectric performance. *Adv. Mater.*, 33(10), p. 2007681 (2021).

[84] S. Roychowdhury, T. Ghosh, R. Arora, M. Samanta, L. Xie, N. K. Singh, A. Soni, J. He, U. V. Waghmare and K. Biswas. Enhanced atomic ordering leads to high thermoelectric performance in AgSbTe$_2$. *Science*, 371(6530), 722–727 (2021).

[85] M. Shen, S. Lu, Z. Zhang, H. Liu, W. Shen, C. Fang, Q. Wang, L. Chen, Y. Zhang and X. Jia. Bi and Sn co-doping enhanced thermoelectric properties of Cu$_3$SbS$_4$ materials with excellent thermal stability. *ACS Appl. Mater. Interfaces*, 12(7), 8271–8279 (2020).

[86] Z. Gao, Q. Yang, P. Qiu, T. R. Wei, S. Yang, J. Xiao, L. Chen and X. Shi. p-type plastic inorganic thermoelectric materials. *Adv. Energy Mater.*, 11(23), p. 2100883 (2021).

[87] N. Ma, Y. Y. Li, L. Chen and L. M. Wu. α-CsCu$_5$Se$_3$: Discovery of a low-cost bulk selenide with high thermoelectric performance. *J. Am. Chem. Soc.*, 142(11), 5293–5303 (2020).

[88] T. Wang, Y. Xiong, H. Huang, P. Qiu, K. Zhao, J. Yang, J. Xiao, X. Shi and L. Chen. Ternary compounds Cu3 R Te3 (R= Y, Sm, and Dy): A family of new thermoelectric materials with trigonal structures. *ACS Appl. Mater. Interfaces*, 12(36), pp.40486–40494 (2020).

[89] A. R. Muchtar, B. Srinivasan, S. L. Tonquesse, S. Singh, N. Soelami, B. Yuliarto, D. Berthebaud and T. Mori. Physical insights on the lattice softening driven mid-temperature range thermoelectrics of Ti/Zr-inserted SnTe – an outlook beyond the horizons of conventional phonon scattering and excavation of heikes' equation for estimating carrier properties. *Adv. Energy Mater.* 11, 2101122 (2021).

[90] T. Xing, C. Zhu, Q. Song, H. Huang, J. Xiao, D. Ren, M. Shi, P. Qiu, X. Shi, F. Xu and L. Chen. Ultralow lattice thermal conductivity and superhigh thermoelectric figure-of-merit in (Mg, Bi) co-doped GeTe. *Adv. Mater.*, 33(17), p.2008773 (2021).

[91] C. Zhou, Y. K. Lee, Y. Yu, S. Byun, Z. Z. Luo, H. Lee, B. Ge, Y. L. Lee, X. Chen, J. Y. Lee and O. Cojocaru-Mirédin. Polycrystalline SnSe with a thermoelectric figure of merit greater than the single crystal. *Nat. Mater.,* 20(10), 1–7 (2021).

[92] G. Kim, D. Byeon, S. Singh, K. Hirata, S. Choi, M. Matsunami and T. Takeuchi. Mixed-phase effect of a high Seebeck coefficient and low electrical resistivity in Ag$_2$S. *J. Phys. D: Appl. Phys.*, 54(11), p.115503 (2021).

[93] D. Byeon, R. Sobota, K. Delime-Codrin, S. Choi, K. Hirata, M. Adachi, M. Kiyama, T. Matsuura, Y. Yamamoto, M. Matsunami and T. Takeuchi. Discovery of colossal Seebeck effect in metallic $Cu_2Se$. *Nature Commun.*, 10(1), 1–7 (2019).

[94] J. Gao, L. Miao, C. Liu, X. Wang,Y. Peng, X. Wei, J. Zhou, Y. Chen, R. Hashimoto, T. Asaka and K. Koumoto. A novel glass-fiber-aided cold-press method for fabrication of n-type $Ag_2Te$ nanowires thermoelectric film on flexible copy-paper substrate. *J. Mater. Chem. A*, 5(47), 24740–24748 (2017).

[95] C. Dun, C. A. Hewitt, H. Huang, J. Xu, C. Zhou, W. Huang, Y. Cui, W. Zhou, Q. Jiang and D. L. Carroll. Flexible n-type thermoelectric films based on Cu-doped $Bi_2Se_3$ nanoplate and polyvinylidene fluoride composite with decoupled Seebeck coefficient and electrical conductivity. *Nano Energy,* 18, pp.306–314 (2015).

[96] Y. Chen, M. He, B. Liu, G. C. Bazan, J. Zhou and Z. Liang. Bendable n-type metallic nanocomposites with large thermoelectric power factor. *Adv. Mater.*, 29(4), p.1604752 (2017).

[97] Q. Jin, W.Shi, Y. Zhao, J. Qiao, J. Qiu, C. Sun, H. Lei, K. Tai and X. Jiang. Cellulose fiber-based hierarchical porous bismuth telluride for high-performance flexible and tailorable thermoelectrics. *ACS Appl. Mater. Interfaces*, 10(2), 1743–1751 (2018).

[98] C. Yang, D. Souchay, M. Kneiß, M. Bogner, H. M. Wei, M. Lorenz, O. Oeckler, G. Benstetter, Y. Q. Fu and M. Grundmann. Transparent flexible thermoelectric material based on non-toxic earth-abundant p-type copper iodide thin film. *Nature Commun.*, 8(1), 1–7 (2017).

[99] L. Wang, Z. Zhang, L. Geng, T. Yuan, Y. Liu, J. Guo, L. Fang, J. Qiu and S. Wang, S., 2018. Solution-printable fullerene/$TiS_2$ organic/inorganic hybrids for high-performance flexible n-type thermoelectrics. *Energy Environ. Sci.*, 11(5), 1307–1317.

[100] J. Choi, K. Cho, J. Yun, Y. Park, S. Yang and S. Kim. Large voltage generation of flexible thermoelectric nanocrystal thin films by finger contact. *Adv. Energy Mater.*, 7(21), p.1700972 (017).

[101] W. Hou, X. Nie, W. Zhao, H. Zhou, X. Mu, W. Zhu and Q. Zhang. Fabrication and excellent performances of $Bi_{0.5}Sb_{1.5}Te_3$/epoxy flexible thermoelectric cooling devices. *Nano Energy,* 50, 766–776 (2018).

[102] Y. Ding, Qiu, K. Cai, Q. Yao, S. Chen, L. Chen. and J. He. High performance *n*-type $Ag_2Se$ film on nylon membrane for flexible thermoelectric power generator. *Nature Commun.*, 10(1), pp.1–7 (2019).

[103] J. A. Perez-Taborda, O. Caballero-Calero, L. Vera-Londono, F. Briones and M. Martin-Gonzalez. High thermoelectric ZT in n-type silver selenide films at room temperature. *Adv. Energy Mater.*, 8(8), p.1702024 (2018).

[104] J. Liang, T. Wang, P. Qiu, S. Yang, C. Ming, H. Chen, Q. Song, K. Zhao, T. R. Wei, D. Ren and Y. Y. Sun. Flexible thermoelectrics: from silver chalcogenides to full-inorganic devices. *Energy Environ, Sci.*, 12(10), 2983–2990 (2019).

[105] Y. Lan, A. J. Minnich, G. Chen and Z. Ren. Enhancement of thermoelectric figure-of-merit by a bulk nanostructuring approach. *Adv. Funct. Mater.*, 20(3), 357–376 (2010).

[106] G. Joshi, H. Lee, Y. Lan, X. Wang, G. Zhu, D. Wang, R. W. Gould, D. C. Cuff, M. Y. Tang, M. S. Dresselhaus and G. Chen. Enhanced thermoelectric figure-of-merit in nanostructured p-type silicon germanium bulk alloys. *Nano Lett.*, 8(12), 4670–4674 (2008).

[107] N. Mingo, D. Hauser, N. P. Kobayashi, M. Plissonnier and A. Shakouri. "Nanoparticle-in-alloy" approach to efficient thermoelectrics: silicides in SiGe. *Nano Lett.*, 9(2), 711–715 (2009).

[108] R. Basu, S. Bhattacharya, R. Bhatt, M. Roy, S. Ahmad, A. Singh, M. Navaneethan, Y. Hayakawa, D. K. Aswal and S. K. Gupta. Improved thermoelectric performance of hot pressed nanostructured n-type SiGe bulk alloys. *J. Mater. Chem. A*, 2(19), 6922–6930 (2014).

[109] S. Bathula, M. Jayasimhadri, B. Gahtori, N. K. Singh, K. Tyagi, A. K. Srivastava and A. Dhar. The role of nanoscale defect features in enhancing the thermoelectric performance of p-type nanostructured SiGe alloys. *Nanoscale*, 7(29), 12474–12483 (2015).

[110] S. Ahmad, A. Singh, A. Bohra, R. Basu, S. Bhattacharya, R. Bhatt, K. N. Meshram, M. Roy, S. K. Sarkar, Y. Hayakawa and A. K. Debnath. Boosting thermoelectric performance of p-type SiGe alloys through in-situ metallic YSi2 nanoinclusions. *Nano Energy,* 27, 282–297 (2016).

[111] S. Bathula, M. Jayasimhadri, B. Gahtori, A. Kumar, A. K. Srivastava and A. Dhar. Enhancement in thermoelectric performance of SiGe nanoalloys dispersed with SiC nanoparticles. *Phys. Chem. Chem. Phys.*, *19*(36), pp.25180–25185 (2017).

[112] K. Xie, K. Mork, U. Kortshagen and M. C. Gupta. High temperature thermoelectric properties of laser sintered thin films of phosphorous-doped silicon-germanium nanoparticles. *AIP Adv.*, 9(1), p.015227 (2019).

[113] J. Wang, J. B. Li, H. B. Yu, J. Li, H. Yang, X. Yaer, X. H. Wang. and H. M. Liu. Enhanced thermoelectric performance in n-type $SrTiO_3$/SiGe composite. *ACS Appl. Mater. Interfaces*, 12(2), 2687–2694 (2019).

[114] M. Omprakash, K. Delime-Codrin, S. Ghodke, S. Singh, S. Nishino, M. Adachi, Y. Yamamoto, M. Matsunami, S. Harish, M. Shimomura and T. Takeuchi. Au and B co-doped p-type Si-Ge nanocomposites possessing ZT= 1.63 synthesized by ball milling and low-temperature sintering. *Japanese J. Appl. Phys.*, 58(12), 125501 (2019).

[115] K. Delime-Codrin, M. Omprakash, S. Ghodke, R. Sobota, M. Adachi, M. Kiyama, T. Matsuura, Y. Yamamoto, M. Matsunami and T. Takeuchi, T. Large figure of merit ZT= 1.88 at 873 K achieved with nanostructured Si0. 55Ge0. 35 (P0. 10Fe0. 01). *Appl. Phys. Exp.*, 12(4), p.045507 (2019).

[116] M. Adachi, S. Nishino, K. Hirose, M. Kiyama, Y. Yamamoto and T. Takeuchi. High dimensionless figure of merit ZT= 1.38 achieved in p-type Si–Ge–Au–B thin film. *Mater. Trans.*, 61(5), 1014–1019 (2020).

[117] O. Muthusamy, S. Ghodke, S. Singh, K. Delime-Codrin, S. Nishino, M. Adachi, Y. Yamamoto, M. Matsunami, S. Harish, M. Shimomura and T. Takeuchi. Enhancement of the thermoelectric performance of Si-Ge nanocomposites containing a small amount of Au and optimization of boron doping. *J. Electron. Mater.*, 49(5), 2813–2824 (2020).

[118] C. R. Ascencio-Hurtado, A. Torres, R. Ambrosio, M. Moreno, J. Álvarez-Quintana and A. Hurtado-Macías. N-type amorphous silicon-germanium thin films with embedded nanocrystals as a novel thermoelectric material of elevated ZT. *J. Alloys Compds,* 890, p.161843 (2022).

[119] M. Omprakash, S. Singh, K. Hirata, M. Adachi, Y. Yamamoto, M. Matsunami, S. Harish, M. Shimomura and T. Takeuchi, Synergetic enhancement of the power factor and suppression of lattice thermal conductivity via electronic structure modification and nanostructuring on a Ni- and B-codoped p-type Si–Ge alloy for thermoelectric application, *ACS Appl. Electron. Mater.* 3 5621–5631 (2021).

[120] Y. Peng, L. Miao, C. Liu, H. Song, M. Kurosawa, O. Nakatsuka, S. Yi Back, J. Soo Rhyee, M. Murata, S. Tanemura, T. Baba,T. Baba, T. Ishizaki and Takao Mori, Constructed Ge quantum dots and Sn precipitate SiGeSn hybrid film with high thermoelectric performance at low temperature *Adv. Energy Mater.*, 12(2), 2103191 (2021).

[121] A. Nag and V. Shubha, Oxide thermoelectric materials: A structure–property relationship, *J. Electron. Mater.* 43, 962 (2014).

[122] J. W. Fergus, Oxide materials for high temperature thermoelectric energy conversion, *J. Euro. Cer. Soc.* 32, 525 (2012).

[123] S. Walia *et al.*, Transition metal oxides–thermoelectric properties, *Prog. Mater. Sci.* 58, 1443 (2013).

[124] M. Shikano and R. Funahashi, Electrical and thermal properties of single-crystalline $(Ca_2CoO_3)_{0.7}CoO_2$ with a $Ca_3Co_4O_9$ structure, *Appl. Phys. Lett.* 82, 1851 (2003).

[125] M. Ohtaki, Recent aspects of oxide thermoelectric materials for power generation from mid-to-high temperature heat source, *J. Cer. Soc. Japan* 119, 770 (2011).

[126] L. Bocher, M. H. Aguirre, D. Logvinovich, A. Shkabko, R. Robert, M. Trottmann and A. Weidenkaff. $CaMn_{1-x}Nb_xO_3$ ($x \leq 0.08$) perovskite-type phases as promising new high-temperature n-type thermoelectric materials. *Inorgan. Chem.*, 47(18), 8077–8085 (2008).

[127] M. Ohtaki, K. Araki and K. Yamamoto. High thermoelectric performance of dually doped ZnO ceramics. *J. Electron. Mater.*, 38(7), 1234–1238 (2009).

[128] K. Park, J. S. Son, S. I. Woo, K. Shin, M. W. Oh, S. D. Park and T. Hyeon. Colloidal synthesis and thermoelectric properties of La-doped $SrTiO_3$ nanoparticles. *J. Mater. Chem. A*, 2(12), 4217–4224 (2014).

[129] Y. Lin, C. Norman, D. Srivastava, F. Azough, L. Wang, M. Robbins, K. Simpson, R. Freer and I. A. Kinloch. Thermoelectric power generation from lanthanum strontium titanium oxide at room temperature through the addition of graphene. *ACS Appl. Mater. Interfaces*, 7(29), 15898–15908 (2015).

[130] S. Singh, R. K. Maurya and S. K. Pandey. Investigation of thermoelectric properties of $ZnV_2O_4$ compound at high temperatures. *J. Phys. D: Appl. Phys.*, 49(42), 425601 (2016).

[131] S. Singh and S. K. Pandey. Understanding the thermoelectric properties of $LaCoO_3$ compound. *Philo. Magazine*, 97(6), 451–463 (2017).

[132] J. Wang, B. Y. Zhang, H. J. Kang, Y. Li, X. Yaer, J. F. Li, Q. Tan, S. Zhang, G. H. Fan, C. Y. Liu and L. Miao. Record high thermoelectric performance in bulk $SrTiO_3$ via nano-scale modulation doping. *Nano Energy*, 35, 387–395 (2017).

[133] L. M. Daniels, S. N. Savvin, M. J. Pitcher, M. S. Dyer, J. B. Claridge, S. Ling, B. Slater, F. Cora, J. Alaria and M. J. Rosseinsky. Phonon-glass electron-crystal behaviour by A site disorder in n-type thermoelectric oxides. *Energy Environ. Sci.*, 10(9), 1917–1922 (2017).

[134] S. Acharya, B. K. Yu, J. Hwang, J. Kim and W. Kim. High Thermoelectric performance of ZnO by coherent phonon scattering and optimized charge transport. *Adv. Funct. Mater.*, 31(43), p.2105008 (2021).

[135] Y. Takashima, Y. Q. Zhang, J. Wei, B. Feng, Y. Ikuhara, H. J. Cho and H. Ohta. Layered cobalt oxide epitaxial films exhibiting thermoelectric ZT= 0.11 at room temperature. *J. Mater. Chem. A*, 9(1), 274–280 (2021).

[136] S. Biswas, S. Singh, S. Singh, S. Chattopadhyay, K. K. H. De Silva, M. Yoshimura, J. Mitra and V. B. Kamble. Selective enhancement in phonon scattering leads to a high thermoelectric figure-of-merit in graphene oxide-encapsulated ZnO nanocomposites. *ACS Appl. Mater. Interfaces*, 13(20), 23771–23786 (2021).

[137] M. Acharya, S. S. Jana, M. Ranjan and T. Maiti2021. High performance (ZT> 1) n-type oxide thermo-electric composites from earth abundant materials. *Nano Energy*, 84, p.105905 (2021).

[138] W. Rahim, J. Skelton and D. Scanlon. $Ca_4Sb_2O$ and $Ca_4Bi_2O$: two promising mixed-anion thermoelectrics, *J. Mater. Chem. A*, 9(36), 20417–20435 (2021).

[139] X. Hu *et al.* Power generation evaluated on a bismuth telluride unicouple module. *J. Electron Mater.* 44, 1785–1790 (2015).

[140] R. Deng *et al.* High thermoelectric performance in $Bi_{0.46}Sb_{1.54}Te_3$ nanostructured with ZnTe. *Energ. Environ. Sci.* 11(6), 1520–1535 (2018).

[141] Technology Inc., H.-z. Thermoelectric device type: The HZ-2 watt module, hwww.hi-z.com (accessed April, 2021).

[142] F. Hao *et al.* High efficiency $Bi_2Te_3$-based materials and devices for thermoelectric power generation between 100 and 300 °C. *Energ. Environ. Sci.* 9(10), 3120–3127 (2016).

[143] B. Zhu *et al.* Realizing record high performance in n-type $Bi_2Te_3$-based thermoelectric materials. *Energ. Environ. Sci.* 13(7), 2106–2114 (2020).

[144] K. L. Company, Thermoelectric device type: KTGM161-18, hwww.kelk.co.jp (accessed April, 2021).

[145] T. Kuroki *et al.* Thermoelectric generation using waste heat in steel works. *J. Electron Mater.* 43, 2405–2410 (2014).

[146] X. Lu *et al.* High-efficiency thermoelectric power generation enabled by homogeneous incorporation of MXene in $(Bi,Sb)_2Te_3$ matrix. *Adv. Energy Mater.* 10(2), 1902986 (2019).

[147] A. Nozariasbmarz *et al.* Bismuth telluride thermoelectrics with 8% module efficiency for waste heat recovery application. *iScience* 23(7), 101340 (2020).

[148] P. Ying *et al.* Towards tellurium-free thermoelectric modules for power generation from low-grade heat. *Nat. Commun.* 12(1), 1121 (2021).

[149] Z. Liu *et al.* Demonstration of ultrahigh thermoelectric efficiency of ~7.3% in $Mg_3Sb_2$/MgAgSb module for low-temperature energy harvesting. *Joule* 5, 1196–1208 (2021).

[150] Z. Bu *et al.* A record thermoelectric efficiency in tellurium-free modules for low-grade waste heat recovery *Nat. Commun.*, 13(1), 237 (2022).

[151] H. Zhu et al. Discovery of ZrCoBi based half Heuslers with high thermoelectric conversion efficiency, *Nat. Commun.*, 9(1), 2497 (2018).

[152] J. Jiang et al. Achieving high room-temperature thermoelectric performance in cubic AgCuTe, *J. Mater. Chem. A*, 8(9), 4790–4799 (2020).

[153] X. Hu et al. Power generation from nanostructured PbTe-based thermoelectrics: comprehensive development from materials to modules, *Energy Environ. Sci.*, 9(2), 517–529 (2016).

[154] C. Fu et al. Realizing high figure of merit in heavy-band p-type half-Heusler thermoelectric materials, *Nat. Commun.* 6(1), 8144 (2015).

[155] B. Jiang et al. Entropy engineering promotes thermoelectric performance in p-type chalcogenides, *Nat. Commun.*, 12(1), 3234 (2021).

[156] W. Li et al. Bismuth telluride/half-heusler segmented thermoelectric unicouple modules provide 12% conversion efficiency, *Adv. Energy Mater*, 10(38), 2001924 (2020).

[157] Z. Bu et al. Realizing a 14% single-leg thermoelectric efficiency in GeTe alloys, *Sci. Adv*, 7(19), eabf2738 (2021).

[158] M. Peyrard, The design of a thermal rectifier, *Europhys. Lett.* 76, 49–55 (2006).

[159] B. Hu, D. He, L. Yang and Y. Zhang, Thermal rectifying effect in macroscopic size, *Phys. Rev. E*, 74(6), 060201(R) (2006).

[160] C. W. Chang, D. Okawa, A. Majumdar and A. Zettel, Solid-state thermal rectifier, *Science* 314(5802), 1121–1124 (2006).

[161] B. Li, L. Wang and G. Casati. Thermal diode: rectification of heat flux. *Phys. Rev. Lett.* 93, 184301 (2004).

[162] F. Giazotto, T. T. Heikkilä, A. Luukanen, A. M. Savin and J. P. Pekola. Opportunities for mesoscopics in thermometry and refrigeration: physics and applications. *Rev. Mod. Phys.* 78, 217–274 (2006).

[163] K. Klinar and A. Kitanovski. Thermal control elements for caloric energy conversion. *Renew. Sustain. Energy Rev.*, 118, p.109571 (2020).

[164] P. R. Gaddam, S. T. Huxtable and W. A. Ducker. A liquid-state thermal diode. *Int. J. Heat Mass Trans.*, **106**, pp.741–744 (2017).

[165] A. L. Cottrill, S. Wang, A. T. Liu, W. J. Wang and M. S. Strano. Dual phase changes thermal diodes for enhanced rectification ratios: theory and experiment. *Adv. Energy Mater.*, 8(11), p.1702692 (2018).

[166] T. Swoboda, K. Klinar, A. S. Yalamarthy, A. Kitanovski and M. Muñoz Rojo. Solid-state thermal control devices. *Adv. Electron. Mater.*, 7(3), p.2000625 (2021).

[167] M. Terraneo, M. Peyrard, and G. Casati, Controlling the energy flow in nonlinear lattices: a model for a thermal rectifier, *Phys. Rev. Lett.* 88(9), 094302 (2002).

[168] B. Li, J. Lan and L. Wang, Interface thermal resistance between dissimilar anharmonic lattices, *Phys. Rev. Lett.* 95(10), 104302 (2005).

[169] W. Kobayashi, Y. Teraoka and I. Terasaki. An oxide thermal rectifier. *Appl. Phys. Lett.*, 95(17), 171905 (2009).

[170] W. Kobayashi, D. Sawaki, T. Omura, T. Katsufuji, Y. Moritomo and I. Terasaki. Thermal rectification in the vicinity of a structural phase transition. *Appl. Phys. Express*, 5(2), p.027302 (2012).

[171] K. Ito, K. Nishikawa, H. Iizuka and H. Toshiyoshi. Experimental investigation of radiative thermal rectifier using vanadium dioxide. *Appl. Phys. Lett.*, 105(25), p.253503 (2014).

[172] X. Zhang, P. Tong, J. Lin, K. Tao, X. Wang, L. Xie, W. Song and Y. Sun2021. Large thermal rectification in a solid-state thermal diode constructed of iron-doped nickel sulfide and alumina. *Phys. Rev. Appl.*, 16(1), p.014031 (2021).

[173] K. Hirata, T. Matsunaga, S. Saurabh, M. Matsunami and T. Takeuchi. Development of high-performance solid-state thermal diodes using unusual behavior of thermal conductivity observed for $Ag_2Ch(Ch= S, Se, Te)$. *Mater. Trans.*, 61(12), 2402–2406 (2020).

[174] K. Hirata, T. Matsunaga, S. Singh, M. Matsunami and T. Takeuchi. High-performance solid-state thermal diode consisting of $Ag_2(S, Se, Te)$. *J. Electron. Mater.*, 49(5), 2895–2901 (2020).

[175] T. Takeuchi, H. Goto, R. S. Nakayama, Y. I. Terazawa, K. Ogawa, A. Yamamoto, T. Itoh and M. Mikami. Improvement in rectification ratio of an Al-based bulk thermal rectifier working at high temperatures. *J. Appl. Phys.*, 111(9), p.093517 (2012).

[176] T. Swoboda, K. Klinar, S. Abbasi, G. Brem, A. Kitanovski and M. Munoz Rojo, Thermal rectification in multilayer phase change material structures for energy storage applications, *iScience* 24(8), 102843 (2021).

[177] A. Kitanovski, Energy applications of magnetocaloric materials, *Adv. Energ. Mater.*, 10(10), 1903741 (2020).

[178] R. A. Kishore and S. Priya, A review on design and performance of thermomagnetic devices, *Renew. Sustain. Energy Rev.* 81, 33 (2018).

[179] R. A. Kishore, D. Singh, R Sriramdas, A. J. Garcia, M. Sanghdasa, and S. Priya, Linear thermomagnetic energy harvester for low-grade thermal energy harvesting, *J. Appl. Phys.* 127, 044501 (2020).

[180] D. Dzekan, A. Waske, K. Nielsch, and S. Fahler, Efficient and affordable thermomagnetic materials for harvesting low grade waste heat, *APL Mater.* 9, 011105 (2021).

[181] S. Kobe, B. Podmiljšak, P. J. McGuiness, M.j Komelj, *CMA's as Magnetocaloric Materials (Pages: 317–363) Complex Metallic Alloys: Fundamentals and Applications* Editor(s): Prof. Jean-Marie Dubois, Prof. Esther Belin-Ferré First published: 24 November 2010

[182] S. Arajs and R. V. Colvin, Thermal conductivity and Lorenz function of gadolinium between 5° and 310°K, *J. Appl. Phys.* 35, 1043 (1964).

[183] T. Takahasi *et al. Electr. Eng. Japan* 148(4), 26–33 (2004).

[184] S. Y. Dan'kov, A. M. Tishin, V. K. Pecharsky and K. A. Gschneidner, Magnetic phase transitions and the magnetothermal properties of gadolinium, *Phys. Rev. B* 57, 3478–3490 (1998).

[185] R. Bjork, C. Bahl and M. Katter, Magnetocaloric properties of $LaFe_{13-x-y}Co_xSi_y$ and commercial grade Gd, *J. Magn. Magn. Mater.* 322, 3882–3888 (2010).

[186] C. Meis, A. K. Froment and D. Moulinier, Determination of gadolinium thermal conductivity using experimentally measured values of thermal diffusivity, *J. Phys. D: Appl. Phys.* 26, 560 (1993).

[187] O. Tegus, E Brück, L Zhang, Dagula, K. Buschow, and F. de Boer, Magnetic-phase transitions and magne-tocaloric effects, *Physica B* 319, 174–192 (2002).

[188] T. Hashimoto, T. Numasawa, M. Shino, and T. Okada, Magnetic refrigeration in the temperature range from 10 K to room temperature: the ferromagnetic refrigerants, *Cryogenics* 21, 647–653 (1981).

[189] F. X. Hu, J. Gao, X. L. Qian, M. Ilyn, A. M. Tishin, J. R. Sun, and B. G. Shen, Magnetocaloric effect in itinerant electron metamagnetic systems $La(Fe_{1-x}Co_x)_{11.9}Si_{1.1}$, *J. Appl. Phys.* 97, 10M303 (2005).

[190] M. Katter, V. Zellmann, G. W. Reppel and K. Uestuener, Magnetocaloric properties of $La, Fe, Co, Si_{13}$ bulk material prepared by powder metallurgy, *IEEE Trans. Magn.* 44, 3044–3047 (2008).

[191] E. Lovell, L. Ghivelder, A. Nicotina, J. Turcaud, M. Bratko, A. D. Caplin, V. Basso, A. Barcza, M. Katter and L. F. Cohen, Low-temperature specific heat in hydrogenated and mn-doped $La(Fe, Si)_{13}$, *Phys. Rev. B* 94, 134405 (2016).

[192] V. Basso, M. Küpferling, C. Curcio, C. Bennati, A. Barzca, M. Katter, M. Bratko, E. Lovell, J. Turcaud, and L. F. Cohen, Specific heat and entropy change at the first order phase transition of La(Fe – Mn – Si)13–H compounds, *J. Appl. Phys.* 118, 053907 (2015).

[193] Z. Ou, N. Dung, L. Zhang, L. Caron, E. Torun, N. van Dijk, O. Tegus, and E. Brück, Transition metal substitution in $Fe_2P$ based $MnFe_{0.95}P_{0.50}Si_{0.50}$ magnetocaloric compounds, *J. Alloys Compd.* 730, 392–398 (2018).

[194] D. T. Cam Thanh, E. Brück, N. T. Trung, J. C. P. Klaasse, K. H. J. Buschow, Z. Q. Ou, O. Tegus, and L. Caron, Structure, magnetism, and magnetocaloric properties of $MnFeP_{1-x}Si_x$ compounds, *J. Appl. Phys.* 103, 07B318 (2008).

[195] J.-Y. Duquesne, J.-Y. Prieur, J. A. Canalejo, V. H. Etgens, M. Eddrief, A. L. Ferreira, and M. Marangolo, Ultrasonic triggering of giant magnetocaloric effect in MnAs thin films, *Phys. Rev. B* 86, 035207 (2012).

[196] L. Mañosa, A. González-Comas, E. Obradó, A. Planes, V. A. Chernenko, V. V. Kokorin, and E. Cesari, Anomalies related to the TA2-phonon-mode condensation in the heusler Ni2mnga alloy, *Phys. Rev. B* 55, 11068–11071 (1997).

[197] Y. K. Kuo, K. M. Sivakumar, H. C. Chen, J. H. Su, and C. S. Lue, Anomalous thermal properties of the Heusler alloy $Ni_{2+x}Mn_{1-x}Ga$ near the martensitic transition, *Phys. Rev. B* 72, 054116 (2005).

[198] X. Zhang, M. Qian, Z. Zhang, L. Wei, L. Geng, and J. Sun, Magnetostructural coupling and magnetocaloric effect in NiMn-Ga-Cu microwires, *Appl. Phys. Lett.* 108, 052401 (2016).

[199] T. Krenke, E. Duman, M. Acet, E. F. Wassermann, X. Moya, L. Mañosa, and A. Planes, Inverse magnetocaloric effect in ferromagnetic Ni-Mn-Sn alloys, *Nat. Mater.* 4, 450 (2005).

[200] V. A. Chernenko, J. M. Barandiarán, J. R. Fernández, D. P. Rojas, J. Gutiérrez, P. Lázpita, and I. Orue, Magnetic and magnetocaloric properties of martensitic $Ni_2Mn_{1.4}Sn_{0.6}$ heusler alloy, J. Magn. *Magn. Mater.* 324, Fifth Moscow International Symposium on Magnetism, 3519–3523 (2012).

[201] T. Krenke, M. Acet, E. F. Wassermann, X. Moya, L. Mañosa, and A. Planes, Martensitic transitions and the nature of ferromagnetism in the austenitic and martensitic states of Ni − Mn − Sn alloys, *Phys. Rev. B* 72, 014412 (2005).

[202] J. Lyubina, Recent advances in the microstructure design of materials for near room temperature magnetic cooling (invited), *J. Appl. Phys.* 109, 07A902 (2011).

[203] V. K. Sharma, M. K. Chattopadhyay, R. Kumar, T. Ganguli, P. Tiwari, and S. B. Roy, Magnetocaloric effect in heusler alloys Ni50Mn34In16 and $Ni_{50}Mn_{34}Sn_{16}$, *J. Phys.: Condens. Matter* 19, 496207 (2007).

[204] K. Buschow, The crystal structures of the rare-earth compounds of the form $R_2Ni_{17}$, $R_2Co_{17}$ and $R_2Fe_{17}$, *J. Less Common Metals* 11, 204–208 (1966).

[205] P. Álvarez-Alonso, P. Gorria, J. A. Blanco, J. Sánchez-Marcos, G. J. Cuello, I. Puente-Orench, J. A. Rodríguez-Velamazán, G. Garbarino, I. de Pedro, J. R. Fernández and J. L. Sánchez Llamazares, Magnetovolume and magnetocaloric effects in $Er_2Fe_{17}$, *Phys. Rev. B* 86, 184411 (2012).

[206] S. Fujieda, Y. Hasegawa, A. Fujita, and K. Fukamichi, Thermal transport properties of magnetic refrigerants $La(Fe_xSi_{1-x})_{13}$ and their hydrides, and $Gd_5Si_2Ge_2$ and mnas, *J. Appl. Phys.* 95, 2429–2431 (2004).

[207] A. M. Tishin and Y. I. Spichkin, *The Magnetocaloric Effect and its Applications* (Institute of Physics Publishing, London, 2003).

[208] I. Zakharov, A. M. Kadomtseva, R. Z. Levitin, and E. G. Ponyatovskii, Magnetic and magnetoelastic properties of a metamagnetic iron-rhodium alloy, *J. Exptl. Theoret. Phys.* 19, 1348–1353 (1964).

[209] R. B. Flippen and F. J. Darnell, Entropy changes of ferromagnetic-antiferromagnetic transitions from magnetic measurements, *J. Appl. Phys.* 34, 1094–1095 (1963).

[210] M. Richardson, D. Melville, and J. Ricodeau, Specific heat measurements on an FeRh alloy, *Phys. Lett. A* 46, 153–154 (1973).

[211] J. S. Kouvel and C. C. Hartelius, Anomalous magnetic moments and transformations in the ordered alloy FeRh, *J. Appl. Phys.* 33, 1343–1344 (1962).

[212] Y. Liu, L. C. Phillips, R. Mattana, M. Bibes, A. Barthélémy, and B. Dkhil, Large reversible caloric effect in FeRh thin films via a dual-stimulus multicaloric cycle, *Nat. Commun.* 7, 11614 (2016).

[213] Y. I. Spichkin, V. K. Pecharsky, and K. A. Gschneidner, Preparation, crystal structure, magnetic and magnetothermal properties of $(Gd_xR_{5-x})Si_4$, where R=Pr and Tb, alloys, *J. Appl. Phys.* 89, 1738–1745 (2001).

[214] F. Holtzberg, R. Gambino and T. McGuire, New ferromagnetic 5: 4 compounds in the rare earth silicon and germanium systems, *J. Phys. Chem. Solids* 28, 2283–2289 (1967).

[215] K. Gschneidner Jr., V. Pecharsky, A. Pecharsky, and C. Zimm, Recent developments in magnetic refrigeration, *Mater. Sci. Forum* 315, 69–76 (1999).

[216] J. M. Elbicki, L. Y. Zhang, R. T. Obermyer, W. E. Wallace, and S. G. Sankar, Magnetic studies of $(Gd_{1-x}Mx)_5Si_4$ alloys (M=La or Y), *J. Appl. Phys.* 69, 5571–5573 (1991).

[217] V. K. Pecharsky and K. A. Gschneidner Jr., Giant magnetocaloric effect in $Gd_5(Si_2Ge_2)$, *Phys. Rev. Lett.* 78, 4494–4497 (1997).

[218] S. Lin, B. S. Wang, J. C. Lin, L. Zhang, X. B. Hu, Y. N. Huang, W. J. Lu, B. C. Zhao, P. Tong, W. H. Song, and Y. P. Sun, Composition dependent-magnetocaloric effect and low room-temperature coefficient of resistivity study of iron-based antiperovskite compounds $Sn_{1-x}Ga_xCFe_3$ ($0 \leq x \leq 1.0$), *Appl. Phys. Lett.* 99, 172503 (2011).

[219] S. Wang, P. Tong, Y. P. Sun, X. B. Zhu, X. Luo, G. Li, W. H. Song, Z. R. Yang, and J. M. Dai, Reversible roomtemperature magnetocaloric effect with large temperature span in antiperovskite compounds $Ga_{1-x}CMn_{3+x}$ (x=0, 0.06, 0.07, and 0.08), *J. Appl. Phys.* 105, 083907 (2009).

[220] W. Zhong, W. Chen, C. Au and Y. Du, Dependence of the magnetocaloric effect on oxygen stoichiometry in polycrystalline $La_{2/3}Ba_{1/3}MnO_{3-\delta}$, *J. Magn. Magn. Mater.* 261, 238–243 (2003).

[221] M. Oumezzine, S. Zemni and O. Peña, Room temperature magnetic and magnetocaloric properties of $La_{0.67}Ba_{0.33}Mn_{0.98}Ti_{0.02}O_3$ perovskite, *J. Alloys Compd.* 508, 292–296 (2010).

[222] D.- L. Hou, C.- X. Yue, Y. Bai, Q.- H. Liu, X.-Y. Zhao, and G.-D. Tang, Magnetocaloric effect in $La_{0.8-x}Nd_xNa_{0.2}Mn$ *Solid State Commun.* 140, 459–463 (2006).

[223] J. H. Lee, S. J. Lee, W. B. Han, H. H. An, and C. S. Yoon, Magnetocaloric effect of $Fe_{64}Mn_{15-x}Co_xSi_{10}B_{11}$ amorphous alloys, *J. Alloys Compd.* 509, 7764–7767 (2011).

[224] X. Li, Y. Pan, and T. Lu, Magnetocaloric effect in Fe-based amorphous alloys and their composites with low boron content, *J. Non-Cryst. Solids* 487, 7–11 (2018).

[225] P. Yu, J. Zhang, and L. Xia, $Fe_{87}Zr_7B_4Co_2$ amorphous alloy with excellent magneto-caloric effect near room temperature, *Intermetallics* 95, 85–88 (2018).

[226] A. Waske, D. Dzekan, K. Sellschopp, D. Berger, A. Stork, K. Nielsch and S. Fähler, Energy harvesting near room temperature using a thermomagnetic generator with a pretzel-like magnetic flux topology, *Nat. Energy* 4, 68–74 (2019).

[227] G. J. Snyder, A. Pereyra and R. Gurunathan, Effective mass from Seebeck coefficient. *Adv. Funct. Mater.*, 32(20), p.2112772 (2022).

# 5 Chemical Aspects of Ligand Exchange in Semiconductor Nanocrystals and Its Impact on the Performance of Future Generation Solar Cells

*Ananthakumar Soosaimanickam, Saravanan Krishna Sundaram, and Moorthy Babu Sridharan*

## 5.1 INTRODUCTION

Recently, inorganic semiconductor nanocrystals (NCs) based optoelectronic devices have attracted considerable interest from researchers. Particularly, semiconductor NCs are showing incredible performance for the fabrication of optoelectronic devices such as light-emitting diodes (LEDs), solar cells, sensors, and photodetectors [1, 2]. Because of the size-dependent band gap, morphology, spectral coverage and solution processing viability, inorganic semiconductor NCs are gaining intensive focus to develop modern electronic devices. Out of other devices, solar cells using inorganic semiconductor NCs seem to be promising candidates [3–6]. Improving the efficiency of semiconductor nanocrystal-based solar cells is a key task in order to integrate them for commercialization. Although the materials play a key role in absorbing sunlight for the energy conversion, the structural assembly of the NCs and their conducting ability determines device efficiency. It is well known that semiconductor NCs are mostly comprised of organic ligands on the surface and depending on the functional properties of these ligands, the NCs could be used for several kinds of applications [7]. In this regard, ligands on the NCs surface play an inevitable role in charge transport properties of a solar cell [8]. The NCs or quantum dot (QD) surface can be passivated using different kinds of ligands as shown in Figure 5.1. In addition to the passivation and morphology variation, it is observed that ligands critically influence on the electronic states of NCs [9, 10]. So, it is essential to choose or select suitable ligands for the NC passivation as well as their assembly for an efficient charge transport. Specifically, when any kind of ligand is employed for a reaction, it is essential to understand how it is useful for the fabrication of an ordered assembly. In short, the conductive or insulating property of a ligand determines the efficiency of a solar cell. Although we use several kinds of ligands for a reaction, only a limited percentage of ligands are quite useful for the fabrication of solar cells, which assist to achieve higher efficiency. Unfortunately, the so-called colloidal hot-injection method compromises use of highly unsaturated fatty acids as the solvent (or) ligands, which are obviously insulative nature [11]. When NCs are purified, several routes such as layer-by-layer deposition, spin coating, dip coating, and spray deposition method are employed to fabricate highly conductive NC thin films. However, native ligands in a NC often hinder the charge transport properties. These ligands therefore have to be eliminated for the efficient charge transport. While removing an insulating ligand, it is imperative to fix a conductive ligand at the same place in order to maintain the structural integrity of the NCs. This specific process of ligands replacement

DOI: 10.1201/9781003340539-5

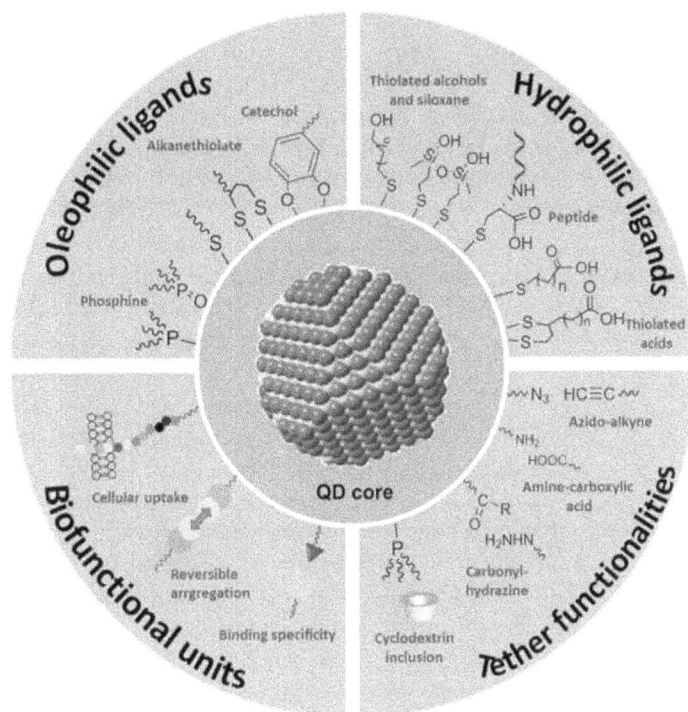

**FIGURE 5.1** Schematic diagram of the different kinds of ligands used for the passivation of QDs. Reprinted from Ref. [7] Copyright 2017 with the permission from Elsevier.

and consequent substitution/modification on the NCs is defined as ligand exchange. This chapter addresses the fundamental aspects of ligand exchange on the semiconductor NCs and its application to the modern solar cell concepts such as dye-sensitized solar cells (DSSCs), hybrid solar cell, quantum dot-sensitized solar cell (QDSSC), and perovskite NC solar cell. In all these cases, the native ligands are replaced from the NCs surface with highly conductive incoming ligands and a significant difference in their performance is notified. There are different chemical aspects which surprise us during this chemical conversion and this should be dealt to understand the mechanism of the ligand exchange.

## 5.2 LIGAND EXCHANGE IN SEMICONDUCTOR NANOCRYSTALS – INTRODUCTION AND TYPES

Ligand exchange in solution is generally carried out between two different biphasic systems in the presence of hydrophilic or hydrophobic ligands. Ligand exchange in semiconductor NCs is generally an equilibrium process. To confirm this, the combined use of Fourier transform infra-red spectroscopy (FTIR) and nuclear magnetic resonance spectroscopy (NMR) could reveal whether the ligand exchange takes place completely or partially. Depending on the peak position of the functional groups in the FTIR and NMR spectrum, the incoming ligand is differentiated from the native ligands the native ligands and it confirms the conversion. The number of ligand molecules coordinating on the NCs surface vary from one to another but usually they orient different directions with respect to the steric influence of the ligands. According to Owen et al. ligands are classified into three different categories namely X-, L- and Z-type depending on their type of bonding on the surface metal atoms [12]. The oleate ligands usually come under the X-type category, which could be replaced through the addition of phosphonic acid. Amines, on the other hand, come under the L-type classification owing to their electron donating ability (ex: ethylenediamine(EDAm), oleylamine (OAm), octadecyalmine(ODAm)). Thus, the L-type ligands donate a pair of electrons which are called as Lewis acids. The Z-type ligands are usually neutral and accept two electrons and therefore they are formally called as Lewis acidic compounds. The replacement of Z-type ligands is acheived by similar kind of Z-type ligands or Lewis acids (ex: $ZnCl_2$, $InCl_3$, $CdCl_2$). These ligands could bind on the different cationic/anionic sites of the NCs and can influence on the optical properties [13]. The selection of incoming ligands for the ligand exchange can be described based on the hard–soft acid base theory (HSAB). According to this concept, a hard acid prefers to bind with a hard base, whereas a soft acid prefers to bind with a soft base. For example, it is visible that in the case of PbS QDs, passivation of $Pb^{2+}$ (soft acid) is mostly accompanied through $I^-$ (soft base) ions [14]. While doing ligand exchange, some of the surface metal atoms could be removed as $ML_x$ (where M = metal atom, L = native ligand) because of the strong bonding and so this vacancy influence on the optical properties. According to the model proposed by Celik et al., a CdSe nanoparticle consists of three kinds of ligand layers, namely strongly bonded ligands, a second outer layer of ligands, and a third layer of weakly bound ligands [15]. This concept is schematically represented in Figure 5.2. Hence, the outer two layers should be removed or exchanged to do a successful ligand exchange. Ligand exchange is carried out either in solution or in solid-state phase. In solution phase, the inorganic salts or organic compounds, which serve as incoming ligands are interacted with the as-synthesized NC solution either directly or after dispersing in a solvent. Here, the incoming ligands are generally shorter chain compound or molecule compared with the native insulating ligands. This obviously promote the charge transport in films. The incoming ligand with a particular penetration depth is reacted with the NC with insulating ligand and the exchange reaction takes place. Also, the degree of ligand exchange depends on the number of cycles of the incoming ligand treatment followed. Organic compounds such as thiols, amines, carboxylic acids, and inorganic salts such as metal halides, metal-complexes, chalcogenide compounds, nitrosonium salts are generally used for the ligand exchange process to remove the native ligands effectively.

In the case of solvents, most of the ligand exchange is carried out using n-hexane, n-octane, toluene etc where the incoming ligands could be dissolved. Here, the native ligands capped NCs are dispersed in a solvent and mixed with a solution which consists of incoming ligands. When this mixture is gently shaken or allowed for a mild heating, the ligand exchange takes place. The rate of ligand exchange depends on the type of native and incoming ligands. For example, it is experimentally observed that the exchange rate of hexadecylamine (HDAm) by 1-hexanethiol (1-HT) is 2 orders of magnitude higher than similar reaction with trioctylphosphine oxide (TOPO) and trioctylphosphine (TOP) [16]. Also, in this case, a higher equilibrium constant value was observed for the HDAm-assisted ligand exchange. In the solid-state ligand exchange, the densely packed NC film is immersed in a solution of incoming ligands for a particular duration and taken out. This

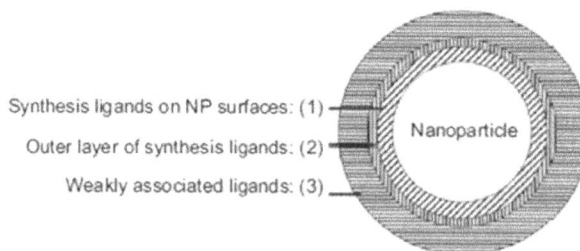

Synthesis ligands on NP surfaces: (1)
Outer layer of synthesis ligands: (2)
Weakly associated ligands: (3)

Nanoparticle

**FIGURE 5.2** Schematic diagram of the different ligand layers of CdSe nanoparticles. Reprinted Ref. [15] Copyright 2012 with the permission from Elsevier.

method is also described in the literature as post-deposition ligand exchange. For the solid-state ligand exchange, there may be several cycles of immersion required depending on the efficacy of the process. Here, the NCs are deposited through potential deposition methods such as dip coating, layer-by-layer deposition (LBL), and doctor blade coating. Out of the others, deposition of QDs through LBL is the mostly adopted approach. Here, for each layer, the NCs are treated with solvents that consist of incoming ligands. Due to the large displacement of ligands from the NC surface, when films are fabricated, formation of cracks takes place owing to the volume shrinkage. To avoid this volume shrinkage effect, each layer is treated with short-chain ligands. High-quality compact NC films could be obtained through this method, which essentially help to attain high efficiency in solar cells. To avoid the waste of a large amount of precursors and solvents in layer-by-layer deposition of NCs, the native ligands are also exchanged before the deposition. For the large area production, doctor blade is preferred due to its versatility. The schematic representation of the solid and solution phase ligand exchange is given in Figure 5.3 [17].

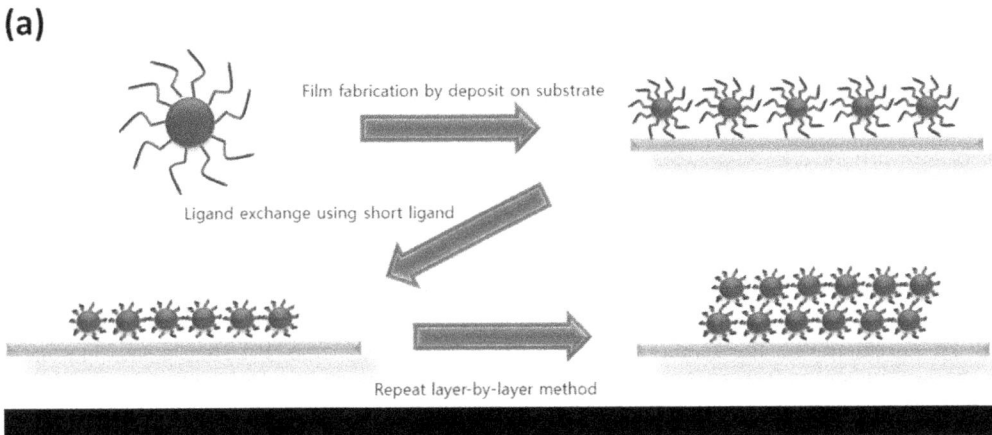

The degree of ligand exchange is important in all these processes since generally not all ligands from the NCs are exchanged and hence the ligands left on the NCs influence on the charge transfer process. Furthermore, when the native ligands are replaced from their position, it is possible to form surface traps, which hardly influence on the surface states of the NCs [18]. Moreover, the NC colloids should be electrostatically stable before and after ligand exchange. Also, the crystal structure and phase of the NCs should be preserved after ligand exchange, which is a key requirement for the modification. The conditions also include the ligand exchanged NCs should not get agglomerated, the incoming ligands should not etch on the surface and the modified NCs should not undergo oxidation. In other words, the

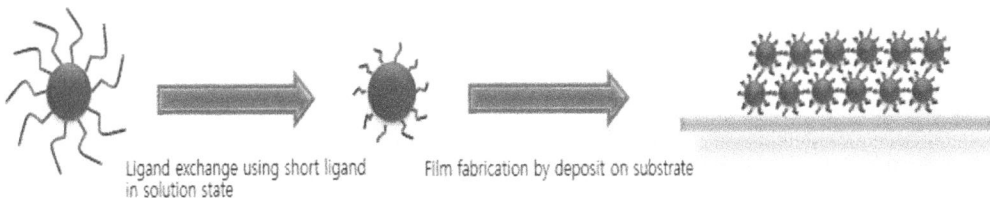

**FIGURE 5.3**   Schematic diagram of (a) solid-phase and (b) solution-phase ligand exchange of NCs. Reprinted from Ref. [17].

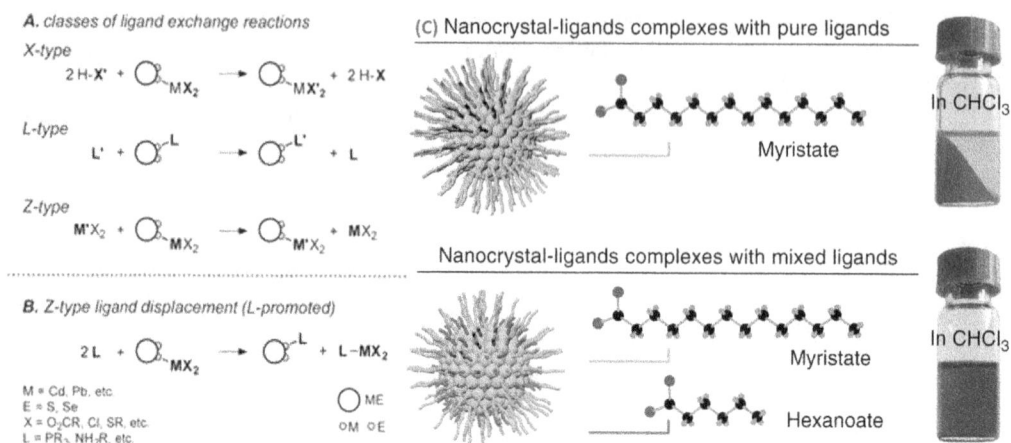

**FIGURE 5.4** (a) Schematic representation of classification of ligands and their exchange reactions. Reprinted with permission from Ref. [12] Copyright @American Chemical Society. (b) Appearance of CdSe NCs in chloroform with native ligands and mixed ligands. Reprinted from Ref. [19].

relative size distribution of the NCs after ligand exchange should not be modified. Any variation in the NC distribution can be explored through analyzing the NC solution using Ultraviolet-visible (UV-visible), X-ray diffraction (XRD), and transmission electron microscopy (TEM) measurements. Also, the reduction of ligand shell could be determined using thermogravimetric analysis (TGA). The position of peaks in the UV-visible spectrum before and after ligand exchange reflects the modification or degradation of NCs in solution. Furthermore, the life-time measurement of the ligand exchanged NCs is another important parameter to conclude the efficiency of the process. Here, the ligand exchanged NCs with less surface traps usually results in higher life-time compared with the NCs with more surface traps. The NCs with high density of surface traps show the shorter life-time, which usually reflects on the poor device performance. The stability of the NCs after ligand exchange is another important parameter to discuss. This is because when the native ligands are replaced by the incoming ligand molecules, the binding ability and the orientation of ligands on the NC surface are different. Thus, ligand exchange is a meticulous process, which is governed by ligands, solvents, and other reaction conditions. All kinds of ligands $L$-, $X$-, and $Z$-type ligands are adopted for the ligand exchange process and the NC surface chemistry could be modified with them. The schematic diagram of this displacement is given in Figure 5.4(a) [12]. Pang et al. have found that when a certain amount of $n$-hexanoic acid is mixed with myristate ligand capped CdSe NCs in chloroform, the stability of the NCs is significantly improved [19]. Here, some of the native ligands are exchanged by the incoming $n$-hexanoic acid and the stability is achieved (Figure 5.4(b)). The functional group associated with the incoming ligand also strongly influences ligand exchange strategy. It is observed that owing to the strong electron withdrawing ability, benzoic acid (BA) exchanged PbS, InP/ZnSe/ZnS and InAs QDs show high colloidal stability compared with the benzenethiol (BT) ligand exchanged QDs [20]. Ligand exchange is also carried out through phase transfer of the prepared NCs [21, 22]. Here, the NCs prepared in organic or aqueous medium are transferred into their counter medium through selective ligand/solvent treatment. That is, it is possible to transfer aqueous synthesized semiconductor NCs to organic medium using suitable ligands. In that case, the pH and functional group of the ligand molecule are considered as potential parameters, which govern the ligand exchange. Short-chain inorganic compounds, which make the NCs temporarily suitable for the coating purpose are widely studied for the fabrication of optoelectronic devices and their electrical properties are evaluated. For example, $Sb_2S_3$ in ethylenediamine and mercaptoethanol (ME) could extract the native ligands from CdSe NCs in toluene without etching and aggregation [23]. Also, this stibanate capped CdSe NCs surprisingly delivered more than 25-fold increase in photocurrent compared with the NCs with native ligands. For the

NCs synthesized through higher boiling point solvents, it is recommended to use dimethyl formamide (DMF) to transfer the NCs [24]. This is because of the fact that DMF has high dielectric constant value ($\varepsilon$ 36.71), which provides effective screening of electrostatic attraction between the oppositely charged ions in order to provide high stability.

## 5.3   WHY IS LIGAND EXCHANGE IMPORTANT FOR THE FABRICATION OF SOLAR CELLS?

In many cases, semiconductor NCs are synthesized at high temperature using higher boiling point solvents and ligands such as OAm, oleic acid (OA), 1-octadecene (1-ODE), TOPO, and TOP [25]. These solvents/ligands could produce highly crystalline, monodispersed nanocrystalline materials due to the decomposition of precursors at high temperature (generally over 200°C). Also, because of the strong bonding with the surface metal atoms, they generally present as tightly bound ligands on the NC surface. Even the non-coordinating solvent, 1-ODE is polymerized at high temperature and form an oligomeric shell [26]. When these ligand capped NCs are deposited as densely packed films for the optoelectronic device applications, the charge transport properties are affected. The as-synthesized colloidal NCs usually consist of these insulating ligands on their surface even after purification. These ligands could hinder the electronic mobility of NCs strongly. Generally, multiple purification steps are required to remove the excess ligands from the NC surface and still the surface ligands after purification could hinder the charge transport. The complete removal of ligands may allow the NCs to fuse together, which deteriorate the performance of the solar cells. Instead, if we could modify the surface by smaller size conductive ligands, it would be beneficial for the fabrication of high-performance solar cell devices through enhanced charge transport [27, 28]. Thus, it is necessary to bring the NCs closer in order to facilitate the charge transport. Ligand exchange is generally used to electronically couple the NCs through reducing the inter-particle distance. Also, the NCs band energy could shift due to the dipole moment of the ligand. Owing to the ligand modification, the LUMO of the semiconductor NCs is altered in favor of the transferring electrons to the other component (ex: polymer) and so the performance of the solar cell is improved. Moreover, this results significant improvement in the charge transport properties and thus carrier enhancement is realized. On the other hand, ligand exchange could influence on the electronic surface states of the NCs. This is because when ligands are replaced, the electronic states of the colloidally prepared NCs are modified depends on the functional groups of the ligands. Kroupa et al. studied the effect of electronic energy levels of PbS QDs through the ligand exchange functionalized cinnamate ligands [29]. The authors tuned the band edge the QDs over 2.0 eV. In general, the charge density of the NCs is altered by the influence of ligands, which additionally modifies the Fermi level. When the Fermi level is varied, consequently the charge transfer kinetics is influenced and the performance of a solar cell is affected. Since the smaller NCs have a large number of surface atoms, the QDs might have more defects due to the ligand exchange. This effect will not only be beneficial for the solar cells, but also other potential area like catalysis. Along with organic ligands, inorganic compounds, weakly bound coordinating compounds (ex: nitrosonium salts), atomic ligands (ex: $NO_3^-$, $Cl^-$) are extensively studied and their influence on the solar cell performance is elucidated. When NC films are treated with these small molecular components, the electron mobility of the films is enhanced and the corresponding efficiency of the solar cell is improved. Thus, in order to improve the exciton dissociation and enhance the carrier mobility of the QDs or NCs in solar cells, ligand exchange is carried out.

## 5.4   LIGAND EXCHANGE IN THE FABRICATION OF MODERN SOLAR CELL TECHNOLOGIES

### 5.4.1   HIGHLY EFFICIENT DYE-SENSITIZED SOLAR CELLS (DSSCs) USING LIGAND EXCHANGE

Dye-sensitized solar cells (DSSCs) have a dye-sensitized layer of metal oxide (typically $TiO_2$) on the fluorine-doped tin oxide (FTO) electrode where the absorption is covered from the visible to the NIR region and the cell assembly further covered with an electrolyte and counter electrode. Here, the counter electrode is typically a hundred nanometer of platinum (Pt) layer on a FTO electrode but due to the higher electrocatalytic activity, copper chalcogenide NCs are considered to use instead of Pt [30, 31]. These NCs are deposited through different approaches like spray deposition, spin coating, doctor blade, etc. In this aspect, ligand exchange on these NCs largely benefit in improving the carrier transport and catalytic activities of the counter electrode layer. To eliminate the insulating ligands and to improve the catalytic properties, the NC layer is annealed. Out of other materials, kesterite group compounds (ex: $Cu_2ZnSnS_4$ and $Cu_2ZnSe_4$) are the frequently studied materials owing to their higher electrocatalytic performance [32, 33]. Since these materials are generally synthesized through colloidal synthetic method in the presence of insulating ligands like OA and OAm [34, 35], it is necessary to remove them for the device applications. Wang et al. carried out solution phase ligand exchange on the OA capped $Cu_2ZnSnSe_4$ NCs using $NH_4S$ in formamide [36]. The finally derived $S^{2-}$ capped $Cu_2ZnSnSe_4$ NCs showed high conductivity without annealing and the fabricated solar cell using $Cu_2ZnSnSe_4$ NC counter electrode showed high performance (7.06% efficiency). The preference of sulphide ligands to remove the native OA, OAm ligands is due to the higher nucleophilic reactivity of $S^{2-}$, which enable to relieve the native ligands from the NC surface. In another study carried out by Guo et al., $CuInS_2$ NCs dispersed in toluene were treated with $(NH_4)S$ dissolved in N-methyl formamide to exchange the ligands [37]. In this case, despite the ligand exchange, the NCs were further allowed sintering to achieve 6.32% efficiency. A similar kind of ligand exchange approach with 1-DDT capped $CuInS_2$ NCs also show that $S^{2-}$ is efficient in removing 1-DDT ligands and with annealing NC layer at 500°C, the resultant solar cell show 5.97% efficiency [38]. All these results are indicating that ligand exchange is a cost-effective approach in fabricating DSSCs using chalcogenide NC layer as the counter electrode.

## 5.4.2  ROLE OF LIGAND EXCHANGE IN QUANTUM-DOT-SENSITIZED SOLAR CELLS (QDSSCs)

Quantum dots (QDs) are used to fabricate high-performance solar cells due to their exciting optical properties such as tunable spectra, higher absorption co-efficient, multiple exciton generation (MEG), etc. [39]. Ligand exchange in QDs to fabricate QDSSCs is a well-studied concept and there are numerous articles investigating about it. Increasing electronic coupling between QDs, high packing density of QDs, and improved charge transport through ligands are the essential parameters to enhance the efficiency of the QDSSCs. Although ligand exchange help to improve the performance of the QDSSCs, purification is important to remove the excessive ligands. In QDSSCs, the QD surface is often treated by thiols to remove insulating long-chain ligands. Along with passivating ligands, solvents play an important role in ligand exchange. It is found that when a low-volatile solvent is used for the ligand exchange, thee QDs order is increased, which leads to the reduction in the trap density. Among other QDs, lead chalcogenides and cadmium chalcogenides are the most studied QD systems. In cadmium chalcogenides, colloidally prepared CdSe QDs are considerably investigated for solar cell applications using ligand exchange strategy. Several ligands are exploited to exchange the native ligands from CdSe QDs and their effect on the optical properties is evaluated [40, 41, 42]. In this, mercaptopropionic acid (MPA) is quite successful in removing the native ligands from the QD surface. It is possible to replace the native ligands of QDs using MPA in solution as well as in solid-state. It is observed that using MPA/methanol mixture, it is possible to precipitate the alloyed $CdSe_xTe_{1-x}$ QDs in chloroform by additionally mixing with water [43]. After this, the QDs are transferred from chloroform to water and this is further precipitated by adding acetone. These QDs are further used to fabricate a solar cell with the structure glass/FTO/$TiO_2$/MPA-CdSeTe QD/electrolyte/$Cu_2S$/glass, which deliver 6.36% efficiency. The absorption spectra of CdSe, CdSeTe, and CdTe QDs and their sensitized films, corresponding $I$-$V$ and IPCE curves of the fabricated devices are given in Figure 5.5. The electronic energy level modification of PbS QDs using several kinds of organic ligands is already demonstrated [44, 45]. Hence, by carefully choosing the

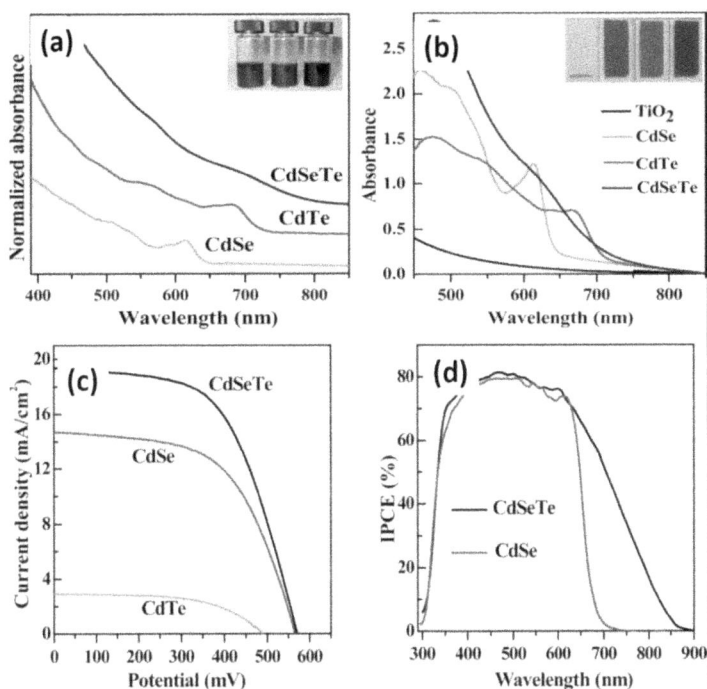

**FIGURE 5.5** (a) UV-visible spectra of the MPA-capped CdSeTe, CdTe, and CdSe QD solutions. (b) UV-visible spectra of the QDs sensitized TiO$_2$ films. (c) *I-V* and (d) IPCE of the different QDs' sensitized solar cell devices. Reprinted with permission from Ref. [43] Copyright@ American Chemical Society.

surface capping ligands, the efficiency of the PbS QD-based solar cells could be tailored. Lin et al. prepared dense, thicker films of PbX ($X$ = S, Se, and Te) nanoinks through ligand exchange strategy [46]. Here, a biphasic system is formed using dimethyl formamide (DMF) or formamide (FA) with $n$-hexane in the presence of halide salts. A simple shaking of this mixture result transformation of QDs from one layer to another layer.

Despite the ligand exchange, it is necessary to eliminate the excess ligands through purification steps and this may form some defects on the QDs depending on the binding strength of the incoming ligands. Similarly, solvents also play important role in improving the performance of a solar cell. For the fabrication of PbS QD solar cells, solvents such as methanol, acetonitrile, propionitrile, dimethyl sulfoxide (DMSO), DMF are used. With respect to the dielectric constant values, these solvents strongly influence on the film quality and corresponding carrier transport of the QDs. It is observed that use of acetonitrile reduces the cracks and this enables the fabrication of thick and dense QD film [47]. Acetonitrile is found to be an efficient solvent for the fabrication of the PbS QD layer because it does not leach out the halides from the surface [48]. However, when comparing the efficiency of the PbS QD layer fabricated through propionitrile and acetonitrile, the ligand exchanged PbS QD layer fabricated using propionitrile shows higher performance (13.3%) and reduced trap density compared with the layer fabricated through acetonitrile (12.5%) [49]. Also, similar to the cadmium chalcogenide nanomaterials, the interaction of MPA with the PbS QD surface is accountable. It is found that ligand exchange using MPA results in fine topography and smooth film surface with the lowest oxidation rate in the case of PbS QDs [50]. It should be noted that most of the ligands are quite comfortable to follow with the physical parameter such as annealing, which also improve the efficiency considerably. Together with MPA, the use of zinc iodide (ZnI) is found to be improving the $V_{oc}$ of the PbS QD-based solar cell [51]. This is due to the suppression of non-radiative channels by MPA and $ZnI_2$. This kind of mixed ligand treatment is sensitive with several factors, which includes solvents, amount and ratio of ligands, rate of ligand exchange etc. Jo et al. used mixed ligand ammonium acetate (AA) and tetrabutyl ammonium acetate (TBAA) to treat PbS QDs [52]. For the ratio 0.2:0.8 (AA:TBA), the resultant solar cell delivered 10.9% efficiency. Other ligands, such as thiols, are considered as one of the best compounds to modify PbS QD surface. This is because compared with amines, both thiols and carboxylic acids are efficiently removing oleate molecules from the PbS QD surface [53]. The stability of the exchanged PbS QDs generally vary with respect to the dispersing solvents. Beygi et al. studied the effect of ligand exchange in PbS QDs using different kinds of amines and their stability in different solvents [50]. It was observed that PbS QDs capped by MPA and butylamine (BA) in tetrachloroethylene resulted high colloidal stability over other ligands due to the strong adsorption of these ligands on the (1 1 1) facets. Furthermore, the solar cell using QDs with these ligands show 2.99% efficiency, which is higher than the solar cell fabricated using native ligands. However, use of BA should be considered with much care because of its high polarity. Interestingly, in the case of PbS QDs, the *n*- or *p*-type nature is dictated by the ligands. When PbS QDs are passivated by EDT, generally *p*-type PbS QDs are achieved. Because of the low carrier mobility of the EDT capped PbS QDs, when the QDs are additionally treated with MPA, the carrier mobility is enhanced [54]. Although EDT and MPA are effective in ligand exchange, for the PbS QDs, which have a larger size (greater than 4 nm), ligand exchange using short-chain carboxylates is found to be effective [55].

Although MPA is widely used to exchange native ligands, there are other potential ligands also investigated. As discussed previously, instead of organic ligands, small inorganic molecules also play important role in boosting the efficiency of QDSSCs. This includes halides, sulphides etc., which actively passivate the QD surface. Considering halides, iodides such as ammonium iodide ($NH_4I$), lead iodide ($PbI_2$) are preferred due to their excellent surface adhering ability with Pb-chalcogenide QDs. This can be explained through soft-acid base chemistry theory, in which the soft-base iodide ($I^-$) ions are prefer to bind with the soft acid ($Pb^{2+}$) ions whereas other halides are not [56]. Ahmad et al. synthesized PbSe QDs through CdSe QDs by cation-exchange process [57]. After this, the OA capped PbSe QDs in octane were ligand exchanged using $PbI_2$ in DMF through a simple shaking process. These $PbI_2$ capped PbSe QDs with EDT-PbSe QDs as the hole-transporter layer delivered 10.68% efficiency with $J_{sc}$=28.11 mA cm$^{-2}$. Crisp et al. studied the effect of metal halides ($PbI_2$, $CdI_2$, $PbCl_2$, $CdCl_2$) treatment on the performance of a solar cell [58]. Here, the metal halides are mixed with DMF and treated with the PbS QD layer and the authors observed that $PbI_2$ remove more oleate molecules compared with the rest of the compounds. After treatment, the $PbI_2$ treated PbS QD layer consists only 1.4% of oleate molecules whereas 26% of oleate molecules are left with the $CdCl_2$ treatment. Through this approach, 7.25% efficiency is achieved for the PbS QDs layer (500 nm thickness). The schematic diagram of the fabricated solar cell structure, the *I-V* curves of the solar cells treated by different ligands, EQE curve of the fabricated solar cell and energy level diagram of the PbS QDs treated by different ligands are collectively given in Figure. 5.6.

Because of the strong interaction between $Pb^{2+}$ and $I^-$, iodide-based ligands deliver promising results due to the capability of suppressing trap states of the QDs. Zhang et al. studied the effect of ligand exchange of PbSe QDs with series of ligands such as tetrabutyl ammonium iodide (TBAI), phenethylethylammonium iodide (PEAI), EDT, MPA and cetyl trimethyl ammonium bromide (CTAB) [59]. Except EDT, all other ligands in methanol were treated with the $FTO/TiO_2$ and the energy levels were modified. Out of the other ligands, TBAI coated PbSe QDs showed high PL quenching due to the high packing density and also delivered high efficiency (3.50%) over other ligands. In the case of PbS QDs, it is demonstrated that TBAI could almost eliminate the insulating OA molecules through multiple washing treatment, which essentially results in high packing density and improved charge transport [60]. The schematic diagram of this treatment and the current-voltage curves of the solar cell with TBAI treated PbS QDs and a reference device is given in Figure. 5.7. It is possible to passivate PbS QDs using multiple steps treatment by different compounds. For

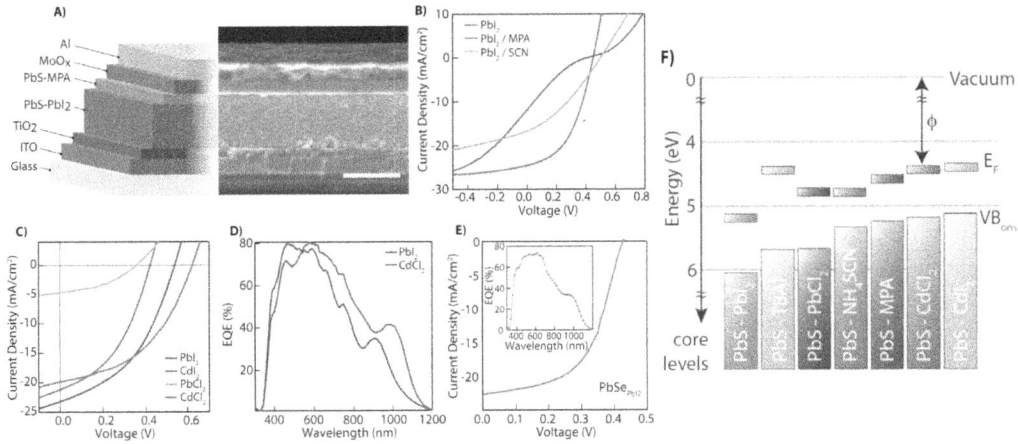

FIGURE 5.6 (a) Schematic representation of the fabricated solar cell with the cross section of false color scanning electron microscope image. (b) *I-V* curve of the solar cells treated with $PbI_2$, $PbI_2$/MPA and $PbI_2$/MCN ligands. (c) *I-V* curve of the solar cells treated with $PbI_2$, $CdI_2$, $PbCl_2$ and $CdCl_2$. (d) External quantum efficiency (EQE) curves of the PbS QD solar cells with $CdI_2$ and $CdCl_2$ ligands treatment and (e) *I-V* curve of the $PbI_2$ treated PbSe QDs solar cell and (f) photoelectron spectroscopy results of the PbS QDs by different surface treatments. Reprinted from Ref. [58].

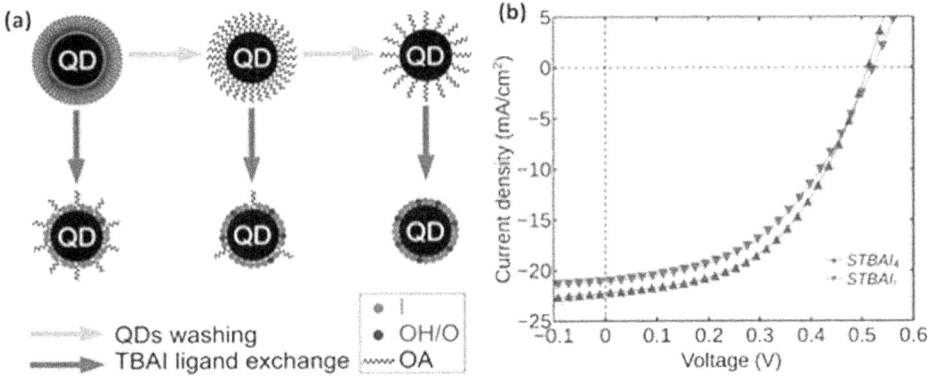

FIGURE 5.7 (a) Schematic diagram of ligand exchange process of PbS QDs using the TBAI assisted washing process. Reprinted from Ref. [60].

example, an experimental study reveals that the ligand exchange in PbS QDs using $NH_4I$ followed with $PbI_2$ result improved efficiency (12.3%) compared with the solar cell treated only with $NH_4I$ (11.5%) [61]. Interestingly, it is found that halide treatment on PbS QDs results in a modification on the QDs packing arrangement, from face-centred (FCC) to body-centred (BCC) [62]. When TOP capped CdSe NCs on $TiO_2$/FTO were immersed in the solution of sodium sulfide ($Na_2S$) in water/methanol mixture under dark conditions, they could efficiently replace the native ligands [63]. Here, the $S^{2-}$ ions are replacing the native TOP ligands and resulting $S^{2-}$ capped CdSe NCs. In this case, through XPS analysis, the authors observed that there is a change in the surface $Cd^{2+}$ atoms before and after ligand exchange. The $S^{2-}$ ions capped CdSe NCs show an impressive 3.17% efficiency with high $J_{sc}$ (10.11 mA/cm²). Sulphide passivation usually improve the performance of the QDSSCs since the surface treatment by $S^{2-}$ is found to be improving the concentration of QDs on the $TiO_2$ surface. Furthermore, it is also experimentally proved that $S^{2-}$ treatment enhancing the electronic coupling and charge transfer rate between the $TiO_2$ and CdSe QDs [64]. However, Ren et al. found that ligand exchange using thiosulfate ($S_2O_3^{2-}$) ions is still beneficial over $S^{2-}$ ions and the solar cell fabricated using thiosulfate ions exchanged PbS QDs showed 6.11% efficiency [65]. It is also possible to achieve a densely packed QD layer through ligand exchange via acids, especially with hydroiodic acid (HI) [66]. Here, the OA ligands are removed through the proton donation by HI, which consequently passivate the surface through iodide. A reduced inter-dot spacing (upto 0.2 nm) was observed in this case owing to the HI treatment. Other than ligands passivation, the performance of QDSSCs could be affected by solvents. It is demonstrated that, use of short-chain alcohols reduce the inter-particle distance between the QDs. However, highly protic solvent, ex: methanol removes excess ligands from the QD surface, which essentially induces aggregation [67]. This is due to the difference in the pKa values of the solvents, which hardly affect the stabilization of QDs in solution.

### 5.4.3  Ligand Exchange in Hybrid Solar Cells

Solution processed organic polymers offer multiple advantages to fabricate highly efficient solar cell devices using flexible substrates. When NCs are mixed with an organic polymer, the ligands have direct contact with the polymeric core and this hybrid blend has importance in fabricating optoelectronic devices. Hybrid solar cells consist of the active layer of a donor/acceptor mixture where the multiple exciton generation of QDs play a crucial role [68]. The important steps namely light absorption, carrier generation, charge separation and recombination are deciding the performance of the hybrid solar cell. Depending on the NCs morphology, the charge transfer is facilitated and the electron-hole separation with consequent charge collection at the respective electrodes takes place. The typical device assembly of a hybrid solar cell is similar to the organic bulk-heterojunction solar cells except the organic acceptor molecules are replaced with the inorganic NCs. Hybrid solar cells possess the active layer, which consists of blends of semiconductor NCs with a low band-gap polymer, for example, poly-3-hexylthiophene (P3HT) [27, 68]. Here, at first, the NCs are ligand exchanged with short-chain conductive ligands. Thereafter, different ratios of the NCs are mixed with a polymer and the corresponding optical properties are evaluated. Generally, a quenching is observed in the PL spectrum of the blend, which indicates the efficient charge transfer from polymer to NCs or QDs. In most of the cases, the native ligands are exchanged through pyridine treatment. Here, the NCs are dispersed in pyridine and the corresponding solution is refluxed overnight. Later, the ligand-exchanged NCs are centrifuged and processed for the film fabrication by redispersing in appropriate solvent(s). For the dispersion of QDs or NCs, solvents like chlorobenzene, dichlorobenzene are widely used to fabricate smooth NCs/polymer hybrid layer. In literature, this process of eliminating native ligands using pyridine is described as liquid-liquid extraction [69]. When a highly insulating TOPO is exchanged by pyridine, improvement in the charge-carrier separation at the interface of the polymer/NC hybrids is realized. However, as indicated earlier, to reduce the ligand shell around the NCs as much as possible, multiple purifications steps are required to reduce the interparticle distance. Lokteva et al. found that even after three times purification using pyridine, some percentage of the OA were left on the CdSe NC surface [70]. Here, the authors observed that for every cycle of pyridine treatment, the agglomeration of NCs was found to be increasing. It seems that although pyridine is good for the removal of native ligands, the preservation of NC monodispersity is not maintained. Celik et al. carried out different methods to execute ligand exchange in CdSe nanorods [15]. The authors observed that when CdSe nanorods were pre-treated (or) washed several times with methanol before pyridine treatment, the efficiency is improved. By this approach, the fabricated device with the structure ITO/PEDOT:PSS/PCPDTBT:CdSe nanorods/Al delivered 3.4% efficiency. Despite this pyridine treatment works well for the ligand exchange of OA/OAm passivated NCs, and there are much better ligands could provide a superior performance. This is because pyridine interaction with the surface metal atoms for ex: $Cd^{2+}$ is quite weak and so the ligands are labile. Also, negatively charged ions, such as phosphonates and stearates, cannot be easily replaced by pyridine molecules. Although pyridine molecules are efficient in removing the native ligands, the efficiency is still affected by the QDs size. For example, it is observed that in the case of InP QDs/P3HT blends, InP QDs with 4.5 nm size provide better efficiency compared with the InP QDs with 2.5 nm size [71]. This is due to the insufficient electron injection at the InP QDs/P3HT interface owing to the modifications in the interfacial energy alignment, which result lower efficiency.

Thiols, which serve as better stabilizing agents for the aqueous synthesized semiconductor NCs, can also serve as better candidates for the ligand exchange. The estimated binding energy value of thiols with the QD surface is between 14 and 23 kJ/mol [15]. The typical thiol compounds used for the ligand exchange of the NCs are tertiary-butyl thiol (TBT), 1,3-benzenedithiol, 1-HT, 3-mercaptopropionic acid (MPA) and 1,2-ethanedithiol (1,2-EDT). In most of the cases, a single thiol is used to exchange the native ligands in the presence of suitable solvents. However, in some cases a mixed ligand treatment is also investigated. Mastria et al. used a mixture of p-methylbenzenethiol/ triethylamine to exchange native ligands of PbS QDs [72]. However, in this case, the fabricated solar cell with QDs/P3HT showed 1.98% efficiency only. Lim et al. prepared CdSe nanotetrapods using hot-injection method and the prepared nanotetrapods were allowed for a two-phase ligand exchange treatment [73]. At first, CdSe tetrapods in n-hexane were treated with tetrafluoroboric acid (HBF$_4$) in water/DMF and after dispersing in OAm/chloroform, tetrapods surface was packed by OAm. To fabricate a solar cell, these CdSe tetrapods were spin-coated and annealed to evaporate OAm. This OAm removed CdSe tetrapods were further treated with 1-hexylamine and pyridine in order to enhance the charge transport between CdSe nanotetrapods and P3HT. Here, compared with pyridine treated CdSe tetrapods, 1-HT treated CdSe nanotetrapods showed higher performance (1.80%). The schematic diagram of the ligand exchange and FTIR, UV-visible spectra of the ligand exchanged tetrapods are represented in Figure 5.8. Not only organic ligands, atomic halide ligands also deliver impressive results. These halide ligands effectively passivate the NCs and reduces the defect sites. Zhang et al. used EDT and CTAB to execute ligand exchange of PbSe QDs in solid-state [74]. Here, the PbSe QD films were firstly treated with CTAB in methanol through the spin-coating method. For the EDT treatment, EDT in acetonitrile was used and the films were treated by the same approach. Out of these two, CTAB treated solar cell (ITO/PEDOT:PSS/P3HT:ZnO/ ZnO/Al) showed 2.90% efficiency. When 1-HT is used to exchange native ligands of CuInS$_2$ NCs, a treatment with acetic acid is required to improve the morphology of the CuInS$_2$ NCs/P3HT hybrids [75]. These evidences clearly describe the active role of ligand molecules in enhancing the performance of the NC-based hybrid solar cells through suppressing surface defects.

Although most of the investigated short-chain thiols are for the aqueous synthesized NCs, long-chain thiol such as 1-dodecanethiol (1-DDT) is also used for the ligand exchange of aqueous synthesized semiconductor NCs [76]. In place of thiols, other organic compounds such as acetic acid [77], m-phenylenediamine [78], butylamine [79], and hexanoic acid [80] are also studied. Nabil et al. prepared CdSe QDs (~ 5 nm) by high-temperature hot-injection method in the presence of TOPO and HDAm [81]. For the ligand exchange, the prepared QDs were dispersed in n-hexane and mixed with TBAI in dimethyl formamide (DMF). The QDs were passivated by iodide (I$^-$) ions through a simple shaking process and prepciptated by isopropanol which subsequently used for the device fabrication. These QDs were further assembled into the device structure ITO/TiO$_x$/ P3HT:CdSe: PC60BM/MoO$_3$/Ag, which showed 2.5% efficiency. Zhou et al. prepared TOPO capped CdSe nanorods and the nanorods were surface treated by pyridine [82]. Then, a hybrid blend was formed with a low band-gap polymer PCPDTBT and used to fabricate the active layer. This active layer was further treated with 1% EDT in acetonitrile and this additional treatment led to 4.7% efficiency. Here, it was proposed that pyridine remove only *L*-type TOPO whereas EDT remove *X*-type phosphonic acid , which help to improve the efficiency. In an another interesting analysis, PbS QDs/ PTBI polymer hybrid blends [1:9 (w/w) ratio] were treated with different ligands, namely, 1,2-EDT,

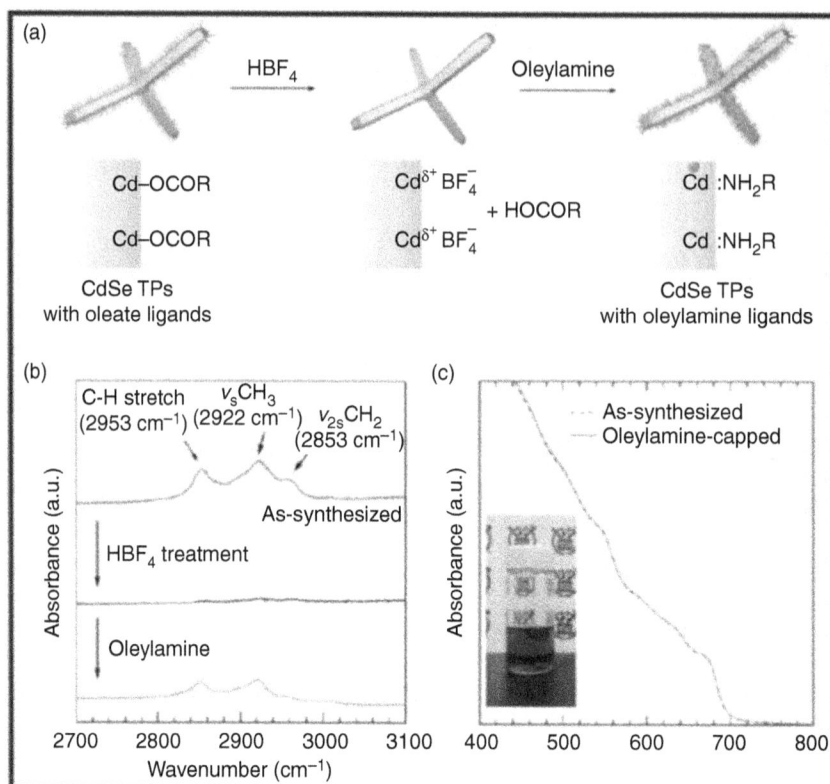

**FIGURE 5.8** (a) Schematic diagram of the ligand exchange of CdSe tetrapods. (b) FTIR spectra of the as-synthesized and ligand exchanged CdSe tetrapods and (c) UV-visible spectra of the as-synthesized and OAm capped CdSe tetrapods. Reprinted with permission from Ref. [73]. Copyright@ American Chemical Society 2014.

3-MPA, malonic acid (MA), and TBAI [83]. Prior to this ligand treatment, the as-synthesized PbS QDs were ligand exchanged with butylamine. Compared with other ligands, the post-deposition-ligand exchange using MPA resulted in higher $V_{oc}$ (0.48 V) and high efficiency (2.49%). A similar kind of approach with CdSe QDs/PCPDTBT blend in the presence of 1-benzenedithiol (1-BDT) result in 4.18% efficiency [84]. Here, an additional CdSe QD layer is fabricated on the active layer and which is further treated with 1-BDT. Importantly, the authors have proposed that it is because of the face-on orientation of 1-BDT, such higher efficiency is achieved. Following this, in an another study, the hybrid blends of P3HT nanowires/CdS QDs were treated with EDT and in this case, 4.1% efficiency was observed [85]. Here, the butylamine capped CdS QDs were ligand exchanged through 1-BDT and improved charge transport was achieved.

Sometimes, the synthesis reaction may generate a by-product, which could serve as a capping ligand for the NCs. For example, Perner et al. observed that when OAm and OA combinely used in a reaction, the resultant dioleamide served as ligand for the CuInS$_2$ NCs and this was further exchanged by 1-HT at 85°C [86]. The fabricated solar cell with a device structure glass/ITO/PEDOT:PSS/PCDTBT-CuInS$_2$ NCs/Ag showed higher performance over the device with NCs without ligand exchange. The schematic diagram of this chemical conversion, PL spectra of the NCs/polymer blend and the I-V curves of the fabricated solar cells are given in Figure 5.9. Since it is difficult to remove the oleate molecules from the NC surface because of the strong biding ability, solvents with different polarity are used to extract them. To transfer the oleate passivated PbS QDs in octane to the water,

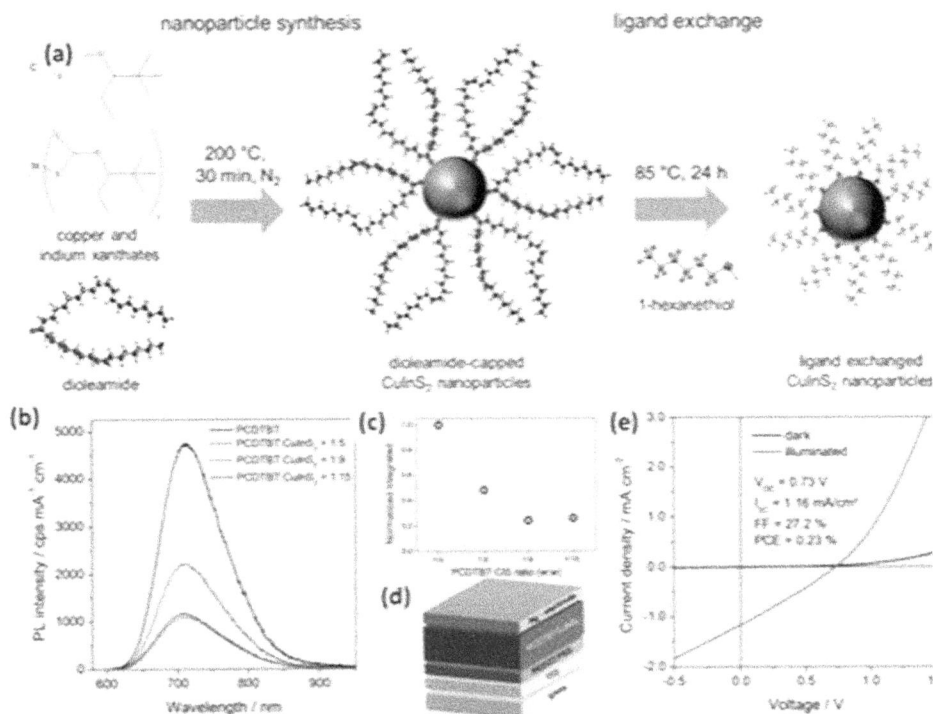

**FIGURE 5.9** (a) Schematic representation of synthesis scheme of CuInS$_2$ NCs using dioleamide and corresponding ligand exchange using 1-hexanethiol. (b) PL spectra of the pure PCDTBT polymer and polymer/NCs blend with different ratios. (c) PL intensity with respect to the ratio of the NCs with the polymer (d) the device structure of hybrid solar cell fabricated using CuInS$_2$ NCs/PCDTBT and (e) the *I-V* curve of the fabricated solar cell. Reprinted from Ref. [86].

the solution is mixed with DMF with MPA/butylamine and ligand exchange in PbS QDs is realized [87]. Nagaoka et al. used butylamine to exchange the oleate molecules from PbS QDs and mixed with PTB1 polymer (1:9 w/w ratio) [88]. This composite layer was further treated with 3-MPA and this post-deposition ligand exchange approach led to 2.8% efficiency. Greaney et al. efficiently carried out the ligand exchange of CdSe NCs using TBT in tetramethyl urea at 250°C [89]. Here, the ligand exchange resulted in extremely higher $V_{oc}$ compared with the NCs with the native ligands, which is about 0.80 V and the fabricated solar cell with CdSe NCs/P3HT active layer showed 1.9% efficiency. It is possible to improve the efficiency of a solar cell through passivating the NC surface by suitable additives. For example, PbI$_2$ passivated PbS NCs with Si-PCPDTBT polymer delivered 4.8% efficiency owing to the trap states passivation [90]. Ligand exchange in the presence of thiol is more sensitive with light since it may produce a disulphide during this conversion [89]. Instead of functionalizing the NCs, the organic donar molecules are functionalized and the active layer is fabricated [91–93]. Compared with OA, removal of TOPO from metal-oxide NCs seems to be difficult. However, it is observed that the Z907 dye molecule could efficiently remove TOPO from TiO$_2$ nanorods (NRs) and could enhance the photocurrent [94]. In this work, interestingly, it is observed that inclusion of methanol into TiO$_2$ NR solution could prevent the recombination of $e^-$-$h^+$ pairs. Although TiO$_2$ NRs could transform high photocurrent than the NCs, the intrinsic defects of the NCs could limit the generation of photocurrent in hybrid blend. These analyses emphasize that the ligand exchange process is helping to achieve high efficiency solar cells through manipulating the NC surface.

### 5.4.4 Ligand Exchange in Perovskite Nanocrystal Solar Cells

All-inorganic lead halide perovskite NCs and hybrid lead halide perovskite NCs ($AMX_3$, where $A = CH_3NH_3^+$, $Cs^+$ $M = Pb^{2+}$, $Sn^{2+}$, $Bi^{2+}$ $X = Cl$, Br and I) are the currently emerging materials which exhibit superior optical properties over traditional cadmium chalcogenide nanomaterials [95]. However, these NCs surface is quite weak for the ligand passivation because of the ionic lattice. Despite this, the field is impressively progressing in the direction of improving the surface properties of metal halide perovskite NCs. Several kinds of halides, metal ions, and organic polymers are explored to passivate the ionic, fragile surface of the perovskite NCs and excellent results are demonstrated [96–98]. Specifically, these surface treatments are quite beneficial in improving the photoluminescent quantum yield (PLQY) of the perovskite NCs, which is useful for the device applications. Considering the role of solvent, it is observed that traditionally used ethanol or methanol deteriorate the perovskite NC surface, whereas mild polar solvents such as methyl acetate, ethyl acetate are found to be suitable for the purification or surface treatment purposes. For the ligand removal or solid-state ligand exchange, the NC films with native ligands are immersed into ethyl (or) methyl acetate several times and the multiple layer assembly of QDs are fabricated. This kind of layer-by-layer assembly is achieved with several hundred nanometers thickness to fabricate potential optoelectronic devices. Through this method, the perovskite QD layer has shown high electron mobility owing to the reduced interparticle distance. Sanehira et al. treated $CsPbI_3$ QD film with the $Pb(NO_3)_2$ in methyl acetate for 10 min and several layers of the QDs (thickness 200–400 nm) were fabricated using this approach [99]. Finally, a solution consisting of formamidinium iodide (FAI) in ethyl acetate was used for the surface passivation. With this approach, the authors achieved impressive 13.43% efficiency for the device structure glass/FTO/TiO$_2$/FAI/CsPbI$_3$ QDs/Spiro-OMeTAD/MoO$_x$/Al. The speculation proposed for this higher efficiency is the introduction of FAI may execute a partial cation exchange (or) could form an alloyed structure with $CsPbI_3$ QDs (or) could form a shell around $CsPbI_3$ QDs (or) it could induce a grain growth on the $CsPbI_3$ QD film.

Among the other halide perovskite QDs, $CsPbI_3$ QDs are much investigated owing to their coverage of photons in the visible region. Potential ligands and inorganic compounds are indeed useful to regulate the surface properties of the metal halide perovskite NCs. Zhang et al. carried out a pseudohalide ligand exchange on the $CsPbI_3$ QDs in order to exploit the QDs for solar cells [100]. In this study, phenylethylamine (PEA) and 2-(4-fluoro phenylethylamine) (FPEA) were used to treat the $CsPbI_3$ QDs in methyl acetate. By optimizing the volume ratio of PEA to methyl acetate as 1:100, the fabricated solar cell with the structure FTO/TiO$_2$/CsPbI$_3$ QDs/PTAA/Ag/MoO$_3$ showed 14.65% efficiency with improved stability. Generally, amines are quite furious with the perovskite NC surface and they are, in fact, providing a very good improvement on the properties of the NCs. For example, when perovskite NCs are treated with ethylenediamine tetraacetic acid (EDTA), a significant improvement in the carrier transport results in efficiency up to 15.25% [101]. Since the halide vacancies contribute to the recombination dynamics in metal halide perovskite QDs, surface passivation using halide compounds are proven to improve the performance of the perovskite QD-based solar cells. In this regard, instead of ligand exchange, most of the studies are describing about the post-surface passivation of halide perovskite QDs using suitable halide compounds. For example, triphenyl phosphite is found to be efficient in promoting the performance of $CsPbI_3$ QD solar cells up to 15.21% [102]. Unlike other semiconductor NCs, perovskite NCs or QDs are generally synthesized using OAm/OA ligands and so these ligands obviously remain on the surface. Since perovskite QD surface is sensitive to the atmospheric moisture, this induces hydrolysis in methyl acetate in which the resultant acetate replaces oleate ions from the $CsPbI_3$ QD surface [103]. However, this methyl acetate mediated exchange should be carried out under optimized cylces/treatment and in such a case, the best solar cell efficiency is 12.85% for the device structure glass/FTO/TiO$_2$/CsPbI$_3$ QDs/spiro-oMeTAD/MoO$_3$/Ag [104].

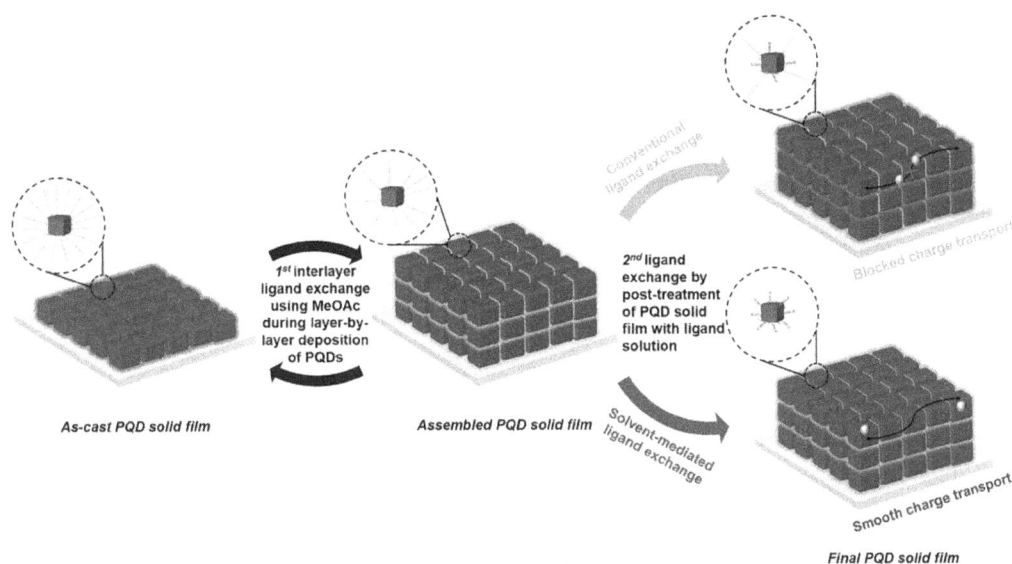

**FIGURE 5.10** (a) Schematic representation of the steps involved in the two-step ligand-exchange process to fabricate CsPbI$_3$ QD films. Reprinted from Ref. [106] Copyright 2022 with the permission from Elsevier.

Interestingly it is further observed that ethyl acetate/butyl acetate mixed solvent system could efficiently stripe OAm/OA ligands and the resultant CsPbBr$_3$ QD solar cell show incredible $V_{oc}$ 1.59 V [105]. Beyond this, when a highly protic solvent for ex: 2-pentanol is employed for the fabrication of CsPbI$_3$ QD-based solar cell, it effectively remove the native OAm ligands and passivate the halide vacancies [106]. As a result, this help to achieve impressive 16.53% efficiency. The schematic representation of this two-step ligand-exchange process is given in Figure 5.10. When amino acids with different chain lengths (glycine, β-alanine, d-alanine, and lysine) mixed with methyl acetate are used to treat CsPbI$_3$ QDs, they could do a dual passivation due to the zwitterionic nature [107]. In this case, the structural, optical properties of the amino acid treated QDs are preserved and due to the short chain, glycine treated CsPbI$_3$ QDs show high intensity in the PL spectra with improved carrier mobility properties. Also, the glycine treated CsPbI$_3$ QDs device (glass/ITO/SnO$_2$/CsPbI$_3$ QDs/spiro-oMeTAD/Ag) show 13.66% efficiency with high fill factor (63.68). These findings clearly emphasize how solvents critically influence on the performance of the QD solar cell. An interesting study with respect to the CsPbI$_3$ QDs size reveals that through maintaining monodispersity, it is possible to achieve higher efficiency. This is because after synthesis, the perovskite QDs are usually resulting in polydispersity owing to the fast reaction kinetics. To solve this, monodispersed CsPbI$_3$ QDs are extracted using gel permeation chromatography (GPC) and these QDs have obviously shown large absorption that results in higher efficiency (15.3%) with ever achieved $V_{oc}$ 1.27 V [108]. Similar to the cadmium chalcogenide nanomaterials, surface treatment on perovskite NCs using MPA improve the carrier transport [109]. Also, MPA is quite useful in removing OAm from the Cu$_{12}$Sb$_4$S$_{13}$ QDs, which function as a hole-transporter layer in the CsPbI$_3$ QD solar cells [110]. Here, the resultant solar cell using Cu$_{12}$Sb$_4$S$_{13}$ QDs delivered higher $J_{sc}$, which was 18.28 mA cm$^{-2}$.

## 5.5 MISCELLANEOUS

Other than the above discussed solar cell assemblies, ligand exchange is also carried out to fabricate heterojunction solar cells. Korala et al. studied the effect of ligand exchange on the Cu$_2$ZnSnS$_4$ NCs using L-, X- and Z-type ligands [111]. Here, to exchange with S$^{2-}$ ligands, the

NC solution in CHCl$_3$ was treated with NH$_4$S in formamide but this was found to be unsuitable to fabricate films using spin-coating. On the other hand, for the solid-state ligand exchange, the NC film was treated with different ligand solutions, namely NH$_4$S, TBAI, ethylenediamine (EDA), and ZnCl$_2$. Interestingly, it was observed that the combined use of TBAI, EDA, and ZnCl$_2$ treated CZTS NC films resulted in high carrier concentration with high conductivity. Also, the CZTS/CdS heterojunction solar cell using this tri-ligand passivated CZTS NC resulted in higher $V_{oc}$ = 391 mV. This reveals each ligand assist the improvement of electrical properties of the NC films. Similarly, CdCl$_2$/methanol treated CdTe NCs in the device structure glass/ITO/CdSe/CdTe/Au show improved grain growth [112]. Since ligand exchange affect the electronic energy level of the QDs, it is possible to fabricate the *n-p* junction in PbS QDs using potential ligand exchange approaches. Aqoma and Jang fabricated a *p-n* junction using PbS QDs, which was prepared using different ligand modifications [87]. For the *p*-QDs, the surface was coated with an organic ligand whereas the *n*-QDs are prepared using phase-transferred MPA-QDs in DMF/*n*-butylamine. Cao et al. have found that when OAm capped CZTS NCs are exchanged using 1-HT in *n*-hexane, the ligand modified NC show high carrier mobility and deliver excellent performance (16.62% efficiency) by serving as hole-transporting layer in perovskite solar cells [113]. Alloyed kesterite NCs such as Cu$_2$ZnSnS$_{4x}$Se$_{4(1-x)}$ NCs are also found to be suitable for the *p-n* junction solar cell in which ligand exchange using 5-amino-1-pentanol result in 7.68% efficiency [114]. When a conjugated system is used to replace the native ligands, the charge transport of the NC film appears to be controversial. Reinhold et al. studied the effect of ligand exchange of TOPO capped CuInS$_2$ NCs in the heterojunction solar cells [115]. Here, the native ligands were exchanged using 4-methylbenzenethiol at 180°C and centrifuged. The NCs were then redispersed in 1,2-dichlorobenzene. The reduction of ligand shell in this case was found to be about 1 nm but however this reduction did not result improvement in the fabricated solar cell. The conclusion of this study is it is the ability to passivate the surface traps of the NCs determine the performance of a solar cell. It is possible to passivate the lead chalcogenide NCs using perovskite compounds. Peng et al. synthesized OAm capped PbS QDs using the hot-injection method and the prepared QDs were ligand exchanged using CH$_3$NH$_3$PbI$_3$ in solid-state [116]. Here, with the as-deposited PbS QDs films, CH$_3$NH$_3$PbI$_3$ in acetonitrile was drop casted with the consecutive rinsing by acetonitrile and *n*-octane. This allowed an improvement in the electrical conductivity to 1.88 × 10$^{-9}$ S cm$^{-1}$ and a layer-by-layer deposition of the prepared QDs together with the EDT-PbS QDs top layer resulted in 5.28% efficiency for the device structure FTO/TiO$_2$/CH$_3$NH$_3$PbI$_3$/PbS QDs/MoO$_3$/Au. Another study on passivating PbS QDs using CH$_3$NH$_3$PbI$_3$ also showed that it reduce recombination and increase the mobility of carriers [50]. Also, this passivation exceeds the performance of traditionally used TBAX (*X*=Cl, Br and I), PbI$_2$, NH$_4$I, and MAI compound, which resulted in 4.41% efficiency in a *p-i-n* junction solar cell. Similar to this, Sun et al. exchanged native ligands of PbS QDs using butylamine and the resultant QDs were fabricated as films after annealing at 70°C [117]. Then formamidiminium halide (FAX where *X* = Br, I) in acetonitrile was used to soak the film to fabricate perovskite coating on the PbS QDs layer. This perovskite layer increase the interdot coupling between the QDs and doubling the exciton diffusion length. As a result, the fabricated solar cell (ITO/ZnO/PbS QDs-FAI/PbS QDs-EDT/Au) delivered 13.8% efficiency. The application of perovskite QDs is extended towards semi-transparent solar cells by exchanging the native OAm/OA ligands using guanidinium thiocyanate in ethyl acetate [118]. In this case, a very high $V_{oc}$ was observed with 5% efficiency together with high transparency. This PbS QDs-CH$_3$NH$_3$PbI$_3$ combination could be made using hexane/N-methyl formamide mediated ligand exchange [119]. The schematic representation of this ligand exchange and the corresponding *I-V* values of the fabricated cells are given in Figure 5.11. All these results are clearly indicating that ethyl acetate is one of the best solvents in removing the native ligands as well as improving the carrier transport properties of perovskite QDs.

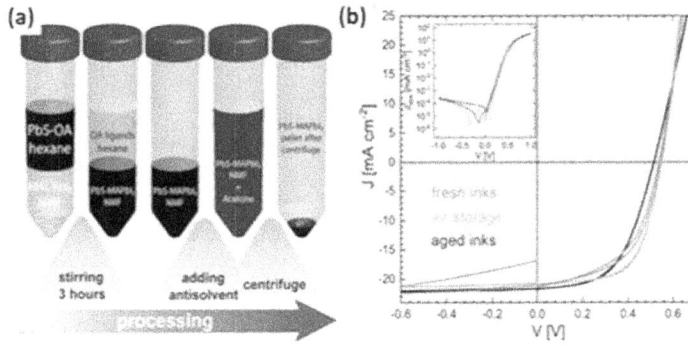

**FIGURE 5.11** (a) Ligand exchange of PbS QDs through phase transfer from *n*-hexane to N-methyl formamide; (b) the current voltage (*I-V*) curves of the fabricated solar cell using ligand exchanged PbS-MAPbI$_3$ inks. Reprinted from Ref. [119]. Copyright @American Chemical Society.

## 5.6  CONCLUSION AND PERSPECTIVES

The discussion in this chapter shows how the ligand exchange process potentially influences on the performance of future generation solar cells. Most of the results indicate that this facile process could be executed even for the large area production of solar cells and further investigation in this direction could deliver promising results for the high-performance solar cells. Furthermore, it is important to minimize the loss of solvents and NCs during ligand exchange and probably the *in situ* ligand exchange approach could solve this problem. However, the *in situ* solid-state ligand exchange approach still generates a volume contraction, which develops cracks in the film and this should be resolved. Furthermore, difficulty associated with the solid-state ligand exchange such as formation of inhomogeneous energy states should be eliminated to execute the same for large-scale applications. Also, the ligands, which withstand oxidation, could be more beneficial for the fabrication of long-term stable, highly efficient NC-based solar cells. Fabrication of NC solar cells through a tandem approach is also interesting to apply the ligand exchange method and extract the charge carriers effectively. Many literature describe that only partial ligand exchange takes place when NCs are treated with ligands, and this may require several steps to replace the native ligands. Also, when ligand exchange takes place, the possibility of agglomeration also arises in several cases and this might be a problem for the large scale. Formation of surface defects in such cases also strongly influence on the optical properties of the NCs. Besides, cracks in the films due to the ligand exchange also lead to major problems for the industrial production. Development of new potential ligands which adopt with different solvents is necessary to execute the ligand exchange strategy effectively or preparation of ligand-free QDs may be another alternative solution for the highly conductive inks. Significant development in defect engineering and molecular modeling of these ligands will additionally strengthen this field. Also, most of the NCs (or) QDs are synthesized through the hot-injection method, which is a challenging technology for upscaling the process. Thus, development of a convenient synthesis process will ease difficulties in processing NCs. Furthermore, a separate protocol or library should be developed for the fabrication of NC-/QD-based solar cells using flexible substrates through ligand exchange strategy. Efforts in these areas may explore other possible directions of the ligand exchange approach in constructing the future generation solar cells.

## ACKNOWLEDGEMENT

One of the authors, Saravanan Krishna Sundaram, sincerely thanks the management of Sri Sairam Institute of Technology, Chennai for their support during this work.

## REFERENCES

1. S. V. Kershaw, L. Jing, X. Huang, M. Gao, A. L. Rogach, Materials aspects of semiconductor nanocrystals for optoelectronic applications, *Mater. Horiz*, 2017, 4, 155–2015.

2. M. Li, C. Wang, L. Wang, H. Zhang, Colloidal semiconductor nanocrystals: synthesis, optical non-linearity, and related device applications, *J. Mater. Chem. C*, 2021, 9, 6686–6721.

3. S. Kumar, G. D. Scholes, Colloidal nanocrystal solar cells, *Microchimica Acta*, 2008, 160, 315–325.

4. C. Steinhagen, T. B. Harvey, C. J. Stolle, J. Harris, B. A. Korgel, Pyrite nanocrystal solar Cells: promising, or fool's gold ?, *J. Phys. Chem. Lett.* 2012, 3(17), 2352–2356.

5. C. Liu, Q. Zeng, H. Wei, Y. Yu, Y. Zhao, T. Feng, B. Yang, Metal Halide perovskite nanocrystal solar cells: progress and challenges, *Small*, 2020, 4(10), 2000419.

6. J. Jasieniak, B. I. MacDonald, S. E. Watkins, P. Mulvaney, Solution-processed sintered nanocrystal solar cells via layer-by-layer assembly, *Nano Lett.* 2011, 11(7), 2856–2864.

7. J. Zhou, Y. Liu, J. Tang, W. Tang, Surface ligands engineering of semiconductor quantum dots for chemosensory and biological applications, *Materialstoday*, 2017, 20, 360–376.

8. A. R. Khabibullin, A. L. Efros, S. C. Erwin, The role of ligands in electron transport in nanocrystal solids, *Nanoscale,* 2020, 12, 23028–23035.

9. M. D. Peterson, L. C. Cass, R. D. Harris, K. Edme, K. Sung, E. A. Weiss, The role of ligands in determining the exciton relaxation dynamics in semiconductor quantum dots, *Ann. Rev. Phys. Chem.*, 2014, 65, 317–339.

10. S. Kilina, K. A. Velizhanin, S. Ivanov, O. V. Prezhdo, S. Tretiak, Surface ligands increase photoexcitation relaxation rates in CdSe Quantum Dots, *ACS Nano*, 2012, 6(7), 6515–6524.

11. E. R. Kennehan, K. T. Munson, G. S. Doucette, A. R. Marshall, M. C. Beard, J. B. Asbury, Dynamic ligand surface chemistry of excited PbS quantum dots, *J. Phys. Chem. Lett.* 2020, 11(6), 2291–2297.

12. N. C. Anderson, M. P. Hendricks, J. J. Choi, J. S. Owen, Ligand exchange and the stoichiometry of metal chalcogenide nanocrystals: spectroscopic observation of facile metal-carboxylate displacement and binding, *J. Am. Chem. Soc.* 2013, 135(49), 18536–18548.

13. M. M. Krause, L. Jethi, T. G. Mack, P. Kambhampati, Ligand surface chemistry dictates light emission from nanocrystals, *J. Phys. Chem. Lett.* 2015, 6(21), 4292–4296.

14. B-S. Kim, J. Hong, B. Hou, Y. Cho, J. I. Sohn, S. N. Cha, J. M. Kim, Inorganic-ligand exchanging time effect in PbS quantum dot solar cell, *Appl. Phys. Lett.* 2016, 109, 063901.

15. D. Celik, M. Krueger, C. Veit, H. F. Schleiermachar, B. Zimmermann, S. Allard, I. Dumsch, U. Scherf, F. Rauscher, P. Niyamakom, Performance enhancement of CdSe nanorod-polymer based hybrid solar cells utilizing a novel combination of post-synthetic nanoparticle surface treatments, *Solar Energy Mat. Solar Cells,* 2012, 98, 433–440.

16. R. Koole, P. Schapotschnikow, C. M. Donega, T. J. H. Vlugt, A. Meijerink, Time-dependent photoluminescence spectroscopy as a tool to measure the ligand exchange kinetics on a quantum dot surface, *ACS Nano*, 2008, 2(8), 1703–1714.

17. H. R. You, J. Y. Park, D. H. Lee, Y. Kim, J. Choi, Recent research progress in surface ligand exchange of PbS quantum dots for solar cell application, *Appl. Sci.* 2020, 10(3), 975.

18. M. J. Greaney, E. Couderc, J. Zhao, B. A. Nail, M. Mecklenburg, W. Thornbury, F. E. Osterloh, S. E.Bradforth, R. L. Brutchey, Controlling the trap state landscape of colloidal CdSe nanocrystals with cadmium halide ligands, *Chem. Mater.* 2015, 27(3), 744–756.

19. Z. Pang, J. Zhang, W. Cao, X. Kong, X. Peng, Partitioning surface ligands on nanocrystals for maximal solubility, *Nat. Comm.*, 2019, 10, 2454.

20. S. Lee, M-J. Choi, G. Sharma, M. Biondi, B. Chen, S-W. Baek, A. M. Najarian, M. Vafaie, J. Wicks, L.L. Sagar, S. Hoogland, F. P. G. Arquer, O. Voznyy, E. H. Sargent, Cascade surface modification of colloidal quantum dot inks enables efficient bulk homojunction photovoltaics, *Nat. Comm.*, 2020, 11,1–8.

21. J. Zylstra, J. Amey, N. J. Miska, L. Pang, C. R. Hine, J. Langer, R. P. Doyle, M. M. Maye, A modular phase transfer and ligand exchange protocol for quantum dots, *Langmuir*, 2011, 27(8), 4371–4379.

22. J. Yang, J. Y. Lee, J. Y. Ying, Phase transfer and its applications in nanotechnology, *Chem. Soc. Rev.* 2011, 40, 1672–1696.

23.   J. J. Buckley, M. J. Greaney, R. L. Brutchey, Ligand exchange of colloidal CdSe nanocrystals with stibanates derived from $Sb_2S_3$ dissolved in a thiol-amine mixture, *Chem. Mater.* 2014, 26(21), 6311–6317.

24.   X. Nie, Y. Zhang, X. Wang, C. Ren, S-Q. Gao, Y-W. Lin, Direct visualization of ligands exchange on the surfaces of quantum dots by a two-phase approach, *Chem. Select*, 2018, 3(8), 2267–2271.

25.   S. Ananthakumar, J. R. Kumar, S. M. Babu, Evolution of non-phosphine solvents in colloidal synthesis of I-III-$VI_2$ and $I_2$-II-IV-$VI_4$ group semiconductor nanomaterials – Current status, *Mater. Sci. Semi. Process.* 2017, 67, 152–174.

26.   E. Dhaene, J. Billet, E. Bennett, I. V. Driessche, J. D. Roo, The trouble with ODE: polymerization during nanocrystal synthesis, *Nano Lett.*, 2019, 19(10), 7411–7417.

27.   S. Ananthakumar, J. Ramkumar, S. M. Babu, Effect of ligand exchange in optical and morphological properties of CdTe nanoparticles/P3HT blend, *Solar Energy*, 2014, 106, 151–158.

28.   S. Ananthakumar, J. Ramkumar, S. M. Babu, Synthesis of thiol modified CdSe nanoparticles/P3HT blends for hybrid solar cell structures, *Mater. Sci. Semi. Process.*, 2014, 22, 44–49.

29.   D. M. Kroupa, M. Voros, N. P. Brawand, B. W. McNichols, E. M. Miller, J. Gu, A. J. Nozik, A. Sellinger, G. Galli, M. C. Beard, Tuning colloidal quantum dot band edge positions through solution-phase surface chemistry modification, *Nat. Comm.*, 2017, 8, 1–8.

30.   J. Xu, X. Yang, Q-D. Yang, T-L. Wong, C-S. Lee, Hierarchical $Cu_2ZnSnS_4$ articles for a low-cost solar cell: morphology control and growth mechanism, *J. Phys. Chem. C*, 2012, 116(37), 19718–19723.

31.   X. Wang, Y. Xie, B. Bateer, K. Pan, Y. Jiao, N. Xiong, S. Wang, H. Fu, Selenization of $Cu_2ZnSnS_4$ enhanced the performance of dye-sensitized solar cells: improved zinc-site catalytic activity for $I_3^-$, *ACS Appl. Mater. Interfaces*, 2017, 9(43), 37662–37670.

32.   S. Ananthakumar, X. Li, A-L. Anderson, P. Yilmaz, S. Dunn, S. M. Babu, J. Briscoe, Photo-enhanced catalytic activity of spray-coated $Cu_2SnSe_3$ nanoparticle counter electrode for dye-sensitised solar cells, *Physica Status Solidi, RRL*, 2016, 10(10), 739–744.

33.   J. Briscoe, S. Dunn, The future of using earth-abundant elements in counter electrodes for dye-sensitized solar cells, *Adv. Mater.* 2016, 28(20), 3802–3813.

34.   S. Ananthakumar, J. R. Kumar, S. M. Babu, Colloidal synthesis and characterization of $Cu_2ZnSnS_4$ nanoplates, *J. Semiconductors*, 2017, 38, 033007.

35.   S. Ananthakumar, J. R. Kumar, S. M. Babu, Influence of co-ordinating and non-coordinating solvents in structural and morphological properties of $Cu_2ZnSnS_4$ (CZTS) nanoparticles, *Optik*, 2017, 130, 99–105.

36.   X. Wang, D-X. Kou, W-H. Zhou, Z-J. Zhou, S-X. Wu, X. Cao, $Cu_2ZnSnSe_4$ nanocrystals capped with $S^{2-}$ by ligand exchange: utilizing energy level alignment for efficiently reducing carrier recombination, *Nano. Res. Lett.* 2014, 9, 1–7.

37.   J. Guo, X. Wang, W-H. Zhou, Z-X. Chang, X. Wang, Z-J. Zhou, S-X. Wu, Efficiency enhancement of dye-sensitized solar cells (DSSCs) using ligand exchanged $CuInS_2$ NCs as counter electrode materials, *RSC Adv.*, 2013, 3, 14731–14736.

38.   R-Y. Yao, Z-J. Zhou, Z-L. Hou, X. Wang, W-H. Zhou, S-X. Wu, Surfactant-free $CuInS_2$ nanocrystals: An alternative counter-electrode material for dye-sensitized solar cells, *ACS Appl. Mater. Interfaces*, 2013, 5(8), 3143–3148.

39.   S. Ananthakumar, D. Balaji, J. R. Kumar, S. M. Babu, Role of co-sensitization in dye-sensitized and quantum dot-sensitized solar cells, *SN Appl. Sci.*, 2019, 1, 1–46.

40.   N. C. Anderson, J. S. Owen, Soluble, chloride-terminated CdSe nanocrystals: ligand exchange monitored by $^1H$ and $^{31}P$ NMR spectroscopy, *Chem. Mater.*, 2013, 25(1), 69–76.

41.   L. Lu, X. Zhang, L. Ji, H. Li, H. Yu, F. Xu, J. Hu, D. Yang, A. Dong, Size-dependent ligand exchange of colloidal CdSe nanocrystals with $S^{2-}$ ions, *RSC Adv.*, 2015, 5, 90570–90577.

42.   B. von Holt, S. Kudera, A. Weiss, T. E. Schrader, L. Manna, W. J. Parak, M. Braun, Ligand exchange of CdSe nanocrystals probed by optical spectroscopy in the visible and mid-IR, *J. Mater. Chem.*, 2008, 18, 2728–2732.

43.   Z. Pan, K. Zhao, J. Wang, H. Zhang, Y. Feng, X. Zhong, Near infrared absorption of $CdSe_xTe_{1-x}$ alloyed quantum dot sensitized solar cells with more than 6% efficiency and high stability, *ACS Nano*, 2013, 7(6), 5215–5222.

44. P. R. Brown, D. Kim, R. R. Lunt, N. Zhao, M. G. Bawendi, J. C. Grossman, V. Bulovic, Energy level modification in lead sulfide quantum dot thin films through ligand exchange, *ACS Nano*, 2014, 8(6), 5863–5872.

45. P. N. Goswami, D. Mandal, A. K. Rath, The role of surface ligands in determining the electronic properties of quantum dot solids and their impact on photovoltaic figure of merits, *Nanoscale*, 2018, 10, 1072–1080.

46. Q. Lin, H. J. Yun, W. Liu, H-J. Song, N. S. Makarov, O. Isaienko, T. Nakotte, G. Chen, H. Luo, V. I.Klimov, J. M. Pietryga, Phase-transfer ligand exchange of lead chalcogenide quantum dots for direct deposition of thick, highly conductive films, *J. Am. Chem. Soc.* 2017, 139(19), 6644–6653.

47. K. Lu, Y. Wang, Z. Liu, L. Han, G. Shi, H. Fang, J. Chen, X. Ye, S. Chen, F. Yan, A. G. Shulga, T. Wu, M. Gu, S. Zhou, J. Fan, M. A. Loi, W. Ma, High-Efficiency PbS quantum-dot solar cells with greatly simplified fabrication processing via "solvent-curing", *Adv. Mater.*, 2018, 30(25), 1707572.

48. A. R. Kirmani, G. H. Cary, M. Abdelsamie, B. Yan, D. Cha, L. R. Rollny, X. Cui, E. H. Sargent, A. Amassian, Effect of solvent environment on colloidal-quantum-dot solar-cell manufacturability and performance, *Adv. Mater.*, 2014, 26(27), 4717–4723.

49. M. Biondi, M-J. Choi, S. Lee, K. Bertens, M. Wei, A. R. Kirmani, G. Lee, H. T. Kung, L. J. Richter, S. Hoogland, Z-H. Lu, F. P. G. Arquer, E. H. Sargent, Control over ligand exchange reactivity in hole transport layer enables high-efficiency colloidal quantum dot solar cells, *ACS Energy Lett.* 2021, 6(2), 468–476.

50. H. Beygi, S. A. Sajjadi, A. Babakhani, J. F. Young, F. C. J. M. Veggel, Solution phase surface functionalization of PbS nanoparticles with organic ligands for single-step deposition of p-type layer of quantum dot solar cells, *Appl. Surf. Sci.* 2018, 459, 562- 571.

51. S. Pradhan, A. Stavrinadis, S. Gupta, Y. Bi, F. Di Stasio, G. Konstantatos, Trap-state suppression and improved charge transport in PbS quantum dot solar cells with synergistic mixed-ligand treatments, *Small*, 2017, 13(21), 1700598.

52. J. W. Jo, Y. Kim, J. Choi, F. P. G. de Arquer, G. Walters, B. Sun, O. Ouellette, J. Kim, A. H. Proppe, R. Quintero-Bermudez, J. Fan, J. Xu, C. S. Tan, O. Voznyy, E. H. Sargent, Enhanced open-circuit voltage in colloidal quantum dot photovoltaics via reactivity-controlled solution-phase ligand exchange, *Adv. Mater.*, 2017, 29(43), 1703627.

53. A. Milam, P. T. Wasdin, H. Turner, M. E. Salyards, A. Clay, M. R. McPhail, Quantum dot thin film imaging enables in situ, benchtop analysis of ligand exchange at the solution-film interface, *Coll. Surf. A: Phys. Eng. Aspects*, 2021, 629, 127457.

54. Z. T. The, L. Hu, Z. Zhang, A. R. Gentle, Z. Chen, Y. Gao, L. Yuan, Y. Hu, T. Wu, R. J. Patterson, S. Huang, Enhanced power conversion efficiency via hybrid ligand exchange treatment of p-type PbS quantum dots, *ACS Appl. Mater. Interfaces* 2020, 12(20), 22751–22759.

55. M. Liu, N. Yazdani, M. Yarema, M. Jansen, V. Wood, E. H. Sargent, Colloidal quantum dot electronics, *Nature Electronics*, 2021, 4, 548–558.

56. R. Wang, Y. Shang, P. Kanjanaboos, W. Zhou, Z. Ning, E. H. Sargent, Colloidal quantum dot ligand engineering for high performance solar cells, *Energy Environ. Sci.*, 2016, 9, 1130–1143.

57. W. Ahmad, J. He, Z. Liu, K. Xu, Z. Chen, X. Yang, D. Li, Y. Xia, J. Zhang, C. Chen, Lead selenide (PbSe) colloidal quantum dot solar cells with >10% efficiency, *Adv. Mater.*, 2019, 31(33), 1900593.

58. R. W. Crisp, D. M. Kroupa, A. R. Marshall, E. M. Miller, J. Zhang, M. C. Beard, J. M. Luther, Metal halide solid-state surface treatment for high efficiency PbS and PbSe QD solar cells, *Sci. Rep.*, 2015, 5, 9945.

59. Y. Zhang, C. Ding, G. Wu, N. Nakazawa, J. Chang, Y. Ogomi, T. Toyoda, S. Hayase, K. Katayama, Q. Shen, Air stable PbSe colloidal quantum dot heterojunction solar cells: ligand-dependent exciton dissociation, recombination, photovoltaic property, and stability, *J. Phys. Chem. C*, 2016, 120(50), 28509–28518.

60. A. E. Tom, A. Thomas, V. V. Ison, Novel post-synthesis purification strategies and the ligand exchange processes in simplifying the fabrication of PbS quantum dot solar cells, *RSC Adv.* 2020, 10, 30707–30715.

61. S. Zheng, J. Chen, E. M. J. Johansson, X. Zhang, PbS colloidal quantum dot inks for infrared solar cells, *iScience*, 2020, 23(11), 101753.

62. W. Chen, J. Zhong, J. Li, N. Saxena, L. P. Kreuzer, H. Liu, L. Song, B. Su, D. Yang, K. Wang, J. Schlipf, V. Korstgens, T. He, K. Wang, P. Muller-Buschbaum, Structure and charge carrier dynamics in colloidal PbS quantum dot solids, *J. Phys. Chem. Lett.* 2019, 10(9), 2058–2065.

63. F. Liu, J. Zhu, J. Wei, Y. Li, J. Lu, Y. Huang, O. Takuya, Q. Shen, T. Toyoda, B. Zhang, J. Yao, S. Dai, *J. Phys. Chem. C*, Ex situ CdSe quantum dot-sensitized solar cells employing inorganic ligand exchange to boost efficiency, 2014, 118(1), 214–222.

64. H. J. Yun, T. Paik, M. E. Edley, J. B. Baxter, C. B. Murray, Enhanced charge transfer kinetics of CdSe quantum dot-sensitized solar cell by inorganic ligand exchange treatments, *ACS Appl. Mater. Interfaces*, 2014, 6(5), 3721–3728.

65. Z. Ren, J. Yu, Z. Pan, J. Wang, X. Zhong, Inorganic ligand thiosulfate-capped quantum dots for efficient quantum dot sensitized solar cells, *ACS Appl. Mater. Interfaces*, 2017, 9(22), 18936–18944.

66. J. W. Jo, J. Choi, F. P. G. Arquer, A. Seifitokaldani, B. Sun, Y. Kim, H. Ahn, J. Fan, R. Quintero-Bermudez, J. Kim, M-J. Choi, S-W. Baek, A. H. Proppe, G. Walters, D-H. Nam, S. Kelly, S. Hoogland, O. Voznyy, E. H. Sargent, Acid-assisted ligand exchange enhances coupling in colloidal quantum dot solids, *Nano Lett.* 2018, 18(7), 4417–4423.

67. J. H. Song, H. Choi, Y-H. Kim, S. Jeong, High performance colloidal quantum dot photovoltaics by controlling protic solvents in ligand exchange, *Adv. Energy Mater.* 2017, 7(15), 1700301.

68. Z. Chen, X. Du, Q. Zeng, B. Yang, Recent development and understanding of polymer–nanocrystal hybrid solar cells, *Mater. Chem. Front.* 2017, 1, 1502–1513.

69. N. T. N. Truong, W. K. Kim, U. Farva, X. D. Luo, C. Park, Improvement of CdSe/P3HT bulk heterojunction solar cell performance due to ligand exchange from TOPO to pyridine, *Solar Energy Mat. Solar Cells,* 2011, 95(11), 3009–3014.

70. Lokteva, N. Radychev, F. Witt, H. Borchert, J. Parisi, J. Kolny-Olesiak, Surface treatment of CdSe nanoparticles for application in hybrid solar cells: The effect of multiple ligand exchange with pyridine, *J. Phys. Chem. C.,* 2010, 114(29), 12784–12791.

71. J. Yin, M. Kumar, Q. Lei, L. Ma, S. S. K. Raavi, G. G. Gurzadyan, C. Soci, Small-Size Effects on Electron Transfer in P3HT/InP Quantum Dots, *J. Phys. Chem. C*, 2015, 119(47), 26783–26792.

72. R. Mastria, A. Rizzo, C. Giansante, D. Ballarini, L. Dominici, O. Inganas, G. Gigli, Role of polymer in hybrid polymer/PbS quantum dot solar cells, *J. Phys. Chem. C*, 2015, 119(27), 14972–14979.

73. J. Lim, D. Lee, M. Park, J. Song, S. Lee, M. S. Kang, C. Lee, K. Char, Modular fabrication of hybrid bulk heterojunction solar cells based on breakwater-like CdSe tetrapod nanocrystal network infused with P3HT, *J. Phys. Chem. C*, 2014, 118(8), 3942–3952.

74. X. Zhang, Y. Zhang, H. Wu, L. Yan, Z. Wang, J. Zhao, W. W. Yu, A. L. Rogach, PbSe quantum dot films with enhanced electron mobility employed in hybrid polymer/nanocrystal solar cells, *RSC Adv,* 2016, 6, 17029–17035.

75. C. Krause, D. Scheunemann, J. Parisi, H. Borchert, Three-dimensional morphology of CuInS2:P3HT hybrid blends for photovoltaic applications, *J. Appl. Phy.* 2015, 118, 205501.

76. S. Ananthakumar, J. Ramkumar, S. M. Babu, Synthesis and efficient phase transfer of CdSe nanoparticles for hybrid solar cell applications, *Conf. Papers in Energy*, 2013, 194638(1–3).

77. W-F. Fu, Y. Shi, L. Wang, M-M. Shi, H-Y. Li, H-Z. Chen, A green, low-cost, and highly effective strategy to enhance the performance of hybrid solar cells: Post-deposition ligand exchange by acetic acid, *Solar Energy Mat. Solar Cells*, 2013, 117, 329–335.

78. F. A. Roghabadi, M. Kokabi, V. Ahmad, G. Abaeiani, Quantum dots crosslinking as a new method for improving charge transport of polymer/quantum dots hybrid solar cells and fabricating solvent-resistant film, *Electrochemical Acta*, 2016, 222, 881–887.

79. J. D. Olson, G. P. Gray, S. A. Carter, Optimizing hybrid photovoltaics through annealing and ligand choice, *Solar Energy Mat. Solar Cells*, 2009, 93(4), 519–523.

80. Y. Zhou, M. Eck, C. Veit, B. Zimmermann, F. Rauscher, P. Niyamakom, S. Yilmaz, I. Dumsch, S. Allard, U. Scherf, M. Kruger, Efficiency enhancement for bulk-heterojunction hybrid solar cells based on acid treated CdSe quantum dots and low bandgap polymer PCPDTBT, *Solar Energy Mat. Solar Cells,* 2011, 95(4), 1232–1237.

81. M. Nabil, S. A. Mohamed, K. Easawi, S. S. A. Obayya, S. Negam, H. Talaat, M. K. El-Mansy, Surface modification of CdSe nanocrystals: Application to polymer solar cell, *Curr. Appl. Phys.,* 2020, 20(3), 470–476.

82. R. Zhou, R. Stalder, D. Xie, W. Cao, Y. Zheng, Y. Yang, M. Plaisant, P. H. Holloway, K. S. Schanze, J.R. Reynolds, J. Xue, Enhancing the efficiency of solution-processed polymer: colloidal nanocrystal hybrid photovoltaic cells using ethanedithiol treatment, *ACS Nano*, 2013, 7(6), 4846–4854.

83. A. E. Colbert, W. Wu, E. M. Janke, F. Ma, D. S. Ginger, Effects of ligands on charge generation and recombination in hybrid polymer/quantum dot solar cells, *J. Phys. Chem. C*, 2015, 119(44), 24733–24739.

84. W. Fu, L. Wang, J. Ling, H. Li, M. Shi, J. Xue, H. Chen, Highly efficient hybrid solar cells with tunable dipole at the donor–acceptor interface, *Nanoscale*, 2014, 6, 10545–10550.

85. S. Ren, L-Y. Chang, S-K. Lim, J. Zhao, M. Smith, N. Zhao, V. Bulovic, M. Bawendi, S. Gradecak, Inorganic–organic hybrid solar cell: Bridging quantum dots to conjugated polymer nanowires, *Nano Lett.*, 2011, 11(9), 3998–4002.

86. V. Perner, T. Rath, F. Pirolt, O. Glatter, K. Wewerka, I. Letofsky-Papst, P. Zach, M. Hobisch, B. Kurnert, G. Trimmel, Hot injection synthesis of CuInS$_2$ nanocrystals using metal xanthates and their application in hybrid solar cells, *New. J. Chem*, 2019, 43, 356–363.

87. H. Aqoma, S-Y. Jang, Solid-state-ligand-exchange free quantum dot ink-based solar cells with an efficiency of 10.9%, *Energy Environ. Sci.*, 2018, 11, 1603–1609.

88. H. Nagaoka, A. E. Colbert, E. Strein, E. M. Janke, M. Salvador, C. W. Schlenker, D. S. Ginger, Size-dependent charge transfer yields in conjugated polymer/quantum dot blends, *J. Phys. Chem. C*, 2014, 118(11), 5710–5715.

89. M. J. Greaney, S. Das, D. H. Webber, S. E. Bradforth, R. L. Brutchey, Improving open circuit potential in hybrid P3HT:CdSe bulk heterojunction solar cells via colloidal tert-butylthiol ligand exchange, *ACS Nano*, 2012, 6(5), 4222- 4230.

90. H. Lu, J. Joy, R. L. Gasper, S. E. Bradforth, R. L. Brutchey, Iodide-passivated colloidal PbS nanocrystals leading to highly efficient polymer:nanocrystal hybrid solar cells, *Chem. Mater.* 2016, 28(6), 1897–1906.

91. M. He, F. Qiu, Z. Lin, Toward high-performance organic–inorganic hybrid solar cells: Bringing conjugated polymers and inorganic nanocrystals in close contact, *J. Phys. Chem. Lett.*, 2013, 4(11), 1788–1796.

92. L. Martinez, S. Higuchi, A. J. MacLachlan, A. Stavrinadis, N. Cates, S. L. Diedenhofen, M.Bernechea, S. Sweetnam, J. Nelson, S. A. Haque, K. Tajima, G. Konstantatos, Improved electronic coupling in hybrid organic–inorganic nanocomposites employing thiol-functionalized P3HT and bismuth sulfide nanocrystals, *Nanoscale*, 2014, 6,10018–10026.

93. K. Yoshida, J-F. Chang, C-C. Chen, T. Higashihara, Thiol-end-functionalized regioregular poly(3-hexylthiophene) for PbS quantum dot dispersions, *ACS Appl. Polym. Mater.* 2021, 3(9), 4450–4459.

94. J. Boucle, S. Chyla, M. S. P. Shaffer, J. R. Durrant, D. D. C. Bradley, J. Nelson, Hybrid solar cells from a blend of poly(3-hexylthiophene) and ligand-capped TiO$_2$ nanorods, *Adv. Fun. Mater.* 2008, 18(4), 622–633.

95. S. Ananthakumar, J. R. Kumar, S. M. Babu, Cesium lead halide (CsPbX$_3$, X=Cl, Br, I) perovskite quantum dots-synthesis, properties, and applications: a review of their present status, *J. Photonics for Energy*, 2016, 6(4), 042001.

96. D. Yang, X. Li, H. Zeng, Surface chemistry of all inorganic halide perovskite nanocrystals: passivation mechanism and stability, *Adv. Mater. Inter.* 2018, 5(8), 1701662.

97. S. R. Smock, Y. Chen, A. J. Rossini, R. L. Brutchey, The surface chemistry and structure of colloidal lead halide perovskite nanocrystals, *Acc. Chem. Res.* 2021, 54(3), 707–718.

98. A. Soosaimanickam, P. J. Rodriguez-Canto, J. P. Martinez-Pastor, R. Abargues, Surface modification of all-inorganic lead halide perovskite nanocrystals, *Nano Tools Devices Enhanced Renew. Energy*, 2021, 61–102.

99. E. M. Sanehira, A. R. Marshall, J. A. Christians, S. P. Harvey, P. N. Ciesielski, L. M. Wheeler, P. Schulz, L. Y. Lin, M. C. Beard, J. M. Luther, Enhanced mobility CsPbI$_3$ quantum dot arrays for record-efficiency, high-voltage photovoltaic cells, *Sci. Adv.*, 2017, 3(10), 1–8.

100. X. Zhang, H. Huang, Y. M. Maung, J. Yuan, W. Ma, Aromatic amine-assisted pseudo-solution-phase ligand exchange in CsPbI$_3$ perovskite quantum dot solar cells, *Chem. Comm.* 2021, 57, 7906–7909.

101. J. Chen, D. Jia, J. Qiu, R. Zhuang, Y. Hua, X. Zhang, Multidentate passivation crosslinking perovskite quantum dots for efficient solar cells, *Nano Energy*, 2022, 96, 107140.

102. Y. Wang, J. Yuan, X. Zhang, X. Ling, B. W. Larson, Q. Zhao, Y. Yang, J. M. Luther, W. Ma, Surface ligand management aided by a secondary amine enables increased synthesis yield of $CsPbI_3$ perovskite quantum dots and high photovoltaic performance, *Adv. Mater.,* 2020, 32(32), 2000449.

103. L. M. Wheeler, N. J. Kramer, U. R. Kortshagen, Thermodynamic driving force in the spontaneous formation of inorganic nanoparticle solutions, *Nano Lett.* 2018, 18(3), 1888–1895.

104. R. Hui, Q. Zhao, J. Su, X. Zhou, X. Ye, X. Liang, J. Li, H. Cai, J. Ni, J. Zhang, Role of methyl acetate in highly reproducible efficient $CsPbI_3$ perovskite quantum dot solar cells, *J. Phys. Chem. C*, 2021, 125(16), 8469–8478.

105. S. Cho, J. Kim, S. M. Jeong, M. J. Ko, J-S. Lee, Y. Kim, High-voltage and green-emitting perovskite quantum dot solar cells via solvent miscibility-induced solid-state ligand exchange, *Chem. Mater.* 2020, 32(20), 8808–8818.

106. D. Jia, J. Chen, J. Qiu, H. Ma, M. Yu, J. Liu, X. Zhang, Tailoring solvent-mediated ligand exchange for $CsPbI_3$ perovskite quantum dot solar cells with efficiency exceeding 16.5%, *Joule*, 2022, 6(7), 1632–1653.

107. D. Jia, J. Chen, M. Yu, J. Liu, E. M. J. Johansson, A. Hagfeldt, X. Zhang, Dual passivation of $CsPbI_3$ perovskite nanocrystals with amino acid ligands for efficient quantum dot solar cells, *Small*, 2020, 16(24), 2001772.

108. S. Lim, G. Lee, S. Han, J. Kim, S. Yun, J. Lim, Y-J. Pu, M. J. Ko, T. Park, J. Choi, Y. Kim, Monodisperse perovskite colloidal quantum dots enable high-efficiency photovoltaics, *ACS EnergyLett.* 2021, 6(6), 2229–2237.

109. J. N. Arenas, A. Soosaimanickam, H. P. Adl, R. Abargues, P. B. Boix, P. J. Rodriguez-Canto, J. P.Martinez-Pastor, Ligand-length modification in $CsPbBr_3$ perovskite nanocrystals and bilayers with PbS quantum Dots for Improved Photodetection Performance, *Nanomaterials*, 2020, 10(7), 1297.

110. Y. Liu, X. Zhao, Z. Yang, Q. Li, W. Wei, B. Hu, W. Chen, $Cu_{12}Sb_4S_{13}$ quantum dots with ligand exchange as hole transport materials in all-inorganic perovskite $CsPbI_3$ quantum dot solar cells, *ACS Appl. Energy Mater.*, 2020, 3(4), 3521–3529.

111. L. Korala, M. B. Braun, J. M. Kephart, Z. Tregillus, A. L. Prieto, Ligand-exchanged CZTS nanocrystal thin films: Does nanocrystal surface passivation effectively improve photovoltaic performance? *Chem. Mater.* 2017, 29(16), 6621- 6629.

112. T. K. Townsend, W. B. Heuer, E. E. Foos, E. Kowalski, W. Yoon, J. G. Tischler, Safer salts for CdTe nanocrystal solution processed solar cells: the dual roles of ligand exchange and grain growth, *J. Mater. Chem. A*, 2015, 3, 13057–13065.

113. Y. Cao, W. Li, Z. Liu, Z. Zhao, Z. Xiao, W. Zi, N. Cheng, Ligand modification of $Cu_2ZnSnS_4$ nanoparticles boosts the performance of low temperature paintable carbon electrode based perovskite solar cells to 17.71%, *J. Mater. Chem. A*, 2020, 8, 12080–12088.

114. J. Embden, A. S. R. Chesman, E. D. Gaspera, N. W. Duffy, S. E. Watkins, J. J. Jasieniak, $Cu_2ZnSnS_{4x}Se_{4(1-x)}$ Solar Cells from Polar Nanocrystal Inks, *J. Am. Chem. Soc.*, 2014, 136, 5237–5240.

115. H. Reinhold, U. Mikolajczak, I. Brand, C. Dosche, H. Borchert, J. Parisi, D. Scheunemann, Shorter is not always better: Analysis of a ligand exchange procedure for $CuInS_2$ nanoparticles as the photovoltaic absorber material, *J. Phys.Chem. C*, 2020, 124(37), 19922–19928.

116. J. Peng, Y. Chen, X. Zhang, A. Dong, Z. Liang, Solid-state ligand-exchange fabrication of $CH_3NH_3PbI_3$ capped PbS quantum dot solar cells, *Adv. Sci.*, 2016, 3(6), 1500432.

117. B. Sun, A. Johnston, C. Xu, M. Wei, Z. Huang, Z. Jiang, H. Zhou, Y. Gao, Y. Dong, O. Ouellette, X. Zheng, J. Liu, M-J. Choi, Y. Gao, S-W. Baek, F. Laquai, O. M. Bakr, D. Ban, O. Voznyy, F. P. G. deArquer, E. H. Sargent, Monolayer perovskite bridges enable strong quantum dot coupling for efficient solar cells, *Joule,* 2020, 4(7), 1542–1556.

118. X. Zhang, Y. Qian, X. Ling, Y. Wang, Y. Zhang, J. Shi, Y. Shi, J. Yuan, W. Ma, α-$CsPbBr_3$ perovskite quantum dots for application in semitransparent photovoltaics, *ACS Appl. Mater. Interfaces*, 2020, 12(24), 27307–27315.

119. N. Sukharevska, D. Bederak, V. M. Goossens, J. Momand, H. Duim, D. N. Dirin, M. V. Kovalenko, B. J. Kooi, M. Loi, Scalable PbS quantum dot solar cell production by blade coating from stable inks, *ACS Appl. Mater. Interfaces*, 2021, 13(4), 5195–5207.

# 6 Tunnel FET-Based Ultra-Low-Power Circuits for Energy Harvesting Applications

*Ramkumar Kannan, Vinodhkumar Nallathambi, Mohanraj Jayavelu, and Valliammai Muthuraman*

## 6.1 INTRODUCTION

Integrated circuit technology has advanced tremendously during the last several decades. Electronic devices have evolved to what we see today because of the scaling of complementary metal-oxide-semiconductor (CMOS) transistors, which has allowed for faster, lower-power, and more complicated processors per unit space. Low power consumption is becoming an increasingly important design measure for analog and digital circuits, as the desire for electronic portability grows [1–2]. Dynamic power consumption reduces quadratically with the scaled power supply voltage, whereas leakage power is limited by the fixed inverse sub-threshold slope (SS) [3].When the inverse sub-threshold slope is fixed, a reduction in power supply leads to an increase in leakage current due to a decrease in threshold voltage with the technology scaling. As a result, traditional thermionic transistors' (MOSFETs and FinFETs)ability to reduce power consumption is limited. At normal temperature, the inverse sub-threshold slope in thermionic-based transistors is restricted to 60 mV/decade [4–5].To alleviate the inverse sub-threshold slope constraint of conventional transistors, devices with different carrier injection processes are needed that are not reliant on the thermal (Boltzmann) distribution of mobile charge carriers. Due to its non-thermal carrier injection process based on the band-to-band tunneling (BTBT) phenomenon, the tunnel field-effect transistor (TFET) is promoted as the most promising post-CMOS technology as it operates with steep sub-threshold slope (SS <60 mV/decade at room temperature) [6]. TFETs have already proven an exceptionally low leakage current, establishing them as serious possibilities for ultra-low power and energy efficient circuit applications.

The utilization of TFETs as an alternative technology for ultra-low power, voltage conversion and management circuits suitable for energy harvesting (EH) sources like rectifiers, charge pumps, power management circuits (PMC), and applications in memory devices were discussed in this chapter.

## 6.2 TUNNEL FET: STATE OF THE ART

The thermionic emission of carriers across a potential barrier generates current in a typical MOSFET. Traditional MOSFETs have one of the most major challenges in advanced technical nodes when it comes to keeping power consumption to a manageable level. As a result of its unique current transport technique, TFETs may be a viable alternative to the existing MOSFETs. This section discusses TFET's basic structure and its operation.

### 6.2.1 DEVICE STRUCTURE

TFETs are essentially gated, reverse-biased PIN transistors. Figure 6.1(a, b) shows an example of how to implement the two types of transistors (*N* and *P*). The drain and source doping of a TFET are

DOI: 10.1201/9781003340539-6

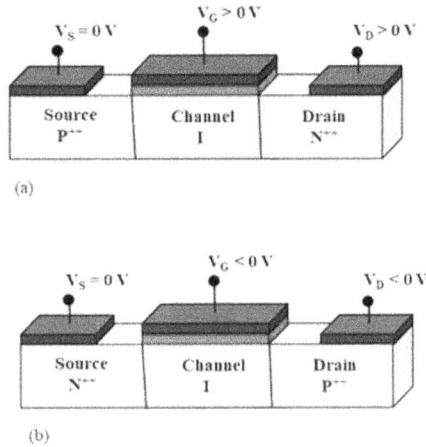

FIGURE 6.1   (a) N-TFET and (b) P-TFET.

the most differentiating structural features. In a TFET, the drain and source are doped differently. In a TFET, the drain and source doping are of completely opposite types, whereas in MOSFET the drain and source doping are identical. The drain of an $n$-type TFET is doped with $n+$, whereas the source is doped with $p+$. The drain of a $p$-type TFET is doped with $p+$, whereas the source is doped with $n+$. The channel is a $p$-type or $n$-type semiconductor that has been intrinsically or minimally doped. Similar to a MOSFET, a dielectric separates the channel from the gate electrode.

Here, the $n$-type and $p$-type biasing methods of the two devices are shown in Figure 6.1. The source of an $n$-type TFET is grounded, while the drain and gate electrodes receive a positive voltage. The source is grounded, and a negative voltage is provided to the drain and gate electrodes in a $p$-type TFET. The dominant carrier in the channel produced under the gate when the TFET is turned on determines whether the TFET is an $n$-type TFET or a $p$-type TFET. When electrons are the majority carriers in the channel, the TFET is referred to as an $n$-type TFET, and when holes are the dominating carriers, the TFET is referred to as a $p$-type TFET. Depending on whether the majority carriers enter or exit the channel through that terminal, the terminals are referred to as source or drain. Electrons enter the channel through the source and exit through the drain in an $n$-type TFET. The holes in a $p$-type TFET enter the channel through the source and exit through the drain. Band-to-band tunneling (BTBT) is the technique through which the dominating carriers enter the channel, and it is explored in depth in the next section.

### 6.2.1.1   Operation

A TFET works by tunneling from one band to another (BTBT) [7–8]. BTBT includes carriers tunneling from the valence band to the conduction band or vice versa via the prohibited bandgap. Figure 6.2 shows the band diagrams of an $n$-type TFET in both the OFF state in Figure6.2(a) and ON state in Figure 6.2(b). The TFET is in the OFFstate when the gate voltage is close to zero. The channel's conduction band is higher than the source's valence band. As a result, BTBT is suppressed, and the TFET is turned off with a very low drain current. When the gate voltage is raised, the carrier density below the gate is modulated, and the conduction band in the channel is lowered. Band bending occurs at the source when a sufficiently high voltage is supplied to the gate, causing the valence band in the source and the conduction band in the channel to align, as seen in Figure 6.2(b). As an outcome, electrons in the source's valence band can tunnel to the channel's conduction band. The positive bias of the drain sweeps the electrons that tunnel through the channel to the drain terminal. An $n$-type TFET operation is based on this principle.

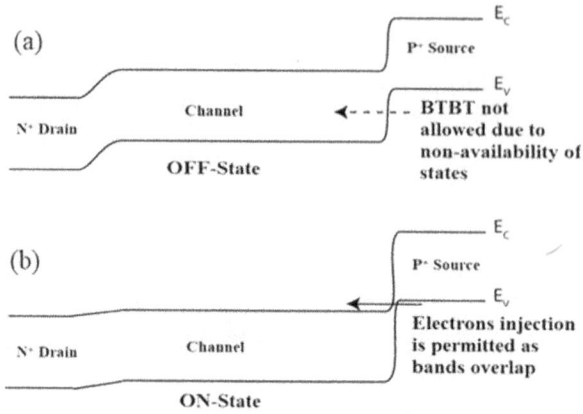

**FIGURE 6.2**  Energy-band of N-TFET (a) OFF state and (b) ONstate.

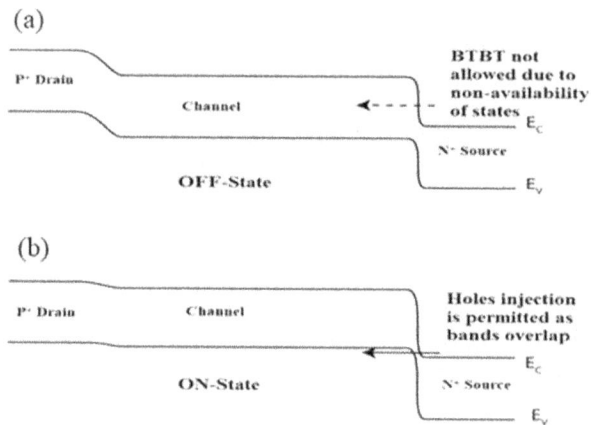

**FIGURE 6.3**  Energy-band of P-TFET (a) OFF state and (b) ON state.

The working principle of a *p*-type TFET is identical to that of an *n*-type TFET. The TFET is in the OFFstate when the gate voltage is close to zero. The source's conduction band is higher than the channel's valence band. As an outcome, BTBT is disabled, and the TFET is turned off as shown in Figure 6.3(a). The valence band in the channel is pushed above the conduction band in the source when a sufficiently enough negative voltage is given to the gate, as seen in Figure 6.3(b). As an outcome, holes are injected into the channel, and the drain's negative bias sweeps the holes to the drain terminal. A *p*-type TFET operation is based on this principle. It is worth noting that, in contrast to a MOSFET, a TFET is an ambipolar device. For instance, when a negative bias is applied to the gate, an *n*-type TFET with a high electron contribution to current transport can display *p*-type behavior with a maximum hole contribution to current transport.

### 6.2.2 TRANSFER CHARACTERISTICS

Figure 6.4 shows the transfer characteristics of double gate TFET (DGTFET) with increasing $V_{GS}$ whose channel length is 50 nm, thickness of 10 nm and gate oxide thickness of 3 nm [9]. The transfer characteristics is divided into three different regions namely offstate, sub-threshold, and super-threshold.

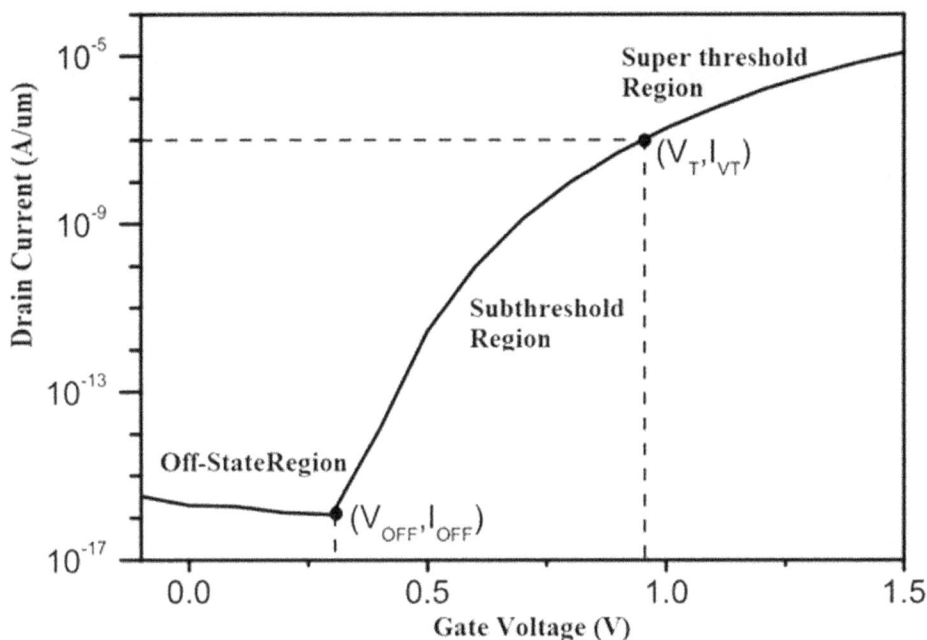

**FIGURE 6.4**  Transfer characteristics of a TFET with increasing $V_{GS}$ and fixed $V_{DS}= 1$ V.

Source: [9].

### 6.2.2.1 OffState

Whenever the gate voltage is less than $V_{OFF}$, the source and channel bands become misaligned. As an outcome, BTBT between the source and channel is prevented, and the TFET's tunneling current is insignificant. Until the gate voltage hits $V_{OFF} = 0.3$ V, the drain current does not begin to increase. The drain current at $V_{GS} = V_{OFF}$ is referred to as the TFET's OFFcurrent and is indicated in this chapter by $I_{OFF}$. The $I_{OFF}$ of the TFET is in the femtoampere range. It is worth noting that when the drain voltage is low, a greater gate voltage may be necessary to align the bands and allow BTBT. As a result, at lower drain voltages, a larger $V_{OFF}$ may be observed [8].

### 6.2.2.2 Sub-Threshold Region

When the gate voltage exceeds off state voltage (i.e. $V_{GS}>V_{OFF}$), the drain current starts to increase super-linearly. The conduction and valance bands at the source and channel junctions are aligned and the BTBT is activated. As a result, the tunneling current begins to flow in this operation zone. The drain current increases rapidly near $V_{GS} = V_{OFF}$ and then progressively declines as the gate voltage decreases. The sub-threshold slope or its counterpart sub-threshold swing quantifies the rate at which the drain current increases as the gate voltage increases. The sub-threshold swing is defined as the change in gate voltage necessary to increase the drain current by a factor of ten, and is expressed in millivolts per decade. A transistor with a least sub-threshold swing is ideal. When a transistor has a small sub-threshold swing, it is possible to boost the drain current from the OFFstate to the ONstate by applying a very low gate voltage, which makes the transistor ideal for operation at low power supply voltage.

### 6.2.2.3 Super-Threshold Region

When gate voltage increases beyond the threshold voltage ($V_T$), the drain current keeps on increasing at a reduced rate. The drain current at $V_{GS}= V_{DS}= V_{DD}$, is known as the ONcurrent ($I_{ON}$) of theTFET.

The TFET has a very low $I_{ON}$. By and large, the $I_{ON}$ produced in a TFET is much less than that required for future low-power applications due to the existing transport method. This is the most significant disadvantage of TFETs, and researchers are currently investigating strategies for increasing the $I_{ON}$ in TFETs.

## 6.3 IMPACT OF DEVICE PARAMETERS ON ELECTRICAL CHARACTERISTICS

This section explores the impact of various device parameters on TFET's electrical characteristics. The gate-source and drain-source voltages, as well as the material and device characteristics, determine the tunneling current in a TFET. An approximate link between tunneling current and device parameters for TFETs is as follows:

$$I_D \propto \exp\left[-\frac{4\sqrt{2m^*}E_g^{*\frac{3}{2}}}{3|e|\hbar\left(E_g^* + \Delta\Phi\right)}\sqrt{\frac{\varepsilon_{Si}}{\varepsilon_{ox}}}t_{ox}t_{Si}\right]\Delta\Phi \tag{6.1}$$

$E_g$ is the material's bandgap, $m$ is its effective carrier mass, $t_{Si}$, $t_{ox}$, $\varepsilon_{ox}$ and $\varepsilon_{Si}$ are the thicknesses and dielectric constants of oxide and silicon films, respectively, $e$ is the charge of an electron, and $\hbar$ is the reduced Planck constant. The below mentioned strategies can be used to improve the tunneling current, according to Equation (6.1):

1. Reduce the gate oxide thickness ($t_{ox}$) or increase the dielectric constant ($\varepsilon_{ox}$) of the gate oxide.
2. Reduce the thickness of the silicon body ($t_{Si}$).
3. Increase the source doping concentration.
4. Choose materials with lower effective carrier mass ($m^*$) and smaller bandgap ($E_g$).

### 6.3.1 GATE DIELECTRIC

We may infer from Equation(1) that the drain current should increase as the dielectric constant of the gate material increases or as the gate dielectric thickness decreases. Due to the fact that the gate oxide's thickness and dielectric constant are exponentially correlated to the tunneling current, the $I_{ON}$ improves superlinearly in a TFET as a result of gate dielectric engineering. It is worth noting that $I_{ON}$ grows linearly with reducing the gate oxide thickness or increasing the dielectric constant of the gate material in a standard MOSFET. As a result, TFETs are projected to exhibit a higher improvement in the $I_{ON}$ owing to gate dielectric engineering than traditional MOSFETs employing the same approach. By using high-k dielectric based TFET as shown in Figure 6.5 tighter gate control could be achieved or by lowering the gate dielectric thickness resulted in an increase in the TFET's $I_{ON}$ [9–16]. For example, it has been discovered that decreasing the gate oxide thickness from 4.5 to 3.5 nm resulted in a six-fold increase in drain current [11].

### 6.3.2 BODY THICKNESS

The influence of silicon body thickness on drain current is determined by two antagonistic effects [10]. With increasing silicon body thickness, the amount of silicon accessible for BTBT rises, resulting in an increase in drain current. However, when the silicon body thickness grows, the coupling between the gate and the channel deteriorates, resulting in a drop in drain current. As a consequence, given a TFET, an optimal body thickness that maximizes drain current may be determined. The drain current of the double gate TFET (DGTFET) is higher when the silicon body thickness is between 10 and 20 nm [12]. It should be emphasized that the sub-thsreshold swing is also heavily

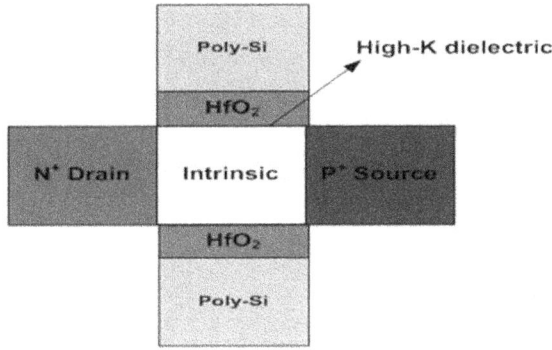

**FIGURE 6.5**   Conventional double gate tunnel FET with high-*k* gate dielectric.

**Source:** [9].

reliant on the TFET body thickness: when the TFET body thickness is lowered, the TFET's sub-threshold swing improves [17]. It has been demonstrated via atomistic modeling that a sub-threshold swing of less than 60 mV/decade can be produced in a single-gate InAs TFET with a body thickness of less than 4 nm and in a double-gate InAs TFET with a body thickness of less than 7 nm [17].

### 6.3.3   SOURCE DOPING PROFILE

When the source doping concentration is increased, the bandgap narrows, resulting in a shorter tunneling distance at the source-channel junction [12, 18]. Experiment findings demonstrate that doubling the source doping concentration increases the $I_{ON}$ of a TFET by a factor of two [12]. The threshold voltage is projected to decrease for high source doping concentrations due to the changed built-in potential at the source–channel junction [12]. It should be noted that increasing the doping concentration of the source causes an increase in the Fermi level into the conduction band, which diminishes the filtering impact on the Fermi tail in the TFET and worsens the sub-threshold swing [19].The tunneling current is also affected by the asymmetry of the source doping profile. The electric field at the source–channel junction increases as the abruptness of the source doping increases, resulting in an increase in tunneling current. Even with a non-abrupt source doping profile, $I_{ON}$ can be increased and sub-threshold swing can be reduced by selecting an optimal gate-source overlap [12].

### 6.3.4   CHANNEL LENGTH

In contrast to a normal MOSFET, the drain current of a TFET is not affected by channel length [20]. Because the tunneling current is mostly determined by the electric field and the band alignment at the source–channel junction, the tunneling current does not vary much as the channel length decreases. When the channel length is less than a critical channel length $L_{crit}$, however, direct source-to-drain leakage takes over. As a result, when the channel length is less than $L_{crit}$ the $I_{OFF}$ increases noticeably. Furthermore, when the channel length is shorter than $L_{crit}$, the $V_T$ of the TFET falls, as does the sub-threshold swing of the TFET [20–22]. For Si-based TFETs, the $L_{crit}$ is about 20 nm [20, 22]. It should be noted that for TFETs designed with materials having a lower bandgap, the effect of channel length variation on drain current may be more noticeable.

### 6.3.5   MULTI GATE STRUCTURES

MOSFET's fundamental purpose for adopting multiple gate architectures like double gate, FinFETs, or gate-all-around is to achieve tighter control over the channel potential and to alleviate short

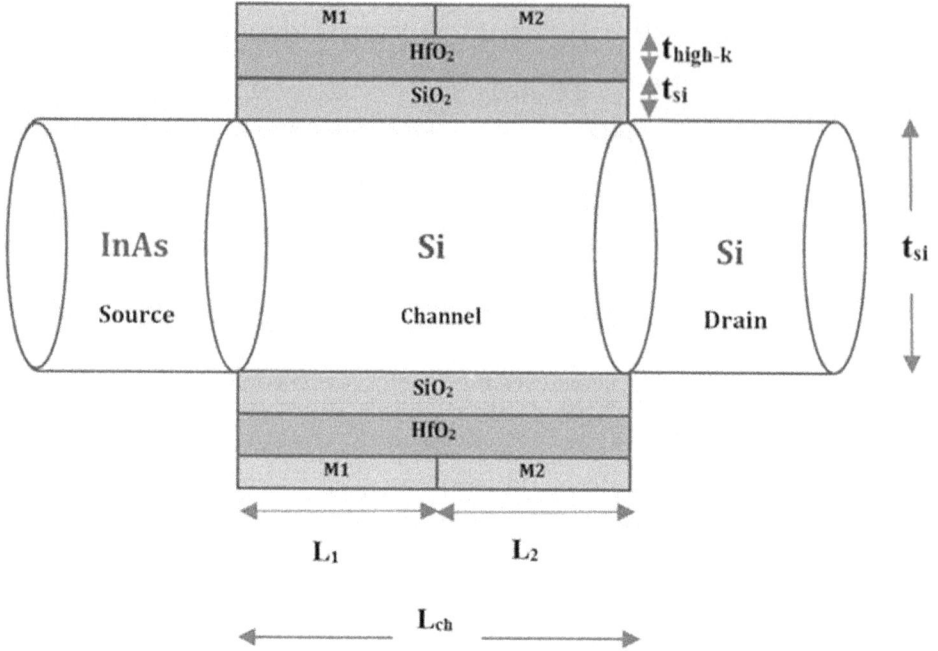

**FIGURE 6.6**   InAs-Si gate-all-around tunnel FET.

**Source:** [25].

channel effects. An increase in effective tunneling area and an improvement in $I_{ON}$ are two benefits of using multiple gate configurations in a tunneling field-effect transistor (TFET). The tunneling area of a TFET is typically located near the source section of the channel. All of the gates come together to produce the channels when using multiple gate setups. As a result, with TFETs with numerous gates, the effective tunneling area grows. The source–channel junction of the DGTFET has two points where tunneling can occur: one beneath each of the two gates. The combined impact of the two gates on the potential over the whole DGTFET body causes an $I_{ON}$ increase of nearly two fold for the DGTFET with a thin body as compared to using a single gate [7]. The engineering gate oxide thickness has less of an influence on a double gate or gate-all-around TFET than a single gate TFET since the gate control over the channel is already tighter in these devices [23]. In addition, a wrap-around gate structure in a TFET has been experimentally proven [24]. By using hetero gate all around structure as shown in Figure 6.6 we can increase the electrical field at the channel edges, the drain current is greatly increased by the wrap-around gate configuration [25].

## 6.4   TUNNEL FET-BASED CHARGE PUMPS

For energy harvesting applications, the performance of charge pumps based on tunnel FETs and FinFETs is examined and compared in this section. The performance of typical charge-pump topologies using TFETs is demonstrated to diminish with increasing power supply voltage and decreasing output current due to the specific electrical properties of TFETs under reverse bias. It is feasible to reduce reverse losses in TFET converters by changing the amplitude of the gate bias of TFETs that are reverse biased. Different circuit topologies have been proposed for TFET-based charge pumps in order to increase their efficiency across a wider range of voltage and load conditions.

The research in power supply circuits that may harvest energy from the environment has grown in recent years [26–29]. Micro-photovoltaic cells (MPV) [26–27] and thermo-electric generators

(TEG) [28–29] are examples of energy harvesting (EH) sources whose output voltage and power are too low for electronic applications. Chargepumps (also known as switched-capacitor converters) are extensively used to enhance the output voltage of EH sources to satisfy the minimal supply voltage constraints of electronic systems. Obtaining good power conversion performance at low voltages (sub-0.25 V) and low powers (sub-W) is difficult due to the large conduction losses of conventional thermionic transistors used in the conversion process and the reverse losses while the transistors are in their offstate [30–31]. Thermionic emission based transistors, like FinFETs, have SS of 60 mV/dec at normal temperature. This property restricts the needed current at low voltage in passive DC-DC converters, increasing forward losses in onstate transistors.

TFET devices (especially heterojunction devices) have better electrical properties than thermionic devices at sub-0.25 V. TFETs, on the other hand, conduct less current at higher levels, making them suitable for low voltage, low-performance applications. So TFETs may be used in power conversion circuits with very low voltages. For example, in [32], TFET-based charge-pump converters outperform FinFET-based converters at sub-0.4 V levels. It is encouraging to see that low voltage energy harvesting may be used in areas where batteries are presently required.

## 6.4.1  CHALLENGES IN CHARGE PUMPS

According to Figure 6.7(a), one may comprehend the limits of TFETs using the gate cross-coupled topology (GCCCP). In comparison to alternative charge-pump topologies [30], this one has been proved to offer the highest performance at low voltage operation; hence it is referred to as the conventional charge pump topology here. The GCCCP converter's operating principle may be broken down into two halves, as seen in Figure 6.7(c). When clock 1 moves from its low to high state, it raises the voltage at node Int1 to $2V_{DD}$-$V_{DS1}$. Node Int2's voltage is lowered to $V_{DD}$-$V_{DS2}$ at the same time. Transistors M1 and M4 are reverse biased in this region, but transistors M2 and M3 function in the onstate. Table 6.1 shows the bias parameters of TFETs in the region-I of operation under steady state circumstances.

In the second period of operation, the high to low (low to high) transition of clock 1 (clock 2) results in forward bias of transistors M1 and M4 and reverse bias of transistors M2 and M3. Table 6.2 shows the bias parameters of TFETs operating in the CGCCP's region-II.

Due to ambipolar conduction, in $n$-TFETs, the reverse current is produced by a reverse BTBT carrier mechanism at the channel-drain interface, as seen in Figure 6.7 (a). With increasing reverse bias (VDS<< 0 V), the BTBT mechanism is reduced, and the reverse current is characterized by excess and drift-diffusion, as illustrated in Figure 6.7(b). $p$-TFETs and positive VDS behave similarly.Due to the fact that the transistors used in charge pumps operate at both forward (onstate) and reverse (offstate) bias throughout successive time periods, it is critical to minimize the reverse current produced by TFETs when they are in their offstate condition.

## 6.4.2  TFET-BASED CHARGE PUMP: SOLUTION

A tunnel FET-based charge pump is shown in Figure 6.8. The auxiliary transistor M1aux and capacitor C1aux are used to bias the gates of the main TFET transistors M1 and M2, while an auxiliary inverter (M2aux and M3aux) and capacitor C2aux are used to bias the gates of the secondary TFET transistors M3 and M4. When Vint1 > Vint2, the auxiliary transistor M1aux and the auxiliary inverter are both turned on, which charges the capacitors at nodes int2* and node 1* to the voltage levels of their respective nodes int1($\approx 2V_{DD}$) and int2 ($\approx V_{DD}$). To put it another way: the VGS is below zero for the transistors M1 and M4 throughout this time period; for the forward biased transistors M2 and M3, it is above zero and below zero for them.

The voltage levels at nodes int1* and int2* are retained while the auxiliary transistors are in the reverse biased (offstate) condition. As a result, the transistors M1 and M4 are forward biased with

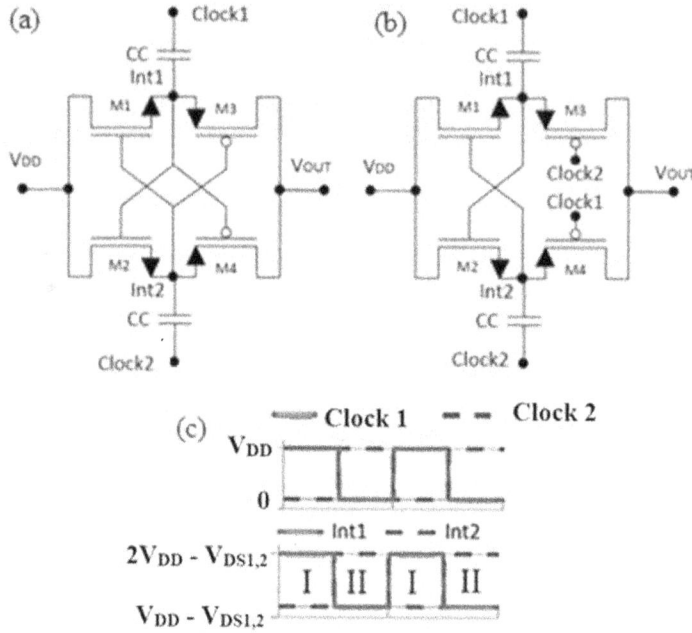

**FIGURE 6.7** (a) Conventional charge pump topology; (b) state-of-the-art (SOA) TFET-based charge-pump and (c) regions of operation.

Source: [32].

**TABLE 6.1**
**Bias Conditions of the TFETs Applied in the GCCCP Considering Region I**

| Reg. I | $V_{GS}$ | $V_{DS}$ | State |
|---|---|---|---|
| M1 (n) | $Int2 - Int1 = -V_{DD}$ | $V_{DD} - Int1 = -V_{DD} + V_{DS1}$ | OFF |
| M2 (n) | $Int1 - Int2 = V_{DD}$ | $V_{DD} - Int2 = V_{DS}2$ | ON |
| M3 (p) | $Int2 - Int1 = -V_{DD}$ | $V_{OUT} - Int1 = -V_{SD3}$ | ON |
| M4 (p) | $Int1 - Int2 = V_{DD}$ | $V_{OUT} - Int2 = V_{DD} - V_{SD3}$ | OFF |

**TABLE 6.2**
**Bias Conditions of the TFETs Applied in the GCCCP Considering Region II**

| Reg. II | $V_{GS}$ | $V_{DS}$ | State |
|---|---|---|---|
| M1 (n) | $Int2 - Int1 = V_{DD}$ | $V_{DD} - Int1 = V_{DS1}$ | ON |
| M2 (n) | $Int1 - Int2 = -V_{DD}$ | $V_{DD} - Int2 = -V_{DD} + V_{DS2}$ | OFF |
| M3 (p) | $Int2 - Int1 = V_{DD}$ | $V_{OUT} - Int1 = V_{DD} - V_{SD4}$ | OFF |
| M4 (p) | $Int1 - Int2 = -V_{DD}$ | $V_{OUT} - Int2 = -V_{SD3}$ | ON |

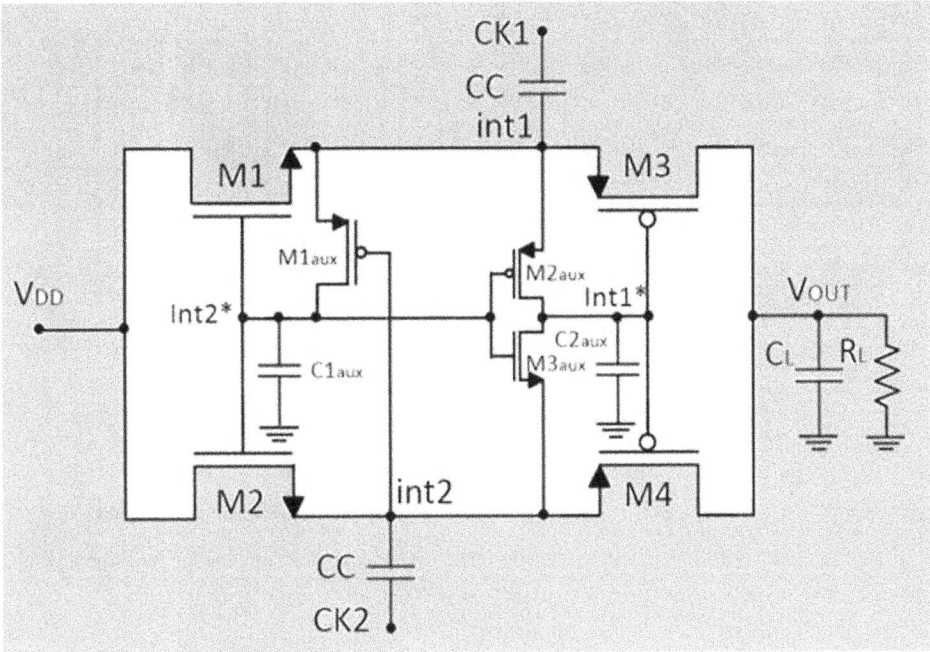

**FIGURE 6.8**  Tunnel FET-based charge pump.

Source: [32].

---

**TABLE 6.3**
**Gate Bias of Tunnel FET-based Charge Pump**

| $V_{GS}$ Region I | | $V_{GS}$ Region II | |
|---|---|---|---|
| M1 ($n$) | **Int2\*–Int1 ≈ 0** | Int2\*–Int1 ≈$V_{DD}$ | |
| M2 ($n$) | Int2\*–Int2 ≈$V_{DD}$ | Int2\*–Int2 ≈$V_{DD}$ | |
| M3 ($p$) | Int1\*–Int1 ≈$-V_{DD}$ | **Int1\*–Int1 ≈ 0** | |
| M4 ($p$) | **Int1\*–Int2 ≈ 0** | Int1\*–Int2 ≈$-V_{DD}$ | |

---

$V_{GS} \approx 0$ V and $V_{GS} \approx -V_{DD}$, respectively, whereas transistors M2 and M3 are reverse biased. Table 6.3 shows the $V_{GS}$ values for the major transistors. Figure 6.9 depicts the transient behavior of the internal nodes within the charge-pump stage at an operating frequency of 100 MHz and a power supply voltage of 160 mV to help clarify how the modified TFET-based charge-pump operation works.

### 6.4.3  Performance of TFET Based Charge Pumps

Two clock frequencies, 1 kHz and 100 MHz, were used in this work to compare the performance of the suggested (HTFET Prop. CP), conventional (HTFET GCCCP) and state-of-the-art (HTFET SOA CP) charge-pump topologies developed using heterojunction TFETs and is shown in Figures 6.10 and 6.11. The performance of the traditional charge pump ($N_{fins}$=14) is presented for comparison purposes. $\Delta K$=2 K ($V_{DD}$=160 mV) and $\Delta K$=6 K ($V_{DD}$=480 mV) are two separate thermogenerator temperature fluctuations that are taken into account. Low supply voltage $V_{DD}$ and the needed output currents show that the FinFET-based charge pump performs poorer than its TFET-based equivalents

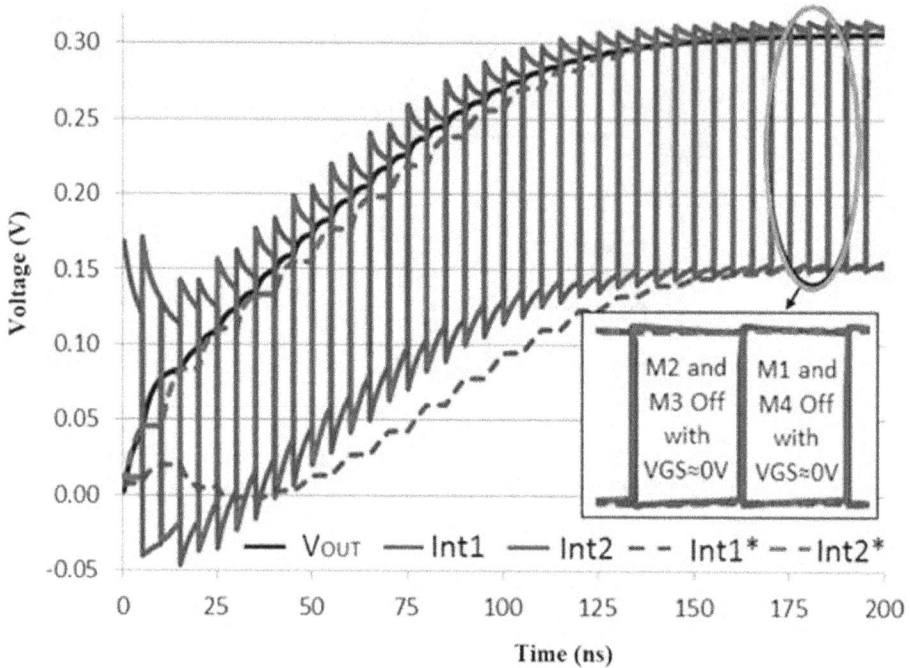

**FIGURE 6.9**   Transient behavior of the tunnel FET-based charge pump.

**Source: [32].**

(at both operating frequencies). This may be observed (due to larger reverse losses of FinFETs in the voltage range considered. The FinFET-based chargepump, on the other hand, performs better at higher voltages ($V_{DD}$=480 mV or more) and higher current demands. Conventional thermionic devices outperform TFETs in terms of driving current at high voltage.

However, TFET-based charge pumps operate better at low voltage and low current. At $V_{DD}$= 160 mV, the HTFET-GCCCP converter has the highest PCE values for both frequencies of operation. Instead, the suggested charge pump is more efficient at higher power supply voltages ($V_{DD}$=480 mV) and lower power consumptions (less than 10 µW). This is directly connected to the proposed charge-pump topology's lowering of reverse losses when the TFETs are in a high reverse bias condition. Due to the auxiliary circuitry's switching losses during clock transitions, the suggested charge-pump design suffers a slight deterioration at low voltage (sub-0.4 V) and low current (sub-µA) .When the output transistors M3 and M4 are subjected to forward bias, the reduced conduction losses in the HTFET SOA CP generate the highest output voltages and power efficiency at low voltages (below 160 mV). However, when the needed output current is low, the switching losses created by the output transistors in this charge-pump design decrease the stage's power conversion efficiency (more than the TFET-based CP equivalents due to high $V_{GS}$).

Figure 6.12 shows the losses in each charge-pump stage for a 1 A load current and the two clock frequencies under consideration for the distribution of power losses. For a minor temperature differential between the plates of the thermo-generator source ($\Delta K$=2 K, $V_{DD}$=160 mV) and for both clock frequencies, the conventional and suggested chargepumps enable the maximum power to the load. For all frequencies of operation, the suggested charge-pump offers the maximum power to the load with a higher temperature variation ($\Delta K$=6 K, VDD=480 mV) (output power). For example, as compared to HTFET-based charge pumps, reverse losses are significantly reduced. The losses

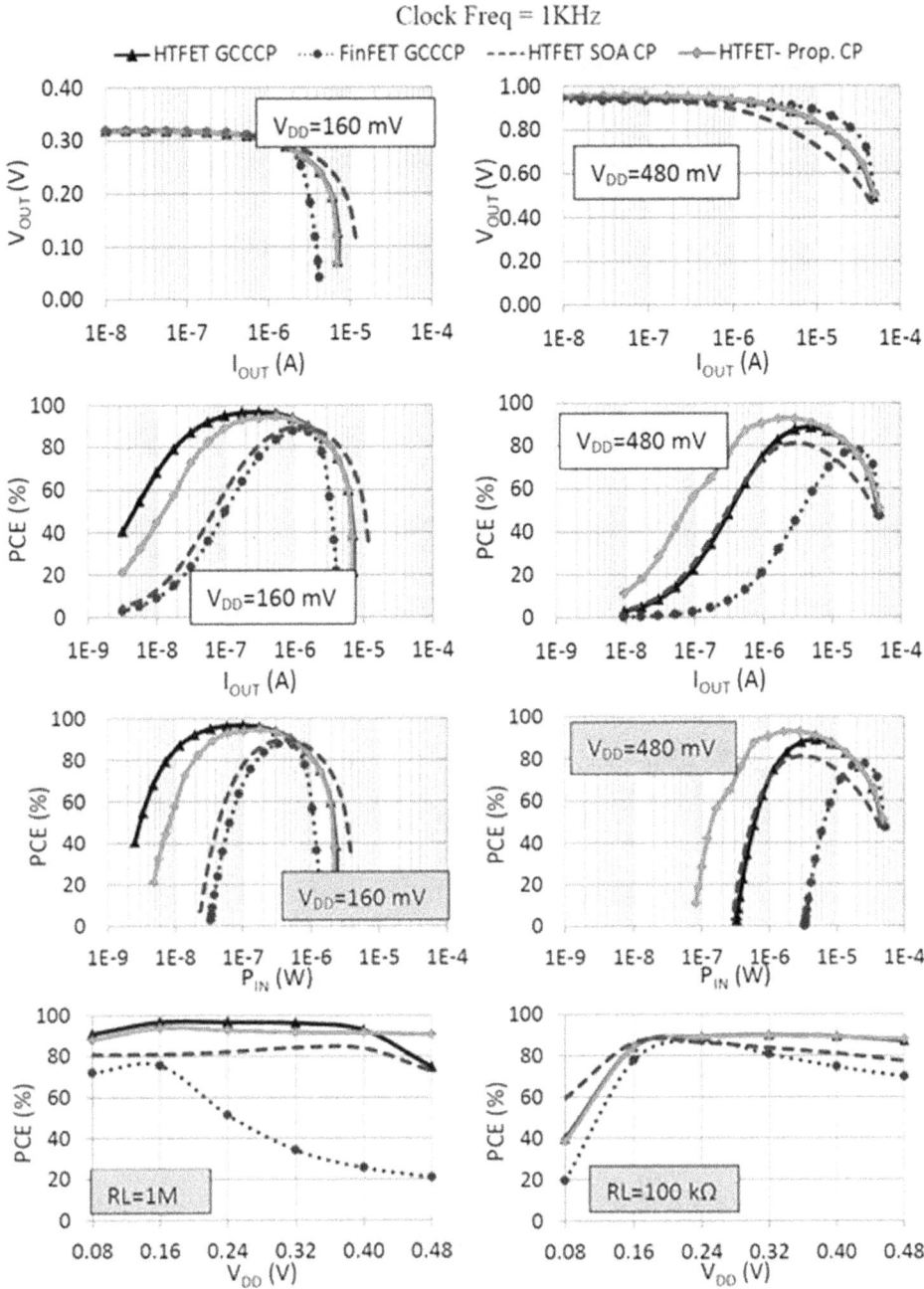

**FIGURE 6.10** Performance comparison of charge pumps at a clock frequency of 1 KHz.

**Source:** [32].

produced by the auxiliary circuitry are shown to grow at a frequency of 100 MHz, lowering the PCE of the converter.

Charge-pump topologies for different power source voltages are summarized in Table 6.4. With an input voltage of more than 400 mV and a power consumption of less than 10 μW, the HTFET-based charge pump is proved to be a viable design option.

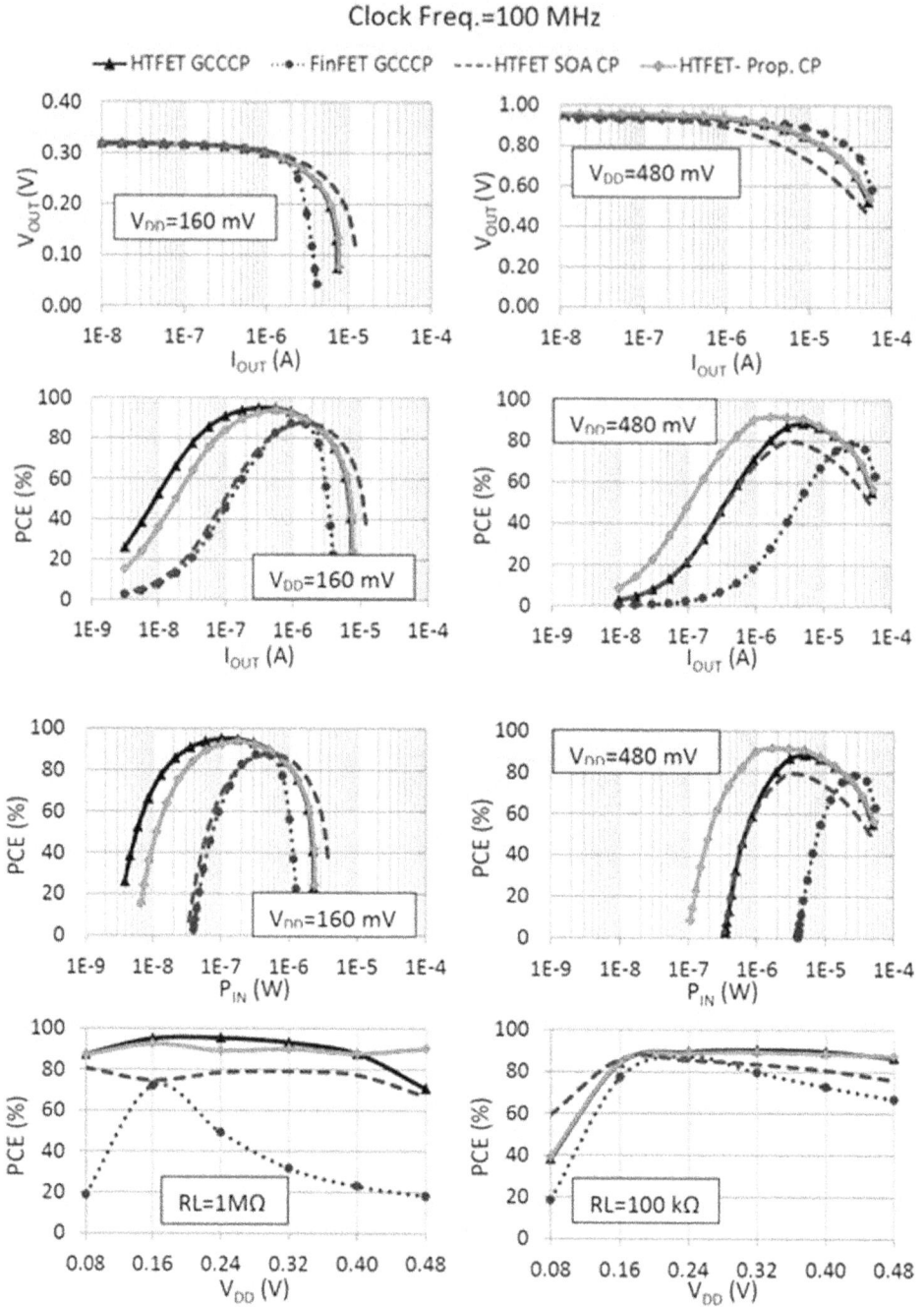

**FIGURE 6.11** Performance comparison of TFET-based charge pumps at a clock frequency of 100 MHz. Source: [32].

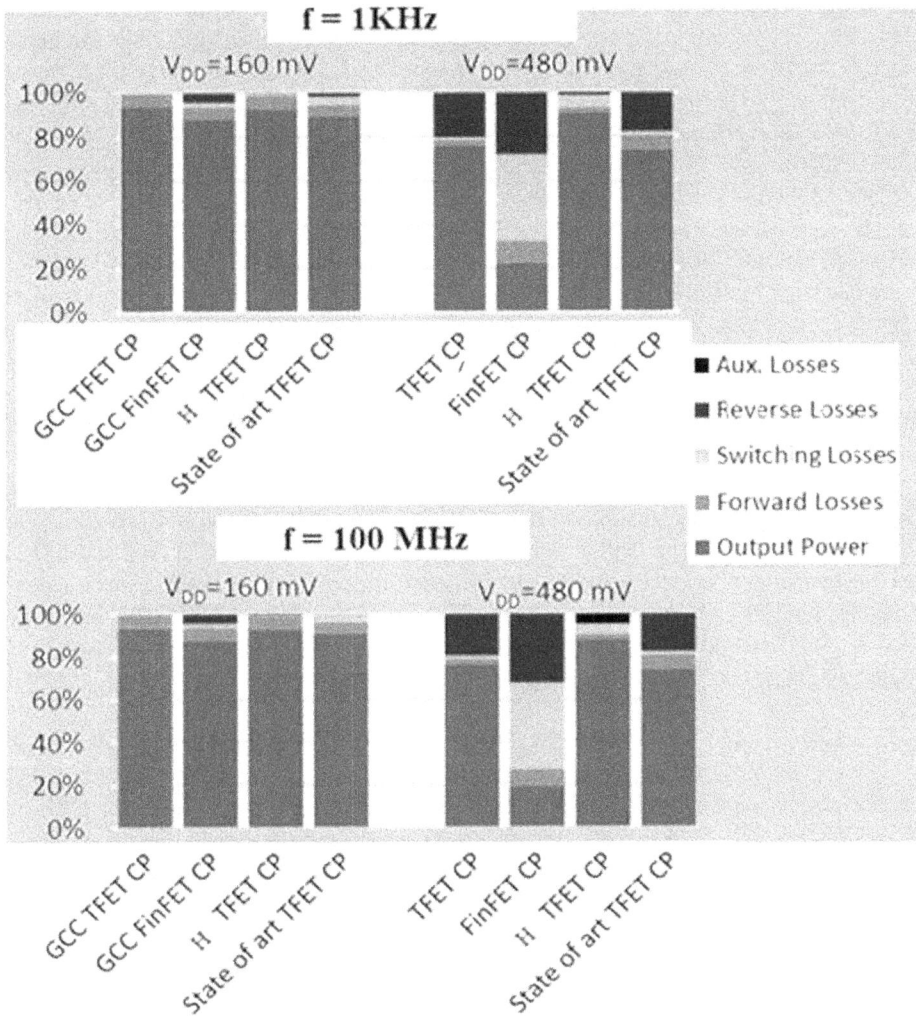

**FIGURE 6.12**   Charge pumps power losses at 1 kHz and 100 MHz.

Source: [32].

## 6.5   TUNNEL FET-BASED RECTIFIERS

**TABLE 6.4**
**Charge-pump Topologies for Different Voltage Range**

|  | Ultra-low voltage | | Lowvoltage | | Medium voltage | |
|---|---|---|---|---|---|---|
| Voltage range | < 160 mV | | 160–480 mV | | > 400 mV | |
| Pin | < 1μW | > 1μW | < 1μW | > 1μW | < 10μW | > 10μW |
| Suitable topology | HTFET-GCCCP | HTFET-SOA CP | HTFET-GCCCP | HTFET-GCCCP | HTFET. Prop. CP | HTFET. Prop. CP |

For ultra-low power applications, the effectiveness of tunnel FET-based rectifiers is examined and compared to that of conventional thermionic device-based rectifiers under similar bias circumstances in this section. Different rectifier topologies are developed and evaluated in order to offset the reverse current conducted by reverse biased TFETs.Several low-power applications can use the radiated energy to power their circuits to profit from the surroundings. When it comes to radio-frequency (RF) power, RFID tags and biomedical implants are two examples of circuits that may be placed in regions that are tough to access. Recent studies have revealed that the area of energy harvesting from ambient has acquired relevance as a result of the continual replacement of their batteries [33–35].

RF-powered circuits suffer from low efficiency at low RF input power levels (sub-μW). This is a direct result of the front-end rectifier's low efficiency at low voltages. The performance of conventional rectifiers decreases when the generated RF voltage at the rectifier input terminals decreases [36–40]. The electrical performance of TFETs in low-power/low-voltage rectifiers is an area of interest because of their increased electrical performance at sub-0.25 V. In [41], it shows that TFET-based passive rectifiers have better rectification efficiency at sub-30 dBm than FinFET-based rectifiers through simulations. Despite the benefits of employing TFETs in rectifiers, it is not always acceptable to use TFET technology in place of traditional thermionic transistors. The electrical properties of TFETs under reverse bias have been discussed in the preceding chapter. Rectifiers' performance can be degraded by reverse losses caused by the BTBT and DD carrier injection mechanisms at low and high reverse bias, respectively. There will always be a second carrier mechanism, but by altering the gate magnitude of the reverse biased TFETs, we may reduce or even eliminate the reversal current that results from the first carrier mechanism. Consequently, in this chapter, we propose and test a TFET-based rectifier that can operate at a wide range of voltages and powers.

The gate cross-coupled rectifier (GCCR) is simple to install and provides good results at low-voltage/power levels [38–40]. For example, the [41] demonstrated that a GaSb-InAs-based GCCR had superior power conversion efficiency (PCE) than other rectifier topologies, which had a PCE of between–40 dBm and–25 dBm, in compared to other rectifier designs. Individual rectifier stage devices can be enhanced by minimizing their reverse losses during their offstate situations, notwithstanding their high performance at low-power operation (reverse biased state).Assuming the RF signal has a sinusoidal pattern, the GCCR operation may be separated into two regions: region I, where the voltage at the node is greater than the voltage at the node, and region II, when the reverse condition holds true. As discussed, the charge-pump architecture, the reverse, conduction, and switching power losses are shown as the primary losses in the rectification process.

The performance of the GCCR with heterojunction TFETs (InAS-GaSb, LG=40 nm) is compared at two different operating frequencies and loads (RL=100 and RL=10 k). This was done for two different operating frequencies and loads. One-micrometer-wide channels are used to simulate TFETs T1 through T4. There are 1pF (10pF) coupling capacitors and 10pF (100pF) load capacitors for a frequency of 915MHz (100MHz). One can see that a load of 100 kΩ allows for far higher PCE values at sub-μW power levels than a load of 10 kΩ does for both frequencies under investigation. It has been shown that the highest efficiency for a 100 kΩ load is around the RF $V_{AC}$ of 0.2 V. PCE is reduced at higher voltages not only because of the increased conduction losses caused by transistors working in the onstate, but also because of the reverse current conduction caused by partially closed transistors running in the off-state.

## 6.5.1 Tunnel FETs in Rectifiers: Advantages

TFET-based rectifiers may operate at ultra-low power (sub-μW) due to the superior electrical properties of tunneling devices at low-voltage operation compared to traditional thermionic devices. TFETs in rectifiers with less reverse and conduction losses and consequently larger PCE at low induced RF voltage have lower reverse current and increased drive current at sub-0.25 V, as will be shown in the following sections, when compared to thermionic devices. For the majority of the period cycle, each GCCR-stage TFET device exhibits an offstate situation. Throughout the time interval t4 to t5, the device T1 is forward biased (onstate), whereas during the rest of the interval, it is reverse biased ($V_{DS}$ is zero). $V_{GS}$= 0 V is reversed throughout the time intervals between time intervals t0–t3 and during time intervals 3–6; during these periods the tunneling device T1 exhibits a positive VGS. An *n*-type tunneling device with a negative differential resistance (NDR) having a non-monotonic reverse current, i.e., the reverse current at low reverse bias increases and subsequently falls at high reverse bias, thereby describing the TFET with the NDR range.

In contrast, the reverse current in thermoelectric grows in amplitude when the reverse bias is increased. The reverse current of a thermionic *n*-FinFET is significantly bigger than that of a heterojunction *n*-TFET. When this characteristic is present, high switching losses (as they are referred to in this context) are incurred in FinFET-based rectifiers. In Figure 6.13, a different topology of tunnel FET-based rectifier is given to avoid the large reverse losses in rectifiers. In this topology, the gate of the transistor T1 & T3 is biased with RF$^+$ when RF$^+$ > RF$^-$ and RF$^-$ otherwise. Similarly the gate of the transistor T2 and T4 is biased with RF$^-$ when RF$^+$ > RF$^-$ and RF$^+$ otherwise. Figure 6.14 shows the active transistors during the first and second regions of operation. The transient response of the TFET-based rectifier is provided in Figure 6.15.

It is shown in Figure 6.16 that the suggested rectifier, which uses GaSb-InAs TFETs and has an LG=40 nm, performs better than the GCCR. The performance of a FinFET-based GCCR is presented for comparison purposes (FinFETs with NFINs=14 and LG=20 nm). There is just one rectifier step examined in both models. For the two frequencies under consideration, the FinFET-based rectifier has the lowest power conversion efficiency at sub-40dBm. For RF VAC magnitudes above 0.3 V, it indicates the maximum output voltage values. Because FinFETs carry a greater current above 0.25 V than heterojunction TFETs, this property may be explained. Using a typical HTFET-GCCR at below–25 dBm and 915 MHz results in the highest power efficiency. This HTFET-efficiency rectifier's degrades at the same RF power levels as the planned HTFET-converter (RF VAC of between 0.2 and 0.6 V) due to increasing losses in the primary transistors (*n*-TFETs placed in an NDR region). Similar results may be achieved at lower frequencies with both HTFET-based rectifiers.

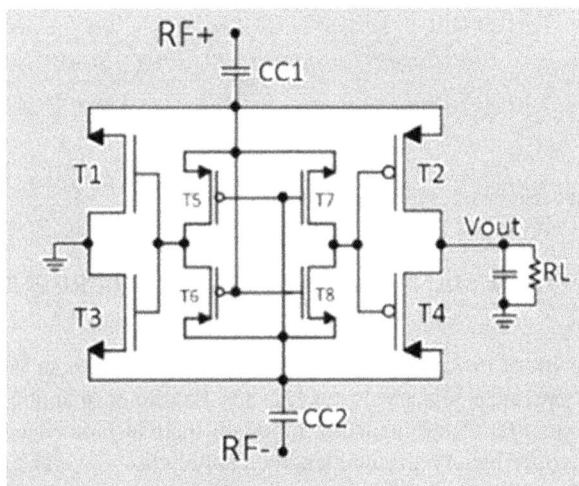

**FIGURE 6.13**   Tunnel FET-based rectifier.

Source: [38].

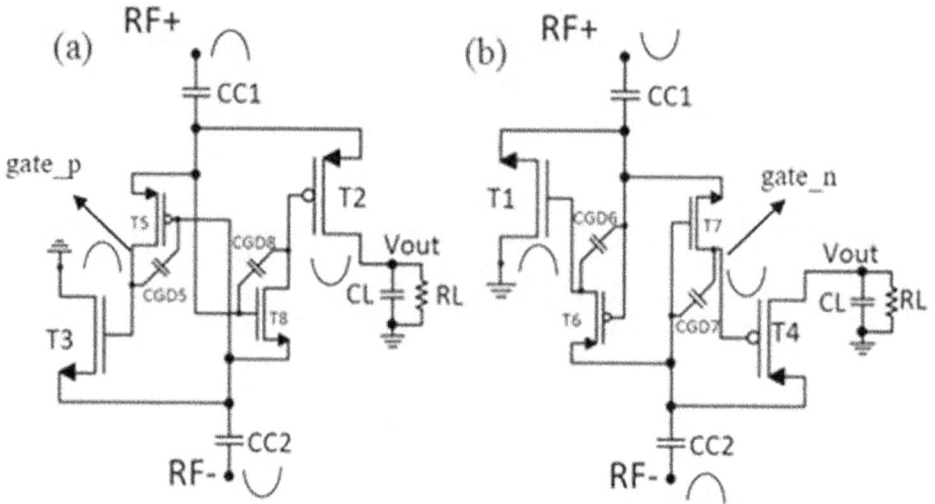

**FIGURE 6.14** Transistors (active) in (a) first and (b) second regions of operation.
Source: [38].

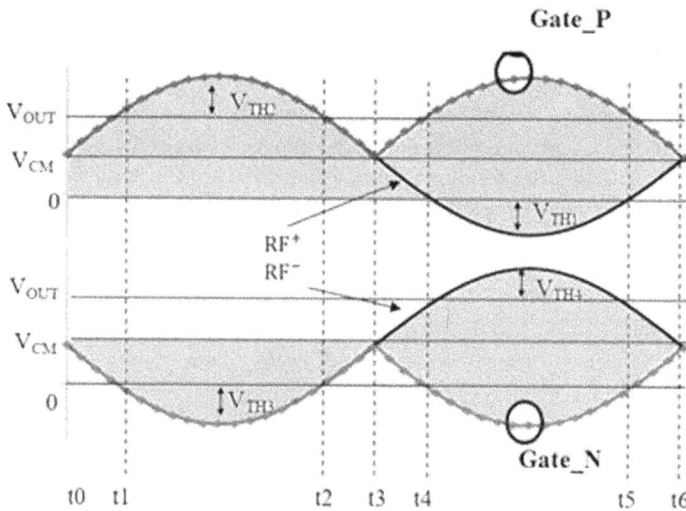

**FIGURE 6.15** Transient behavior of TFET-based rectifier.

## 6.6 TFET-BASED POWER MANAGEMENT CIRCUIT FOR RF ENERGY HARVESTING

In recent years, the growth of low-power embedded systems has led to an increase in the creation of efficient circuits that operate at lower voltages [42–43]. Examples of applications that can benefit from collecting environmental electromagnetic radiation include biomedical implants and wearable devices. This can reduce battery size and lengthen battery life. Several studies have previously shown that a load may be powered wirelessly over short distances using ultra-high frequency (UHF) radiation at levels of power that can be legallytransmitted [44–52]. As a result, RF energy harvesters

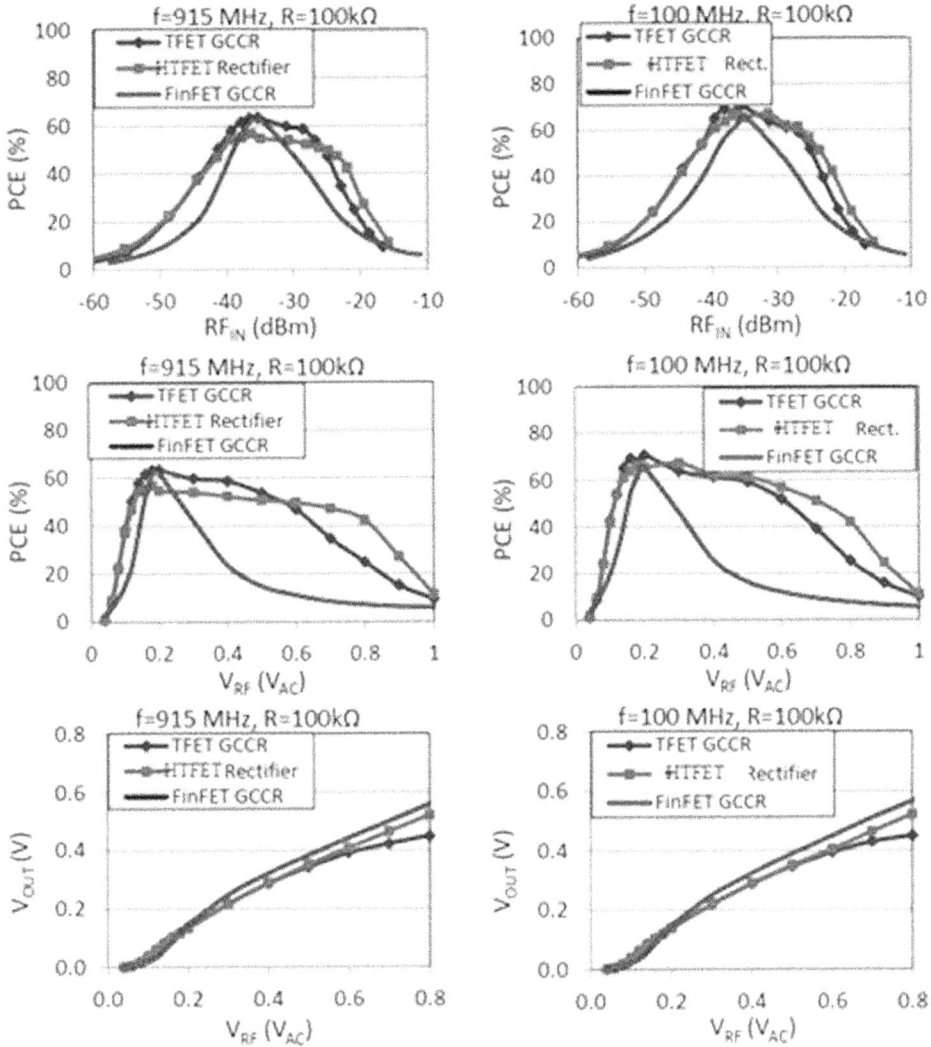

**FIGURE 6.16** Various rectifier's performance at two different frequencies.

Source: [38].

can only operate over short distances due to the weakening of received radiation power through time and space, as well as the low-power conversion efficiency (PCE) shown by front-end rectifiers at low RF power levels (below–20 dBm).

For a good system operation, efficient rectifiers are needed since low-power levels of electromagnetic radiation result in low output voltage values in the receiving antenna. In extreme low-voltage/power circumstances, simulations showed that TFET-based rectifiers outperform thermionic device-based equivalents (sub-20 dBm). Because of tunneling devices' unique electrical properties, it is possible to reduce the amount of energy required for each switch operation, enabling for the development of more energy-efficient digital cells at lower voltages [53–55]. The ability to construct efficient PMCs at low voltages is made possible by this property (sub-0.25 V). As a result, tunneling devices in ultra-low voltage PMCs for RF energy harvesting applications should be studied for their performance and limits. An ultra-low power RF energy harvesting power management circuit

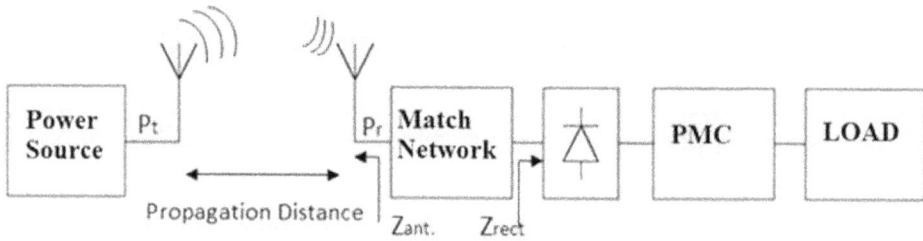

**FIGURE 6.17**   RF power transfer system.

**Source: [56].**

(PMC) based on tunnel FET technology and the benefits of utilizing tunneling devices in RF PMCs are discussed.

### 6.6.1 RF POWER TRANSFER SYSTEM CHALLENGES

In Figure 6.17, a schematic diagram of the RF power transfer system is given [56]. Following a conventional impedance antenna, the receiver has a lumped matching network between the antenna and the rectifier. Using the rectifier's output voltage to supply power to a load necessitates the inclusion of a power management circuit (PMC).

Long distances between the receiver and transmitter circuits cause power density attenuation, which must be overcome in the RF power system's receiver. As shown in Equation (6.2), the power received at the rectifier's input depends on transmitting power $P_t$, antenna gain $G_t$, wavelength of the transmitting signal $\lambda$, and propagation distance $R$. Equation (6.3) expresses the relationship between the peak amplitude of the antenna $V_A$ and the received power if the receiver antenna is adequately matched with the rectifier.

$$P_R = P_t G_t G_r \left( \frac{\lambda_0}{4\pi R} \right)^2 \tag{6.2}$$

$$V_A = \sqrt{8 R_A P_r} \tag{6.3}$$

### 6.6.2 TFET-BASED PMC

In Figure 6.18, the RF TFET-based PMC's building components are shown. It is assumed that the RF power source has an impedance of 50 and operates at 915 MHz. Thereafter the PMC is needed to enhance the low output voltage of the rectifier and enable a load, after matching and rectification. Startup, controller, and boost circuit are three different parts of the PMC. When the boost converter is activated, the first module is responsible for pre-charging the power capacitors linked to nodes $VDD_{INT}$, $VDD_{STARTUP}$ (by rectifier), and input and output $C_{BOOST}$ and $C_{OUT}$ capacitors to acceptable voltage levels.

An external controller is used to provide control signals to switches in the boost converter, which are powered by startup capacitors. Self-sustaining mode (SSM) of operation occurs when the boost converter is activated and the load connected for the first time. This means that the power capacitors are directly charged by output capacitance $C_{OUT}$, and not via the rectifier. The controller module is responsible for ensuring that the boost converter (which interacts with the rectifier's output) receives enough voltage to enhance rectifier efficiency.

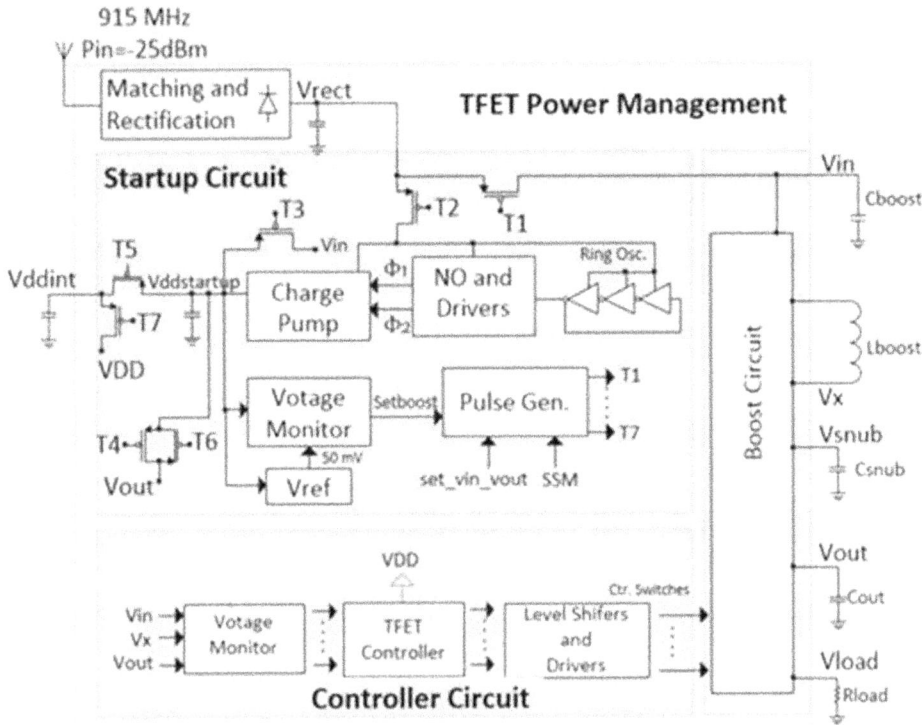

**FIGURE 6.18**   Power management circuit using TFET for energy harvesting applications.

Source: [56].

### 6.6.3   TFETs in PMCs and Boost Converters: Advantages

To assess a transistor's performance in a PMC or boost converter, the internal resistance measured as a function of |VDS| can be examined. It is demonstrated in Figure 6.19 that under forward bias circumstances of sub-0.25 V, a heterojunction TFET device has the lowest internal resistance. To put it another way, as compared to ordinary MOSFETs, this feature reduces the conduction losses in both the input transistor S2 and the output transistor T4 throughout the time interval t1 to t2 and time interval t2 to t3, respectively. With the TFET-based circuitry's lower static and dynamic power consumption, appropriate boost controller functioning may be minimized as compared to the usage of conventional thermoelectric technologies. Because of the higher current carried by TFETs during sub-0.2 V operation, smaller buffers may be designed. Tunnel FETs have a lower reverse loss than Si-FinFETs because of their higher internal resistance ($R_{OFF}$) and a narrower range of reverse bias settings. Boost converter output transistors (S4) with low reverse bias magnitudes, such as |VDS| <0.3 V for homojunction TFETs and |VDS|< 0.6 V for heterojunction TFETs, have this property, as a benefit. The reverse current carried by TFETs (greater than Si-FinFETs) increases the boost converter's reverse losses at higher magnitudes of reverse bias, hence TFET-based boost converters can only operate at low voltage.

Figure 6.20, the circuit's transient behavior is demonstrated for an–25 dBm RF power. The input $C_{BOOST}$ and output $C_{OUT}$ capacitors are shown to be pre-charged to 200 mV prior to the boost conversion procedure. Once charged, the controller's power supply node ($V_{DD}$) is activated, and the boost converter enters synchronous mode.Once the load is connected, the circuit enters self-sustaining mode (SSM), in which the output capacitor is responsible for charging the capacitors necessary to

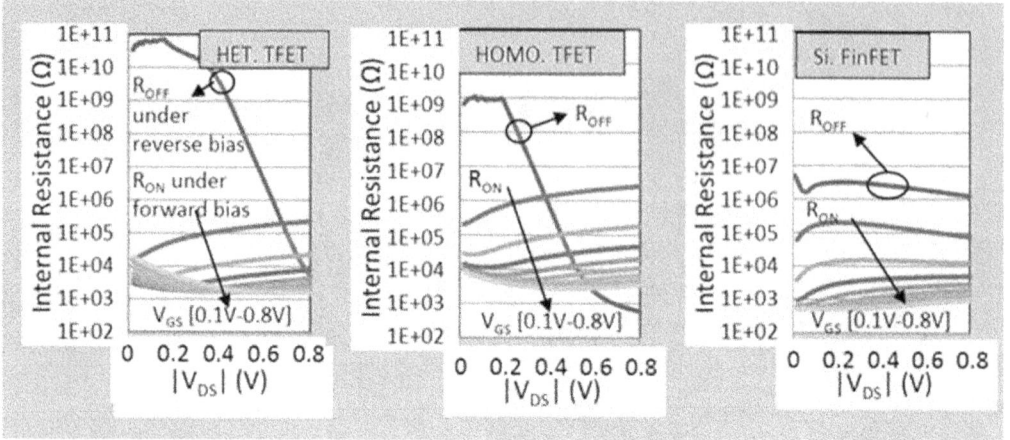

**FIGURE 6.19**   Internal resistance of various technologies (*n*-type).
Source: [56].

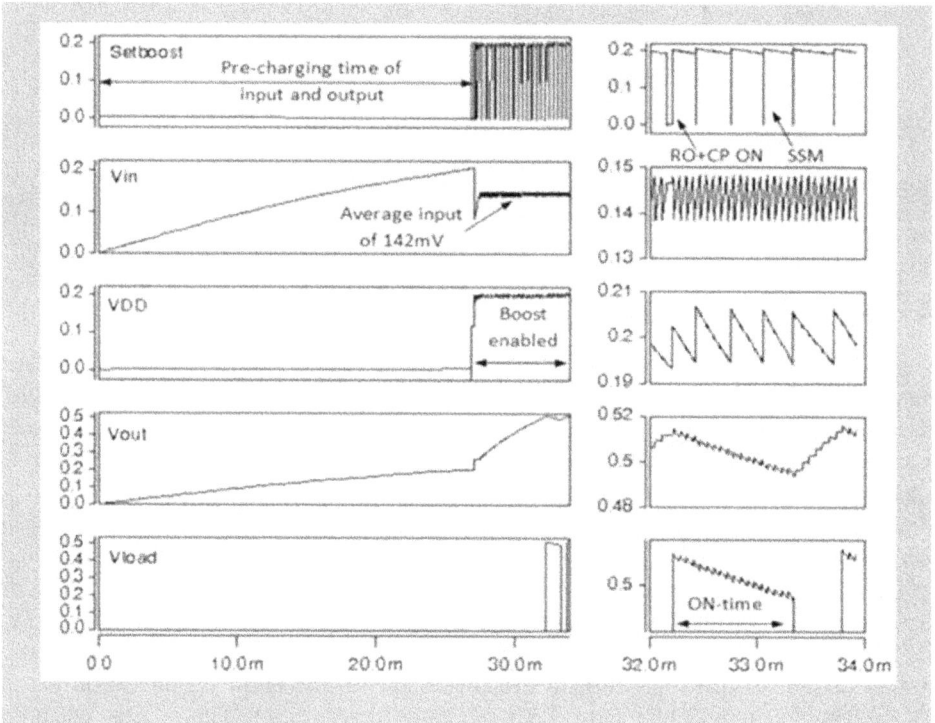

**FIGURE 6.20**   Transient simulation of tunnel FET-based power management circuit.
Source: [56].

power the load. When SSM is enabled, the ring oscillator and charge pump are disabled, and the charge rate of the capacitors in the startup module is increased, lowering the boost conversion's offtime. The voltage at the boost converter's input node ($V_{IN}$) is controlled to an average of roughly 142 mV, allowing for maximum power transfer from the rectifier (1.28 μW).

The inductor current and boost frequency are proportional to the capacitance value $C_{BOOST}$. Increased current values caused by increased capacitance necessitate longer channel widths for the input and output transistors in order to decrease forward losses and raise the PCE of the boost converter. While increasing the size of the output transistor can help minimize forward losses, increasing reverse current conduction and the resulting reverse losses diminish the boost converter's PCE.

Figure 6.21 illustrates the power loss distribution of a TFET-based PMC operating at its maximum boost conversion efficiency: PCE=86 percent, WS4=25 μm, $C_{OUT}=C_{BOOST}=0.05$ μF, $L=10$ mH. The startup circuit consumes 41.9 nW, the controller circuit 11.88 nW, and the boost converter consumes 116 nW, with the TFET switches S1 to S5 accounting for the majority of the losses. It is demonstrated that the input and output transistors S2 and S4 are responsible for more than 85% of the boost converter losses, respectively (S2 due to forward losses and S4 due to reverse losses). Despite these losses, the proposed TFET-based PMC exhibits promising performance for μW applications when compared to current RF-PMC. The startup and controller circuits' low-power consumption enables high DC-DC conversion efficiencies at low-voltage/power levels, whereas the good rectification performance at–25 dBm (approximately 40%) is due to the improved electrical characteristics of TFETs at sub-0.25 V in comparison to conventional thermionic devices.

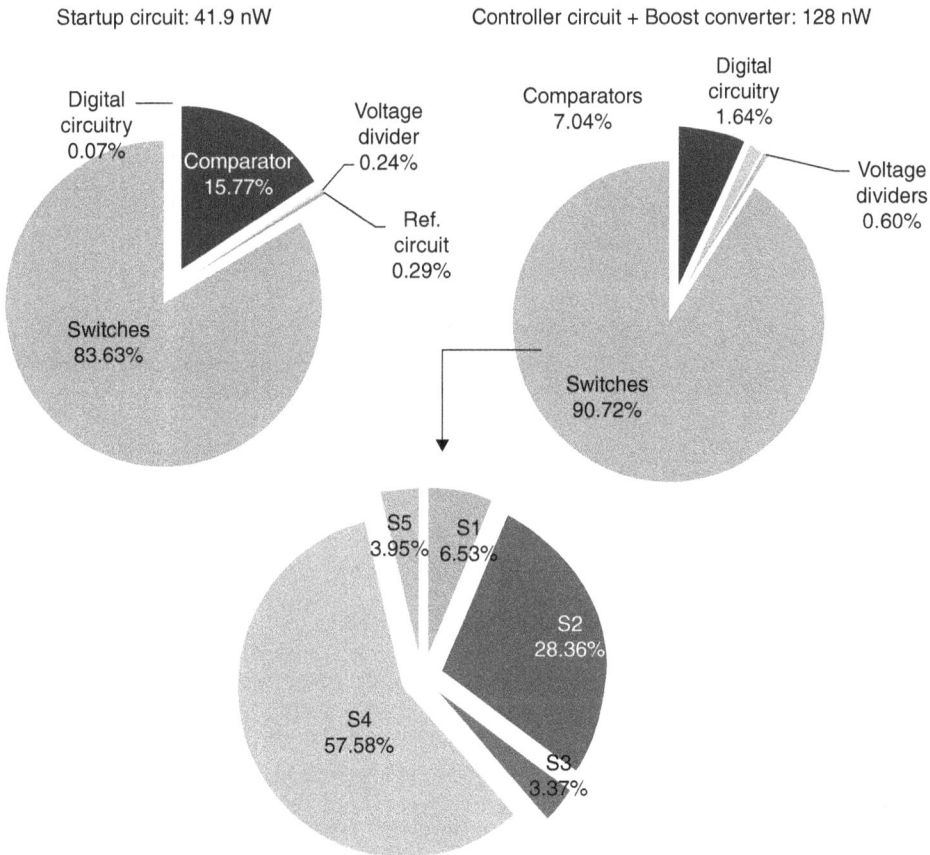

**FIGURE 6.21** Power losses distribution in the power management circuit.

**Source:** [56].

## 6.7   CONCLUSION AND FUTURE WORK

Front-end charge pumps and rectifiers based on tunnel FETs were discussed in this chapter. Traditional front-end topology limitations were recognized, and circuit-level improvements were proposed to extend the voltage/power range functioning of such circuits (PMC) for RF energy harvesting applications (W). Techniques for increasing the efficiency of PMCs based on TFETs have been discussed. The following are a few areas where tunnel FET-based circuit design for energy harvesting applications should be improved.

The suggested circuits' performance must be further evaluated using an upgraded tunnel FET compact model. For ultra-low power TFETs, process and temperature fluctuations, electrical noise and layout parasitics must be taken into account. Verilog-A lookup tables would be replaced with an analytical model that describes the TFET's behavior in all areas of operation for both $n$- and $p$-type configurations.

In order to validate the TFET-based circuit design methodologies suggested in this chapter, an experiment is needed. To achieve the device-level performance predicted by simulations, the leakage current of heterojunction III-V TFETs must be reduced by overcoming numerous phenomena that reduce leakage current. In order to create III-V based TFETs with the same degree of maturity as silicon-based devices, additional research is needed into III-V and innovative materials with significant bulk and interface defects.

## REFERENCES

[1]   G. E. Moore, "Cramming more components onto integrated circuits," *Electronics*, vol. 87, pp. 114–117, Apr. 1965.
[2]   W. F. Brinkman, D. E. Haggan, and W. W. Troutman, "A history of theinvention of the transistor and where it will lead us," *IEEE Journal of Solid-State Circuits,* vol. 32, pp. 1858–1865, Dec. 1997.
[3]   C. A. Mack, "Fifty years of Moore's law," *IEEE Transactions on Semiconductor Manufacturing,* vol. 24, pp. 202–207, May 2011.
[4]   S. M. Sze, Physics of Semiconductor Devices. New York: John Wiley and Sons, 2nd ed., 1981.
[5]   D. A. Neaman, Semiconductor Physics and Devices. Asia: Tata McGraw-HillEducation, 3rd ed., 2002.
[6]   A. M. Ionescuand H. Riel, "Tunnel field-effect transistors as energyefficient electronic switches," *Nature,* vol. 479, pp. 329–337, Nov. 2011.
[7]   E.-H. Toh, G. H.Wang, G. Samudra, and Y.-C.Yeo, "Device physics anddesign of double-gate tunneling field-effect transistor by silicon film thickness optimization," *Applied Physics Letters*, vol. 90, no. 263507, 2007.
[8]   J. Knoch, S. Mantl, and J. Appenzeller, "Impact of the dimensionalityon the performance of tunneling FETs: Bulk versus one-dimensional devices," *Solid-State Electronics,* vol. 51, pp. 572–578, Apr. 2007.
[9]   K. Boucart and A. M. Ionescu, "Double-gate tunnel FET with high-ĸgatedielectric," *IEEE Transactions on Electron Devices*, vol. 54, pp. 1725–1733, July 2007.
[10]   W. Lee and W. Y. Choi, "Influence of inversion layer on tunneling fieldeffect transistors," *IEEE Electron Device Letters*, vol. 32, pp. 1191–1193, Sept. 2011.
[11]   A. S. Verhulst, W. G. Vandenberghe, D. Leonelli, R. Rooyackers, A. Vandooren, G. Pourtois, S. D. Gendt, M. M.Heyns, and G. Groeseneken, "Boosting the on-current of Si-based tunnel field-effect transistors," *ECS Transactions*, vol. 33, no. 6, pp. 363–372,2010.
[12]   A. Sandow, J. Knoch, C. Urban, Q.- T. Zhao, and S. Mantl, "Impact of electrostatics and doping concentration on the performance of silicontunnel field-effect transistors," *Solid-State Electronics,* vol. 53, pp. 1126–1129, Oct. 2009.
[13]   S. Mookerjea, D. Mohata, R. Krishnan, J. Singh, A. Vallett, A. Ali,T. Mayer, V. Narayanan, D. Schlom, A. Liu, and S. Datta, "Experimental demonstration of 100 nm channel length $In_{0.53}Ga_{0.47}As$-basedvertical inter-band tunnel field effect transistors (TFETs) for ultra lowpower logic and SRAM applications," in *IEEE International Electron Devices Meeting (IEDM)*, pp. 1–3, IEEE, Dec. 2009.

[14] H. Zhao, Y. Chen, Y. Wang, F. Zhou, F. Xue, and J. Lee, "In$_{0.7}$Ga$_{0.3}$Astunneling field-effect transistors with an Ion of 50 µA/µm and a subthreshold swing of 86 mV/dec using HfO$_2$ gate oxide," *IEEE Electron Device Letters,* vol. 31, pp. 1392–1394, Dec. 2010.

[15] R. Iida, S.-H.Kim, M. Yokoyama, N. Taoka, S.-H. Lee, M. Takenaka, and S. Takagi, "Planar-type In$_{0.53}$Ga$_{0.47}$As channel band-to-band tunneling metal-oxide-semiconductor field-effect transistors," *Journal of Applied Physics*, vol. 110, no. 124505, 2011.

[16] S. K. Kim and W. Y. Choi, "Impact of gate dielectric constant variation on tunnel field-effect transistors (TFETs)," *Solid-State Electronics,* vol.116, pp. 88–94, 2016.

[17] M. Luisier and G. Klimeck, "Atomistic full-band design study of InAsband-to-band tunneling field-effect transistors," *IEEE Electron DeviceLetters,* vol. 30, pp. 602–604, June 2009.

[18] W. G. Vandenberghe, A. S. Verhulst, K.-H. Kao, K. D. Meyer, B. Sore, W. Magnus, and G. Groeseneken, "A model determining optimal dopingconcentration and material's band gap of tunnel field-effect transistors," Applied *Physics Letters*, vol. 100, no. 193509, 2012.

[19] J. Knoch, S. Mantl, and J. Appenzeller, "Impact of the dimensionalityon the performance of tunneling FETs: Bulk versus one-dimensional devices," *Solid-State Electronics,* vol. 51, pp. 572–578, Apr. 2007.

[20] S. Saurabh and M. J. Kumar, "Impact of strain on drain current andthreshold voltage of nanoscale double gate tunnel field effect transistor: Theoretical investigation and analysis," *Japanese Journal of AppliedPhysics*, vol. 48, p. 064503, June 2009.

[21] Y. Gao, T. Low, and M. Lundstrom, *"Possibilities for VDD = 0.1 V logicusing carbon-based tunneling field effect transistors,"* in *Symposium onVLSI Technology (VLSIT),* pp. 180–181,IEEE, 2009.

[22] K. Boucart and A. M.Ionescu, "Length scaling of the double gate tunnel FET with a high-k gate dielectric," *Solid-State Electronics*, vol. 51, pp. 1500–1507, Nov. 2007.

[23] F. Xue, H. Zhao, Y.-T. Chen, Y. Wang, F. Zhou, and J. Lee, "In$_{0.7}$Ga$_{0.3}$As tunneling field-effect-transistors with LaAlO$_3$ and ZrO$_2$ high-κ dielectrics," *ECS Transactions*, vol. 41, no. 3, pp. 249–253, 2011.

[24] Y. Morita, T. Mori, S. Migita, W. Mizubayashi, A. Tanabe, K. Fukuda,T. Matsukawa, K. Endo, S. O'uchi, Y. X. Liu, M. Masahara, and H. Ota,"Performance enhancement of tunnel field-effect transistors by syntheticelectric field effect," *IEEE Electron Device Letters*, vol. 35, pp. 792–794, July 2014.

[25] M. Sathishkumar, T. A. Samuel, K. Ramkumar, I. V. Anand and S. B. Rahi. Performance evaluation of gate engineered InAs–Si heterojunction surrounding gate TFET. Superlattices and Microstructures, vol 162, 107099, Feb 1 2022.

[26] D. Brunelli, C. Moser, L. Thiele and K. Benini,"Design of a Solar-Harvesting Circuit for Batteryless Embedded Systems," in *IEEE Transactions on Circuits and Systems I: Regular Papers*, vol. 56, no. 11, pp. 2519–2528, Nov. 2009.

[27] S. W. Hsu, E. Fong, V. Jain, T. Kleeburg and R. Amirtharajah, "Switched-capacitor boost converter design and modeling for indoor optical energy harvesting with integrated photodiodes," *International Symposium on Low Power Electronics and Design*, pp. 169–174, 2013.

[28] L. Mateu, C. Codrea, N. Lucas, M. Pollak and P. Spies, "Human body energy harvesting thermogenerator for sensing applications," *International Conference on Sensor Technologies and Applications*, pp. 366–372, 2007.

[29] G. Bassi, L. Colalongo, A. Richelli and Z. Kovács-Vajna, "A 150mV-1.2V fully-integrated DC-DC converter for Thermal Energy Harvesting," *International Symposium on Power Electronics Power Electronics, Electrical Drives, Automation and Motion*, pp. 331–334, 2012.

[30] O. Y. Wong, H. Wong, W. S. Tam and C. W. Kok, "A comparative study of charge pumping circuits for flash memory applications," *Microelectronics Reliability*, vol. 52, no.4, pp. 670–687, 2012.

[31] G. Palumbo and D. Pappalardo, "Charge pump circuits: An overview on design strategies and topologies," in *IEEE Circuits and Systems Magazine*, vol. 10, no. 1, pp. 31–45, 2010.

[32] D. Cavalheiro. Ultra-low power circuits based on tunnel FETs for energy harvesting applications. PhD Thesis, University of Catalonia, 2017.

[33] T. Paing et al., "Custom IC for ultralow power RF energy scavenging," *IEEE Transactions in Power Electron.,* vol. 26, no. 6, pp. 1620–1626, Jun. 2011.

[34]  S.-E. Adami et al., "Ultra-low power autonomous power management system with effective impedance matching for rf energy harvesting," in *International Power Systems (CIPS), International Confernce on*, pp.1–6, Feb. 2014.

[35]  G. Sain et al., "A battery-less power management circuit for RF energy harvesting with input voltage regulation and synchronous rectification," in *Circuits and Systems, International Midwest Symposium on*, pp. 1–4, Aug. 2015.

[36]  S. Jinpeng et al. "Design and implementation of an ultra-low power passive UHF RFID tag," *Journal of Semiconductors*, vol. 33, no. 11, pp.115011, Nov., 2012.

[37]  D.-S. Liu et al. "New analysis and design of arf rectifier for rfid and implantable devices," *Sensors*, vol. 11, no. 7, pp. 6494–6508, 2011.

[38]  D. Cavalheiro, F. Moll and S. Valtchev, "Insights into tunnel FET-based charge pumps and rectifiers for energy harvesting applications," *IEEE Transactions on Very Large Scale Integration (VLSI) Systems*, vol 25, no. 3, pp. 988–997, 2017.

[39]  P. Burasa, N. G. Constantin and K. Wu, "High-efficiency wideband rectifier for single-chip batteryless active millimeter-wave identification (MMID) tag in 65-nm bulk CMOS technology," *Microwave Theory and Techniques, IEEE Transactions on*, vol. 62, no. 4, pp. 1005, 1011, April 2014.

[40]  S. Mandal and R. Sarpeshkar, "Low-power CMOS rectifier design for RFID applications," *Circuits and Systems, IEEE Transactions* vol. 54, no. 6, pp. 1177–1188, 2007.

[41]  H. Liu et al., "Tunnel FET RF rectifier design for energy harvesting applications," Emerging and Selected Topics in Circuits and Systems, *IEEE Journal on*, vol. 4, no. 4, pp. 400–411, Dec. 2014.

[42]  H. J. Visser and R. J. M. Vullers, "RF energy harvesting and transport for wireless sensor network applications: principles and requirements," *Proc. IEEE*, vol. 101, no. 6, pp. 1410–1423, June 2013.

[43]  T. Soyata, L. Copeland and W. Heinzelman, "RF energy harvesting for embedded systems: A survey of tradeoffs and methodology," in *IEEE Circuits and Systems Magazine*, vol. 16, no. 1, pp. 22–57, 2016.

[44]  V. Marian V., B. Allard, C. Vollaire and J. Verdier, "Strategy for microwave energy harvesting from ambient field or a feeding source," in *IEEE Transactions on Power Electronics*, vol. 27, no. 11, pp. 4481–4491, Nov. 2012.

[45]  C. Yao and W. Hsia, "A 21.2 dBm dual-channel UHF passive CMOS RFID tag design," in *IEEE Transactions Circuits and Systems I*, vol. 61, no. 4, pp. 1269–1279, 2014.

[46]  C. Mikeka, H. Arai, A. Georgiadis and A. Collado, "DTV band micropower RF energy harvesting circuit architecture and performance analysis," RFID-Technologies and Applications (RFID-TA), *2011 IEEE International Conference on*, pp. 561–567, 2011.

[47]  M. Stoopman, S. Keyrouz, H. J. Visser, K. Philips and W. A. Serdijn, "Co-design of a CMOS rectifier and small loop antenna for highly sensitive rf energy harvesters," in *IEEE Journal of Solid-State Circuits*, vol. 49, no. 3, pp. 622–634, March 2014.

[48]  G. Papotto, F. Carrara and G. Palmisano, "A 90-nm CMOS threshold-compensated RF energy harvester," *IEEE Journal of Solid-State Circuits*, vol. 46, no. 9, pp. 1985–1997, Sept. 2011.

[49]  S. Scorcioni, L. Larcher, A. Bertacchini, L. Vincetti and M. Maini, "An integrated RF energy harvester for UHF wireless powering applications," *Wireless Power Transfer, 2013 IEEE*, pp. 92–95, 2013.

[50]  P. Theilman, C. C. Presti, D. Kelly and M. Asbeck, "Near zero turn-on voltage high-efficiency UHF RFID rectifier in silicon-on-sapphire CMOS," *IEEE Radio Frequency Integrated Circuit Symposium*, pp. 105–108, May 2010.

[51]  K. Kotani, "Highly efficient CMOS rectifier assisted by symmetric and voltage-boost PV-cell structures for synergistic ambient energy harvesting," *Proceedings of IEEE Custom Integrated Circuits Conference*, pp. 1–4, Sep. 2013.

[52]  M. Stoopman, S. Keyrouz, H. J. Visser, K. Philips, and W. A. Serdijin, "A self-calibrating RF energy harvester generating 1 V at –26.3 dBm," in VLSI Circuits (VLSIC), Symposium on, pp. C226–C227, 2013.

[53]  D. H. Morris, U. E. Avci, R. Rios and I. A. Young, "Design of low voltage tunneling-FET logic circuits considering asymmetric conduction characteristics," in *IEEE Journal on Emerging and Selected Topics in Circuits and Systems*, vol. 4, no. 4, pp. 380–388, Dec. 2014.

[54]  D. Esseni, M. Guglielmini, B. Kapidani, T. Rollo, and M. Alioto, "Tunnel FETs for ultralow voltage digital VLSI circuits: part I – device–circuit interaction and evaluation at device level," *Very Large Scale Integration (VLSI) Systems, IEEE Trans. on*, vol. 22, no. 12, pp. 2488–2498, Dec. 2014.

[55]  M. Alioto and D. Esseni, "Tunnel FETs for ultra-low voltage digital VLSI circuits: part II–evaluation at circuit level and design perspectives," Very Large Scale Integration (VLSI) Systems, *IEEE Trans. on*, vol. 22, no. 12, pp. 2499–2512, Dec. 2014.

[56]  D. Cavalheiro, F. Moll and S. Valtchev, "TFET-based power management circuit for RF energy harvesting," *IEEE Journal of the Electron Devices Society*, 5(1), pp.7–17, 2017.

# 7 Biodegradable Energy Storage Devices

*Anushya Ganesan, Sarika Raj, Amala Mithin Minther Singh Amirthaiah, and Kalai Kumar Kamaraj*

## 7.1 INTRODUCTION

The advancements in the other class of electronics encourage the contrary, where they can completely disintegrate, decompose, degrade, or disappear in a well-regulated way after a desired lifetime, in contrast to the developments in conventional electronics, which emphasize robust, consistent, and long-lasting operation. These devices, which are generally used as environmental monitors, hardware security electronics, biomedical implants, sensors, and stimulators, as well as bioelectronic medications or drug delivery systems, are referred to as transitory or biodegradable electronics in a wide sense [1]. Electronic device development for sustainable energy generation and storage is vital for ecologically responsible technology [2, 3]. Despite the latest major advancements in this field, insufficient attention has been paid to the pollution caused by connected waste streams, as well as the toxic and damaging consequences on the human body [4, 5]. In this context, emerging device innovations based on bio/ecoresorbable active and passive materials are critical. After a defined period of stable operation, the hardware dissolves totally into ground water or biofluids. Applications include temporary biological implants, secure data systems and environmental monitoring, in addition to "green" electronics [6–8]. The key to creating biodegradable energy storage devices is to employ biocompatible and biodegradable materials rather than hazardous or non-degradable ones. As an illustration, biocompatible macromolecular polymers and naturally occurring biodegradable substances including cellulose, poly (1,8-octanediol-co-citrate), polycaprolactone, silk protein, and spider silk have been employed as substrates and encapsulating materials [9]. Systematic study has shown the dissolving behavior of water-soluble metals such as W, Fe, Mg, Mo, and Zn in thin films or foils, which may be used as bioresorbable current collectors or circuit in biodegradable electronic devices. Several metal oxides can also function as bioresorbable semiconductors in degradable devices and can be hydrolyzed in aqueous solutions, including ZnO, MoO and MgO [10]. Transient supercapacitors (TSCs), a new class of advanced supercapacitors that can completely degrade with biologically and environmentally benign by products in vivo after serving their intended purpose, have a wide range of potential applications in the fields of personalized medicine, green electronics, implantable technology, military security, and other areas. Tian *et al.* demonstrated an easy super assembly manufacturing process for a three-dimensional network Zn@PPy hybrid electrode that is entirely biodegradable and implantable. An enhanced biodegradable supercapacitors may be entirely decomposed *in vivo* in 30 days without having any negative effects on the host organism and has a maximum energy density of $0.394 \text{ mW h cm}^{-2}$ [11]. According to Figueroa-Gonzalez *et al.* biodegradable supercapacitors are environmentally safe and sustainable energy sources that have the potential to replace conventional technologies like Li or alkaline batteries that are created with hazardous materials and pollute the environment [12]. Although they are necessary, biodegradable and

DOI: 10.1201/9781003340539-7

rechargeable batteries that are also adaptable and secure for power sources *in vivo* are still lacking. Biodegradable fiber conductors, polydopamine/polypyrrole composite material as the anode and $MnO_2$ as the cathode, biodegradable chitosan as the separator, and bodily fluid as the electrolyte were used by Mei *et al.* to create a biocompatible battery. It integrates effectively with biological tissues and may be injected directly into the body with little invasiveness without triggering an immune reaction. It provided a specific capacity of 25.6 mA h $g^{-1}$ with a retention of 69.1% after 200 charge/discharge cycles to power numerous biomedical devices. It has been shown, for instance, to successfully power biosensors within the body [13]. In the transportation industry, focus has switched to renewable energy and fuel cells due to the unsustainable nature and environmental effects of fossil fuels. Microbial fuel cell (MFC), which achieves upto 50% chemical oxygen demand elimination and power densities in the range of 34–980 mW/$m^2$, provides an environmentally beneficial method of producing energy while simultaneously cleaning wastewater. The device makes use of bacteria's metabolism to generate electricity [14]. Recent environmental considerations about fossil fuels and carbon emissions have generated a lot of interest in the feasibility of employing microorganisms to generate electrical power. Bacteria have a strong potential in microbial fuel cells due to their rapid growth rate during cultivation. The performance of the bacteria-imprinted polymer-coated electrode was evaluated by using it as the anode in a microbial fuel cell. Enhanced bacterial interaction with the anode and increased electron transfer via a respiratory enzyme are likely responsible for the higher output discovered [15]. This chapter describes in detail various biodegradable energy storage devices, such as fuel cells, solar cells, supercapacitors, and batteries.

## 7.2 SUPERCAPACITORS

Production of renewable energy from the available solar and wind energy resources has been increased rapidly day by day. In this generation, due to the uneven nature of solar and wind energy, efficient energy storage systems are found to be highly required to store electrical energy produced from the above said sources. Supercapacitors and batteries were the outstanding candidates for high performance of energy storage devices for their low-cost and eco-friendly nature. Supercapacitors (SCs) also known as ultracapacitors or electrochemical capacitors. SCs are governed the same as by capacitors. But it utilizes higher surface area electrodes, thinner dielectrics to achieve greater capacitances, high power density, long life cycle (>105 cycles), high reliability and has the ability to store and release the energy within the time frame of a few seconds [16–20]. As a result, supercapacitors may become an attractive power solution or an increasing number of applications ideal for industrial, UPS, medical, automotive, hybrid battery packs, power supply, actuators, emergency lighting, telematics, security equipment, RAID systems, portable instruments, server blade cards, camera flash systems, notebook computers, and energy harvesting applications. On the other hand, the green natural-derived materials, biodegradable resources, and raw materials, such as bioplastics or cellulose, are increasingly used as building for the development of biodegradable electronics and robotics in the field such as precision agriculture, biomedicine, wearable electronics, environmental monitoring [21–29].

Migliorini *et al.* showed interest in energy storing devices using polymeric substrates. This polymeric substrate shows interest in the field of producing complex devices. Mainly authors focused on the eco-friendly supercapacitors based on biodegradable poly(3-hydroxybutyrrate) PHB, ionic liquids and cluster-assembled gold electrodes. The energetic performance of the supercapacitor was analyzed from electrochemical characterization based on the amount of ionic liquid employed. And also, cyclic operation and resistant to moisture exposure analysis was carried out for the novel hydrophobic plastic materials [30].

Currently, used tea leaves were used for the preparation of carbon with high surface area and electrochemical properties. The electrode derived from the tea leaves has 100% capacitance retention up to 5000 cycles. The amount of varying the activating agent KOH, the surface area and pore

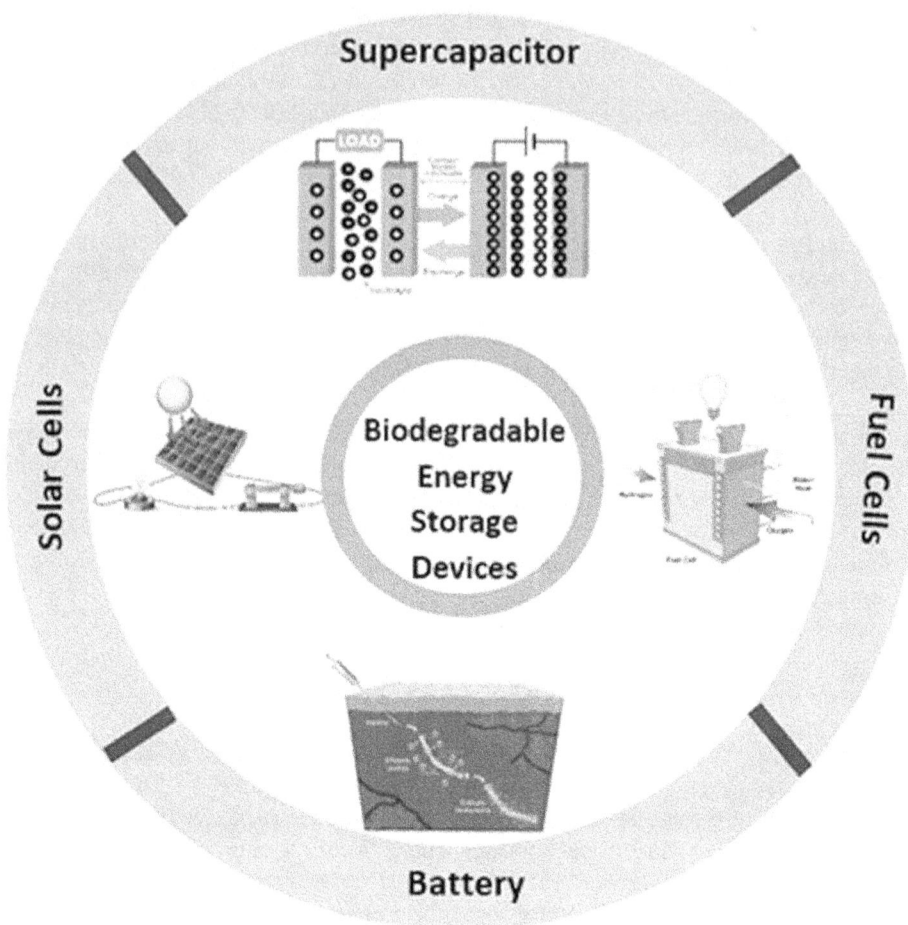

**FIGURE 7.1** Biodegradable energy storage devices.

size of the carbon derived from tea leaves were controlled. The maximum surface is of the activated tea leaves was 2532 m$^2$ g$^{-1}$, which is much higher than the inactivated tea leaves – 3.6 m$^2$ g$^{-1}$. Hence the electrolyte ions have a deep effect on the energy storage capacity. The specific capacitance of 292 F g$^{-1}$ is observed in 3 m KOH electrolyte with outstanding cyclic stability. Over 30% of the charge storage capacity improved, when the device temperature ranges from 10 to 80° C. This study proves used tea leaves can be used for the fabrication of environment-friendly high-performance supercapacitor devices at a low cost [31]. The following Table 7.1, shows the comparison of various tea leaves derived from carbon to carbon derived from various another biomass.

Biodegradable polymer gel electrolyte based on chitosan–lithium perchlorate complex has been investigated and the conductivity was found to be 5.5 × 10$^{-3}$ S cm$^{-1}$. Solid burial degradation analysis of the PGE suggested that the material has good biodegradation characteristics. The biodegradable material shows capacitance of 80 F g$^{-1}$ and is considered to be stable with a coulombic efficiency for about 99% [39].

Carbon-based supercapacitors with eco-friendly binders like, gelatin, chitosan, casein, guar gum, and carboxymethyl cellulose have been fabricated by low-cost water-soluble method [40]. Gelatin and chitosan-based devices shows a dielectric response similar to an electrochemical double layer capacitor. Equally, carboxymethyl cellulose, casein, guar gum–carbon-based electrodes also show dominant capacitance results.

**TABLE 7.1**

**Comparison of the Tea Leaves Derived Carbon to Carbon from Another Biomass**

| Name | Activating agent | Surface area | Highest specific capacitance | Reference |
|------|------------------|--------------|------------------------------|-----------|
| Tri-doped carbon | Microwave with ammonium polyphosphate | 228 | 170 | 32 |
| Corn straw | KOH | 894 | 166 | 33 |
| Corn straw-nitrogen doped | KOH | 1252 | 181 | 33 |
| Bamboo @MnO$_2$ | HCl | - | 174 | 34 |
| Cellulose-based activated nanofibers (ACNF) | Steam | 865 | 130 | 35 |
| Multiwall carbon nanotubes (MWCNT)/ACNF | Steam | 1120 | 160 | 35 |
| Banana peel | Hydrothermal | 1358 | 126 | 36 |
| N-doped banana peel CF | Hydrothermal | 648 | 186 | 36 |
| Activated waste tire | KOH | 1625 | 135 | 37 |
| Oil palm kernel shell | KOH | 462 | 210 | 38 |
| Oil palm kernel shell-P doped | Steam | 727 | 123 | 38 |

Nano cellulose has extensive efforts to develop electrochemical energy storage devices like supercapacitors. The abundant hydroxyl groups present in nano cellulose has reactive surfaces, which enable a variety of hybridization possibilities with diverse active materials to construct nano cellulose-based composite and separators. The high aspect ratio along with good mechanical properties makes nano cellulose highly attractive for flexible energy storage devices [41]. The good thermal and structural stability, and good wettability in various electrolytes over a wide potential window, positions it as a competitive candidate for the electrochemical energy storage applications. Table 7.2 shows the summary of nano cellulose-based supercapacitor electrode performances.

Colherinhas *et al.* used self-organized charged polypeptide nanosheets as electrodes to an electric double-layer supercapacitor [52]. This proposed device gives high potential for the energy storage using only biodegradable organic matter. This biodegradable supercapacitor device serves as a basis in the field of bio-integrated electronics. Table 7.3 summarizes the active performances of activated carbon derived from biowaste for supercapacitors.

## 7.3 MICROSUPERCAPACITORS (MSCS)

For the advanced miniaturized energy storage, following SCs, microsupercapacitors (MSCs) were considered to be the primary choice due to their satisfactory power density and a fast maintained frequency response. The size of these MSCs ranges from few microns to centimeters and it has a sandwich structure. The loading capacity and the dispersion state of the active nano or micron particles were responsible for the electrochemical properties, which is available in MSCs. Compared to the bulky structures, the 1D structure has numerous advantages, like high mechanical flexibility (which helps in the long term and repetitive deformation), easy self-integration extension, fit into small spaces of different shapes (helps in versality of the designs), integration of other 1D devices (helps in multi-functional wearable systems) [65].

Implantable medical electronic devices such as pacemakers, neurostimulators, and other sensor systems have seen rapid progress to realize real-time monitoring and accurate diagnosis and treatment. Similarly, implantable MSCs are a great pursuit area in the medical field. Recently, Sim *et al.* [66] reported a novel electrochemical material having high flexibility with outstanding electrochemical

**TABLE 7.2**
**Summary of the Nano-Cellulose-Based Supercapacitor Electrode Performance**

| Nano Cellulose based electrode | Electrolyte/ Conductivity (S/cm) | Separator | Specific capacitance (F/g) | Power density | Energy density | Capacitance retention | References |
|---|---|---|---|---|---|---|---|
| PPy/CNF (soak and polymerization) | 23.77 | | 2.26 (F/cm$^2$) | 1.2 mW/cm$^3$ | 16.95 mWh/cm$^3$ | 70.5% after 5000 cycles | 42 |
| CNF/VGCF/PPy (in-situ polymerization) | 11.25 | | 678.66 @ 1.875 mA/cm$^2$ | 334.57 W/kg | 15.08 Wh/kg | 91.38 after 2000 cycles | 43 |
| PANI-CNF (sol-gel and electrospinning) | | | 234 | 500 W/kg | 32 Wh/kg | 90% after 1000 cycles | 44 |
| GO/CNC (non-liquid-crystal spinning and followed by chemical reduction) | H$_2$SO$_4$-PVA gel / 64.7 | | 141.1 | 496.4 mW/cm$^3$ | 5.1 mWh/cm$^3$ @ 0.123 A/cm$^3$ | 92% after 1000 cycles | 45 |
| BP/CNT@CNF | HQ-H2SO4 | | 380.8 @ 1A/g | 3.975 kW/kg | 28.2 Wh/kg | 91% after 12,000 cycles | 46 |
| CNF/porous Co$_3$O$_4$ (LBLA) | KOH | | 594.8 mF/cm$^2$ | 799.97 kW/kg | 18.75 Wh/kg | | 47 |
| NC-Metal (oxides) CNF/MnO$_x$ | | | 269.7 | 2.75 kW/kg | 37.5 Wh/kg | 80% after 1000 cycles | 48 |
| MnCo$_2$O$_4$ (carbonization) | | | 871.5 | 125 W/kg | 30.26 Wh/kg | 89.3% after 5000 cycles | 49 |
| Gallium Oxynitride (GON) | | Sulfonation film | 152 mF/cm$^2$ at 10 mA/cm$^2$ | 1383 W/kg | 58 Wh/kg | 100% after 20,000 cycles | 50 |
| RuO2 | 1 M H$_2$SO$_4$ | - | 650-735 | | | | 51 |
| PANI | H$_2$SO$_4$ Aqueous | | 120-1530 | | | | |
| | Non-aqueous | | 100-670 | | | | |

**TABLE 7.3**

**Summary of Key Performance Metrics for Activated Carbon Derived from Biowaste for Supercapacitors**

| Biowaste | Energy Density (Wh/kg) | Power Density (W/kg) | Cycles | Percentage Retention (%) | References |
|---|---|---|---|---|---|
| Banana peel | - | - | 5000 | ~100 | 53 |
| Banana peel waste | 0.75 | 31 | - | - | 54 |
| Bradyrhizobium japonicum with a Soybean Leaf as a Separator | - | - | 8000 | 91 | 55 |
| Coconut shell | | | 3000 | 97.2 | 56 |
| Dead Neem leaves (Azadirachta indica) | 55 | 569 | - | - | 57 |
| Eucalyptus tree leaves | - | - | 15,000 | 97.7 | 58 |
| Garlic peel | - | - | 100 | 95-98 | 59 |
| Garlic Skin | 14.65 | 310.67 | 5000 | 94 | 60 |
| Lemon peel | 6.61 | 425.26 | 3000 | 92 | 61 |
| Banana-peel | 40.7 | 8400 | 1000 | 88.7 | 62 |
| Rice husk | 5.11 | - | 10,000 | 90 | 63 |
| Wood sawdust | 5.7-7.8 | 250-5000 | 10,000 | 94.2 | 64 |

behavior. Author trapped poly (3,4-ethylenedioxythiophene): poly (styrenesulfonate) and ferritin on multiwalled carbon nanotube. Here the fiber electrodes showed electrochemical stability and electrical conductivity. This novel electrochemical material shows 16% capacitance loss in the mouse and hence, the electrodes achieve excellent biocompatibility for *in vivo* and cell response tests.

Biodegradable smart implantable MSCs have attracted numerous investigations in the field of medical electronics. Lately, the research group of Sheng [67] described the fabrication of $MoO_x$ flake on water-soluble Mo foil by an electrochemical oxidation method. The fabricated $MoO_x$ supercapacitor has high specific capacitance of 112.5 mF cm$^{-2}$ at 1 mA cm$^{-2}$ and a recoverable energy density of 15.64 Wh cm$^{-2}$. In the stimulated body fluid environment at pH–7.4, the reported supercapacitor effectively works for 1 month with the controllable working life cycle. The electrochemical performance of $MoO_x$ flake shows good biodegradation and biocompatibility behavior, respectively, and it was systematically studied by the research group. Figure 7.2(a, b) represents the schematic representation of the synthesis and its biodegradable applications of MSCs and SEM image and crystal structure of $MoO_x$ electrode, respectively. The greatest common divisor (GCD) curve was shown in Figure 7.2(c). The biodegradability study was studied by placing the $MoO_x$ electrode material in a neutral phosphate buffer filling solution [Figure 7.2(d)]. When the time proceeded, the $MoO_x$ flakes were dissolved completely. Then the exposed Mo foil appeared to be cracked and corroded and decomposed into black powder. Figure 7.2(e) shows 6 months after implanting the encapsulated biodegradable supercapacitor into a living mouse, the entire MSC device was completely dissolved, and the mouse did not show any inflammatory response. As we all know, molybdenum is an indispensable trace element to maintain the normal life activities of the human body, so the dissolution of electrode materials can also further provide the human body's daily demand for Mo element.

Another research group [68], developed an edible and biocompatible MSC, which has a highly flexible and excellent electrochemical performance in bioenvironment area using scribing strategy. This edible MSC composed of gelatin, edible gold, active carbon, and agar aqueous electrolyte. MSCs charged to 2 V even after soaking in simulated gastric fluid for 28 mins and then it starts to degrade completely in the simulated gastric fluid after 40 mins. The high safety of this edible MSC

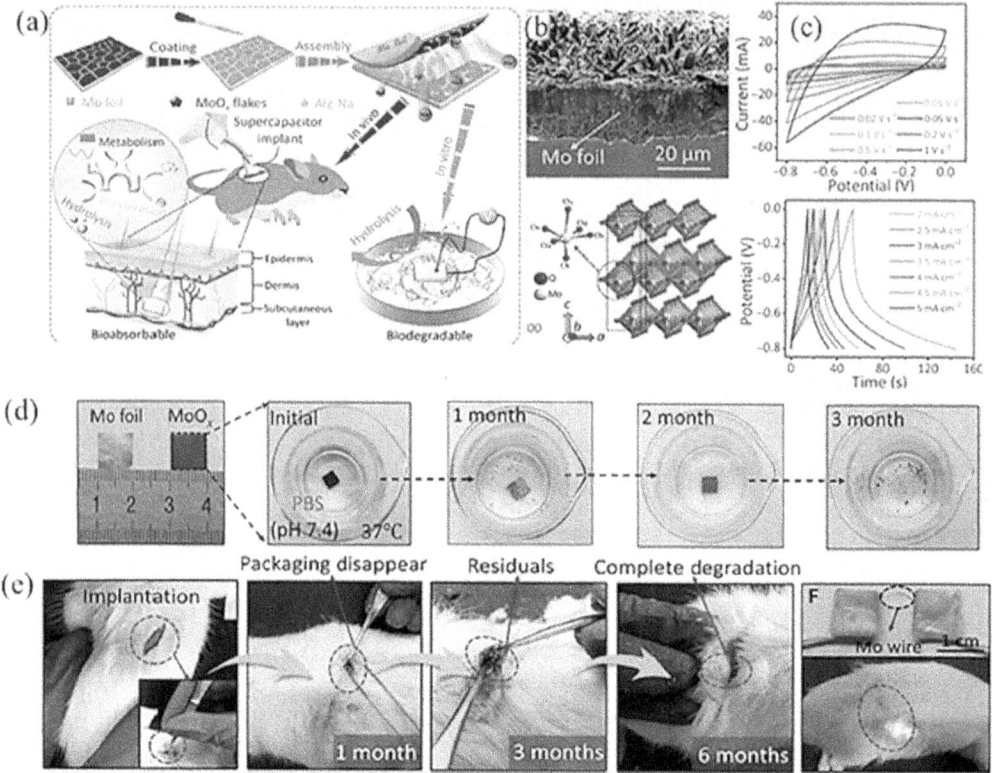

**FIGURE 7.2** (a) Schematic representation of the synthesis procedure and applications of the biodegradable MSCs. (b) SEM image and crystal structure of $MoO_x$ electrode. (c) CV and GCD curves of the $MoO_x$ electrode. (d) Digital pictures of the time-sequential of a single $MoO_x$ electrode (1 cm $X$ 1 cm) immersed in PBS (pH 7.4) at 37° C. (e) *In vivo* degradation evaluation of the supercapacitor implant in the subcutaneous of SD rats.

**Source:** [67].

capsule was tested by the testers by swallowing it and they showed no adverse effect on them and hence it proved its safety.

Lee *et al.* [69] fabricated a biodegradable MSC material using water-soluble metal electrodes such as tungsten (W), iron (Fe), and molybdenum (Mo), a biopolymer, agarose hydrogel electrolyte and a biodegradable poly (lactic-co-glycolic acid) substrate, encapsulated with polyanhydride. The electrochemical performance of these unusual MSCs is intensely enhanced during repetitive charge/discharge cycles. This enhanced performance was followed by the role of pseudo capacitance that originates from metal-oxide coatings generated by electrochemical corrosion at the interface between the water-soluble metal electrode and the hydrogel electrolyte. These electrodes also serve as biocompatible current collectors, whereas Fe and Mo are essential to maintain normal biological function in the human body. The ideal capacitive behavior of the fabricated MSCs was 0.05 mA cm$^{-2}$.

## 7.4 BATTERY

Biodegradable battery generates electricity from renewable fuels (glucose, sucrose, fructose, etc.) providing a sustained, on-demand portable power source. When enzymes in our bodies break down glucose, several electrons and protons are released. Therefore, by using enzymes to breakdown

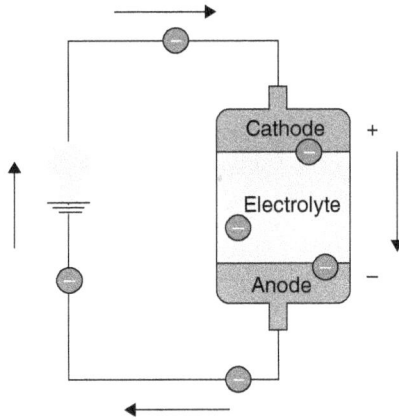

**FIGURE 7.3** Schematic representation of a biodegradable battery.

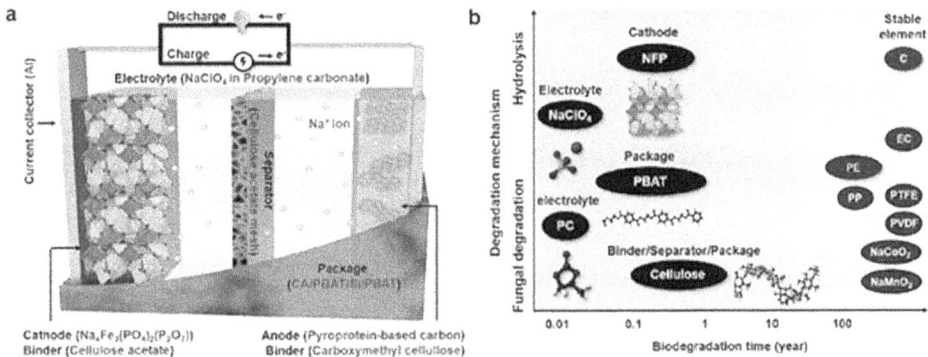

**FIGURE 7.4** (a) Schematic illustration of the materials and battery structure. (b) Biodegradation mechanism after brief biodegradation times for various battery components.

**Source: [70].**

glucose, biobatteries directly receive energy from glucose. These batteries then store this energy for future use. Biodegradable batteries use biocatalysts, either biomolecules such as enzymes or even whole living organisms to catalyse oxidation of biomass-based materials for generating electrical energy. A biodegradable battery consists of two types of metals suspended in an acidic solution. They contain an anode, cathode, separator, and electrolyte, which are the basic components of a battery. Each component is layered on top to another component. Anodes and cathodes are the negative and positive areas on a battery. The anode is located at the top of the battery and the cathode is located at the bottom of the battery (see Figure 7.3). Anodes are the components that allow electrons to flow outside the battery, whereas cathodes are devices that allow current to flow out from the battery.

An eco-friendly and biodegradable sodium-ion secondary battery (SIB) is developed through extensive material screening followed by the synthesis of biodegradable electrodes and their continuous assembly with an unconventional biodegradable separator, electrolyte, and package. Each battery component decomposes in nature into non-toxic compounds or elements via hydrolysis and/ or fungal degradation, with all of the biodegradation products naturally abundant and eco-friendly [70]. The following Figure 7.4 explains the clear setup of biobattery with its biomaterials and with its biodegradation times.

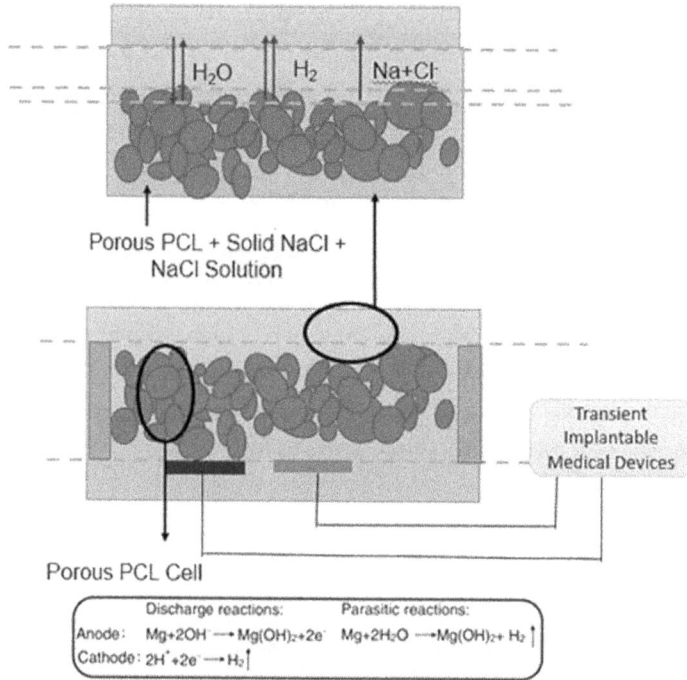

**FIGURE 7.5** Schematic illustration of principle operation of biodegradable battery.

Esquivel *et al.* [71] fabricated low-cost batteries for single use application using exclusive organic materials like, cellulose, carbon, and wax. This organic material features an integrated quinone based redox chemistry to generate electricity within a compact factor. Once the device depleted, the battery can be disposed without any need for a recycling process. And also, its components are non-toxic and it shows to be biotically degradable in a standard test.

A biodegradable battery containing a solid electrolyte of sodium chloride and polycaprolactone (PCL) was reported by the group She *et al.* [72]. This approach harnesses the body fluid that diffuses into the cell as an element of the electrolyte; however, the large excess of sodium chloride suspended in the polycaprolactone holds intracell ionic conditions constant. This biodegradable battery, has long shelf life and desirable functional life span. The polymeric skeleton of the solid electrolyte system acts as an insulating layer between electrodes, preventing the metallic structure from short-circuit during discharge. Figure 7.5 shows the principle operation of biodegradable battery. Liquid absorbed into the cell activates the battery by converting solid NaCl into a super saturated NaCl solution. Na+ and Cl− and as well as hydrogen gas, leach out from PCL film. The chemistry in the electrolytic cell is the cathode protection of Fe through the oxidation of Mg anode and parasitic corrosion of the Mg.

Natural gel–electrolyte with light weight and flexible batteries between textile-based electrodes was reported with the discharge capacity of 100 mAh g$^{-1}$ at 14 mA g$^{-1}$ with respect to the anode [73]. The aging process of the gel–matrix was examined and confirms that this device can refresh by re-wetting the gel–electrolyte. This reported textile-based batteries can bend up to 180° with minor influence on the battery voltage.

Li-ion electrodes were fabricated by using low-cost cellulose fibers (2 mm length) as the binder and supporter for the electrode, by eliminating heavy and inactive current collectors as substrates (aluminium/copper) and expensive binders [74]. To improve the electrochemical performances, water-soluble carbon was utilized and also to reduce the preparation time. Flexible and resistant

**TABLE 7.4**
**Summary of Key Parameters of Various Wood-Based Materials for Battery**

| Wood species | Electrode | Treatment condition | Surface area ($m^2 g^{-1}$) | Active mass loading ($mg\ cm^{-2}$) | Electrochemical performance | Electrolyte | Reference |
|---|---|---|---|---|---|---|---|
| Bass wood | TARC | $N_2$ (950°C,2h); $NH_3$ activation | 1438 | - | 241 mW $cm^{-2}$ at 330 mA $cm^{-2}$; 945.2 Wh $kg_{Zn}^{-1}$ at 10 mA $cm^2$; over 100 h after 4 mechanical charges | KOH (6M) | 75 |
| Poplar wood | CW | Ar (1400°C,2h) | - | 7 | 330 mAh/g; initial Coulombic efficiency of 88.3% | - | 76 |
| Balsa wood | CNT/ Rucoated F-Wood | chemical delignification | - | - | 0.85 V at 100 mA $g^1$; 220 cycles | LiTFSI/ TEGDME (1 M) | 77 |
| Alsa wood | - | chemical delignification | - | - | over 200 cycles; capacity of 1000 mAh $g^1$; low overpotential of 1.5 V | Lithium bis (trifluoromethan) sulfonimide (1 M) | 78 |
| Bass wood | 3D-wood | (1000°C, 6h) | 216.77 | - | 91.77% at 10mA $cm^{-2}$; energy efficiency is 75.44% | 0.1M $VOSO_4$+ $3MH_2SO_4$ | 79 |

$LiFePO_4$ (LFP), $Li_4Ti_5O_{12}$ (LTO), organic 3,4,9,10-perylenetetracarboxylic dianhydride (PTCDA), and graphite electrodes are obtained with active mass loadings similar to those obtained by the current casting method. The initial discharge capacity of approximately 130 mAh·$g^{-1}$ at 2°C is obtained for an LFP/LTO paper battery with an approximately 91.6% capacity retention after 1000 cycles. An all-organic prelithiated PTCDA/graphite cell without a transition metal is prepared and electrochemically tested. It is one of the first self-standing batteries that is composed of organic redox active molecules and biodegradable components reported in the literature.

Wood is a biodegradable and renewable material that naturally has a hierarchical porous structure, excellent mechanical performance and versatile physicochemical properties. Wood-based eco-friendly materials were fabricated and used for the energy storage and flexible electronics. Table 7.4 displays various key parameters of various wood-based materials for batteries.

## 7.5 TRANSIENT ELECTRONICS

Transient electronics is an emerging technology, the key attribute of which is an ability to physically disappear, entirely or in a part, in a controlled manner after a period of stable operation [80]. Biodegradable transient electronic device play an important role in the medical/diagnostic processes, such as wound healing and tissue regeneration. The fully biodegradable and biocompatible magnesium–molybdenum trioxide battery, which is an alternative method for an *in vivo* on-board power supply was reported by Huang *et al.* [81]. This high-performance biodegradable battery delivers a stable high output voltage as well as prolongs the lifespan. This reported biodegradable and biocompatible battery provides a promising candidate to the advanced energy harvesters for self-powered transient bioresorbable implants as well as eco-friendly electronics.

Yin *et al.* [82] reported water-activated primary batteries, which involve constituent materials that are degradable, environmentally benign, and biocompatible. Magnesium foils serve as the anodes, while metal foils based on Fe, W, or Mo serve as the cathodes, the packages are formed with polyanhydrides. The performance of single cell batteries that consist of metal foils can be evaluated most conveniently by the use of a PDMS liquid chamber filled with phosphate buffered saline as the electrolyte. For the constant current density – 0.1 mA cm$^{-2}$, the discharging behavior of an anode–cathode spacing was considered to be ~2 mm. Hence, the voltage is stable for 24 h.

## 7.6  FUEL CELLS

Fuel cells use a one-step energy extraction cycle to convert chemical energy into electrical energy. Solid oxide fuel cells (SOFCs) have sparked interest in green energy research due to their ability to generate electrical energy through non-combustion chemical reactions and mechanical processes. The release of electrons via an external circuit generates electricity. Interconnect serves as a separator between the fuel and the mixing oxidants, as well as connecting the SOFC components. Protective interconnect coating is critical because it protects against chromium poisoning [83]. The conversion of ammonia from a pollutant to an energy-rich, carbon-free fuel provides an opportunity to achieve net-zero wastewater services. However, little is known about how the product quality of ammonia recovered from real wastewater may affect its downstream use in fuel cells. The energy balance from this study shows that, despite the high latent heat demand for separation, the low cost of heat combined with the power produced by ammonia results in a favorable economic return when compared to conventional biological treatment [84].

In terms of high efficiency and clean electrochemical energy conversion devices, SOFCs offer numerous advantages. Due to the high operating temperature, however, this technology is limited to stationary applications and causes component degradation and long-term stability issues. The development of new designs and modifications to improve the electrochemical performance at intermediate temperatures and the durability of SOFC components is critical to bringing this technology one step closer to the market [85]. Hydrogen energy is the future mainstream replacement for fossil-based fuels because it is environmentally friendly, renewable, efficient, and cost-effective. However, the extremely low volumetric density presents the main challenge in hydrogen storage, and thus, developing effective storage techniques is a critical step toward achieving a sustainable hydrogen economy. Hydrogen physically or chemically stored into nanomaterials in the solid state is a promising prospect for effective large-scale hydrogen storage, with applications in both reversible on-board storage and regenerable off-board storage. Its appealing features include safety, compactness, lightness, reversibility, and the ability to produce sufficient pure hydrogen fuel under mild conditions [86].

Long-distance, heavy-duty transportation applications that are beyond the capabilities of electric vehicle technology require hydrogen fuel cells. Recent advances in membrane science have allowed alkaline fuel cells, which do not require precious metal catalysts, to compete with acidic membrane fuel cells. We use cryo-electron microscopy, electrochemistry, and numerical modeling to investigate the cathode of alkaline fuel cells, where oxygen combines with water to form hydroxide anions [87]. Hydrogen fuel cell vehicles (FCVs) are seen as a promising solution to energy security and environmental pollution issues. However, the technology is still being developed, and the hydrogen consumption is unknown. The quantitative evaluation of current and future FCV's life cycle energy consumption and pollution emissions in China involves complex processes and parameters. As a result, this study focuses on the life cycle assessment (LCA) of FCV and the key parameters of FCV production as well as different hydrogen production methods such as steam methane reforming, catalysis decomposition, methanol steam reforming, electrolysis-photovoltaic (PV), and electrolysis-Chinese electricity grid mix (CN). While accounting for various assumption scenarios, sensitivity analysis of the bipolar plate, glider mass, power density, fuel cell system efficiency, and

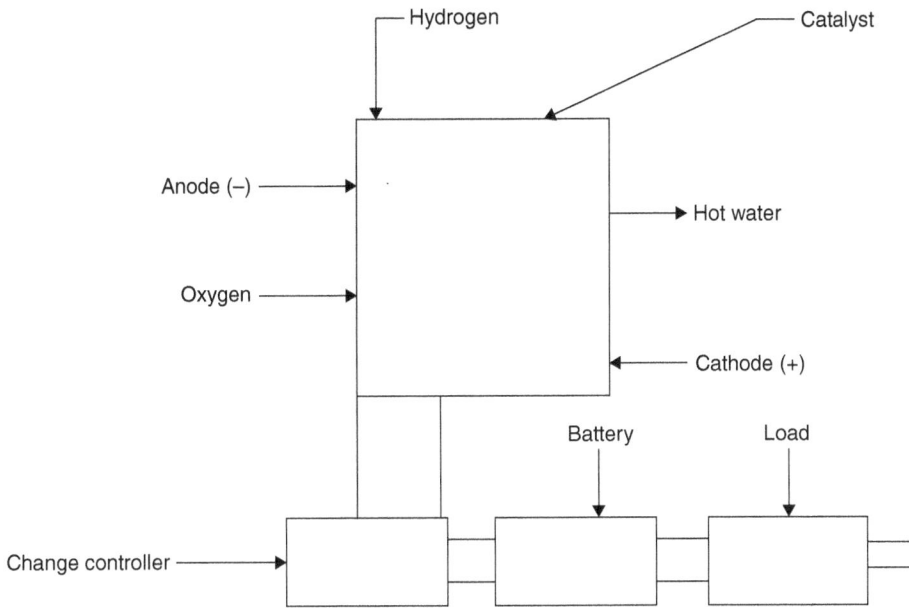

**FIGURE 7.6**   Schematic representation of fuel cell.

energy control strategy is performed. Fuel cell durability is one of the technical nuts and bolts of technology industrialization in the automotive sector, so methods for improving durability are especially important. By detecting and correcting fuel cell faults in real time, the fault tolerant control process increases fuel cell durability. Fuel cells are prone to failure due to their sensitivity to operating conditions. Because dynamic conditions are frequently encountered in vehicle applications, fault risk is heightened. Dynamic conditions complicate fuel cell control because they affect reactant supply, thermal management, and water management. If not corrected, these flaws degrade the fuel cell and shorten its useful lifetime [88–89].

The schematic diagram of the fuel cell is illustrated in Figure 7.6. In this, it contains hydrogen and oxygen as major inorganic elements as source, catalyst, anode, cathode, charge controller, battery, and load. The hydrogen and oxygen are passed in the chamber and catalyst is used as the separation space between the hydrogen and oxygen. Hydrogen acts as the anode and oxygen acts as the cathode. The two terminals are connected with the charge controller and then to the battery. The produced charges were controlled by the controller and stored in the battery.

In today's economic and environmental climate, the sustainable use of energy is critical to the development of our society. Fuel cells, particularly proton exchange membrane fuel cells (PEMFC), are one of the most promising portable power generation systems. PEMFCs are being considered as a technology to reduce greenhouse gas emissions in transportation applications, as an alternative to internal combustion engines. The extent to which PEMFCs are used commercially is heavily dependent on the efficiency and robustness of the PEMFC operation. Because of the complexity of these electrochemical devices, early detection of PEMFC failure modes is critical to improving their efficiency and robustness, as it can prevent or mitigate fault-induced damages, lowering maintenance costs. While the failures have not yet caused irreversible damage to the cell, early detection allows for the restoration of optimal operating conditions. The following are two desirable characteristics of an early diagnosis method that would be well suited for implementation in commercial portable PEMFC systems. To begin, the diagnosis process should be carried out in real time on the operating cell, without influencing or interfering with the cell's normal operation. Second, the diagnosis system should be built with lightweight, non-voluminous, and low-cost hardware [90].

Advanced modeling tools are widely used to improve proton exchange membrane fuel cell control strategies. Existing dynamic fuel cell models, on the other hand, are plagued by a conflict between computational speed and simulation accuracy. To address this contradiction, a novel discretized fuel cell model based on existing segmented modeling approaches is presented. This model accurately investigates the dynamic temperature, two-phase water, and current-density distributions using a new twofold subdivision method and a newly designed interpolation algorithm. When compared to the finite-element method, the simulation time is reduced to minutes while the high-fidelity characteristic is maintained. Model validations are performed at both the cell and stack levels to demonstrate the model's high accuracy and real-time simulation capability, demonstrating the model's high potential for use in the development of fuel cell control strategies [91].

Hydrogen energy and fuel cell technology are critical clean energy pathways for achieving carbon neutrality. Because of its simple structure, portability, and quick start-up, the proton exchange membrane fuel cell (PEMFC) has a wide range of commercial application prospects. The cost and durability of the PEMFC system, on the other hand, are the main barriers to commercial applications of fuel cell vehicles. The core hydrogen recirculation components of fuel cell vehicles, such as mechanical hydrogen pumps, ejectors, and gas-water separators, are reviewed in this chapter in order to understand the problems and challenges in their simulation, design, and application. Mechanical pumps used in PEMFC systems are summarized in terms of their types and operating characteristics. Furthermore, corresponding design suggestions are provided based on an analysis of the mechanical hydrogen pump's design challenges [92].

SOFC technology is gaining popularity among researchers as a promising power generation system with high energy efficiency, increased fuel flexibility, and low environmental impact when compared to conventional power generation systems. SOFCs are devices that convert chemical energy directly into electrical energy with negligible emissions Because of their high energy conversion efficiency when compared to other types of fuel cells, SOFCs are regarded as a promising power generation system. Internal gas reforming, which is highly effective for improving efficiency, is the source of fuel flexibility. However, ceramics limit the materials used in SOFCs, and a long start-up period is required due to the high operating temperature. As a result, many efforts are currently being made to operate SOFCs at intermediate temperatures, which is also effective in expanding the application area of SOFCs [93–94].

SOFCs are electrochemical energy conversion devices that generate electricity directly from the chemical energy in fuels. SOFCs can offer high power generation efficiency and low $CO_2$ and air pollutant emissions because the intermediate conversion of chemical energy to thermal and mechanical energies is avoided. Furthermore, SOFCs can operate on a wide range of fuels, including hydrogen, carbon monoxide, and common hydrocarbons like methane, propane, butane, and ethanol. The majority of SOFC research has concentrated on two geometric designs: planar and tubular. Despite their low current collection losses and ease of stacking, planar SOFCs have issues, such as large sealing areas and high thermal stress. Tubular SOFCs, on the other hand, require significantly less sealing and are mechanically robust. Tubular SOFCs also have a high thermal cycling endurance and allow for quick start-up and shut-down operations [95].

Rising energy demand and environmental concerns have prompted research and development of novel technologies for clean and efficient electricity generation. SOFCs continue to garner attention as one of the most promising power generation technologies, owing to their ability to convert chemical energy directly to electricity while emitting little pollution. The cutting-edge nickel-yttrium-stabilized zirconia (Ni-YSZ) anode provides excellent electrochemical performance, but coking quickly deactivates Ni, resulting in rapid performance degradation under direct hydrocarbon fuel-cell operation. The development of alternative anode materials with coking-resistant properties would be critical to the advancement of fuel cell technology [96].

SOFCs use nickel-based cermet anodes, which have good compatibility with electrolytes made of stabilized zirconia or doped ceria, as well as high activity toward hydrogen electrocatalytic oxidation and high electronic conductivity. By directly feeding methane to a nickel cermet anode-based SOFC, the anode can play two roles: catalyzing methane reforming reactions and promoting electrochemical oxidation of hydrogen, carbon monoxide, and methane. However, because nickel-based anodes are prone to carbon deposition in methane fuel, quick carbon accumulation over the anode is common, resulting in rapid degradation of the fuel cell's performance. Furthermore, sulphur in the fuel gas could seriously poison the nickel cermet anodes. However, nickel cermet anodes are susceptible to reoxidation by the oxidant, which can have a significant impact on cell integrity [97]. While electrolyte-supported cell-based solid oxide cell technologies for stationary applications have been developed, anode-supported cell (ASC) and metal-supported cell (MSC) configurations are thought to be more suitable for mobile and high-power density applications. Both cell types share a very thin electrolyte layer applied at the cell's core, which greatly reduces ohmic losses and thus allows for a lower operating temperature. However, in the case of an MSC, the mechanical support is made of a porous metal layer, which is typically formed by the same material as the metallic interconnector, whereas in the case of an ASC, it is a ceramic layer. Though several research groups and companies have recently considered the MSC to be the most promising candidate for mobile applications due to its high mechanical robustness and redox stability, world records in fuel conversion efficiency to electricity of >70 percent [with respect to the lower heating value (LHV)] [98].

SOFCs are promising renewable energy sources. Many works are devoted to the investigation of materials, specific aspects of SOFC operation, and the creation of devices based on them. There is, however, no work that covers the entire spectrum of SOFC concepts and designs. The current review makes an attempt to collect and structure all types of SOFC that exist today. The structural characteristics of each type of SOFC have been described, as have their advantages and disadvantages. A comparison of the designs revealed that the anode supported design is the most suitable for operation at temperatures below 800°C among the well-studied dual-chamber SOFC with oxygen-ion conducting electrolyte. SOFC with proton-conducting electrolyte and electrolyte-free fuel cells are two other SOFC types that show promise for low-temperature operation. SOFCs have piqued the interest of researchers due to their high efficiencies in converting chemical energy to electricity, as well as their flexibility in utilizing conventional hydrocarbon-based fuels such as methane and ammonia, and have seen a surge in the research and deployment of commercial systems ranging from stationary to transport applications in recent years. Durability and performance stability over long-term operation have been the focus of R&D efforts in both academic and industrial sectors; however, lowering system costs is a persistent key issue required for widespread commercialization of SOFC technology [99–100].

Electrolytes with high proton conduction and low activation energy are appealing for lowering the high operating temperature of solid-oxide fuel cells below 600°C. We created a semiconducting electrolyte $SrFeTiO^{3-}$ (SFT) material with high ionic conduction and exceptionally high protonic conduction at low operating temperature but low electronic conduction to avoid short-circuiting [101]. Because of its high energy density, clean production from water and clean electricity, and zero emissions when converted into usable energy, hydrogen is the sustainable energy carrier. More than 30 countries, including developed countries such as Germany, the United Kingdom and Japan, have issued hydrogen energy-related strategies by 2022. They have developed their own hydrogen energy strategies for 2020, with the goal of elevating the development of the hydrogen energy industry to the level of national energy strategy. As a result, it is critical to accelerate the development of hydrogen energy 50, fuel cell technology, and industrial development. Because of its many advantages, such as no pollution, high energy conversion efficiency, low noise, and fast start-up, proton exchange membrane fuel cell (PEMFC) is the most widely used of the various fuel cell technologies [102].

Fabrication of electrodes for polymer electrolyte fuel cells is an intriguing process that must optimize a balance of gas transport, electrical conductivity, proton transport, and water management. In this study, four different electrospray deposition electrodes were studied using various catalytic inks, including nafion and epoxy doped with graphene-nanoplatelets as binders. After studying the behavior of those electrodes in a single open cathode fuel cell proton electrolyte membrane, it is clear that the addition of epoxy as a binder doped with graphene improves the performance of the fuel cell and increases the mechanical stability of the electrode, avoiding catalyzts loose during electrode manipulation in the fuel cell assembly process and increasing the fuel cell's durability [103].

To function properly, hybrid fuel cell battery electric vehicles require complex energy management systems (EMSs). A hybrid system with a poor EMS can have a low efficiency and a high rate of degradation of the fuel cell and battery pack. Many different types of EMSs have been reported in the literature, such as equivalent consumption minimization strategies and fuzzy logic controllers, which typically focus on a single goal optimization, such as $H_2$ usage minimization. Different vehicle and system specifications make comparing EMSs difficult and can frequently result in false claims about system performance. Fuel cells that run on hydrogen fuel are a less advanced technology, but they have advantages over batteries in terms of range and refueling speed. Fuel cells, on the other hand, cannot accept regenerative braking, are less dynamically responsive, and lack a sufficiently developed refueling infrastructure. Unfortunately, fuel cells and batteries are frequently portrayed as rival technologies, with one "winning" over the other. Instead, the two should be viewed as complementary technologies, each deployable from stable electrochemical power sources (including supercapacitors and a variety of battery and fuel cell types) to meet the needs of different applications, either individually or in combination (hybridized) to extract the best features from each [104].

Environmental pollution caused by the excessive use of fossil fuels to generate energy has become one of the world's most serious issues, owing to the release of polluting gases ($CO_2$, $CH_4$, $N_2O$, and others). On rare occasions, pollution levels can rise above normal, reducing the quality of our environment. Despite efforts to reduce the impact of pollution, the problem is worsening and is not being addressed adequately. Furthermore, it is an issue that is overlooked in many international development and global health agendas, as well as in many countries' planning strategies. Organic waste generated in rural and agro-industrial areas, on the other hand, is a type of contamination that is not adequately managed. In some cases, organic waste accounts for half of all municipal solid waste, and it is frequently disposed of in open dumps or "landfills," where decomposition releases some greenhouse gases. At various stages of agricultural and industrial operations, agro-industrial waste (pulps, stalks, seeds, peels, etc.) is generated [105]. Fuel cells provide sustainable energy by converting the chemical energy of fuels into electrical energy in a clean and efficient process; however, some fuels are not environmentally friendly. For example, hydrogen fuel is primarily derived from fossil fuels and poses safety risks due to the fact that hydrogen is flammable and combustible, and hydrogen fuel cells must operate at high pressures. Direct liquid fuel cells (DLFCs) are a greener alternative that use renewable biofuels and have enhanced safety features. Acidic DLFCs also have a highly efficient cation exchange membrane (CEM) that allows cations to be transferred from the anode to cathode. Because of their high-power densities, these fuel cells have gained commercial interest in powering portable electronic devices with methanol or formic acid fuel. Due to the corrosive acidic medium, acidic DLFCs are limited to expensive noble metal catalysts, suffer from CO catalyzt poisoning, and experience fuel crossover as protons flow from anode to cathode [106].

In the last few decades, the urgent need to reduce global carbon emissions from fossil fuels by switching to clean energy through the use of alternative energy has inspired the investigation of novel portable sources of renewable and sustainable energy production. Fuel cells, in conjunction with green hydrogen, are being used to meet energy demands and to replace fossil fuel-powered combustion engines and turbines. Microfuel cells are also being researched to meet the energy demands of various MEMS devices and portable electronics. A fuel cell is typically an assembly consisting of an anodic and a cathodic chamber separated by a proton exchange membrane (PEM) that allows only ions ($H^+$ or other cations) to travel from the anodic chamber to the cathodic chamber. The anodic and cathodic chambers house the anode and cathode electrodes, which are connected via a conductive load to form a close circuit. The high cost of PEM fuel cells is tolerable in relatively expensive and bulky applications, such as fuel cell electric vehicles, and hydrogen fuel can be conveniently stored in conventional pressurized cylinders or tanks. The higher cost of PEM, on the other hand, is unsustainable for low-power portable electronic devices. As a result, a novel membrane-free or membrane less microfluidic fuel cell (MMFC) concept has recently been introduced in which reactant stream lamination allows the removal of PEM [107]. Because fuel cells have stringent air cleanliness requirements, oil-free dry compressors are commonly used. To avoid oil pollution, female and male rotors in twin screw compressors are driven by synchronous gears. Twin screw compressors have played an important role in industry in recent decades; however, the twin screw compressor rotor profiles were mostly for oil compressors, such as SMR, GHH, Hitachi, and so on. Many dry twin screw compressor rotor profiles directly referred to the above profiles. Because of the benefits of environmental protection, pollution-free operation, and high energy conversion efficiency, hydrogen-oxygen vehicle fuel cells have a bright future in research and development [108].

## 7.7 SOLAR CELL

The primary goal of solar panels and other renewable energy generation technologies is sustainability. Unfortunately, solar panels have a paradox that must be resolved soon: they use silicon and other metals. While these are renewable materials that require recycling and proper manufacturing practices, they are non-biodegradable and require more energy to break down into raw materials than natural resources. Every manufacturer appears to be moving in both directions. Recently, scientists discovered the efficiency of perovskite, a much less expensive alternative to silicon, in the fabrication of solar films that aid in solar energy absorption. However, preliminary testing has yet to demonstrate perovskite's widespread adoption in existing solar panel systems. It is very likely that it will become a significant part of overall solar technology in the future. As a result, solar technology is at a fork in the road that leads to excellent sustainability practices. First, solar energy manufacturers can continue to improve their existing technologies, which will allow them to use fewer materials by improving efficiency, design, and longevity. As an alternative to silicon, plastic, and steel, natural and easily recyclable materials can be used.

Anaerobic-aerobic treatment solutions, such as upflow anaerobic sludge blanket (UASB) reactors followed by high-rate algal ponds (HRAP), have already proven to be effective for pollutant and micropollutant removal, as well as energy recovery from raw sewage and microalgal biomass co-digestion. Pre-treatment techniques may be used to improve microalgal biomass solubilization and methane yield because microalgae cells have complex structures that make them resistant to anaerobic digestion. Solar energy for biomass solubilization is one of the thermal pre-treatments [109]. The ability to remove (1) antibiotic residues; (2) their transformation products (TPs); (3) antibiotic resistance determinants; and (4) genes identifying the indicator bacteria in treated wastewater using solar light-driven photolysis and $TiO_2$ based photocatalysis (secondary effluent). The study included 16 antimicrobials from various classes and 45 by-products of their transformation. Tetracycline was the most susceptible to photochemical decomposition, being completely removed in the photocatalysis process and more than 80% removed in the solar light-driven photolysis. In the case of the tested genes, 83.8 percent removal (on average) was observed using photolysis and 89.9 percent removal using photocatalysis, with the genes sul1, uidA, and intI1 showing the highest degree of removal by both methods [110]. When petroleum-based plastic debris enters the ocean, it releases dissolved organic carbon (DOC) and has the potential to leach fluorescent dissolved organic matter (FDOM). Although DOC is available for microbial uptake, FDOM bioavailability has received little attention. Although petrol-based plastic is the most common type of plastic found in the ocean, the use of biodegradable plastic has grown significantly [111]. Natural pigments derived from the peels of *Musa paradisiaca, Mangifera indica, Punica granatum,* and *Ananas comosus,* as well as their conversion efficiency in dye-sensitized solar cells (DSSCs). The properties of the zinc oxide (ZnO) photo-anode film were investigated. Natural extracts were used as photosensitizers in the fabrication of DSSCs. Various spectroscopy techniques were used to examine the properties of natural pigments. The solar to electrical energy efficiencies of natural dye-based on ZnO DSSCs were determined to be 0.009 percent, 0.024 percent, 0.010 percent, and 0.002 percent for *Musa paradisiaca*, *Mangifera indica*, *Punica granatum*, and *Ananas comosus*, respectively. It was possible to convert visible light into electricity, resulting in improved photoelectric characteristics [112]. Because flexible solar cells are required for portable applications, materials that are light and soft are required.

One of the generations of solar cells that generate clean energy and require low-cost fabrication is the dye-sensitized solar cell. Natural dye-based solar cells perform better than synthetic dye-based solar cells. The green synthesis technique is used to produce zinc oxide (ZnO). The ZnO film's structural property exhibits good behavior with preferred (101) orientation. The x-ray diffraction (XRD) peak of (101) yielded the lattice parameters *a* and *c* of 2.9214 Å and 5.3600 Å, respectively. A UV-Vis spectrophotometer was used to measure the dyes' absorption spectra, and the dyes had high average absorption values of 0.2872, 1.6585, and 0.5886 for beetroot, Eosin y, and yombotumtum, respectively. The current voltage (*I-V*) measurements of the cells revealed that solar cells based on beetroot, yombotumtum, and Eosin y dyes have power conversion efficiencies (PCEs) of 0.75 percent, 1.17 percent, and 2.65 percent, respectively. This demonstrates that a readily available and environmentally friendly black mineral dye can be used as an active material in DSSCs [113]. With the world's population growing and energy resources dwindling as a result of pollution-caused global warming, a clean, green, low-cost, biodegradable, and compostable type of solar cell is urgently needed in the near future to save the human habitat on this planet. President Obama's Clean Power Act provides strong political backing for this line of work. By actively contributing to this goal, the scientific community can help make this policy a success. The US Department of Energy reports that significant progress has been made in the development of clean coal, low-cost photovoltaic cells, and solar cell conversion efficiencies of 43.5 percent. Solar cells that are highly efficient, clean, green, low cost, compostable, or biodegradable, on the other hand, are desperately needed [114]. Bioresorbable electronic materials are used as the foundation for implantable devices that provide active diagnostic or therapeutic function over a timeframe that corresponds to a biological process, then disappear within the body to avoid secondary surgical extraction. Power supply approaches in these physically transient systems are critical. This chapter describes a fully biodegradable monocrystalline silicon photovoltaic (PV) platform based on microscale cells (microcells) designed to operate at wavelengths with long penetration depths in biological tissues (red and near infrared wavelengths), allowing for realistic levels of power to be provided by external illumination [115]. Because of its complex composition, low biodegradability index, and high toxicity, mature landfill leachate is one of the most difficult effluents to treat. The performance of Galvanic Fenton (GF) and solar cells was investigated in this study. For the remediation of landfill leachate, galvanic Fenton (SGF) systems coupled to a biological reactor were evaluated [116].

Flexible perovskite solar cells (PSCs) on paper are promising biodegradable power sources for wearable and portable electronics, but fabrication of paper-based PSCs with high mechanical flexibility and efficiency remains difficult due to paper-based PSCs' limited optimization. The typical structure of cellophane/OMO/C60 pyrrolidinetris-acid (CPTA)/$CH_3NH_3PbI_3$/Spiro-OMeTAD/Au was demonstrated using 25 m cellophane paper substrates combined with $TiO_2$/ultrathin Ag/$TiO_2$ (OMO) electrodes. PSCs grown on cellophane paper demonstrated a high-power conversion efficiency (PCE) of 13.00%. More interestingly, the performance of paper-based PSCs showed minor degradation after bending with radiuses ranging from 0.3 to 10 mm: they retained 97.6% of the initial PCE after bending with a radius of 0.3 mm and even 95.8% of the initial PCE after bending with a radius of 1 mm for 1000 cycles. PSCs ultraflexibility was attributed to ultrathin substrates to relieve strain in bended devices and OMO electrodes superior flexibility [117]. A new biopolymer electrolyte for solar cells has been created and tested. To create a solid electrolyte, potassium iodide was added to a biopolymer matrix. Impedance spectroscopy reveals that salt doping increases ionic conductivity, with conductivity maxima obtained at 60 wt percent KI concentration. The formation

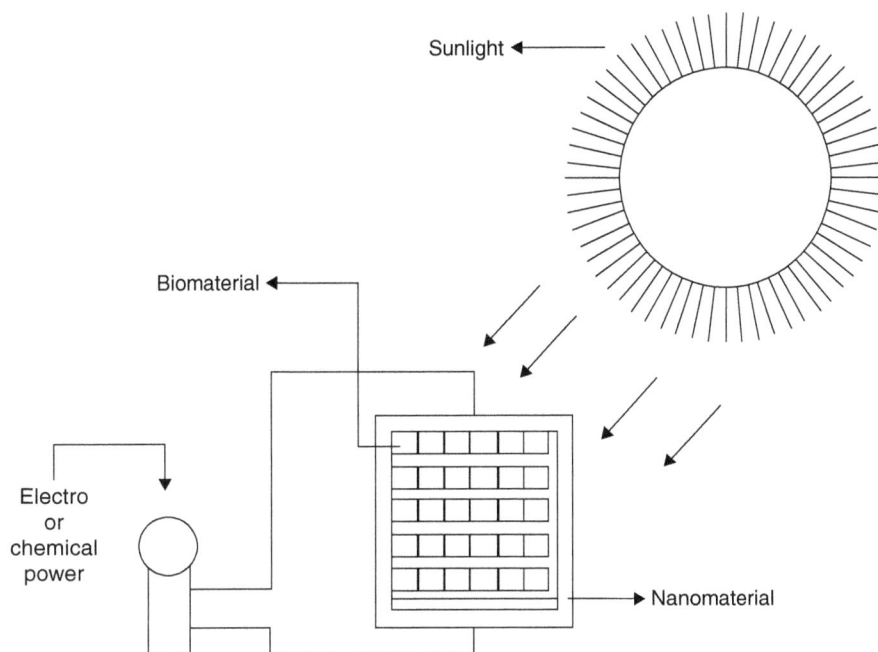

**FIGURE 7.7**   Schematic layout of biosolar cell.

of composite nature is confirmed by infrared spectroscopy, while x-ray diffraction and optical microscopy confirm the reduction of crystallinity of biopolymer electrolyte by salt doping. A dye sensitized solar cell with $J_{sc}$ of $5.68 \times 10^{-4}$ mA/cm$^2$, $V_{oc}$ of 0.56 volt, and efficiency of 0.63 percent was created using a maximum ionic conductivity electrolyte [118]. Weathering tests under outdoor conditions and accelerated degradation tests are carried out to simulate ageing processes over the lifetime of materials and to select appropriate stabilisers. Oxo- and biodegradable polymers have been studied and used in agricultural products over the last few decades. With recent advancements in the efficiency of PSCs, diverse functionalities for next-generation charge-transport layers are required. The hole-transport layer (HTL) in various synthesized materials modified with functional groups is specifically investigated. Organic photovoltaics, organic light-emitting diodes, and organic field effect transistors with environmentally friendly functions can be used [119–120].

Figure 7.7 shows the schematic layout of a biosolar cell. In this, the biomaterials' surface over the nano material surface absorbs the rays of sunlight (heat) and stores it in the solar cell. Two terminals were connected with the bulb from the solar cell. It converts the heat energy into electro or chemical power.

Modification of molecular properties of metal-free organic dyes is necessary in order to present sensitizers with competing electronic and optical properties for DSSCs. The effects of additional donors on the photophysical properties of the dyes were investigated using density-functional theory (DFT) and time-dependent density-functional theory (TDDFT). The donor subunit containing 2-hexylthiophene ($P_2$) demonstrated the highest chemical hardness, open circuit voltage ($V_{oc}$), and other comparable electronic properties among the optimized dyes [121]. Researchers have been inspired to develop renewable energy technologies as a result of the depletion of fossil fuel-based resources, their skyrocketing prices, and current environmental issues. Solar energy is one of the most important renewable resources. We created highly porous and conductive nanocomposites photoanodes of ZnO and $TiO_2$ for the first time by combining nanocellulose fibres (i.e. ZnO + LC and $TiO_2$+LC) obtained from the self-growing plant (Lily), which are promising for DSSC applications. This porous structure of lignocellulose (LC) based nanocomposites photoanodes not only simplifies the fabrication process towards low-cost techniques, but it also improves device efficiency by absorbing more dye and addressing the electrolyte leakage and stability issues of DSSCs. Furthermore, the additional structural advantages of these photoanodes reduce backward scattering and allow electron manipulation within the DSSCs [122].

The development of solar-powered evaporation techniques offers a promising solution to freshwater scarcity. Developing a highly efficient evaporation system, on the other hand, remains a challenge. This study used two-dimensional heterostructure nanosheets as a photothermal agent. We used the nanosheets to improve graphene absorption and facilitate heat localization due to their higher visible light absorption and ultralow thermal conductivity. Graphene also contributed low surface reflection and a broad absorption spectrum. DSSCs are a low-cost, easy-to-manufacture photovoltaic device in the hybrid organic-inorganic solar cell family. The DSSC's basic architecture consists of a semiconductor photoanode with dye molecules adsorbed on its surface, a counter electrode, and an electrolyte that fills the space between these electrodes. The electrolyte typically serves as a carrier for the iodide/triiodide redox species and is critical for internal charge transfer between the electrodes to ensure continuous dye replenishment. The performance of solar cells using these three-layer electrodes is compared to that of the same devices using ITO or FTO on plastic substrates as bottom electrodes or Au and Ag silver as top electrodes in the traditional configuration. It results in a 15% variation in efficiency performance as a function of materials or technologies [123–124].

Paper-based microbial fuel cells have received a lot of attention among many paper-based batteries and energy storage devices because bacteria can harvest electricity from any type of organic matter that is readily available in those difficult regions. The promise of this technology, however, has not been translated into practical power applications due to its short power duration, which is insufficient to fully operate those systems for a relatively long period of time. In this paper, we demonstrate for the first time a simple and long-lasting paper-based biological solar cell that employs photosynthetic bacteria as biocatalysts. Bacterial photosynthesis and respiration generate power continuously and self-sustainably by converting light energy into electricity [125]. Natural resource utilization also aims to increase the value of the natural resource itself. This review describes the applications of various natural sources for DSSC components, as well as the modifications that have been made to improve their performance. The following topics are covered in the discussion: the use of natural dyes as sensitizer dyes in liquid DSSC applications: (1) biopolymers for quasi-solid DSSC electrolytes; (2) green synthesis methods for photoanode semiconductors; and (3) natural carbon counter electrode development [126].

1D nanostructures such as rods, tubes, and fibers significantly improve charge collection efficiency by providing direct transport pathways to DSSCs. Furthermore, a mesoporous structure with a high specific surface area greatly increases dye loading, which improves photovoltaic current performance. Photovoltaic dye-sensitized solar cells have demonstrated the use of highly photochemically and thermally stable upconversion nanoparticles (UCNPs) (DSSCs). Lanthanide-based UCNPs have tunable near-infrared (NIR) absorption and a strong ability to transfer photons from higher wavelengths (NIR light) to lower wavelengths (visible light) via the UC procedure [127]. Polymer photovoltaic devices were created using a biodegradable poly-L-lactic acid (PLLA) substrate loaded with nanoclay to improve thermal properties and possibly reduce water and oxygen permeation. Compounding and extrusion were used to create the nanoclay-loaded PLLA substrates. It was discovered that PLLA as a substrate has potential, but there are several challenges beyond photovoltaics that must be overcome before widespread application in this field can be considered. The planarity of the PLLA surface, the mechanical stresses induced by the extrusion process, the processing temperature limitation, and the available range of solvents for solution processing are the most important aspects [128].

Eco-friendly printing is critical for the mass production of thin-film photovoltaic (PV) devices in order to protect human safety and the environment while also reducing energy consumption and capital expenditure. However, it is difficult for perovskite PVs due to a lack of environmentally friendly solvents for ambient fast printing. We present for the first time in this study an eco-friendly printing concept for high-performance perovskite solar cells. Under ambient conditions, both the perovskite and charge transport layers were created using eco-friendly solvents and scalable fast blade coating. By controlling film formation during scalable coating, the insights enable high-quality coatings for both the perovskite and charge transport layers. These coatings' excellent optoelectronic properties translate to a power conversion efficiency of 18.26 percent for eco-friendly printed solar cells, which is comparable to conventional devices fabricated via spin coating from toxic solvents in an inert atmosphere [129].

Organic solvents used in DSSC electrolytes are not only toxic and explosive, but they are also prone to leakage due to volatility and low surface tension. DSSC dyes are ruthenium-complex molecules that are expensive and require a complicated synthesis process. The effect of device aging time and electrolyte composition on the photovoltaic performance of eco-friendly DSSCs was investigated. $TiO_2$ plasma treatment was used to improve dye adsorption and the wettability of water-based electrolytes on $TiO_2$. The plasma treatment proved to be an effective method of improving the photovoltaic performance of the eco-friendly DSSCs by increasing efficiency by 3.4 times. The organic-synthetic dye was replaced with chlorophyll extracted from spinach for more environmentally friendly DSSCs [130].

Solar energy is arguably the most abundant renewable energy source at our disposal, but significant advances are required to make solar cells economically viable. Biodegradable and flexible solar cells are being researched extensively for environmentally friendly electronic applications. Biomaterials-based solar cells are gaining popularity due to their ability to produce energy that is sustainable, scalable, abundant, renewable, and environmentally friendly. This review highlights recent research progress in the emerging group of biomaterials and their integration for flexible solar cell devices. The greater the emphasis on absolute recyclable solar cell technology, processing conditions, and optimized processing conditions to produce a large amount of energy [131]. Solar energy has the potential to be the most abundant renewable energy source at our disposal, but significant advances are needed to make photovoltaic technologies economically viable, environmentally friendly, and thus scalable. Cellulose nanomaterials are new high-value nanoparticles derived from plants that are plentiful, renewable, and sustainable. The first demonstration of efficient polymer solar cells fabricated on optically transparent cellulose nanocrystal (CNC) substrates is presented here. Solar cells manufactured on CNC substrates exhibit good rectification in the dark and achieve a power conversion efficiency of 2.7 percent. Furthermore, we show that these solar cells can be easily separated and recycled into their major components at room temperature using low-energy processes, paving the way for truly recyclable solar cell technology. Organic solar cells on CNC substrates that are efficient and easily recyclable are expected to be an appealing technology for sustainable, scalable, and environmentally friendly energy production [132–133].

## 7.8  CONCLUSION

One of the key research areas addressing the rising ecological challenge associated with electronic waste is the creation of green and biodegradable electrical components. The risk associated with electronic waste would be significantly reduced for the specific instance of energy storage if energy storage technologies were biodegradable after their service lifetime. Furthermore, biodegradable and non-toxic materials must be the foundation of any energy storage systems created for biointegrated electronics. For the effective storage of electrochemical energy with direct applications in biomedical electronics, research integrating natural biopolymers and tiny biological molecules has shown outstanding results. Recent researches have offered several ideas for temporary, biodegradable energy systems and devices that not only give other electrical devices a steady and reliable power supply, but also lower environmental costs and biological hazards when the supply job is done. Even in the absence of any external power sources, transient, biodegradable, rechargeable electrochemical energy storage devices with superior packaging designs may offer a crucial alternative. Beyond a technological perspective, society is more concerned than ever before with the sustainability challenges surrounding energy gadgets. Since scientists have made significant progress in this area over the past 10 years, it is critical to close the big gap between academia and industry, as well as to simplify the regulatory environment. This would speed the transition of different innovations from labs to commercial applications and encourage the use of transient, biodegradable energy sources for a sustainable future. It is undeniable that transient, biodegradable energy systems have enormous potential to address many of the present bounds in both basic and practical research.

## REFERENCES

1. Rajaram K., S.M. Yang and S.W. Hwang, 2022. Transient, biodegradable energy systems as a promising power solution for ecofriendly and implantable electronics, *Advanced Energy & Sustainability Research*, 1–13.
2. Larcher, D and J.M. Tarascon, 2015. Towards greener and more sustainable batteries for electrical energy storage, *Nature Chemistry*, 7, 19–29.
3. Wang, H., J.D. Park and Z.J. Ren, 2015. Practical energy harvesting for microbial fuel cells: a review, *Environmental Science and Technology*, 49, 3267–3277.
4. Lim, S.R and J.M. Schoenung, 2010. Human health and ecological toxicity potentials due to heavy metal content in waste electronic devices with flat panel displays, *Journal of Hazardous Materials*, 177, 251–259.
5. Lisbana, D and T. Snee, 2011. A review of hazards associated with primary lithium and lithium-ion batteries, *Process Safety and Environmental Protection*, 89, 434–442.
6. Hwang, S.W., H, Tao, D. H. Kim, H. Cheng, J.K. Song, E., Rill, M.A. Brenckle, B., Panilaitis, S.M. Won, Y.S. Kim, Y.M. Song, K. J. Yu, A., Ameen, R., Li, Y., Su, M., Yang, D.L. Kaplan, M.R. Zakin, M.J. Slepian, Y., Huang, F.G. Omenetto and J.A. Rogers, 2012. A physically transient form of silicon electronics, *Science*, 337, 1640–1644.
7. Kang, S.K., R. K.J. Murphy, S.W. Hwang, S. M. Lee, D.V. Harburg, N. A. Krueger, J., Shin, P., Gamble, H., Cheng, S., Yu, Z., Liu, J. G. McCall, M., Stephen, H., Ying, J., Kim, G., Park, R.C. Webb, C. H. Lee, S., Chung, D.S. Wie, A. D. Gujar, B., Vemulapalli, A.H. Kim, K.M. Lee, J., Cheng, Y., Huang, S. H. Lee, P. V. Braun, W. Z. Ray and J. A. Rogers, 2016. Bioresorbable silicon electronic sensors for the brain, *Nature* 30, 71–76.
8. Yu, K.J., D., Kuzum, S.W. Hwang, B. H. Kim, H., Juul, N. H. Kim, S. M. Won, K., Chiang, M. Trumpis, A. G. Richardson, H., Cheng, H., Fang, M., Thompson, H., Bink, D.,Talos, K. J. Seo, H. N. Lee, S.K. Kang, J.H. Kim, J.Y. Lee, Y., Huang, F. E. Jensen, M. A. Dichter, T. H. Lucas, J., Viventi, B., Litt and J. A. Rogers, 2016. Bioresorbable silicon electronics for transient spatiotemporal mapping of electrical activity from the cerebral cortex, *Nature Materials*, 15, 782–791.
9. Chen, G., N., Matsuhisa, Z., Liu, D., Qi, P., Cai, Y., Jiang, C.,Wan, Y.,Cui, W.R. Leow, Z., Liu, S., Gong, K.Q. Zhang, Y., Cheng and X. Chen, 2018. Plasticizing silk protein for on-skin stretchable electrodes, *Advanced Materials*, 30, e1800129.
10. Shou, W., B.K. Mahajan, B.,Ludwig, X.,Yu, J. Staggs, X. Huang and H. Pan, 2017. Low-cost manufacturing of bioresorbable conductors by evaporation-condensation-mediated laser printing and sintering of Zn nanoparticles, *Advanced Materials*, 29, 1700172.
11. Tian, W., Y., Li, J., Zhou, T., Wang, R. Zhang, J., Cao, M., Luo, N., Li, N., Zhang, H., Gong, J., Zhang, L., Xie and B. Kong, 2021. Implantable and biodegradable micro-supercapacitor based on a superassembled three-dimensional network Zn@PPy hybrid electrode, *ACS Applied Materials and Interfaces*, 13, 8285–8293.
12. Figueroa-Gonzalez, E., R., Mendoza, A.I. Oliva, C., Gomez-Solis, L.S. Valle-Garcia, E., Oliva, V., Rodriguez-Gonzalez, A., Encinas and J. Oliva, 2022. Highly efficient and biodegradable flexible supercapacitors fabricated with electrodes of coconut-fiber/graphene nanoplates, *Journal of Physics D: Applied Physics*, 55, 035501.
13. Mei, T., C., Wang, M., Liao, J., Li, L., Wang, C., Tang, X., Sun, B., Wang and H. Peng, 2021. A biodegradable and rechargeable fiber battery, *Journal of Materials Chemistry A*, 9, 10104–10109

14. Obileke, K., H., Onyeaka, E.L. Meyer and N. Nwokolo, 2021. Microbial fuel cells, a renewable energy technology for bio-electricity generation: A mini-review, *Electrochemistry Communications*, 125, 107003.
15. Lee, M.H.., J.L. Thomas, W.J. Chen, M.H. Li, C.P. Shih and H.Y. Lin, 2015. Fabrication of bacteria-imprinted polymer coated electrodes for microbial fuel cells, *ACS Sustainable Chemistry & Engineering*, 3, 1190–1196.
16. Nyholm, L., G. Nystr, A. Mihranyan and M. Stromme., 2011. Toward flexible polymer and paper-based energy storage devices. *Advanced Materials*, 23(33), 3751–3769.
17. Lota, G., K. Fic and E. Frackowiak E., 2011. Carbon nanotubes and their composites in electrochemical applications. *Energy & Environmental Science*, 4(5), 1592–1605.
18. Zhang, L.L and X.S Zhao., 2009. Carbon-based materials as supercapacitor electrodes. *Chemical Society Reviews*, 38(9), 2520–2531.
19. Kumar, U.N., S. Ghosh and T. Thomas., 2019. Metal oxynitrides as promising electrode materials for supercapacitor applications. *ChemElectroChem*, 6(5), 1255–1272.
20. Zhai, Z., W. Yan., L. Dong., J. Wang., C. Chen., J. Lian., X. Wang., D. Xia and J. Zhang., 2020. Multi-dimensional materials with layered structures for supercapacitors: advanced synthesis, supercapacitor performance and functional mechanism. *Nano Energy*, 78, 105193.
21. Agate, S., M. Joyce., L. Lucia, and L. Pal., 2018. Cellulose and nanocellulose-based flexible-hybrid printed electronics and conductive composites-A review. *Carbohydrate polymers*, 198, 249–260.
22. Svorc, L., M. Rievaj and D. Bustin., 2013. Green electrochemical sensor for environmental monitoring of pesticides: Determination of atrazine in river waters using a boron-doped diamond electrode. *Sensors and Actuators B: Chemical*, 181, 294–300.
23. Xu, W., Q. Xu., Q. Huang., R. Tan. W. Shen and W. Song., 2015. Electrically conductive silver nanowires-filled methylcellulose composite transparent films with high mechanical properties. *Materials Letters*, 152, 173–176.
24. Boutry, C.M., Y. Kaizawa., B.C Schroeder., A. Chortos., A. Legrand., Z. Wang., J. Chang., P. Fox and Z. Bao., 2018. A stretchable and biodegradable strain and pressure sensor for orthopaedic application. *Nature Electronics*, 1(5), 314–321.
25. Rossiter, J., J. Winfield and I. Ieropoulos., 2016, April. Here today, gone tomorrow: biodegradable soft robots. In *Electroactive polymer actuators and devices (EAPAD) 2016* (Vol. 9798, pp. 312–321). SPIE.
26. Liu, H., H. Qing., Z. Li., Y. L Han., M. Lin., H. Yang., A. Li., T. J Lu., F. Li., and F. Xu., 2017. A promising material for human-friendly functional wearable electronics. *Materials Science and Engineering: R: Reports*, 112, 1–22.
27. Walker, S., J. Rueben., T.V Volkenburg., S. Hemleben., C. Grimm., J. Simonsen and Y. Menguc., 2017. Using an environmentally benign and degradable elastomer in soft robotics. *International Journal of Intelligent Robotics and Applications*, 1(2), 124–142.
28. Amjadi, M., K. U Kyung., I. Park and M. Sitti., 2016. Stretchable, skin-mountable, and wearable strain sensors and their potential applications: a review. *Advanced Functional Materials*, 26(11), 1678–1698.
29. Pedersen, S.M., S. Fountas, H. Have and B. S Blackmore., 2006. Agricultural robots-system analysis and economic feasibility. *Precision agriculture*, 7(4), 295–308.
30. Migliorini, L., T. Santaniello, F. Borghi, P. Saettone, M. Comes Franchini, G. Generali, and P. Milani, P., 2020. Eco-friendly supercapacitors based on biodegradable poly (3-hydroxy-butyrate) and ionic liquids. *Nanomaterials*, 10(10), 2062.
31. Bhoyate, S., C.K. Ranaweera, C. Zhang, T. Morey, M. Hyatt, P.K. Kahol, M. Ghimire, S.R. Mishra, and R.K. Gupta, 2017. Eco-friendly and high performance supercapacitors for elevated temperature applications using recycled tea leaves. *Global Challenges*, 1(8), 1700063.
32. Ramasahayam, S.K., Z. Hicks, and T. Viswanathan, 2015. Thiamine-based nitrogen, phosphorus, and silicon tri-doped carbon for supercapacitor applications. *ACS Sustainable Chemistry & Engineering*, 3(9), 2194–2202.
33. Xie, Q., R. Bao, A. Zheng, Y. Zhang, S. Wu, C. Xie, and P. Zhao, 2016. Sustainable low-cost green electrodes with high volumetric capacitance for aqueous symmetric supercapacitors with high energy density. *ACS Sustainable Chemistry & Engineering*, 4(3), 1422–1430.

34.  Huang, T., Z. Qiu, D. Wu, and Z. Hu, 2015. Bamboo-based activated carbon@ $MnO_x$ nanocomposites for flexible high-performance supercapacitor electrode materials. *International Journal of Electrochemical Science*, 10, 6312–6323.

35.  Deng, L., R.J. Young, I.A. Kinloch, A.M. Abdelkader, S.M. Holmes, D.A. De Haro-Del Rio, and S.J. Eichhorn, 2013. Supercapacitance from cellulose and carbon nanotube nanocomposite fibers. *ACS Applied Materials & Interfaces*, 5(20), 9983–9990.

36.  Liu, B., L. Zhang, P. Qi, M. Zhu, G. Wang, Y. Ma, X. Guo, H. Chen, B. Zhang, Z. Zhao, and B. Dai, 2016. Nitrogen-doped banana peel-derived porous carbon foam as binder-free electrode for supercapacitors. *Nanomaterials*, 6(1), 18.

37.  Boota, M., M.P. Paranthaman, A.K. Naskar, Y. Li, K. Akato, and Y. Gogotsi, 2015. Waste tire derived carbon-polymer composite paper as pseudocapacitive electrode with long cycle life. *ChemSusChem*, 8(21), 3576–3581.

38.  Misnon, I.I., N.K.M. Zain, R. Abd Aziz, B. Vidyadharan, and R. Jose, 2015. Electrochemical properties of carbon from oil palm kernel shell for high performance supercapacitors. *Electrochimica Acta*, 174, 78–86.

39.  Kumar, M.S. and D.K. Bhat, 2009. $LiClO_4$- doped plasticized chitosan as biodegradable polymer gel electrolyte for supercapacitors. *Journal of Applied Polymer Science*, 114(4), 2445–2454.

40.  Landi, G., L. La Notte, A.L. Palma, A. Sorrentino, M.G. Maglione and G. Puglisi, G., 2021. A comparative evaluation of sustainable binders for environmentally friendly carbon-based supercapacitors. *Nanomaterials*, 12(1), 46.

41.  Chen, C. and L. Hu, 2018. Nanocellulose toward advanced energy storage devices: structure and electrochemistry. *Accounts of Chemical Research*, 51(12), 3154–3165.

42.  Fu, Q., Y. Wang, S. Liang, Q. Liu, and C. Yao, 2020. High-performance flexible freestanding polypyrrole-coated CNF film electrodes for all-solid-state supercapacitors. *Journal of Solid State Electrochemistry*, 24(3), 533–544.

43.  Chen, Y., S. Lyu, S. Han, Z. Chen, W. Wang, and S. Wang, 2018. Nanocellulose/polypyrrole aerogel electrodes with higher conductivity via adding vapor grown nano-carbon fibers as conducting networks for supercapacitor application. *RSC Advances*, 8(70), 39918–39928.

44.  Yanilmaz, M., M. Dirican, A.M. Asiri, and X. Zhang, 2019. Flexible polyaniline-carbon nanofiber supercapacitor electrodes. *Journal of Energy Storage*, 24, 100766.

45.  Chen, G., T. Chen, K. Hou, W. Ma, M. Tebyetekerwa, Y. Cheng, W. Weng, and M. Zhu, 2018. Robust, hydrophilic graphene/cellulose nanocrystal fiber-based electrode with high capacitive performance and conductivity. *Carbon*, 127, 218–227.

46.  Santos, M.C., D.R. da Silva, P.S. Pinto, A.S. Ferlauto, R.G. Lacerda, W.P. Jesus, T.H. da Cunha, P.F. Ortega, and R.L. Lavall, 2020. Buckypapers of carbon nanotubes and cellulose nanofibrils: foldable and flexible electrodes for redox supercapacitors. *Electrochimica Acta*, 349, 136241.

47.  Xiao, L., H. Qi, K. Qu, C. Shi, Y. Cheng, Z. Sun, B. Yuan, Z. Huang, D. Pan, and Z. Guo, 2021. Layer-by-layer assembled free-standing and flexible nanocellulose/porous $Co_3O_4$ polyhedron hybrid film as supercapacitor electrodes. *Advanced Composites and Hybrid Materials*, 4(2), 306–316.

48.  Guo, X., Q. Zhang, Q. Li, H. Yu, and Y. Liu, Y., 2019. Composite aerogels of carbon nanocellulose fibers and mixed-valent manganese oxides as renewable supercapacitor electrodes. *Polymers*, 11(1), 129.

49.  Cai, N., J. Fu, V. Chan, M. Liu, W. Chen, J. Wang, H. Zeng, and F. Y, 2019. MnCo2O4 nitrogen-doped carbon nanofiber composites with meso microporous structure for high-performance symmetric supercapacitors. *Journal of Alloys and Compounds*, 782, 251–262.

50.  J. Wang, F. Zhang, Z. Xu, W. Hu, H. Jiang, L. Liu, L. Gai, 2021. Gallium oxynitride @ carbon cloth with impressive electrochemical performance for supercapacitors, *Chemical Engineering Journal*, 411, 128481.

51.  Muzaffar, M.B., A. M. Basheer, K. Deshmukh, J. Thirumalai, 2019. A review on recent advances in hybrid supercapacitors: Design, fabrication and applications, *Renewable and Sustainable Energy Reviews*, 101 123–145.

52.  Colherinhas, G., T. Malaspina, T. and E.E. Fileti, 2018. Storing energy in biodegradable electrochemical supercapacitors. *ACS Omega*, 3(10), 13869–13875.

53.   Liu, B., L. Zhang, P. Qi, M. Zhu, G. Wang, Y. Ma, X. Guo, H. Chen, B. Zhang, Z. Zhao and B. Dai, 2016. Nitrogen-doped banana peelerived porous carbon foam as binder-free electrode for supercapacitors. *Nanomaterials*, 6(1), 18.
54.   Taer, E., R.Taslim, Z. Aini, S.D. Hartati, W.S. Mustika. 2017. Activated carbon electrode from banana-peel waste for supercapacitor applications. *AIP Conference Proceedings*, 1801, 40004.
55.   Yao, Q., H. Wang, H., C. Wang, C. Jin and Q. Sun, 2018. One step construction of nitrogen-carbon derived from bradyrhizobium japonicum for supercapacitor applications with a soybean leaf as a separator. *ACS Sustainable Chemistry & Engineering*, 6(4), 4695–4704.
56.   Sun, K., C.Y. Leng, J.C. Jiang, Q. Bu, G.F. Lin, X.C. Lu, and G.Z. Zhu, 2017. Microporous activated carbons from coconut shells produced by self-activation using the pyrolysis gases produced from them, that have an excellent electric double layer performance. *New Carbon Materials*, 32(5), 451–459.
57.   Biswal, M., A. Banerjee, M. Deo and S. Ogale, 2013. From dead leaves to high energy density supercapacitors. *Energy & Environmental Science*, 6(4), 1249–1259.
58.   Mondal, A.K., K. Kretschmer, Y. Zhao, H. Liu, C. Wang, B. Sun and G. Wang, G., 2017. Nitrogen-doped porous carbon nanosheets from eco-friendly eucalyptus leaves as high performance electrode materials for supercapacitors and lithium ion batteries. *Chemistry-A European Journal*, 23(15), 3683–3690.
59.   Selvamani, V., R. Ravikumar, V. Suryanarayanan, D. Velayutham and S. Gopukumar, 2016. Garlic peel derived high capacity hierarchical N-doped porous carbon anode for sodium/lithium ion cell. *Electrochimica Acta,* 190, 337–345.
60.   Zhang, Q., K. Han, S. Li, M. Li, J. Li and K. Ren, 2018. Synthesis of garlic skin-derived 3D hierarchical porous carbon for high-performance supercapacitors. *Nanoscale*, 10(5), 2427–2437.
61.   Senthilkumar, S.T., N. Fu, Y. Liu, Y. Wang, L. Zhou and H. Huang, 2016. Flexible fiber hybrid supercapacitor with $NiCo_2O_4$ nanograss@ carbon fiber and bio-waste derived high surface area porous carbon. *Electrochimica Acta,* 211, 411–419.
62.   Zhang, Y., Z. Gao, N. Song and X. Li, 2016. High-performance supercapacitors and batteries derived from activated banana-peel with porous structures. *Electrochimica Acta,* 222, 1257–1266.
63.   Teo, E.Y.L., L. Muniandy, E.P. Ng, F. Adam, A.R. Mohamed, R. Jose and K.F. Chong, 2016. High surface area activated carbon from rice husk as a high performance supercapacitor electrode. *Electrochimica Acta,* 192, 110–119.
64.   Huang, Y., L. Peng, Y.L. Liu, G. Zhao, J.Y. Chen and G. Yu,, 2016. Biobased nano porous active carbon fibers for high-performance supercapacitors. *ACS Applied Materials & Interfaces*, 8(24), 1520–15215.
65.   Zhai, Z., W. Yan, L. Dong, J. Wang, C. Chen, J. Lian, X. Wang, D. Xia, and J. Zhang, 2020. Multidimensional materials with layered structures for supercapacitors: Advanced synthesis, supercapacitor performance and functional mechanism. *Nano Energy,* 78, 105193.
66.   Sim, H.J., C. Choi, D.Y. Lee, H. Kim, J.H. Yun, J.M. Kim, T.M. Kang, R. Ovalle, R.H. Baughman, C.W. Kee, and S.J. Kim, 2018. Biomolecule based fiber supercapacitor for implantable device. *Nano Energy,* 47, 385–392.
67.   Sheng, H., J. Zhou, B. Li, Y. He, and C. Yu. 2021. A thin, deformable, high-performance supercapacitor implant that can be biodegraded and bioabsorbed within an animal body. *Scientific Advances* 7, eabe3097.
68.   Gao, C., C. Bai, J. Gao, Y. Xiao, Y. Han, A. Shaista,Y. Zhao and L. Qu, 2020. A directly swallowable and ingestible micro-supercapacitor. *Journal of Materials Chemistry A*, 8(7), 4055–4061.
69.   Lee, G., S.K. Kang, S.M. Won, P. Gutruf, Y.R. Jeong, J. Koo, S.S. Lee, J.A. Rogers and J.S. Ha, 2017. Fully biodegradable microsupercapacitor for power storage in transient electronics. *Advanced Energy Materials*, 7(18), 1700157.
70.   Lee, M. H., J. Lee, S.K. Kang, M. S Park, Y.S Yun and S. J Kim, 2021. A biodegradable secondary battery and its biodegradation mechanism for Eco-Friendly energy storage systems. *Advanced Materials*, 33(10), 2004902.
71.   Esquivel, J.P., P. Alday, O.A. Ibrahim, B. Fernandez, E. Kjeang and N. Sabate, 2017. A metal-free and biotically degradable battery for portable single-use applications. *Advanced Energy Materials*, 7(18), 1700275.

72.   She, D., M. Tsang and M. Allen, 2019. Biodegradable batteries with immobilized electrolyte for transient MEMS. *Biomedical Microdevices*, 21(1), 1–9.

73.   Gellner, S., A. Schwarz-Pfeiffer and E. Nannen, 2021. Textile-based battery using a biodegradable gel-electrolyte. *Multidisciplinary Digital Publishing Institute Proceedings*, 68(1), 17.

74.   Delaporte, N., G. Lajoie, S. Collin-Martin and K. Zaghib, 2020. Toward low-cost all-organic and biodegradable Li-ion batteries. *Scientific Reports*, 10(1), 1–18.

75.   Tang, Z., Z. Pei, Z. Wang, H. Li, J. Zeng, Z. Ruan, Y. Huang, M. Zhu, Q. Xue, J. Yu and C. Zhi, 2018. Highly anisotropic, multichannel wood carbon with optimized heteroatom doping for supercapacitor and oxygen reduction reaction. *Carbon*, 130, 532–543.

76.   Zheng, Y., Y. Lu, X. Qi, Y. Wang, L. Mu, Y. Li, Q. Ma, J. Li and Y.S. Hu, 2019. Superior electrochemical performance of sodium-ion full-cell using poplar wood derived hard carbon anode. *Energy Storage Materials*, 18, 269–279.

77.   Li, L., W. Han, L. Pi, P. Niu, J. Han, C. Wang, B. Su, H. Li, J. Xiong, Y. Bando and T. Zhai, 2019. Emerging in-plane anisotropic two-dimensional materials. *InfoMat*, 1(1), 54–73.

78.   Jiao, M., T. Liu, C. Chen, M. Yue, G. Pastel, Y. Yao, H. Xie, W. Gan, A. Gong, X. Li, and L. Hu, 2020. Holey three-dimensional woodbased electrode for vanadium flow batteries. *Energy Storage Materials*, 27, 327–332.

79.   Xia, Y., H. Zhang, Y. Sun, L. Sun, F. Xu, S. Sun, G. Zhang, P. Huang, Y. Du, J. Wang and S.P. Verevkin, 2019. Dehybridization effect in improved dehydrogenation of $LiAlH_4$ by doping with two-dimensional $Ti_3C_2$. *Materials Today Nano*, 8, 100054.

80.   Huang, X., D. Wang, Z. Yuan, W. Xie, Y. Wu, R. Li, Y. Zhao, D. Luo, L. Cen, B. Chen and H. Wu, 2018. Biodegradable batteries: a fully biodegradable battery for self-powered transient implants (small 28/2018). *Small*, 14(28), 1870129.

81.   Huang, X., D. Wang, Z. Yuan, W. Xie, Y. Wu, R. Li, Y. Zhao, D. Luo, L. Cen, B. Chen and H. Wu, 2018. A fully biodegradable battery for self-powered transient implants. *Small*, 14(28), 1800994.

82.   Y.N. Lowrance, M. Azham Azmia, H.A. Rahman, H. Zakaria and S. Hassan, 2021. A review on the solid oxide fuel cells (SOFCs) interconnect coating quality on electrophoretic deposition (EPD) processing parameters, *AIP Conference Proceedings* 2339, 020182.

83.   C.J. Davey, B. Luqmani, N. Thomas and E.J. McAdam, 2022. Transforming wastewater ammonia to carbon free energy: Integrating fuel cell technology with ammonia stripping for direct power production, *Separation and Purification Technology*, 289, 120755.

84.   M. Zubair Khan, A. Iltaf, H.A. Ishfaq, F.N. Khan, W.H. Tanveer, R.-H. Song, M.T. Mehran, M. Saleem, A. Hussain and Z. Masaud, 2021. Flat-tubular solid oxide fuel cells and stacks: A review, *Journal of Asian Ceramic Societies*. 23,152–159.

85.   J. Zheng, C.-G. Wang, H. Zhou, E. Ye, J. Xu, Z. Li and X.J. Loh, 2021. Current research trends and perspectives on solid-state nanomaterials in hydrogen storage, *AAAS Research*, 2021, 3750689.

86.   Z. Yan, M. Colletta, A. Venkatesh, L.F. Kourkoutis, H.D. Abruna and T.E. Mallouk, 2022. Managing gas and ion transport in a PTFE fiber-based architecture for alkaline fuel cells, *Cell Reports, Physical Science*.

87.   Y. Chen, L. Lan, Z. Hao and P. Fu, 2022. Cradle-grave energy consumption, greenhouse gas and acidification emissions in current and future fuel cell vehicles: Study based on five hydrogen production methods in China, *Energy Reports*, 8, 7931–7944.

88.   J. Aubry, N. Yousfi Steiner, S. Morando, N. Zerhouni and D. Hissel, 2022. Fuel cell diagnosis methods for embedded automotive applications, *Energy Reports*, 8, 6687–6706

89.   M.A. Rubio, D.G. Sanchez, P. Gazdzicki, K.A. Friedrich and A. Urquia, 2022. Failure mode diagnosis in proton exchange membrane fuel cells using local electrochemical noise, *Journal of Power Sources*, 541, 231582.

90.   Y. Liu, S. Dirkes, M. Kohrn, M. Wick, S. Pischinger, 2022. A high-fidelity real-time capable dynamic discretized model of proton exchange membrane fuel cells for the development of control strategies, *Journal of Power Sources*, 537, 231394.

91.   J. Han, J. Feng, P. Chen, Y. Liu and X. Peng, 2022. A review of key components of hydrogen recirculation subsystem for fuel cell vehicles, *Energy Conversion and Management: X* 15, 100265.

92.   S. Hussain and L. Yangping, 2020. Review of solid oxide fuel cell materials: cathode, anode, and electrolyte, *Energy Transitions*, 4, 113–126.

93. Y.-W. Ju, J.-E. Hong and J. Hyodo., 2012. New buffer layer material La(Pr)CrO$_3$ for intermediate temperature solid oxide fuel cell using LaGaO$_3$-based electrolyte film, *Journal of Material Research*, 27, 15.

94. D. Panthi1, N. Hedayat, T. Woodson, B.J. Emley and Y. Du, 2019. Tubular solid oxide fuel cells fabricated by a novel freeze casting method, *Journal of American Ceramic Society*, 1–11.

95. X. Yang, J. Chen, D. Panthi, B. Niu, L. Lei, Z. Yuan, Y. Du, Y. Li, F. Chen and T. He, 2018. Electron doping of Sr$_2$FeMoO$_6$_d as high performance anode materials for solid oxide fuel cells, *Journal of Material Chemistry A*, 7, 733.

96. W. Wang, C. Su, Y. Wu, R. Ran and Z. Shao, 2012. Progress in solid oxide fuel cells with nickel-based anodes operating on methane and related fuels, *Chemical Reviews*, ACS, Publications.

97. L. Wehrle, Y. Wang, P. Boldrin, N.P. Brandon, O. Deutschmann and A. Banerjee, 2022. Optimizing solid oxide fuel cell performance to re-evaluate its role in the mobility sector, *ACS Environmental Au*, 2, 42–64.

98. K.A. Kuterbekov, A.V. Nikonov, K.Zh. Bekmyrza, N.B. Pavzderin, A.M. Kabyshev, M.M. Kubenova, G.D. Kabdrakhimova and N. Aidarbekov, 2022. Classification of solid oxide fuel cells, *Nanomaterials,* 2022(12), 1059.

99. K. Develos-Bagarinao, T. Ishiyama, H. Kishimoto, H. Shimada and K. Yamaji, 2021. Nanoengineering of cathode layers for solid oxide fuel cells to achieve superior power densities, *Nature Communications*, 12, 3979.

100. M.A.K. Yousaf Shah, Y. Lu, N. Mushtaq, S. Rauf, M. Yousaf, M.I. Asghar, P.D. Lund and B. Zhu, 2022. Demonstrating the potential of iron-doped strontium titanate electrolyte with high performance for low temperature ceramic fuel cells, *Renewable Energy*, S0960-1481(22), 00995-8.

101. T. Niu, W. Huang, C. Zhang, T. Zeng , J. Chen, Y. Li and Y. Liu, 2022. Study of degradation of fuel cell stack based on the collected high-dimensional data and clustering algorithms calculations, *Energy and AIS* 2666- 5468(22), 00033-7.

102. M.A. Gomez, A.J. Navarro, J.J. Giner-Casares, M. Cano, A.J. Fernandez-Romero and J.J. Lopez-Cascales, 2022. Electrodes based on nafion and epoxy-graphene composites for improving the performance and durability of open cathode fuel cells, prepared by electrospray deposition, *International Journal of Hydrogen Energy,* 47, 13980e13989.

103. R. Luca, M. Whiteley, T. Neville, P.R. Shearing and D.J.L. Brett, 2022. Comparative study of energy management systems for a hybrid fuel cell electric vehicle - A novel mutative fuzzy logic controller to prolong fuel cell lifetime, *International Journal of Hydrogen Eenergy* (online).

104. S. Rojas-Floresa, M. De La Cruz-Noriegaa, R. Nazario-Navedaa, Santiago M. Benitesa, D. Delfin-Narcisob, W. Rojas-Villacortac, Cecilia V. Romero, 2022. Bioelectricity through microbial fuel cells using avocado waste, *Energy Reports*, 8, 376–382.

105. Kimberly Waters, Diana Nguyen, Lauren Hernandez, Kelly Vu, Allyson Fry-Petit, Stevan Pecic, John L. Haan, 2022. The electrochemical oxidation of butanediol isomers in an alkaline direct liquid fuel cell, *Journal of Power Sources*, 535, 231401.

106. Muhammad Tanveer, Tehmina T. Ambreen, H. Khan, G. Man Kim and C. Woo Park, 2022. Paper-based microfluidic fuel cells and their applications: A prospective review, *Energy Conversion and Management*, 264, 115732.

107. M. Geng, L. Wang, D. Cui, X. Li, J. Li and H. Jiang, 2022. Profile design of twin screw air compressor for fuel cell, *Energy Reports,* 8, 21–26.

108. L. Vassalle, F. Passos, A. Trindade Rosa-Machado, C. Moreira, M. Reis, M. Pascoal de Freitas, I. Ferrer and C. Rossas Mota, 2022. The use of solar pre-treatment as a strategy to improve the anaerobic biodegradability of microalgal biomass in co-digestion with sewage, *Chemosphere,* 286, 131929.

109. E. Felis, M. Buta-Hubeny, W. Zielinski, J. Hubeny, M. Harnisz, S. Bajkacz and E. Korzeniewska, 2022. Solar-light driven photodegradation of antimicrobials, their transformation by-products and antibiotic resistance determinants in treated wastewater, *Science of the Total Environment,* 836, 155447.

110. R.-C. Cristina, M.-. Rebeca, S.-Y. Marola, A.-S.Xos'e Anton, 2022. Leaching and bioavailability of dissolved organic matter from petrol-based and biodegradable plastics *Marine Environmental Research*, 176, 105607.

111. O. Adedokun, O. Lydia Adedeji, I. Taiwo Bello, M. Kofoworola Awodele, A. Oladiran Awodugba, 2021. Fruit peels pigment extracts as a photosensitizer in ZnO-based dye-sensitized solar cells, *Chemical Physics Impact*, 3, 100039.

112. G.F. Gemeda, H.F. Etefa, C.-C. Hsieh, M.A. Kebede, T. Imae, Y.-W. Yen, 2022. Preparation of ZnO/NiO-loaded flexible cellulose nanofiber film electrodes and their application to dye-sensitized solar cells, *Carbohydrate Polymer Technologies and Applications*, 3, 100213.

113. A.N. Ossai, A.B. Alabi, S.C. Ezike, A.O. Aina, 2020. Zinc oxide-based dye-sensitized solar cells using natural and synthetic sensitizers, *Current Research in Green and Sustainable Chemistry*, 3, 100043.

114. C.V. Kumar and C. Baveghems, 2015. Biodegradable, biocompatible, bioinspired and bioabsorbale (edible) *Functional Materials for Solar Cell Applications*, 6, 1259–1268

115. L. Lu, Z. Yang, K. Meacham, C. Cvetkovic, E.A. Corbin, A. Vazquez-Guardado, M. Xue, L. Yin, J. Boroumand, G. Pakeltis, T. Sang, K. Jun Yu, D. Chanda, R. Bashir, R.W. Gereau IV, X. Sheng and J.A. Rogers, 2018. Biodegradable monocrystalline silicon photovoltaic microcells as power supplies for transient biomedical implants, *Advances in Energy Materials*, 2018, 1703035

116. L.A. Castillo-Suareza, V. Lugo-Lugob, I. Linares-Hernandeza, V. Martinez-Mirandaa, M. Esparza-Sotoa and M. de los Angeles Mier-Quirogaa, 2019. Biodegradability index enhancement of landfill leachates using a solar Galvanic-Fenton and Galvanic-Fenton system coupled to an anaerobic-aerobic bioreactor, *Solar Energy*, 188, 989–1001.

117. H. Lia, X. Lia, W. Wanga, J. Huanga, J. Lia, S. Huanga, B. Fane, J. Fanga and W. Songa, 2019. Ultraflexible and biodegradable perovskite solar cells utilizing ultrathin cellophane paper substrates and TiO2/Ag/TiO2 transparent electrodes, *Solar Energy*, 188, 158–163.

118. R. Singh, B. Bhattacharya, H.-W. Rhee, P.K. Singh., 2014. New biodegradable polymer electrolyte for dye sensitized solar cell, *International Journal of Electrochemical Science*, 9, 2620–2630.

119. P. Rizzarelli, M. Rapisarda, L. Ascione, F. D. Innocenti and F. Paolo La Mantia, 2021. Influence of photo-oxidation on the performance and soil degradation of oxo- and biodegradable polymer-based items for agricultural applications, *Polymer Degradation and Stability*, 188, 109578

120. J. Lee, G.-W. Kim, M. Kim, S. Ah Park, and T, Park, 2020. Nonaromatic Green-solvent-processable, dopant-free, and lead-capturable hole transport polymers in perovskite solar cells with high efficiency, *Advance in Energy Materials*, 2020, 1902662

121. S. Olusegun Afolabi, B. Semire and M. Abidemi Idowu, 2021. Electronic and optical properties' tuning of phenoxazine-based D-A2-n-A1 organic dyes for dye-sensitized solar cells. DFT/TDDFT investigations, *Heliyon*, 7, e06827

122. M. Saleem, M. Irfan, S. Tabassum, M.D. Albaqami, M.S. Javed, S. Hussain, M. Pervaiz, I. Ahmad, A. Ahmad and M. Zuber, 2021. Experimental and theoretical study of highly porous lignocellulose assisted metal oxide photoelectrodes for dye-sensitized solar cells, *Arabian Journal of Chemistry*, 14, 102937.

123. Y. Huang, Y. Zhao, Y. Liu, B. Xu, S. Xu and G. Bai, 2021. Lanthanide doped two dimensional heterostructure nanosheets with highly efficient harvest towards solar energy, *Materials & Design*, 210 110023.

124. V. Selvanathan, M.H. Ruslan, A.A. Nasser Alkahtani, N. Amin, K. Sopian, G. Muhammad and M. Akhtaruzzaman, 2021. Organosoluble, esterified starch as quasi-solid biopolymer electrolyte in dye-sensitized solar cell, *Journal of Materials Research and Technology*, 12, 1638e1648.

125. M. Girtan and B. Negulescu, 2022. A review on oxide/metal/oxide thin films on flexible substrates as electrodes for organic and perovskite solar cells, *Optical Materials: X*, 13, 100122.

126. L. Liu and S. Choi, 2020. A paper-based biological solar cell, *SLAS Technology 2020*, 25(1), 75–78.

127. Y. Kusumawati, A.S. Hutama, D.V. Wellia and R. Subagyo, 2021. Natural resources for dye-sensitized solar cells, *Heliyon* 7, e08436.

128. A.A. Ansari, M.K. Nazeeruddin and M.M. Tavakoli, 2021. Organic-inorganic upconversion nanoparticles hybrid in dye-sensitized solar cells *Coordination Chemistry Reviews*, 436, 213805.

129. M. Strange, D. Plackett, M. Kaasgaard and F.C. Krebs, 2008. Biodegradable polymer solar cells, *Solar Energy Materials & Solar Cells*, 92, 805–813.

130. X. Chang, Y. Fan, K. Zhao, J. Fang, D. Liu, M.-C. Tang, D. Barrit, D.-M. Smilgies, R. Li, J. Lu, J. Li, T. Yang, A. Amassian, Z. Ding, Y. Chen, S. (Frank) Liu, and W. Huang, 2021. Perovskite solar cells toward eco-friendly printing, *AAAS Research,* 2021, 9671892.

131. J.-H. Kim, S.-Y. Park, D.-H. Lim, S.-Y. Lim, J. Choi and H.-J Koo, 2021 Eco-friendly dye-sensitized solar cells based on water-electrolytes and chlorophyll, *Materials,* 14, 2150.

132. K. Kumar Sadasivuni, K. Deshmukh, T.N. Ahipa, A. Muzaffar, M. Basheer Ahamed, S.K. Khadheer Pasha and M. Al-Ali Al-Maadeed, 2018. Flexible, biodegradable and recyclable solar cells: a review, *Journal of Materials Science: Materials in Electronics.* 24, 1256–1267.

133. Y. Zhou, C. Fuentes-Hernandez, T.M. Khan, J.-C. Liu, J. Hsu, J. Won Shim, A. Dindar1, J.P. Youngblood, R.J. Moon and B. Kippelen, 2013. Recyclable organic solar cells on cellulose nanocrystal substrates, *Scientific Reports*, 42, 154–167.

# 8 High-Energy-Density Lithium-Ion Batteries for Future Power Systems

*Shunli Wang, Xiaoyong Yang, Chunmei Yu,*
*Josep M. Guerrero, and Yanxin Xie*

## 8.1 INTRODUCTION – OVERVIEW OF LITHIUM-ION BATTERIES

The applicaition scope of lithium-ion batteries, as one of the most important electrochemical energy storage devices, has been small capacity batteries in consumer electronic products, and power tool applications, and has gradually extended to new energy electric vehicles, electric ships, electric aircraft, robots, and other emerging fields; these areas not only require lithium-ion batteries to have a greater capacity but also put forward higher requirements for their energy density. Initially, Sony commercialized lithium-ion cell energy density of only 80 W·h/kg in 1991; nowadays, the energy density of lithium-ion batteries has reached 300 W·h/kg.

In the past, the energy density of lithium-ion batteries has satisfied the linear relationship with the increase over time, but in recent years, the increase in energy density has gradually slowed down. At present, the lithium-ion battery development plans proposed by countries around the world are mostly based on the linear increase in energy density to develop research and development goals, the Chinese, American, and Japanese governments plan to develop a battery prototype device with an energy density of 400 ~ 500 W·h / kg and achieve mass production in 2025–2030.

The improvement of the energy density of lithium-ion batteries is based on continuous optimization of existing materials and the search for new material combinations [1], as shown in Figure 8.1.

The choice of material determines the theoretical value of the energy density of lithium-ion batteries. The positive and negative electrode materials are the active energy storage materials of lithium-ion batteries, and the essence of increasing the energy density lies in improving the potential difference between the positive and negative electrodes and the specific capacity of the material. Due to the diversity of applications leading to the diversity of performance indicators, lithium-ion batteries will continue to be the common development of a variety of materials in the future. From the perspective of improving energy density, anode materials have developed from graphite anode materials to nano-silicon carbon anodes, and lithium composite anode materials may appear in the future; the cathode material is mainly the further optimization and replacement of the existing lithium cobalt oxide ($LiCoO_2$) [2–4], ternary layered (NCM/NCA)[5], lithium-rich manganese group (Li-riched or OLO) [6–9], lithium manganese oxide ($LiMn_2O_4$), lithium nickel manganese oxide ($LiNi_{0.5}Mn_{1.5}O_4$) cathode material. In addition, the selection of electrolyte, diaphragm, and conductive additive materials needs to be considered based on the interface compatibility with the positive and negative electrode materials and the performance of the material itself.

In the design process of the actual battery, the calculation of energy density needs to consider inactive substances, and the comprehensive technical specifications of the battery must be considered while achieving high energy density. The cell design of lithium-ion batteries is of great significance to the actual performance of the battery.

DOI: 10.1201/9781003340539-8

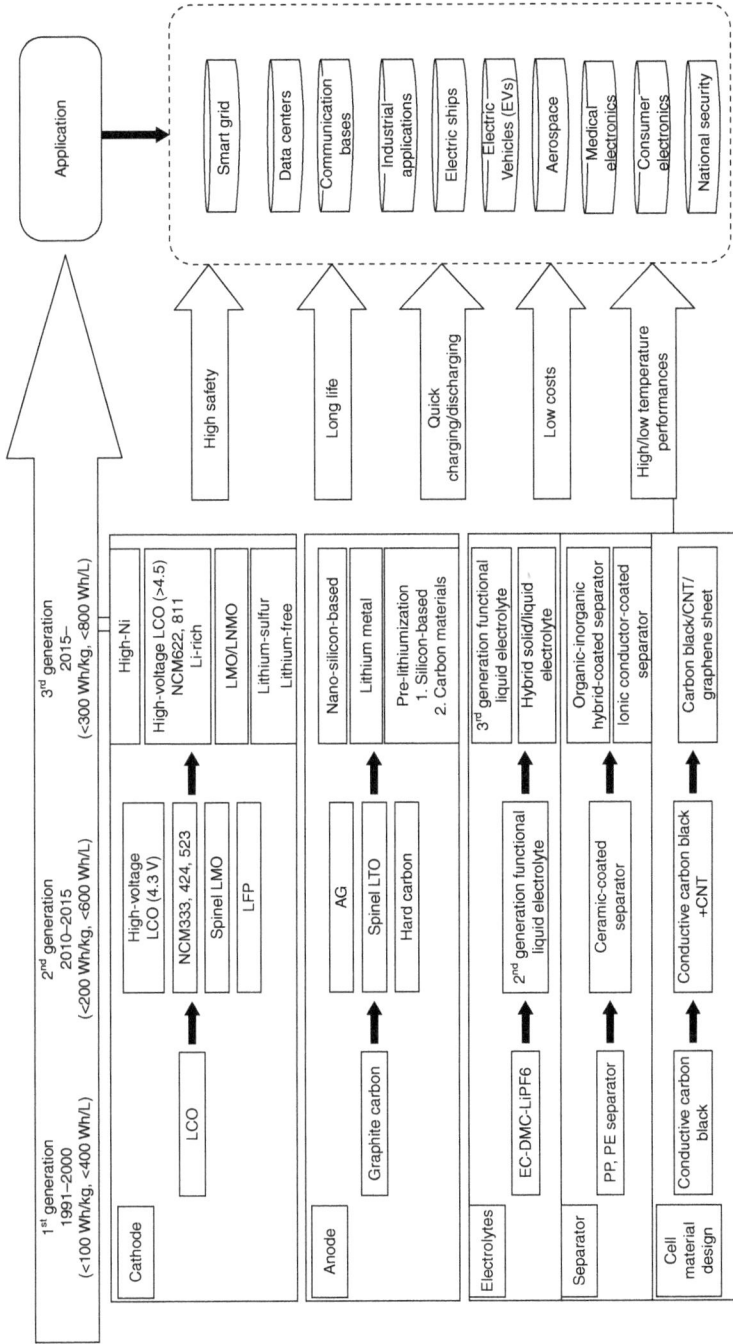

**FIGURE 8.1** Progress and upgrade of critical materials used for various generations of lithium-ion batteries, driven by performances, cost, and safety.

At the same time, the application of the battery not only energy density, but also the power density, charging rate, cycle life, service life, energy efficiency, safety index, single battery cost, and other technical indicators, whether the battery can be applied depends on whether a technical indicator can meet the minimum requirements of the application, called the battery "barrel effect."

Achieving a high-energy-density lithium-ion battery is a systematic project, which not only involves the improvement of the battery chemical system but also involves the upgrading of the whole industry chain system of the battery. The energy density of the battery is related to the voltage and capacity of the battery, and the simple pursuit of the high energy density of the battery cannot only not improve the performance of the battery, but also may affect the life and safety of the battery. Therefore, the upgrade of high-energy-density batteries not only represents the upgrading of lithium-ion batteries but also represents the progress and development of the power system and energy system.

## 8.2    INTRODUCTION OF LITHIUM-ION BATTERIES

### 8.2.1    LITHIUM-ION BATTERY WORKING PRINCIPLE

The internal chemical reaction of the lithium-ion battery is a basic redox reaction, which is also the working principle of a lithium-ion battery in the actual use process. It converts electric energy into heat energy through a chemical reaction. It can be observed from the chemical reaction equation that the charge-discharge process of lithium-ion batteries is the intercalation and deintercalation process of lithium ions. When a lithium-ion battery is charged, the lithium atoms in the positive electrode oxidize to lose electrons, thus becoming lithium-ions. A large number of lithium ions produced by the oxidation reaction to the positive electrode pass through the electrolyte solution to the carbon layer of the negative electrode of the batteries. The capacity of the battery is related to the number of lithium ions produced by the reaction to the positive electrode, on the one hand, and the number of lithium-ions exchanged with the negative electrode by the electrolyte, on the other hand.

During the discharging process, an oxidation reaction occurs in the negative electrode, and the lithium ions embedded in the carbon layer of the negative electrode come out and move back to the positive electrode. The more lithium ions return to the positive electrode, the higher the discharge capacity is. Similarly, during charging, lithium ions are generated in the positive electrode of the battery, and the generated lithium ions move to the negative electrode through the electrolyte. The lithium ions are embedded in the negative electrode and the micropores of the carbon layer. The more lithium ions embedded in the carbon layer, the higher the charging capacity. The internal chemical reaction process in the lithium-ion battery is shown in Figure 8.2.

Most of the cathode materials of lithium-ion batteries are lithium compounds, such as lithium cobalt oxide, lithium manganate, lithium iron phosphate, and ternary materials. Because of this, they are called lithium-ion batteries. The anode materials of lithium-ion batteries were originally made of alloys and metal lithium, but eventually became graphite for its better performance. The electrolyte is an organic solution that dissolves lithium salts. In general, the internal electrochemical reaction process of a lithium-ion battery is the exchange of lithium ions between the positive and negative poles. The positive and negative reactions and the overall reaction equations are described as follows [10]:

The positive electrode reaction is shown in Equation (8.1).

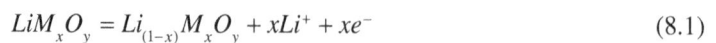

$$LiM_xO_y = Li_{(1-x)}M_xO_y + xLi^+ + xe^-$$    (8.1)

The negative electrode reaction is shown in Equation (8.2).

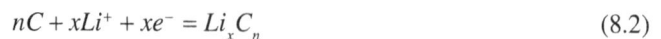

$$nC + xLi^+ + xe^- = Li_xC_n$$    (8.2)

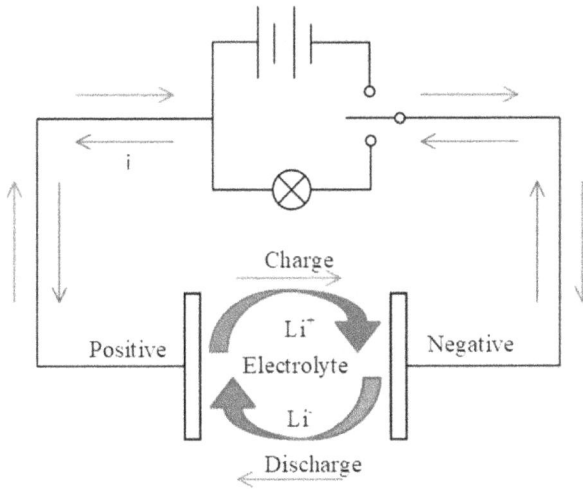

**FIGURE 8.2**   The working principle of lithium-ion batteries.

The total reaction of the battery is shown in Equation (8.3).

$$LiM_xO_y + nC = Li_{(1-x)}M_xO_y + Li_xC_n \qquad (8.3)$$

In Equation (8.3), $M$ can be Co, Mn, Fe, and Ni, which represent lithium cobalt oxide, lithium manganate, lithium iron phosphate, and lithium nickel oxide batteries, respectively. In the process of charging and discharging lithium-ion batteries, there is no metal lithium, only lithium ions. When the battery is charged, lithium ions are generated on the positive electrode of the battery, and the resulting lithium ions travel through the electrolyte to the negative electrode. The carbon as the negative electrode is layered and has many micropores. The lithium ions that reach the negative electrode are embedded in the pores of the carbon layer. The more lithium ions embedded in the carbon layer, the higher the charging capacity. Similarly, when a battery is discharged, the lithium ions embedded in the carbon layer of the negative electrode break out and move back to the positive electrode. The more lithium ions repolarize, the higher the discharge capacity.

The battery capacity refers to the discharge capacity. In the process of charge-discharge of lithium ion, it is in a state of motion from positive electrode to negative electrode to positive electrode. The working principle of lithium-ion batteries is different from the oxidation-reduction process of general batteries, but the insertion-deintercalation process of lithium ions. That is, they can be reversibly inserted or extracted from the host material. In the two stages of charge-discharge, intercalation and intercalation are carried out back and forth between the positive and negative electrodes. During charging, lithium ions are first intercalated from the positive electrode, reaching the negative electrode through the electrolyte, and inserting lithium ions in the negative electrode.

The negative electrode of the battery is rich in lithium and lithium-rich is a positive electrode material made with a small amount of lithium doped, which is combined with positive electrode active materials, such as $LiMn_2O_4$, etc. The lithium-rich condition can cause the unit cell to shrink and change from the cubature of charging-discharging and improve the structural stability of the material cycle performance. The process of discharging and charging is opposite to each other. The cathode material of a lithium-ion battery is composed of a lithium-intercalation compound. If there is an external electric field, the $Li^+$ in the cathode material can be extracted and inserted into the crystal lattice under the action of the electric field. Taking $LiCoO_2$ as an example, its positive electrode is shown in Equation (8.4).

$$LiCoO_2 \rightarrow xLi^+ + Li_{(1-x)}CoO_2 + xe^-  \qquad (8.4)$$

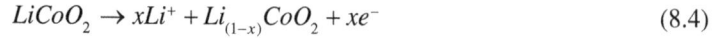

The negative electrode is shown in Equation (8.5).

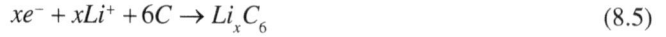

$$xe^- + xLi^+ + 6C \rightarrow Li_xC_6  \qquad (8.5)$$

The general equation of battery reaction is shown in Equation (8.6).

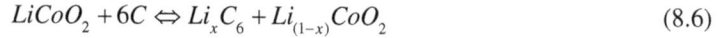

$$LiCoO_2 + 6C \Leftrightarrow Li_xC_6 + Li_{(1-x)}CoO_2  \qquad (8.6)$$

The lithium-ion battery has three important components: a diaphragm and positive and negative electrodes. Its operation mainly depends on the back-and-forth movement of ions between the positive and negative electrodes. The working principle of the lithium-ion battery mainly depends on the lithium-ion concentration difference between ions at both ends. During charging, $Li^+$ is de-embedded from the positive electrode and embedded into the negative electrode through the corresponding electrolyte. After this series of chemical reactions, the positive electrode is in the low lithium state and the negative electrode is in the multi lithium state. At the same time, its compensation charge is supplied to the negative electrode from the external circuit. During discharging, $Li^+$ is removed from the negative electrode and embedded into the positive electrode again through the action of the electrolyte.

### 8.2.2  Lithium-Ion Battery Composition

Lithium-ion batteries are mainly composed of four parts: positive electrode material, negative electrode material, diaphragm, and electrolyte. The cathode material determines the capacity and voltage of the batteries, which provides lithium ions for the batteries and determines the capacity and voltage. Commonly used materials of positive electrodes are lithium manganese oxide, and conductive fluid collection using 10 to 20-micron thickness of electrolytic aluminum foil. The anode material is mainly graphite, or carbon with a graphitic structure, and the conductive collector uses electrolytic copper foil with a thickness of 7 to 15 microns. The anode material at the anode acts as a current through an external circuit while allowing reversible absorption or emission of lithium ions released from the cathode. The main function of the anode is to store lithium ions and realize the insertion and intercalation reaction to lithium ions during the charge-discharge process.

The diaphragm is a special composite mold that prevents electrons from free shuttle between the positive and negative electrodes, while the lithium ions in the electrolyte can pass freely. The electrolytes are mainly materials with high ionic conductivity, which can make it easy for lithium ions to move back and forth, it is generally composed of lithium salt and organic solvent, and it has the function of conducting ions.

The diaphragm is an insulator essentially and free electrons cannot pass through it, so it does not conduct electricity. In the battery, the element exists in the form of ions, and the ions can easily pass through the separator, and the electrons leave the element on the new carrier, positive electrode material, or negative electrode material. When it is in contact with the separator, the separator cannot absorb the free electrons on the electrode, thereby preventing the passage of electrons. The common materials of the separator are single-layer PP film, PE film, and PP/PE/PP three-layer composite film. The electrolyte realizes the conduction of lithium ions between the positive and negative electrodes of the batteries. The cathode and anode determine the basic performance of the battery, while the electrolyte and diaphragm determine the safety of the battery. Currently, $LiPF_6$ is the most widely used electrolyte. There are lithium ions, metal ions, oxygen ions, and carbon layers in the batteries. The composition of the lithium-ion battery is made up of some compounds. The reaction

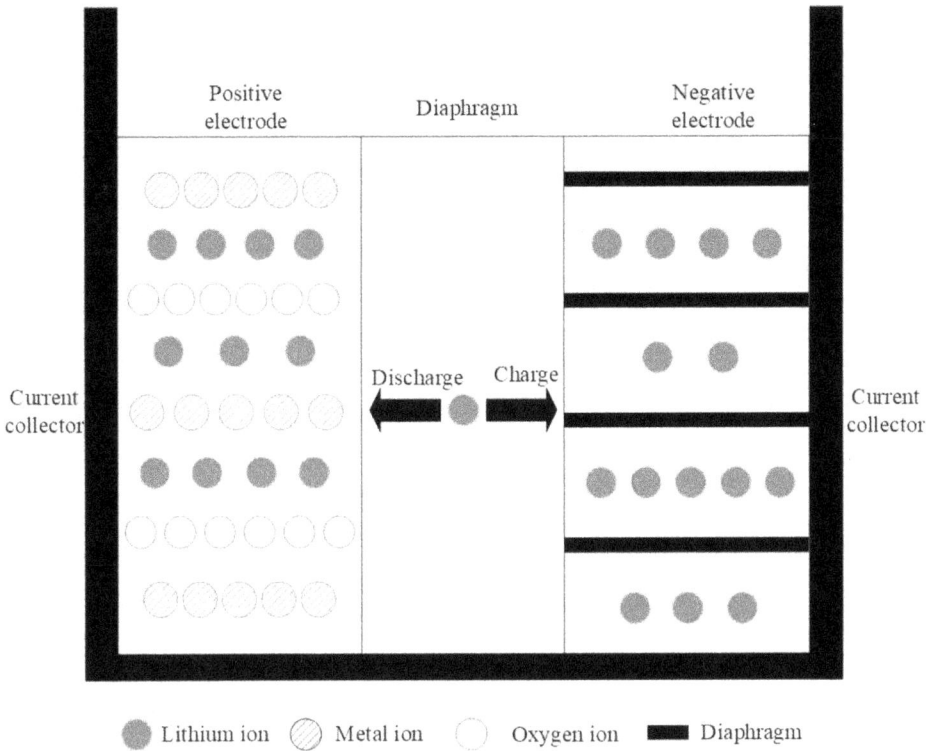

FIGURE 8.3 Schematic diagram of the lithium-ion battery structure.

is done by the movement of ions inside the cell. The cell's diaphragm acts as a barrier, separating the battery's poles. The internal structure of the battery is objective, as shown in Figure 8.3.

In Figure 8.3, the lithium-ion battery is an indispensable portable energy storage element. The performance of the lithium-ion battery is characterized by many external parameters, such as voltage, current, and internal resistance [11].

Table 8.1 shows the composition of lithium-ion batteries and the functions of each part. Detailed requirements for lithium-ion batteries are as follows:

### 8.2.2.1 Cathode Material

The cathode material of lithium-ion batteries is an important part of secondary lithium-ion batteries, which not only participates in electrochemical reactions as electrode materials but also acts as a lithium-ion source. When designing and selecting the cathode material for lithium-ion batteries, it is necessary to comprehensively consider the specific energy, cycle performance, safety, cost, and its impact on the environment.

The ideal lithium-ion battery cathode material should meet the following conditions: (1) large specific capacity: the cathode material is required to have a low relative molecular mass, and a large amount of $Li^+$ can be inserted into its host structure; (2) high working voltage: the negative value of Gibbs free energy of the system discharge reaction is required to be large; (3) good high rate performance of charge and discharge: the interior and surface of the electrode material are required to have a high diffusion rate; (4) good safety performance: the material is required to have high chemical stability and thermal stability; (5) easy to prepare, environmentally friendly, inexpensive.

Lithium-ion battery cathode materials are generally lithium-containing transition metal oxides and polyanionic compounds. Because transition metals tend to have multiple valence states, they

**TABLE 8.1**

**The Composition of Lithium-Ion Batteries and the Functions of Each Part**

|  | Composition | Notes |
|---|---|---|
| Cathode | Cathode materials, fluid collectors (aluminum foil conductive agents, binders) | The cathode is generally a transition metal oxide with a high redox potential, such as lithium iron phosphate materials, ternary materials, etc. The cathode binder is generally PVDF (polyatomic) |
| Anode | Anode material, collector (copper foil) conductive agent, binder | The negative electrode is an active material with a low redox potential that can embed lithium, such as graphite, Si, Sn alloy, etc. |
| Separator | Substrate + coating | The current diaphragm substrate is generally PP (polypropylene) or PE (polyethylene), and the coating is ceramic (alumina) particles. |
| Electrolyte | Lithium salt + organic solvent + additive | The electrolyte is the transmission medium between the positive and negative electrodes of lithium-ion batteries, generally, a carbonate organic solvent dissolved in lithium salts, and the lithium salts mainly include $LiPF_6$, $LiClO_4$ and the like. |
| Cell material design | Connectors, packages | The soft-packed battery core is an aluminum-plastic film and polar ear; aluminum shell battery cores are internal connection tape, internal protective film, external insulation film, housing, top cover, pole, and so on. |

can maintain the electrical neutrality of the lithium-ion embedding and shedding process; the lithium compound has a higher electromotive force relative to lithium, which can ensure that the battery has an open-circuit voltage. In general, relative to the potential of lithium, the transition metal oxide is greater than the transition metal sulfide.

At present, the commercial lithium-ion battery cathode material generally uses lithium compounds, such as $LiCoO_2$, which has a theoretical specific capacity of 274 mA·h·g$^{-1}$ and an actual specific capacity of about 146 mA·h·g$^{-1}$. $Li(NiCoMn)O_2$ ternary material, the theoretical specific capacity is similar to $LiCoO_2$, but the actual specific capacity varies slightly according to the composition. $LiMn_2O_4$ materials specific capacity 148 mA·h·g$^{-1}$, actual specific capacity 115 mA·h·g$^{-1}$; the theoretical specific capacity of $LiFePO_4$ materials is 170 mA·h·g$^{-1}$, and the actual specific capacity can reach about 150 mA·h·g$^{-1}$.

Nowadays, the main development idea of cathode materials is to develop related various derivative materials based on $LiCoO_2$, $LiMnO_2$, $LiFePO_4$ and other materials, of which the application of ternary material NCM is more extensive.

### 8.2.2.2 Anode Material

As an important part of lithium-ion batteries, the ideal negative electrode material should meet the following conditions: (1) the embedding Li reaction has a low redox potential, so that the lithium-ion battery has a higher output voltage; (2) in the process of Li embedding and shedding, the electrode potential changes less to ensure that the voltage fluctuation is small when charging and discharging; (3) the structural stability and chemical stability in the process of incarcetion and de-Li are better, so that the battery has a high cycle life and safety; (4) has a high reversible specific capacity; (5) good lithium ion and electronic conductivity to obtain higher charge-discharge rate and low-temperature charge-discharge performance; (6) if the embedded Li potential is below 1.2 V, the negative electrode surface should be able to generate a dense and stable solid electrolyte film (SEI), thereby preventing the electrolyte from continuously reducing on the negative electrode surface and irreversibly consuming the positive Li; (7) the preparation process is simple, easy to scale, and the manufacturing and use cost is low; (8) rich in resources and friendly in environment.

According to the reaction mechanism of the negative electrode and lithium, many negative electrode materials can be divided into three categories: insertion reaction electrode, alloy reaction electrode, and conversion reaction electrode. Among them, the insertion reaction electrode mainly refers to carbon anode and $TiO_2$-based anode material; the alloy reaction electrode specifically refers to tin or silicon-based alloys and compounds; the conversion reaction electrode refers to the metal oxide, metal sulfide, metal hydride, metal nitride, metal phosphide, metal fluoride and the like that are active against lithium through the conversion reaction.

At present, the negative electrode is mainly concentrated in carbon anode, lithium titanate, and silicon-based alloy materials, the use of traditional carbon anode meets the requirements of consumer electronics. Power batteries, energy storage batteries, and the use of lithium titanate as the negative electrode can meet the requirements of high power density and long cycle life of the battery, which is expected to further improve the battery energy density.

There are currently two types of commercialized lithium-ion battery anodes. One type is carbon materials, such as natural graphite, synthetic graphite, intermediate phase carbon microspheres (MCMB [12, 13]), and so on. Compared with natural graphite, MCMB electrochemical performance is superior, mainly because the outer surface of the particles in the edge surface of the graphite structure, the reactivity is uniform, and it is easy to form a stable SEI film [14, 15], which is conducive to the embedding and de-embedding of Li. There is also a class of $Li_4Ti_5O_{12}$ anode materials with spinel structures [16–21], which have a theoretical specific capacity of 175 mA·h·g$^{-1}$ and an actual specific capacity of up to 160 mA·h·g$^{-1}$. Although the $Li_4Ti_5O_{12}$ operating voltage is high, due to the particularly excellent cycle performance and magnification performance, it has safety advantages compared to carbon materials, so this material has a strong application demand for power and energy storage lithium-ion batteries. But the susceptibility to chemical reactions in the electrolyte causes flatulence to cause the battery to bulge.

The next generation of high-capacity anode materials includes Si anode and Sn-based alloys. However, alloy anode materials face the problem of high capacity with the high volume change, to solve the problem of material powdering caused by volume expansion, alloy, and carbon composite materials are often used, and composite materials can improve the energy density of existing lithium-ion batteries to a certain extent, but they are not as expected.

### 8.2.2.3  Electrolyte

The liquid electrolyte of lithium-ion batteries is generally composed of two parts: a non-aqueous organic solvent and an electrolyte lithium salt. The role of the electrolyte is to form a good ion-conductive channel between the positive and negative electrodes inside the battery. The following conditions should be met when the non-aqueous electrolyte is used in the lithium-ion battery system: (1) high conductivity, generally $3\times10^{-3}$~$2\times10^{-2}$ S·cm$^{-1}$; (2) good thermal stability, no decomposition reaction occurs in a wide temperature range; (3) the electrochemical window is wide and should be stable in the range of 0~4.5 V; (4) high chemical stability, do not react with the positive electrode, negative electrode, fluid collector, diaphragm, binder, etc.; (5) it has good solvation properties for ions; (6) no toxicity, low vapor pressure, safe to use; (7) it can promote the reversible reaction of the electrode as much as possible, which is easy to prepare and is low cost. Chemical stability, safety, and reaction rate are the main factors.

### 8.2.2.4  Separator

The main role of the separator is to separate the positive and negative electrodes of the battery, to avoid the contact between the positive and negative electrodes and a short circuit when the battery is inside due to the short circuit temperature rising to exceed the temperature tolerated by the separator. The commonly used PP/PE will melt, and close the pores to prevent Li + from passing, which prevents the battery from burning and exploding.

Requirements for lithium-ion battery separators: good chemical stability and certain mechanical strength in the electrolyte, which can withstand the oxidation/reduction of the active substance of the electrode, and resist the corrosion of the electrolyte; the resistance of the diaphragm to the movement of electrolyte ions is small, thereby reducing the internal resistance of the battery so that the energy loss of the battery is reduced when discharging at a large current, which requires a certain pore size and porosity; it should be a good insulator of electrons and be able to block the growth of particles and dendrites that shed material from the electrode; good thermal stability and automatic shutdown protection. Of course, there must be a wealth of materials and low prices.

The main performance requirements of lithium-ion battery separator materials are thickness uniformity, mechanical properties, air permeability, physical and chemical properties, and four other major performance indicators. According to different physical and chemical characteristics, lithium-ion battery separator materials can be divided into woven film, non-woven fabric, microporous film, composite film, diaphragm paper, rolling film, and other categories. Due to the excellent mechanical properties, chemical stability, and relatively inexpensive characteristics of polyolefin materials, commercial lithium-ion battery separator materials are still mainly used in polyolefin microporous films such as polyethylene and polypropylene. To improve the safety of power batteries, functional composite separators such as ceramic separators are prepared based on polyolefin microporous films.

### 8.2.3  Lithium-Ion Battery Basic Parameters

#### 8.2.3.1  Voltage

The monomer voltage mainly depends on the type of monomer positive and negative electrode materials, the general lithium cobalt oxide, ternary positive electrode with graphite anode can obtain a full charge voltage of about 4.2 V, while lithium iron phosphate can only reach 3.6 V. The voltage here should be precisely the potential depending on the material properties, and the potential value is equal to the battery open-circuit voltage after standing for a long time. The single-terminal voltage in the closed-loop is the voltage value detected by us with an external instrument, and its value is equal to the battery potential minus the resistance occupancy voltage inside the battery. The internal resistance of the battery is not constant and will be affected by a variety of factors. In addition to being determined by the material, the monomer voltage changes with the change of charge and is a one-to-one correspondence. So in many cases, the battery charge (SOC) that cannot be directly and simply measured is often speculated with the battery open-circuit voltage [22, 23]. The monomer voltage is related to the degree of activity of the active substance inside the battery, so the temperature that affects the activity can also affect the level of the monomer voltage in a small range. The higher the monomer voltage, the more energy the battery of the same capacity contains. So under the premise of ensuring safety, increasing the upper limit of the monomer voltage is a technical route to improve the energy density of the system.

#### 8.2.3.2  Capacity

Lithium-ion battery capacity, measurable capacity, is the maximum amount of power that can be charged and discharged within the maximum and minimum voltage range reasonably available to the battery. Before being mounted on a vehicle, the capacity of the individual can be measured by charging and discharging. Once in the car, the battery capacity can only be estimated by algorithms. In a battery management system (BMS), accurately estimating the battery state of charge (SOC) is an important indicator of its design level [22, 24, 25]. The current well-known practice is to integrate the loop current timing in dynamic operating conditions, and in the non-operating state, check the battery power with the battery open-circuit voltage. Although there are many kinds of other methods, they are either poorly stable or the amount of computation is too large, and it is rare to be applied to batches. The capacity of the monomer is affected by the degree of aging, and the capacity decay to a limit value is when the battery is eliminated, which shows that the two have an absolute correlation. Secondly, the capacity is also affected by temperature, at low temperatures, the activity of the active substance decreases, the ions that can be provided become less, and the capacity will inevitably decline, so the capacity is also an important factor affecting the energy density of the battery.

### 8.2.3.3 Internal Resistance

Inside a lithium-ion battery, lithium ions move from one electrode to another through the electrolyte, and the factors that hinder the movement of ions in the process constitute the internal resistance. Due to the internal resistance of the battery, the terminal voltage of the battery is lower than the electromotive force and open-circuit voltage during discharge, and the terminal voltage of the battery is higher than the electromotive force and open-circuit voltage during charging. The essence of electric current is the directional movement of electric charges, and the motion of the electrons is opposed by the material itself and the magnetic field. Internal resistance is one of the important parameters of lithium-ion batteries that are not always static. The magnitude of the internal resistance directly affects the battery operating voltage, operating current, and battery capacity. In the process of battery charge-discharge, changes in temperature, charge-discharge rate, charge-discharge time, electrolyte concentration, and active material quality cause changes in the internal resistance of the battery.

The internal resistance of the battery includes the ohmic resistance $R_0$ and the polarization of internal resistance $R_p$. The ohmic resistance comes from the resistance of the electrolyte, diaphragm, and the contact resistance of various parts. During the internal oxidation-reduction reaction of the battery, an electric field is generated. Under the action of the electric field, the dielectric generates polarization charges due to the polarization effect. The polarization resistance is the resistance of the polarization charge to the current. During the charge-discharge process of lithium-ion batteries, the internal resistance generated by the polarization reaction is mainly electrochemical and concentration polarization. The nature of the active material, the structure of the electrode, the manufacturing process of the battery, and the operating conditions of the battery lead to the difference in the polarization internal resistance. The main influencing factor of the polarization internal resistance is the operating conditions of the batteries. The total internal resistance $R$ of the battery is equal to the ohmic resistance $R_0$ plus the polarization internal resistance $R_p$. The calculation equation is shown in Equation (8.7).

$$R = R_0 + R_p \tag{8.7}$$

In Equation (8.7), different types of lithium-ion batteries show different internal resistance characteristics. Even the same type of lithium-ion battery, due to the difference in internal material composition, operating environment, and aging degree, also show different internal resistances. The internal resistance is an important indicator to measure the battery performance. The internal resistance is most sensitive to temperature, and the internal resistance value can vary greatly at different temperature conditions. An important reason for the degradation of the battery performance at low temperatures is that the internal resistance increases drastically at low temperatures. Under normal circumstances, lithium-ion batteries are used as power sources. From the external analysis, the internal resistance is usually as small as possible. Especially in the case of high-power applications, small internal resistance is a necessary condition.

### 8.2.3.4 Power Density

The power output per unit massed or cubature of the battery is called power density, also known as specific power, and the unit is W/kg or W/L. The specific power is the magnitude of the working current that the battery can withstand. Higher specific power means that the battery can withstand high current discharge. Specific power is an important indicator for evaluating whether batteries and battery packs meet the rapid acceleration and climbing capabilities of new energy vehicles.

### 8.2.3.5 Energy Density

The energy density of a battery refers to the amount of energy stored in a given space unit or material mass unit. The battery energy density is generally divided into two dimensions: weight energy density (Wh/kg) and cubature energy density (Wh/L), also known as mass-specific energy or cubature ratio energy. Specific energy is an important indicator for evaluating whether high-power batteries can meet the application requirements of new energy vehicles. It is also an important indicator for comparing the performance of different types of batteries.

In the application process of new energy vehicles, the installation of battery packs requires corresponding components such as battery boxes, connecting wires, and current and voltage protection devices. The actual specific energy of the battery pack is less than the rated specific energy of the batteries. The smaller the gap between the actual specific energy and the rated specific energy, the higher the design level and integration degree of the battery pack. Therefore, the specific energy serves as an important parameter used to evaluate the performance of the battery pack for new energy vehicles.

In the conventional detection of lithium-ion batteries, the charge and discharge state of the battery is constant current charge and discharge, the median voltage is used for the estimation of the working voltage, and the calculation formula of energy density is Equation (8.8):

$$ED_m = \frac{w}{m} = \frac{I\int_0^t Vdt}{m} = \frac{C_0 V_{(0,mid)}}{m_{ea} + m_{an} + m_{ele}} \qquad (8.8)$$

where $ED_m$ represents the mass-energy density of the monomer, in units of Wh/kg; $C_0$ is the design capacity of the full battery, the unit is usually Ah; $V_{(0,mid)}$ is the median voltage of the whole battery, generally considered to be the difference between the median voltage of the positive and negative poles $V_{ca,mid} - V_{ac,mid}$, a deformation of Equation (8.9):

$$ED_m = \frac{C_0\left(V_{ca,mid} - V_{ca,mid}\right)}{C_0/C_{ca} + C_0/C_{an} + C_0 \cdot \delta_{ele}} = \frac{V_{ca,mid} - V_{ca,mid}}{1/C_{ca} + 1/C_{an} + \delta_{ele}} \qquad (8.9)$$

$C_{ca}$ and $C_{an}$ are the gram capacity of the positive and negative electrodes of the full battery, respectively, in mAh/g; it is the rehydration coefficient or amount of liquid injection of the whole battery, which is usually g/Ah; therefore, starting from the chemical system, the following aspects can be considered:

(1) Increasing $V_{ca,mid} - V_{ac,mid}$. It means increasing the median voltage of the positive electrode and reducing the median voltage of the negative electrode; measures that need to be taken include cathode materials using high charging cut-off voltage, high discharge platform type oxides, negative electrode materials using lower charging platform species (alloyed substances, silicon, etc.), and this is also one of the important reasons why graphite is superior to most metal oxides and reduced graphene oxides. Side effects are that the high voltage of the positive electrode makes it easy to cause interface damage and electrolyte consumption.

(2) Increasing the gram capacity of the cathode material $C_{ca}$. Using traditional high specific capacity, multi-variable valence, multi-lithium-deliberation type compounds. Side effects are that high gram capacity cathode material makes it easy to destroy the cycle life.

(3) Increasing the gram capacity of the negative electrode material $C_{an}$. The use of high specific capacity silicon anode, conversion materials such as ferric oxide, etc. Side effects are the physical expansion of the battery, battery power damage, and cycle life attenuation.

(4) Reducing the amount of liquid injection in the whole battery. Changing the shape and space design of the battery, reducing the capacity of the battery, etc.

Moreover, to improve the energy density of the battery, it is necessary to consider not only the energy density of the cell but also the system energy density of the battery. One is the energy density of the single cell, and the other is the energy density of the battery system. A battery cell is the smallest unit of a battery system. M cells form a module, and N modules form a battery pack, which is the basic structure of the vehicle power battery. It can be seen that improving the energy density of the battery also needs to consider the structure of the module and the structural design of the battery pack.

System energy density refers to the charge of the entire battery system after the completion of a single combination of the weight or volume of the entire battery system. Because the battery system contains a battery management system, a thermal management system, a high and low voltage loop, etc., occupying part of the weight and internal space of the battery system, the energy density of the battery system is lower than that of the single body.

Of course, in addition to the chemical system, the level of production processes such as compaction density, foil thickness, etc., will also affect the energy density. In general, the greater the compaction density, the higher the capacity of the battery in a limited space, so the compaction density of the main material is also regarded as one of the reference indicators of the battery energy density. The battery is a very comprehensive product, you want to improve the performance of one aspect, may sacrifice the performance of other aspects, which is the basis of the understanding of battery design and development. Power batteries belong to the vehicle, so energy density is not the only measure of battery quality.

## 8.3   REALIZATION PATH OF HIGH-ENERGY DENSITY LITHIUM-ION BATTERY

How to increase energy density? The use of new material systems, the fine-tuning of lithium-ion battery structure, and the improvement of manufacturing capabilities are the three stages for R&D engineers to "dance with long sleeves." The energy density of the monomer mainly relies on the breakthrough of the chemical system, mainly including two aspects:

(1) Increase the battery size. Battery manufacturers can achieve the effect of power expansion by increasing the original battery size. The most familiar example is that Tesla, a well-known electric vehicle company that first used Panasonic 18650 batteries [26–30], will be equipped with a new 21700 battery. But the battery cell "fattening" or "growing" is only a symptom, not a cure. The method of extracting money from the bottom of the kettle is to find the key technology to improve the energy density from the positive and negative electrode materials and electrolyte components that make up the battery cell.

(2) Chemical system change. As mentioned earlier, the energy density of the battery is subject to the positive and negative electrodes of the battery. Since the energy density of the current negative electrode material is much greater than that of the positive electrode, it is necessary to continuously upgrade the positive electrode material to improve the energy density.

The energy density of the system mainly lies in improving the group efficiency of the battery pack: the group test of the battery pack is the ability of the battery to deploy troops for the individual cells and modules, and it is necessary to use every inch of space to the greatest extent on the premise of safety.

### 8.3.1 IMPROVEMENT OF THE ENERGY DENSITY OF LITHIUM-ION BATTERY CELLS – THE TRANSFORMATION OF CHEMICAL SYSTEMS

#### 8.3.1.1 Cathode Material

At present, the cathode materials of commercial lithium-ion batteries are mainly lithium cobalt oxide (LCO) [31, 32], lithium iron phosphate (LFP), lithium manganate (LMO), nickel-cobalt-manganese oxide (NCM), and nickel-cobalt aluminum oxide (NCA). Since 2012, the energy density of commercial lithium-ion battery monomers has increased from 120 W·h/kg to 300 W·h/kg, and the energy density has benefited from the development of high-specific energy cathode materials.

To further improve the energy density of lithium-ion batteries, cathode materials need to be considered from three aspects:

(1) Develop a cathode material that can achieve high specific capacity at low potentials, such as high nickel cathode materials.

NCM and NCA are also known as ternary materials, according to the proportion of Ni, Co, Mn/Al elements to divide, NCM and NCA cathode material theoretical capacity of about 275 mA · h/g, has been replaced by the original NCM111 step by step to NCM424, NCM523, NCM622, NCM721, NCM811, NCM90.0.5, etc., such a development trend is because in NCM materials Ni and Co are the main active materials, Mn maintains the stability of the material during charge and discharge, and generally does not participate in electrochemical reactions. Considering the price of Co and the inactive or weak activity of Mn, the content of Co and Mn should be gradually reduced, the content of Ni should be continuously improved, and its content should be continuously improved, but the increase of Ni content will cause an increase in surface residuals, an intensification of cationic mixing, an increase in $Ni^{3+}$ and $Ni^{4+}$ and other problems. At present, the main precursor process, sintering process, doping and coating, and other aspects to improve the problem of high nickel NCM and NCA, of which Ni content near 0.8 high nickel cathode material has been commercialized. Some enterprises use NCM811 positive electrodes with a graphite anode, the cell energy density has exceeded 270 W·h/kg, and even reached 300 W·h/kg, much higher than NCM622/graphite anode of 230 W·h/kg and NCM523/graphite.

(2) Increase the de-embedding lithium potential of cathode materials to achieve higher capacities, such as high-voltage LCO, NCM, and lithium-rich manganese-based oxide cathode materials.

Researchers are currently developing LCO cathode materials with higher voltages such as 4.50 V, 4.53 V, 4.55 V, 4.60 V, etc. As the voltage is further increased to 4.5 to 4.6 V, the specific capacity of the LCO positive electrode and the energy density of the corresponding battery cells will continue to increase.

Lithium-rich manganese-based cathode materials have the advantages of high specific capacity (250~400 mA·h/g), high operating voltage, environmental friendliness, and low cost, and are expected to become the preferred cathode materials for the next generation of high-specific energy batteries. Under the known lithium-ion battery material system, only lithium-rich manganese-based cathode materials with silicon-carbon anodes are expected to make the energy density of lithium-ion battery cells reach more than 400 W·h/kg

(3) Development of cathode materials with high operating voltages, such as the lithium spinel cathode (LNMO) of nickel manganese oxide.

At present, in the mainstream cathode materials, the current specific capacity of LFP and manganese iron phosphate (LFMP) can reach 160~165 mA·h/g, close to the theoretical limit of 170 mA·h/g, and the specific capacity has no obvious room for improvement; the specific capacity of high-voltage LMO cathode materials is theoretically possible to further improve, and the specific capacity of LCO, NCM, NCA and lithium-rich manganese group (Li-rich) cathode materials has a relatively large room for improvement. Of course, while improving the specific capacity of the above materials, it is also necessary to solve the problems of material stability, material and electrolyte interface stability, and further optimize the polar plate design.

Existing oxide cathode materials are limited by their lower theoretical capacity, to achieve higher energy density, it is necessary to develop higher capacity cathode materials, such as lithium-sulfur [33–37] battery systems using elemental sulfur (or sulfur-containing compounds) as the positive electrode, the sulfur elemental is a "conversion reaction" based on the positive electrode material, its theoretical specific capacity of 1675 mA·h/g, The theoretical specific energy can reach 2600W·h/kg. In addition, the low cost and environmental friendliness of elemental sulfur also make it have commercial potential. In recent years, lithium-sulfur batteries have also made great progress. In 2016, the United States Sion Power Company successfully developed a new lithium-sulfur battery (20 A·h@400 W·h/kg) that can be used on drones. Lithium-sulfur battery volume energy density is low, the current rate characteristics, cycle characteristics, high- and low-temperature characteristics, and the safety of the whole life cycle need to be significantly improved, especially since the solution of high facial volume metal lithium anode is still very immature. With the maturity of solid electrolyte technology, the safety and service life problems faced by existing liquid electrolyte systems, as well as the problem of a metallic lithium anode, are expected to be further solved, which will open up the application of lithium-sulfur batteries and other high-capacity lithium-free cathode materials (such as $MnO_2$, $FeS_2$, $MoS_2$, $CuF_2$, $FeF_3$, etc.).

### 8.3.1.2 Anode Material

At present, the commercial lithium-ion battery anode material is still dominated by graphite, graphite anode has the advantages of high conductivity and high stability, but its theoretical specific capacity is low, the specific capacity of the current graphite anode reaches 365 mA·h/g, which is close to its theoretical maximum of 372 mA·h/g, if you want to further improve the energy density of the power battery, the negative electrode material must also have a breakthrough, the development of a new high-capacity anode material has been another important research direction for the development of high specific energy lithium-ion batteries.

(1) Nano-silicon-based anode material [38–40]. Silicon is the current theoretical specific capacity of the highest negative electrode material, lithium in silicon to form $Li_{4.4}Si$, the specific capacity of silicon up to 4200 mAh /g, much higher than the theoretical specific capacity of a graphite anode, silicon oxide $SiO_x$ (0 $<x<$2) also has a high specific capacity (theoretical specific capacity greater than 2000 mA·h/g). In addition, the silicon-based anode material also has the advantages of low lithium potential, stable discharge platform, rich reserves, etc., so it is one of the most promising materials to replace the graphite anode. However, from the current research and practice, no matter what form of the materials, the structural unit of silicon is accompanied by 300% volume expansion and contraction during the charge and discharge process, and the repeated volume change makes it easy to lead to cracks and pulverization of silicon-containing anode material particles, and the electrical contact and ion conduction become worse; at the same time, cracks and new fracture surfaces caused by volume changes will contact the electrolyte and form an unstable solid electrolyte membrane (SEI membrane), resulting in reversible capacity loss; SEI films repeatedly rupture and form with uncontrollable thickness and uniformity, further limiting ion transport and affecting the material's conduction network, resulting in reversible capacity attenuation, affecting the magnification and cycle performance. Designing the right micro-nano structure is critical for silicon-based anode materials and can improve the above problems to some extent.

At present, two methods are suitable for applications. One is silicon-based material. Nanoscale silicon-based particle size directly affects whether it occurs in the lithiumization process, silicon-based anode material in the nano dimension can release stress faster. And compared to the larger size of the silicon anode, the material does not make it easy to produce cracks under the constraint of the surrounding medium, and effectively improves the magnification performance and specific capacity of the silicon-based material. The other is nano-silicon carbon composite materials. Because nano-silicon is easy to agglomerate, the volume, the local electron, and ion conductivity change greatly after lithium embedding, and the silicon surface does not make it easy to produce a stable SEI film. So the development of nano-silicon carbon composite materials is the key to promoting the application of silicon anode.

(2) Lithium metal anode [41–44]. Lithium metal is regarded as the ultimate material for lithium-ion batteries due to its theoretical capacity of up to 3860 mAh/g and the lowest reduction potential (−3.040 V) relative to the standard hydrogen electrode. However, while the metal lithium anode brings high energy density, its shortcomings are also very obvious. First of all, the uncontrollable growth of lithium dendrites during charging and discharging may eventually puncture the diaphragm and cause the battery to short circuit, causing huge safety problems; secondly, the surface unstable SEI film is easy to rupture during the charging and discharging process, resulting in direct contact and reaction between metal lithium and electrolyte, and finally the metal lithium and electrolyte are consumed in large quantities, which greatly shortens the life of the battery; at the same time, it also leads to the active material being easy to fall off the collector, forming "dead lithium," the combined effect of the two makes the Coulomb efficiency of the battery reduced, the cycle performance further deteriorates, the pulverization of lithium and the direct reaction with the electrolyte leads to safety also becoming a serious problem. In addition, the area of the metal lithium electrochemical reaction is the geometric area, which affects the high magnification characteristics. At present, the research modification of the negative electrode of lithium metal mainly is about one is combined with a suitable carrier, two is the design of an artificial protective layer on the surface of metal lithium.

At present, large-scale commercial anode materials are mainly graphite carbon materials and lithium titanate (LTO), with the gradual improvement of battery energy density, traditional anode materials are difficult to meet the demand for high specific energy negative electrodes of the next generation of lithium-ion batteries. In the development direction of the silicon-based anode, researchers control the agglomeration of nano-silicon through nano-silicon-carbon composite, inhibit the side reaction, and thus inhibit the expansion of the silicon-based negative electrode to a certain extent, improve its circulation, and make it have a certain application scenario.

Nowadays, the technical route of high nickel ternary with the nano-silicon-carbon anode has been widely recognized. With the gradual maturity of the nano-silicon-carbon anode preparation process, the nano-silicon-carbon anode with a large increase in energy is expected to be widely used [38–40]. However, in a higher energy density (such as 400 W· h/kg) of the battery, the application of the silicon-carbon anode is facing greater challenges, with the increase in specific capacity, the volume expansion of silicon-carbon materials will become more and more serious, even if the battery cell reaches a higher energy density, its volume expansion is difficult to meet the requirements of the actual application. Compared with the nano-silicon-carbon anode, the composite metal lithium anode has a higher specific capacity, and the control of volume expansion is relatively easy, so for batteries with an energy density of more than 400 W·h/kg, the composite metal lithium anode material is an ideal choice.

In addition to the improvement in materials, it is also important to improve the method of performing the lithium anode. The formation of a composite lithium anode by pre-lithiumization of the anode material is one of the effective ways to achieve a high specific energy anode.

Such as (1) to control its volume change in the reasonable range, the silicon-based material needs to control its specific capacity. And in terms of higher specific capacity requirements, by pre-lithiumization, the efficiency of the first Coulomb can be improved. (2) Carbon materials (soft carbon, hard carbon, graphite) have excellent cycling properties and low reduction potential, and the negative electrode material alone can make the battery cell have good performance. Because of the advantages of adjustable carbon material structure and adjustable porosity, the loading of lithium metal as a porous carrier can increase the lithium deposition point and alleviate the volume change during the deposition/peeling process of lithium metal.

Above, by pre-lithiumization of the traditional anode material to form a composite lithium anode, the development of a high specific energy anode can be further promoted. Compared with the lithium compensation method of the first irreversible capacity of the negative electrode from the positive electrode or electrolyte additive, the lithium capacity that can be compensated by direct pre-lithification of the negative electrode can be higher.

### 8.3.1.3  Electrolytes

Lithium-ion battery organic electrolyte consists of a high-purity organic solvent, an electrolyte lithium salt, and necessary additives. At present, the commonly used organic solvents are vinyl carbonate, which has relatively high molecular symmetry, higher melting point, higher ionic conductivity, better interfacial properties, and can form a stable SEI film, which solves the problem of solvent co-embedding of graphite anode. However, it must be added and used in conjunction with the co-solvent. These co-solvents mainly include propylene carbonate and some chain carbonates with low viscosity, low boiling point, and low dielectric constant, such as dimethyl carbonate. In addition, other chain carbonates are gradually being applied to lithium-ion batteries.

At present, $LiPF_6$ is commercially applied, and the single nature of $LiPF_6$ is not optimal, but its comprehensive performance is the most advantageous. $LiPF_6$ has a relatively moderate number of ion migration, good oxidation resistance, and good aluminum foil passivation ability in commonly used organic solvents so that it can match various positive and negative electrode materials. However, the chemical and thermodynamic stability of $LiPF_6$ is not good enough, and the reaction Equation (6.10) occurs at room temperature, and the demarcation at high temperatures is particularly severe.

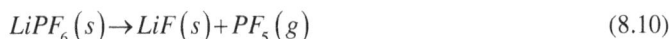

$$LiPF_6(s) \rightarrow LiF(s) + PF_5(g) \tag{8.10}$$

In Equation (8.10), $PF_5$ is a strong Lewis acid that easily attacks oxygen atoms in organic solvents, resulting in open-loop polymerization of the solvent and cracking of ether bonds. Secondly, $LiPF_6$ is more sensitive to water, the presence of trace water will lead to its decomposition, and the product causes the interfacial resistance to increase, affecting the cycle life of lithium-ion batteries, corroding electrodes, and collectors, and seriously affecting the electrochemical properties of the battery.

In recent years, the development of electrolytes with higher energy density has become an inevitable trend. The development of new antioxidant solvents (such as fluoride solvents, and sulfone solvents), the improvement of the interface compatibility of electrodes and electrolytes, and the development of electrolytes with wide electrochemical windows, high thermal stability, and good flame resistance have become the focus of research in the electrolyte industry. Although through the optimization of lithium salts, solvents, and additives, the performance of the battery can be appropriately improved to meet the development needs of high-energy-density batteries. But with the introduction of high-capacity positive and negative electrode materials such as metal lithium anode and lithium-rich manganese-based cathode, the modification and optimization of the electrolyte always have greater limitations on the improvement of battery performance, and it is difficult to take into account the control interface side reactions to meet the requirements of high- and low-temperature magnification.

Although unconventional electrolytes such as ionic liquids have the advantages of high oxidation resistance, flame retardancy, and relatively stable interface, they also face problems such as high cost, high density, and large viscosity, and cannot meet the practical application needs of high-energy-density batteries. Therefore, the introduction of higher stability of the solid electrolyte in the cell, and the gradual development of mixed solid-liquid electrolyte batteries and all-solid-state electrolyte batteries become the key to the future development of practical high-energy-density lithium-ion batteries.

The true density of the solid electrolyte will generally be significantly higher than that of the liquid electrolyte, and because the solid electrolyte does not have fluidity, the volume of the solid electrolyte in the battery may be higher, so under the same positive and negative material system, the battery energy density of the battery directly using the solid electrolyte is lower than that of the traditional liquid electrolyte battery.

The use of solid-state electrolytes [17, 45–47] to achieve high-energy-density lithium-ion battery technical approach, according to theoretical calculations, solid-state electrolyte and metal lithium negative electrode, high voltage or no lithium positive electrode matching use, can make solid-state batteries achieve significantly higher than the energy density of traditional liquid lithium-ion batteries. However, starting from the technical angle, the poor fluidity and wettability of the solid electrolyte will bring huge process problems to the preparation of solid batteries. Because of the many problems in the material and process of solid-state batteries, as a transitional technology that is expected to achieve high energy density, mixed solid-liquid electrolyte lithium-ion batteries containing a small amount of liquid electrolyte, composite metal lithium-ion batteries with negative electrode solids, it is possible to gradually improve energy density and safety based on existing liquid electrolyte batteries, and take into account the comprehensive performance of magnification performance and cycle performance.

In addition to lithium salts and solvents, additives are also an integral part of the electrolyte. Additives are characterized by small amounts but significantly improve the performance of a certain aspect of the electrolyte. Different additives have different roles, which can be divided into flame retardant additives, film-forming additives, and some additives that can improve the conductivity of the electrolyte and the efficiency of battery circulation. At present, the functional additives studied mainly include flame retardant additives that improve battery safety, overcharge-resistant additives, high-voltage electrolytes for high-voltage batteries, etc., and there are also special additives for problems such as flatulence bulges.

### 8.3.1.4 Separator

Lithium-ion battery separators play an important role in isolating the positive and negative electrodes, preventing internal electrical short circuits, and allowing fast transport of ion charge carriers between the positive and negative electrodes. PP, PE separator is the earliest commercial use in lithium-ion battery separator, polyolefin separator with its high strength, excellent chemical stability, and low price occupy the current 3C battery main market. However, these polyolefin diaphragms have problems such as low porosity, poor wettability of the electrolyte, and serious heat shrinkage at high temperatures. Poor wettability of the electrolyte affects the magnification performance and cycle stability of the battery, and brings a series of drawbacks in the production process; severe heat shrinkage of polyolefin separators can lead to severe internal short circuits in the battery, which eventually causes the battery to fire or explode under abnormal conditions. By coating, impregnating, spraying, etc., a new material with high-temperature resistance and hydrophilic properties is added to the single-layer polyolefin diaphragm to obtain a diaphragm with better performance.

At present, the market is used to coat nano-oxides and PVDF on one or both sides of the polyolefin diaphragm to improve the high-temperature resistance and hydrophilic properties of the diaphragm, and the modified diaphragm has higher safety and better cycling performance; however, with the development of science and technology, people's capacity requirements and safety requirements for lithium-ion batteries is increasing, and while the positive and negative electrode materials of lithium-ion batteries are updated, the separator also needs to find new material systems to match the performance requirements. To this end, great efforts have been made to develop high-performance diaphragms with good electrolyte wettability, thermal stability, and electrochemical properties.

For lithium-ion battery separators, their main role is to prevent internal short circuits and provide ion channels, the thickness of the diaphragm, porosity, pore size, pore uniformity, diaphragm flatness, chemical stability, wettability, mechanical properties, and thermal shutdown performance are directly affecting the performance and safety of the battery. For Li-S battery separators, its main role in addition to preventing internal short circuits and providing ion channels, the effective suppression of the polysulfide "shuttle effect" will directly affect the capacity retention rate, cycle life, and utilization of active substances in the battery. The main role of the diaphragm and electrolyte is to block the electron transmission between the positive and negative electrodes of the battery and to provide a guarantee for the transmission of lithium ions.

At present, it is generally believed in the industry that the gradual transformation from liquid batteries to all-solid-state batteries is an inevitable way to improve energy density and safety, and it is also the future development trend of lithium-ion batteries. To this end, the development of the diaphragm is expected to go through the following three stages.

Phase 1: improve the performance of existing diaphragms. Mainly based on the physical properties of PP and PE diaphragms, the development of diaphragms with higher thermal stability, interfacial stability, and magnification characteristics.

Phase 2: the introduction of solid electrolytes in the pores of porous diaphragms. The use of solid electrolytes will be conducive to the advantages of high-temperature resistance, high mechanical strength, and prevention of lithium dendrite puncture while improving the charging cut-off voltage of the battery and promoting the development and application of high-voltage cathode materials (such as lithium nickel manganate, lithium-rich manganese group, etc.).

Phase 3: development of lithium-ion conductor membranes [48,49]. Similar to the perfluorosulfonic acid proton exchange membrane (Nafion membrane), it is possible to develop a successfully dense lithium-ion conductor membrane, and take into account the advantages of high-temperature resistance, high voltage resistance, high ion migration number, stability for lithium metal, and stability for air, to achieve polymer inorganic composite all-solid-state lithium-ion batteries, which is crucial for the industrialization of all-solid batteries.

## 8.3.1.5 Cell Material Design

In the design of the energy density of lithium-ion batteries, on the one hand, it is to focus on the eigenvalue characteristics of the positive and negative electrode materials, that is, to enhance the specific energy of the active substance; on the other hand, it is also necessary to further optimize the energy density of the battery cell from the perspective of the auxiliary material, that is, the proportion of compressed inactive substances. In the selection of auxiliary materials, first, the proportion of quality is constantly decreasing; second, the auxiliary materials also reflect a significant gain effect in the electrochemical properties of the battery cell, such as the ceramic coating of the diaphragm not only enhances the mechanical properties and safety of the diaphragm but also helps to play the capacity of the battery cell.

It can be achieved from three aspects: coil, that is, by reducing the thickness of the copper foil/aluminum foil/diaphragm [50–52], reducing the dough density of the substrate; packaging, that is, increasing the size of the specifications and improving the utilization rate of space; reduce the content of additives/conductive agents.

To reduce the content of inactive substances, the selection of low density, the thin thickness of the positive and negative foil is conducive to the improvement of the energy density of the battery cell; under the premise of meeting the electrochemical stability, the existing lithium-ion battery foil system mainly uses aluminum foil, and the negative electrode adopts copper foil. In commercial batteries, the design of high-energy-density cells will also choose thinner positive and negative foils.

In the case of a given size, the energy density of the battery cell is closely related to the surface capacity, thickness, porosity, $N/P$ ratio, and other conditions of the positive and negative pole sheets. In addition, the design of positive/negative pole/diaphragm size, winding/lamination mode, inner string structure design, and diversion terminal structure also has a certain impact on energy density.

### 8.3.2  IMPROVEMENT OF THE ENERGY DENSITY OF LITHIUM-ION BATTERY GROUPS–SYSTEM EQUALIZATION

In general, the "slimming" of the lithium-ion battery pack mainly has the following ways: (1) Optimize the layout structure, from the aspect of the overall size, you can optimize the layout of the system inside so that the internal parts of the battery pack are more compact and efficient. (2) Topology optimization, through simulation calculations to ensure the rigidity and structural reliability of the premise, to achieve weight reduction design. With this technology, topology optimization and topography optimization can be achieved, ultimately helping to reduce the weight of the battery box. (3) Material selection, you can choose low-density materials, such as the battery pack cover has been gradually transformed from the traditional sheet metal cover to the composite material cover, which can reduce weight by about 35%. For the box under the battery pack, it has gradually changed from the traditional sheet metal scheme to the aluminum profile scheme, with a weight reduction of about 40% and a significant lightweight effect. (4) System integration design and system structure design are considered as a whole, and structural parts are shared and shared as much as possible to achieve the ultimate lightweight.

Among them, the battery module is formed by several battery cells in series and parallel combinations, which is the element that makes up the battery pack. The battery module is rarely evaluated as a subject in actual operation, and occasionally its voltage value is detected in some systems. People tend to think of the module as a big battery. The difference is that the module has a single body consistency problem, and the internal cell voltage difference is the focus of the balance function investigation. The performance of the battery module is often subject to the lowest performance battery cells in the composition battery module and is mainly reflected in the capacity of this indicator, so it also has an impact on the energy density of lithium-ion batteries. When charging, the single body with high voltage is the first to be filled; when discharging, the low-voltage battery is discharged first. Likely, the two batteries are not the same. Therefore, the consistency of the cells' parameters within the module has a decisive impact on the performance of the module [53–59].

Therefore, consistency is a parameter that needs to be considered at the module level more than the monomer. This parameter is guaranteed by screening the battery by various means at the beginning of the module; after the module is produced, consistency is an important indicator of its acceptance; during operation, it can only rely on the balancing function of the BMS [60–63].

Battery balancing is a difficult point of BMS, which is essentially through human intervention to make all of the batteries in the battery pack consistent. Common balancing techniques are divided into passive equalization and active equalization. When the active balancing technology is used, the monomer battery is generally added to the DC/DC circuit, and the purpose of battery balancing is achieved using energy replenishment or energy transfer, and the balance is achieved during charging and discharging; passive equalization is the use of external resistors to drain a battery with higher energy to a set value. Active equalization requires high voltage acquisition accuracy and complex circuit structure. Although the passive equalization structure is simple, it can only achieve the effect of equalization when the battery is charged, and the energy utilization rate is low.

In contrast, due to the high energy utilization rate of active equalization technology, the effect of battery balancing can be achieved when charging and discharging, so it is the future development

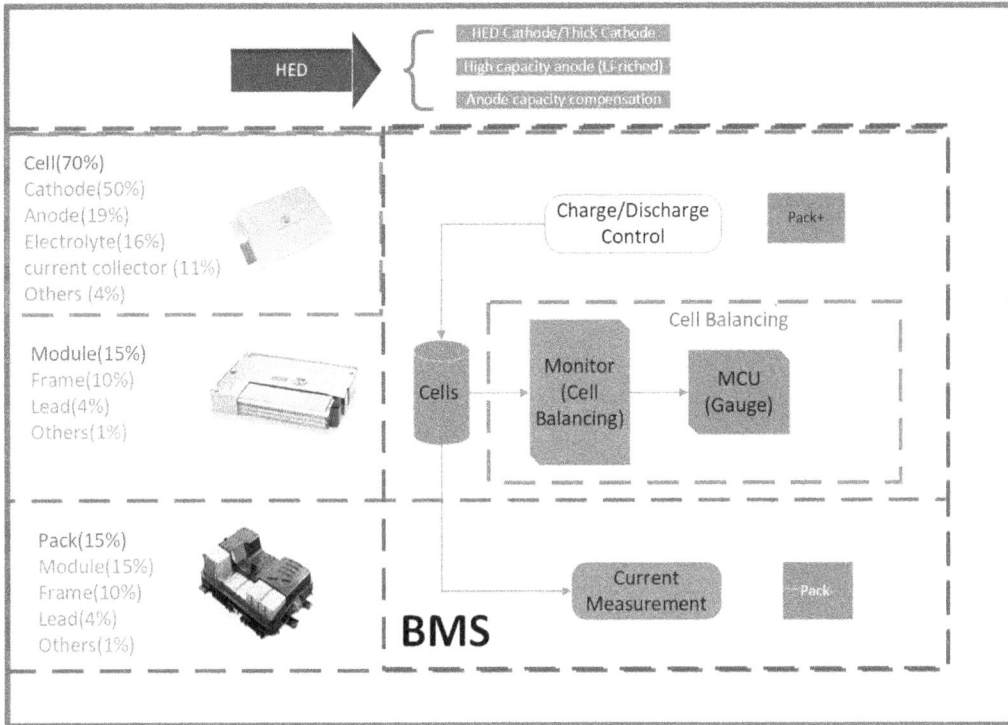

**FIGURE 8.4** Balanced solutions for lithium-ion batteries.

direction. In addition, the equilibrium technology and the type of battery also have a certain relationship, it is generally believed that LFP is more suitable for active equalization, and ternary batteries are suitable for passive equalization.

The application of the battery needs to comprehensively balance the mass-energy density, volume energy density, safety, recyclability, charge and discharge rate, high- and low-temperature characteristics, manufacturing costs, and other aspects of the indicators. The targeted solution must be implemented in a cell system, avoid conflict, and support large-scale engineering applications. Figure 8.4 lists the battery pack architecture technical strategies to be adopted to achieve each indicator.

## 8.4 REALIZATION OF FUTURE POWER SYSTEMS

The power industry is the basic pillar industry of the country, which has strict requirements for the safety and reliability of equipment, and we need to have a rigorous and respectful attitude towards the promotion of power storage. Many lithium-ion battery companies will take electric energy storage as the target market.

Lithium-ion batteries have multiple advantages in the field of energy storage.

(1) With relatively high density, strong endurance, and the application of cathode materials, the life and safety of power batteries have been greatly improved, and they are preferred to be used in the field of energy storage.
(2) The cycle life of lithium-ion batteries is long, and the shortcomings of improving energy density, weak endurance, and high price in the future make the application of lithium-ion batteries in the field of energy storage possible.

(3) Lithium-ion battery magnification performance is good, preparation is relatively easy, in the future to improve high-temperature performance and poor cycle performance and other shortcomings are more conducive to the application of energy storage field.

(4) Driven by national policies, the demand for lithium-ion batteries in the energy storage field is also growing rapidly, becoming a follow-up force to promote the growth of the lithium-ion battery market.

At present, lithium energy storage batteries are mainly used in the field of power and communication energy storage. Among them, the power system energy storage lithium-ion battery market concentration is relatively high. The business layout of energy storage battery manufacturers is related to the company's advantages: the main energy storage technology providers, energy storage converter providers, and energy storage system integrators are mainly consumption, powerful lithium-ion battery leaders, inverter leaders, and photovoltaic or new energy vehicle terminal application leaders.

At present, the power grid has become an important boost to stimulate the development of the lithium-ion battery energy storage industry, and the demand is very obvious. As we all know, the power grid is configured according to the maximum electricity load, and the rapid load growth needs to be configured with the corresponding power supply and power grid, but these power supplies and power grids built to meet the peak load are idle during the trough and peak hours of electricity consumption, so that not only the power generation efficiency will be reduced, but also a large waste of resources, lithium-ion battery energy storage can just solve this problem.

Lithium-ion battery energy storage has many application scenarios in the power industry, which are generally divided into three major links: power generation side, transmission and distribution side, and electricity consumption side, and there are dozens of application scenarios. On the power generation side, the role of energy storage is mainly loaded regulation, smoothing intermittent energy, improving new energy consumption, increasing the backup capacity of the power grid, and participating in frequency regulation. On the transmission and distribution side, the configuration of energy storage is to improve circuit quality, reduce line losses, improve the backup capacity of the power grid, improve the utilization efficiency of transmission and distribution equipment, and delay the demand for capacity increase. In terms of distributed energy storage on the electricity side, energy storage can improve distributed energy consumption, peak shaving and valley filling, load transfer, reduce electricity costs, and improve power supply reliability and power quality.

With the new construction and expansion of 5G base stations, the demand for communication iron lithium-ion battery energy storage backup power supply has risen, and communication operators and tower companies have purchased communication backup lithium-ion batteries in bulk in the form of bidding. Overall, in the global market, domestic lithium-ion battery factories have evolved a mature technical route and a strong cost control capability, and is the most powerful competitor in the global lithium-ion battery energy storage market, the domestic lithium-ion battery industry chain will fully benefit from the rapid development of the global electrochemical energy storage industry.

The lithium-ion battery is undoubtedly an excellent invention in the history of human science and technology, although the energy density of the first generation of lithium-ion batteries is only 80 W·h/kg and 200 W·h/L, after more than 20 years of effort, the comprehensive performance of lithium-ion batteries has been significantly improved, the lithium-ion battery industry has surpassed nickel-cadmium batteries, nickel-metal hydride batteries, and other products to leap to the second place in the energy storage industry. In the future, the development of high-energy-density lithium-ion batteries that take into account the comprehensive performance of high safety and high stability is still full of challenges, and lithium-ion battery enterprises and research institutions need to cooperate more closely to jointly solve the scientific and engineering problems faced in new materials and new systems. It is expected that in the near future, with the breakthrough of new materials and new technologies, a new generation of high-energy-density lithium-ion batteries will surely enter thousands of households and bring a new dawn to the future power energy storage system.

## REFERENCES

1. Liu, L., J. Xu, S. Wang, F. Wu, H. Li, and L. Chen, Practical evaluation of energy densities for sulfide solid-state batteries. *Transportation*, 2019. 1: p.1–49.
2. Zhang, J.N., Q.H. Li, C.Y. Ouyang, X.Q. Yu, M.Y. Ge, X.J. Huang, E.Y. Hu, C. Ma, S.F. Li, R.J. Xiao, W.L. Yang, Y. Chu, Y.J. Liu, H.G. Yu, X.Q. Yang, X.J. Huang, L.Q. Chen, and H. Li, *Trace doping of multiple elements enables stable battery cycling of* LiCoO2 at 4.6V. *Nature Energy*, 2019. 4(7): p. 594–603.
3. Wei, A.J., J.P. Mu, R. He, X. Bai, Z. Liu, L.H. Zhang, Z.F. Liu, and Y.J. Wang, *Preparation of* $Li_4Ti_5O_{12}$/carbon nanotubes composites and $LiCoO_2$/$Li_4Ti_5O_{12}$ full-cell with enhanced electrochemical performance for high-power lithium-ion batteries. *Journal of Physics and Chemistry of Solids*, 2020. 138: p. 1–36.
4. Wei, C.X., Y.S. Hong, Y.C. Tian, X.A. Yu, Y.J. Liu, and P. Pianetta, *Quantifying redox heterogeneity in single-crystalline* $LiCoO_2$ cathode particles. *Journal of Synchrotron Radiation*, 2020. 27: p. 713–719.
5. Wang, S.L., S.M. Chen, W.Q. Gao, L.L. Liu, and S.J. Zhang, A new additive 3-Isocyanatopropyltriethoxysilane to improve electrochemical performance of Li/NCM622 half-cell at high voltage. *Journal of Power Sources*, 2019. 423: p. 90–97.
6. Zou, W., F.J. Xia, J.P. Song, L. Wu, L.D. Chen, H. Chen, Y. Liu, W.D. Dong, S.J. Wu, Z.Y. Hu, J. Liu, H.E. Wang, L.H. Chen, Y. Li, D.L. Peng, and B.L. Su, Probing and suppressing voltage fade of Li-rich $Li_{1.2}Ni_{0.13}Co_{0.13}Mn_{0.54}O_2$ cathode material for lithium-ion battery. *Electrochimica Acta*, 2019. 318: p. 875–882.
7. Zhu, J., G.L. Cao, Y.J. Li, S.L. Wang, S.Y. Deng, J. Guo, Y.X. Chen, T.X. Lei, J.P. Zhang, and S.H. Chang, Nd2O3 encapsulation-assisted surface passivation of Ni-rich LiNi0.8Co0.1Mn0.1O2 active material and its electrochemical performance. Electrochimica Acta, 2019. 325: p. 1–49.
8. Zhou, Z.K., Y.Z. Kang, Y.L. Shang, N.X. Cui, C.H. Zhang, and B. Duan, Peak power prediction for series-connected LiNCM battery pack based on representative cells. *Journal of Cleaner Production*, 2019. 230: p. 1061–1073.
9. Ran, Q.W., H.Y. Zhao, Y.Z. Hu, Q.Q. Shen, W. Liu, J.T. Liu, X.H. Shu, M.L. Zhang, S.S. Liu, M. Tan, H. Li, and X.Q. Liu, Enhanced electrochemical performance of dual-conductive layers coated Ni-rich $LiNi_{0.6}Co_{0.2}Mn_{0.2}O_2$ cathode for Li-ion batteries at high cut-off voltage. *Electrochimica Acta*, 2018. 289: p. 82–93.
10. Wang, S., K. Liu, Y. Wang, D.-I. Stroe, C. Fernandez, and J.M. Guerrero, *Multidimensional Lithium-Ion Battery Status Monitoring, CRC Press*, 2022 p. 1–469.
11. Jiang, C., S. Wang, B. Wu, C. Fernandez, X. Xiong, and J. Coffie-Ken, A state-of-charge estimation method of the power lithium-ion battery in complex conditions based on adaptive square root extended Kalman filter. *Energy*, 2021. 219: p. 1–25.
12. Tang, Y.W., L.J. Wu, W.F. Wei, D.Q. Wen, Q.W. Guo, W.C. Liang, and L. Xiao, Study of the thermal properties during the cyclic process of lithium ion power batteries using the electrochemical-thermal coupling model. *Applied Thermal Engineering*, 2018. 137: p. 11–22.
13. Smolianova, I., L.H. Jin, Y.Z. Xin, V. Dementiev, and Z.Z. Ling, A high-capacity graphene/mesocarbon microbead composite anode for lithium-ion batteries. *Journal of Zhejiang University-Science A*, 2020. 21(5): p. 392–400.
14. Liu, Y., H.J. Bai, Q.Z. Zhao, J.G. Yang, Y.J. Li, C.M. Zheng, and K. Xie, Storage aging mechanism of $LiNi_{0.8}Co_{0.15}Al_{0.05}O_2$/graphite Li-ion batteries at high state of charge. *Journal of Inorganic Materials*, 2021. 36(2): p. 175–180.
15. Choi, W., H.C. Shin, J.M. Kim, J.Y. Choi, and W.S. Yoon, Modeling and applications of electrochemical impedance spectroscopy (EIS) for lithium-ion batteries. *Journal of Electrochemical Science and Technology*, 2020. 11(1): p.1–13.

16.  Xiang, Y., W.F. Zhang, B. Chen, Z.Q. Jin, H. Zhang, P.C. Zhao, G.P. Cao, and Q.Q. Meng, Nano-$Li_4Ti_5O_{12}$ particles in-situ deposited on compact holey-graphene framework for high volumetric power capability of lithium ion battery anode. *Journal of Power Sources*, 2020. 447: p. 1–12.

17.  Zhu, G.L., C.Z. Zhao, H. Yuan, B.C. Zhao, L.P. Hou, X.B. Cheng, H.X. Nan, Y. Lu, J. Zhang, J.Q. Huang, Q.B. Liu, C.X. He, and Q. Zhang, Interfacial redox behaviors of sulfide electrolytes in fast-charging all-solid-state lithium metal batteries. *Energy Storage Materials*, 2020. 31: p. 267–273.

18.  Shi, X.Y., S.S. Yu, T. Deng, W. Zhang, and W.T. Zheng, Unlock the potential of Li4Ti5O12 for high-voltage/long-cycling-life and high-safety batteries: Dual-ion architecture superior to lithium-ion storage. *Journal of Energy Chemistry*, 2020. 44: p. 13–18.

19.  Magdaline, T.B. and A.V. Murugan, Microwave-assisted hydrometallurgical extraction of $Li_4Ti_5O_{12}$ and $LiFePO_4$ from ilmenite: effect of PPy-Br-2 derived C-coating with N, Br, and $Nb^{5+}$ co-doping on electrodes for high-rate energy storage performance. *Dalton Transactions*, 2020. 49(19): p. 6227–6241.

20.  Lacroix, R., J.J. Biendicho, G. Mulder, L. Sanz, C. Flox, J.R. Morante, and S. Da Silva, Modelling the rheology and electrochemical performance of $Li_4Ti_5O_{12}$ and $LiNi_{1/3}Co_{1/3}Mn_{1/3}O_2$ based suspensions for semi-solid flow batteries. *Electrochimica Acta*, 2019. 304: p. 146–157.

21.  Huddleston, W., F. Dynys, and A. Sehirlioglu, Nickel percolation and coarsening in sintered $Li_4Ti_5O_{12}$ anode composite. *Journal of the American Ceramic Society*, 2020. 103(8): p. 4178–4188.

22.  Wang, S., P. Takyi-Aninakwa, Y. Fan, C. Yu, S. Jin, C. Fernandez, and D.I. Stroe, A novel feedback correction-adaptive Kalman filtering method for the whole-life-cycle state of charge and closed-circuit voltage prediction of lithium-ion batteries based on the second-order electrical equivalent circuit model. *International Journal of Electrical Power and Energy Systems*, 2022. 139: p. 1–13.

23.  Liu, Y., S. Wang, Y. Xie, C. Fernandez, J. Qiu, and Y. Zhang, A novel adaptive H-infinity filtering method for the accurate SOC estimation of lithium-ion batteries based on optimal forgetting factor selection. *International Journal of Circuit Theory and Applications*, 2022.50(10): p. 3372–3386.

24.  Wang, S., S. Jin, D. Bai, Y. Fan, H. Shi, and C. Fernandez, A critical review of improved deep learning methods for the remaining useful life prediction of lithium-ion batteries. *Energy Reports*, 2021. 7: p. 5562–5574.

25.  Wang, S., C. Fernandez, C. Yu, Y. Fan, W. Cao, and D.-I. Stroe, A novel charged state prediction method of the lithium ion battery packs based on the composite equivalent modeling and improved splice Kalman filtering algorithm. *Journal of Power Sources*, 2020. 471(228450): p. 1–13.

26.  Zhu, J.G., M. Knapp, D.R. Sorensen, M. Heere, M.S.D. Darma, M. Muller, L. Mereacre, H.F. Dai, A. Senyshyn, X.Z. Wei, and H. Ehrenberg, Investigation of capacity fade for 18650-type lithium-ion batteries cycled in different state of charge (SoC) ranges. *Journal of Power Sources*, 2021. 489: p. 1–12.

27.  Zhu, J.G., M. Knapp, D.R. Sorensen, M. Heere, M.S.D. Darma, M. Muller, L. Mereacre, H.F. Dai, A. Senyshyn, X.Z. Wei, and H. Ehrenberg, Investigation of capacity fade for 18650-type lithium-ion batteries cycled in different state of charge (SoC) ranges. *Journal of Power Sources*, 2021. 489(7): p. 1–12.

28.  Zhao, C.P., J.H. Sun, and Q.S. Wang, Thermal runaway hazards investigation on 18650 lithium-ion battery using extended volume accelerating rate calorimeter. *Journal of Energy Storage*, 2020. 28: p. 1–9.

29.  Willenberg, L., P. Dechent, G. Fuchs, M. Teuber, M. Eckert, M. Graff, N. Kurten, D.U. Sauer, and E. Figgemeier, The development of jelly roll deformation in 18650 lithium-ion batteries at low state of charge. *Journal of the Electrochemical Society*, 2020. 167(12): p. 1–9.

30.  Kim, J., A. Mallarapu, D.P. Finegan, and S. Santhanagopalan, Modeling cell venting and gas-phase reactions in 18650 lithium ion batteries during thermal runaway. *Journal of Power Sources*, 2021. 489: p. 1–16.

31.  Yu, C.Y., J. Choi, V. Anandan, and J.H. Kim, High-temperature chemical stability of $Li_{1.4}Al_{0.4}Ti_{1.6}(PO4)$ (3) solid electrolyte with various cathode materials for solid-state batteries. *Journal of Physical Chemistry C*, 2020. 124(28): p. 14963–14971.

32.  Kannan, D.R. and M.H. Weatherspoon, The effect of pulse charging on commercial lithium cobalt oxide (LCO) battery characteristics. *International Journal of Electrochemical Science*, 2021. 16(4): p. 1–9.

33. Zhou, G., A. Yang, G. Gao, X. Yu, J. Xu, C. Liu, Y. Ye, A. Pei, Y. Wu, Y. Peng, Y. Li, Z. Liang, K. Liu, L.W. Wang, and Y. Cui, Supercooled liquid sulfur maintained in three-dimensional current collector for high-performance Li-S batteries. *Science Advances*, 2020. 6(21): p. eaay5098.

34. Zhang, R.H., C. Chi, M.C. Wu, K. Liu, and T.S. Zhao, A long-life Li-S battery enabled by a cathode made of well-distributed B4C nanoparticles decorated activated cotton fibers. *Journal of Power Sources*, 2020. 451: p. 1–9.

35. Yuan, H., H.X. Nan, C.Z. Zhao, G.L. Zhu, Y. Lu, X.B. Cheng, Q.B. Liu, C.X. He, J.Q. Huang, and Q. Zhang, Slurry-coated sulfur/sulfide cathode with Li metal anode for all-solid-state lithium-sulfur pouch cells. *Batteries & Supercaps*, 2020. 3(7): p. 596–603.

36. Polrolniczak, P., D. Kasprzak, J. Kazmierczak-Razna, M. Walkowiak, P. Nowicki, and R. Pietrzak, Composite sulfur cathode for Li-S batteries comprising hierarchical carbon obtained from waste PET bottles. *Synthetic Metals*, 2020. 261: p. 1–6.

37. Marinescu, M., L. O'Neill, T. Zhang, S. Walus, T.E. Wilson, and G.J. Offer, Irreversible vs reversible capacity fade of lithium-sulfur batteries during cycling: The effects of precipitation and shuttle. *Journal of the Electrochemical Society*, 2018. 165(1): p. A6107–A6118.

38. Wang, F., Z.L. Hu, L.M. Mao, and J. Mao, Nano-silicon @ soft carbon embedded in graphene scaffold: High-performance 3D free-standing anode for lithium-ion batteries. *Journal of Power Sources*, 2020. 450: p. 1–10.

39. Liu, C.Y., H. Li, X.B. Kong, and J.B. Zhao, Modeling analysis of the effect of battery design on internal short circuit hazard in $LiNi_{0.8}Co_{0.1}Mn_{0.1}O_2/SiO_x$ graphite lithium ion batteries. *International Journal of Heat and Mass Transfer*, 2020. 153: p. 1–12.

40. Hu, X.Q., S.M. Huang, X.H. Hou, H.D. Chen, H.Q. Qin, Q. Ru, and B.L. Chu, A double core-shell structure silicon carbon composite anode material for a lithium ion battery. *Silicon*, 2018. 10(4): p. 1443–1450.

41. Zheng, H.W., H.L. Zhang, Y. Fan, G. Ju, H.B. Zhao, J.H. Fang, J.J. Zhang, and J.Q. Xu, A novel Mo-based oxide beta-$SnMoO_4$ as anode for lithium ion battery. *Chinese Chemical Letters*, 2020. 31(1): p. 210–216.

42. Su, Y.B., L.H. Ye, W. Fitzhugh, Y.C. Wang, E. Gil-Gonzalez, I. Kim, and X. Li, A more stable lithium anode by mechanical constriction for solid state batteries. *Energy & Environmental Science*, 2020. 13(3): p. 908–916.

43. Shishvan, S.S., N.A. Fleck, R.M. McMeeking, and V.S. Deshpande, Growth rate of lithium filaments in ceramic electrolytes. *Acta Materialia*, 2020. 196: p. 444–455.

44. Liu, D.H., Z.Y. Bai, M. Li, A.P. Yu, D. Luo, W.W. Liu, L. Yang, J. Lu, K. Amine, and Z.W. Chen, Developing high safety Li-metal anodes for future high-energy Li-metal batteries: strategies and perspectives. *Chemical Society Reviews*, 2020. 49(15): p. 5407–5445.

45. Zhou, L., M.K. Tufail, L. Yang, N. Ahmad, R.J. Chen, and W. Yang, Cathode-doped sulfide electrolyte strategy for boosting all-solid-state lithium-ion batteries. *Chemical Engineering Journal*, 2020. 391: p. 1–27.

46. Zheng, C., L.J. Li, K. Wang, C. Wang, J. Zhang, Y. Xia, H. Huang, C. Liang, Y.P. Gan, X.P. He, X.Y. Tao, and W.K. Zhang, Interfacial reactions in inorganic all-solid-state lithium-ion batteries. *Batteries & Supercaps*, 2021. 4(1): p. 8–38.

47. Yan, L., J.H. Huang, Z.W. Guo, X.L. Dong, Z. Wang, and Y.G. Wang, Solid-state proton battery operated at ultralow temperature. *ACS Energy Letters*, 2020. 5(2): p. 685–691.

48. Kitajima, S., W. Choi, and D. Im, Simple scalable processing method for a polymer/inorganic hybridized electrolyte. *Composites Science and Technology*, 2020. 197: p. 1–8.

49. Husmann, S., A.J.G. Zarbin, and R.A.W. Dryfe, High-performance aqueous rechargeable potassium batteries prepared via interfacial synthesis of a Prussian blue-carbon nanotube composite. *Electrochimica Acta*, 2020. 349: p. 1–11.

50. Zhang, F., K.M. Wu, X. Xu, W.Z. Wu, X. Hu, K.F. Yu, and C. Liang, 3D printing of graphite electrode for lithium-ion battery with high areal capacity. *Energy Technology*, 2021. 9(11): p. 1–21.

51. Chen, J.Q., Y. Zhao, H.T. Gao, S.D. Chen, W.J. Li, X.H. Liu, X.L. Hu, and S. Yan, Rolled electrodeposited copper foil with modified surface morphology as anode current collector for high corrosion resistance in lithium-ion battery electrolyte. *Surface & Coatings Technology*, 2021. 421: p. 1–9.

52.  Chen, J.Q., X.G. Wang, H.T. Gao, S. Yan, S.D. Chen, X.H. Liu, and X.L. Hu, Rolled electrodeposited copper foil with modified surface morphology as anode current collector for high performance lithium-ion batteries. *Surface & Coatings Technology*, 2021. 410: p. 1–9.

53.  Zhang, Y.L., Y. Hong, and K. Choi, Optimal energy-dissipation control for SOC based balancing in series connected Lithium-ion battery packs. *Multimedia Tools and Applications*, 2020. 79(23–24): p. 15923–15944.

54.  Yu, Y.Q., R. Saasaa, A.A. Khan, and W. Eberle, A series resonant energy storage cell voltage balancing circuit. *IEEE Journal of Emerging and Selected Topics in Power Electronics*, 2020. 8(3): p. 3151–3161.

55.  Yan, Y.D., K.Y. Hu, and C.H. Tsai, Digital battery management design for point-of-load applications with cell balancing. *IEEE Transactions on Industrial Electronics*, 2020. 67(8): p. 6365–6375.

56.  Wang, Y.J., J.Q. Tian, Z.H. Chen, and X.T. Liu, Model based insulation fault diagnosis for lithium-ion battery pack in electric vehicles. *Measurement*, 2019. 131: p. 443–451.

57.  Wang, W.Z. and M. Preindl, Dual cell links for battery-balancing auxiliary power modules: A cost-effective increase of accessible pack capacity. *IEEE Transactions on Industry Applications*, 2020. 56(2): p. 1752–1765.

58.  Wang, B., F.F. Qin, X.B. Zhao, X.P. Ni, and D.J. Xuan, Equalization of series connected lithium-ion batteries based on back propagation neural network and fuzzy logic control. *International Journal of Energy Research*, 2020. 44(6): p. 4812–4826.

59.  Uno, M. and K. Hasegawa, Modular equalization system based on star-connected phase-shift switched capacitor converters with inherent constant current characteristics for electric double-layer capacitor modules. *IEEE Transactions on Power Electronics*, 2020. 35(10): p. 10271–10284.

60.  Yang, X., S. Wang, W. Xu, J. Qiao, C. Yu, P. Takyi-Aninakwa, and S. Jin, A novel fuzzy adaptive cubature Kalman filtering method for the state of charge and state of energy co-estimation of lithium-ion batteries. *Electrochimica Acta*, 2022. 415: p. 1–13.

61.  Qiao, J., S. Wang, C. Yu, X. Yang, and C. Fernandez, A novel intelligent weight decreasing firefly–particle filtering method for accurate state-of-charge estimation of lithium-ion batteries. *International Journal of Energy Research*, 2022. 46(5): p. 6613–6622.

62.  Shi, H., L. Wang, S. Wang, C. Fernandez, X. Xiong, B.E. Dablu, and W. Xu, A novel lumped thermal characteristic modeling strategy for the online adaptive temperature and parameter co-estimation of vehicle lithium-ion batteries. *Journal of Energy Storage*, 2022. 50: p. 1–41.

63.  Hu, Y., S. Wang, J. Huang, P. Takyi-Aninakwa, and X. Chen, A novel seasonal autoregressive integrated moving average method for the accurate lithium-ion battery residual life prediction. *International Journal of Electrochemical Science*, 2022. 17(5): p. 1–17.

# 9 Recent Advances in Dielectric Materials for Energy Storage Devices
## *A Comprehensive Overview*

*Vaishali Misra, Saleem Khan, Manisha Yadav, Ajay Singh, and Vishal Singh*

## 9.1 INTRODUCTION

The non-renewable energy shortage, environmental issues, and environmental quality all have prompted major attempts in recent years to develop cleaner and more sustainable sources, i.e., wind, solar power, heat, and water energy, in tandem with the economic growth and community. Consequently, there is a significant demand for technologies that can efficiently store, absorb, and deliver power. Commercial systems for the storage of electric energy are typically split into long-term and short-term based on their energy storage time. A battery is generally used for long time storage, whereas a capacitor is used for short time storage. Rechargeable batteries, cells, electrolytic supercapacitors, and dielectric capacitors are examples of energy storage systems [1, 2]. Batteries and fusion reactors, for example, have large energy densities (10–300 Wh/kg and 200–1000 Wh/kg, respectively), but minimum power density (typically below 500 W/kg), preventing their use across high-power applications. Long charge-discharge operations limit the power as well as energy density (0.04–30 Wh/kg) of electrochemical supercapacitors. Contrastingly, the dielectric capacitors exhibit rapid charge-discharge rates and a substantially greater power density, meeting the demands of super high-power electronics. High-power energy storage systems have applications in aerospace, military, defence, telecommunication, nuclear, electric vehicles, grids, and medical fields as shown in Figure 9.1. Unfortunately, the poor energy density of dielectrics, which is an order less than fuel cells, batteries, and supercapacitors, restricts their potential applications. Power electronics as well as power systems contains roughly 25% of the volume and weight of dielectric capacitors contains roughly 25% of volume and weight of dielectric capacitors. As a result, an innovative material that can considerably boost the energy density of dielectric capacitors is urgently needed [3]. The rapid development of this equipment and devices demands new dielectric materials with high storage capability. The research into dielectrics with large temperature stability and high energy density has garnered a lot of attention [4, 5]. Dielectrics have become the most significant element for electrical and electronic insulating utilization because they may protect equipment or devices from damaging factors such as dielectric loss, sparking shorting, and dielectric breakdown, among others [6]. As a result, it is possible to suggest that if the dielectric capacitors storage density would be enhanced to compete with electrochemical supercapacitors and batteries than their utility would be substantially broadened. Dielectric capacitors with high storage density, for example, can help to compact electronic and electrical systems and go even closer to nanotechnology, lightness, and integration. Because of the greater need for high-power capacitors, research in dielectric materials with greater electrical storage capability has gained attention. However, an overall assessment of the research state of dielectric materials for high storage has yet to be completed. As a result, several

DOI: 10.1201/9781003340539-9

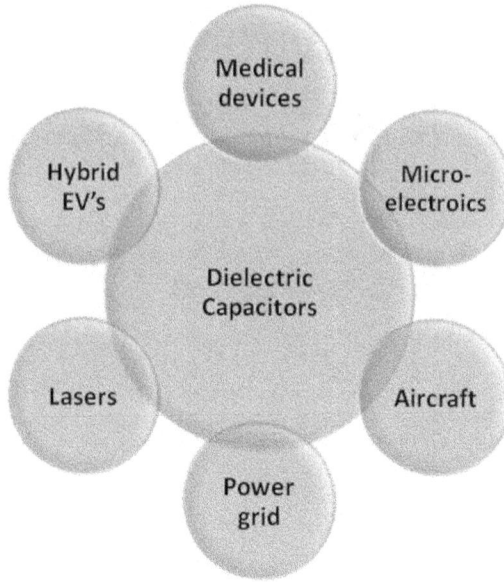

**FIGURE 9.1** Applications of dielectric capacitors.

physical concepts are initially implemented in the current study to emphasize the basic criteria for exploring interesting materials. Dielectric glass-ceramics, relaxor ferroelectrics, and polymer-based ferroelectrics, antiferroelectrics are all summarized. Out of all these, antiferroelectrics receive extraordinary emphasis. Finally, some broad prospects for future studies are presented. The goal of this chapter is to provide complete information on high energy storage dielectric materials at the research level, as well as to encourage their future implementation in practice.

## 9.2 GENERAL PRINCIPLE OF DIELECTRIC CAPACITORS

As shown in Figure 9.2, a conventional dielectric capacitor comprises electrodes sandwiched between dielectric materials. The role of capacitors in electronic devices is to store electric energy. Capacitance, which measures the capacity of capacitors to store energy, may be computed as follows [7, 8]:

$$C = \epsilon_0 \epsilon_r \frac{A}{d} \tag{9.1}$$

where $C$ denotes capacitance, $\epsilon_0$ denotes vacuum permittivity, $\epsilon_r$ denotes relative permittivity, the overlaying region of the two plates is denoted by the letter $A$, and $d$ is spacing among them. In the charging cycle, the electrical field is formed throughout space, allowing dipoles in the dielectric medium to rearrange in the field's orientation, leading a charged ion to be induced on parallel sheets. When the electric potential difference induced by the collected ions ($Q$) equals the applied voltage, the charging cycle is completed, and the electrostatical field energy is stored in the capacitor. The derivative of charge concerning voltage may also be used to determine capacitance $C$:

$$C = \frac{dQ}{dV} \tag{9.2}$$

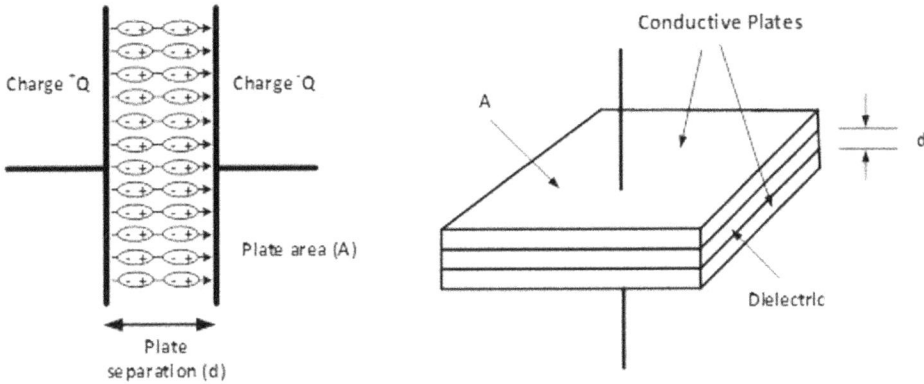

**FIGURE 9.2**　Schematic of the dielectric capacitor.

During the charging process, external bias transfers charges across conductor plates, indicating the work needs to be done and also that electric energy is being consumed. At the same time, energy is stored in the dielectrics. As a result, the following equations might be used to calculate the quantity of the stored energy $U$ [7, 8]:-

$$U = \int_0^Q V dq = \int_0^Q \frac{q}{C} dq = \frac{1}{2}\frac{Q^2}{C} = \frac{1}{2}CV^2 = \frac{1}{2}VQ \tag{9.3}$$

## 9.2.1　Energy Density (Stored and Recoverable)

Energy storage density ($J$) is defined as the amount of the energy in unit volume of a material.

$$J = \frac{U}{Ad} = \frac{\int_0^{Q_{max}} V dq}{Ad} = \frac{\int_0^{\delta_{max}} V d\delta}{d} = \int_0^{D_{max}} E dD \tag{9.4}$$

Here, $E = \frac{V}{d}$ signifies the exerted electric field, $d$ represents the charge density of surface, and $D$ represents the electric displacement (in Maxwell's equations). The dielectric constant of linear dielectrics (such as quartz, glass) stays static while the applied electric field ($E$) varies, allowing the storage density ($J$) to be determined [9, 10].

$$J = \int_0^{D_{max}} E dD = \int_0^{E_{max}} \epsilon_0 \epsilon_r E dE = \frac{1}{2}\epsilon_0 \epsilon_r E^2 \tag{9.5}$$

The relative permittivity and the square of the electric field are dependent on the energy density, as indicated in Equation (9.5). As the dielectric constant depends on the electric field, Equation (9.5) is inapplicable to non-linear dielectric materials. The electrical conductivity of dielectric materials having large dielectric constant is high, if the displacement is extremely near to polarization, then Equation (9.4) assumes the form of [7]

$$J = \int_0^{P_{max}} E dP \qquad (9.6)$$

The spontaneous and max polarization of dielectric material is denoted by $P$ and $P_{max}$, respectively. According to Equation (9.6), the storage density of the material is equal to the area covered by the vertical axis and the hysteresis loop during the charging process (i.e., the sum of the blue and red sections in Figure 9.3). Figures 9.3 (a–d) depict the $P$–$E$ field hysteresis loops for linear dielectrics, ferroelectrics, relaxor ferroelectrics, antiferroelectrics, respectively. As demonstrated in Figure 9.3, not all of the stored energy can be released due to the energy loss ($J_{loss}$) caused by electric hysteresis during the discharging process and the hysteresis loops while discharging. During discharging, the recoverable energy-storage density ($J_{rec}$), which is equal to the region surrounded by the vertical axis and the hysteresis loops,

$$J_{rec} = - \int_{P_{max}}^{P_r} E dP \qquad \text{(upon discharging)} \qquad (9.7)$$

As illustrated in Figure 9.3 and Equation (9.7), strong electric field breakdown strength (BDS), high max polarization, low residual polarization, and narrow $P$–$E$ loops all contribute to high $J_{rec}$. Despite having a higher BDS and lower energy loss, linear dielectric materials are unsuitable for high energy storage applications because of their small maximum polarization [8]. As a result, materials for energy storage applications are the emphasis of this chapter, including ceramics, thin and thick films, and polymer-based nanocomposites.

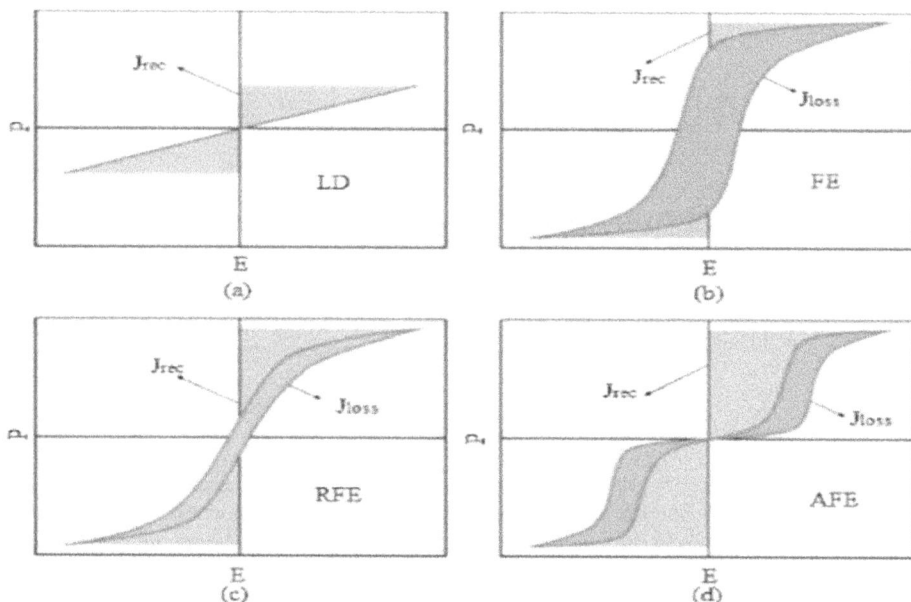

**FIGURE 9.3** $P$–$E$ loops for (a) linear dielectrics; (b) ferroelectrics; (c) relaxor ferroelectrics; (d) antiferroelectrics with their energy storage parameters. Reprinted and adapted with permission from [11]. Copyright (2019) Materials Research Bulletin.

## 9.2.2 Energy Storage Efficiency

The storage efficiency ($\eta$) of a dielectric capacitor is just as important as its energy density in real applications. During the depolarization process, some of the stored energy is normally dissipated. $J_{loss}$ denotes the amount of loss in energy density as the area of energy loss of the hysteresis loop $P$–$E$, as shown in Figure 9.3. Though charging in a strong electric field can result in higher energy densities, increased conduction of the material causes polarization loss, which reduces the amount of energy that can be recovered during a discharge [12]. Such a loss of energy, i.e., the energy squandered by the hysteresis loss, will cause temperature rise/heat generation in the capacitor, reducing its temperature stability and longevity [7, 13, 14]. The energy storage efficiency could be stated as [12]

$$\eta = \frac{J_{rec}}{J_{rec} + J_{loss}} \tag{9.8}$$

To achieve the highest storage efficiency, the dielectric materials should have low dielectric/hysteresis loss, low remnant polarization, and delayed saturation polarization (electric displacement). It is worth noting that dielectrics high energy storage efficiency translates to less waste heat, improved dependability, and a longer capacitor lifetime in real-world applications [13].

## 9.2.3 Dielectric Breakdown Field

Electric field based functioning of dielectric capacitors are determined by their intended uses. High electric fields can cause dielectric materials to break down, thereby reducing their electrical resistance dramatically. The dielectric breakdown field is the maximum electric field ($E_{BD}$). The $E_{BD}$ has a significant impact on dielectric capacitor reliability and performance. Equations (9.3)–(9.5) show that a dielectric capacitor's energy density is highly connected to its $E_{BD}$, which is time as well as field dependent [14]. The bandgap and permittivity of the material influence the $E_{BD}$ of a dielectric capacitor. Dielectric material performance is affected by crystal structures, size distribution, imperfection chemistry, sample thickness, microstructural homogeneity, and pore size distribution, and also electrode configuration and working (voltage, frequency, temperature, pressure, and aging time) and environmental conditions. To attain better capacitance sustainability under high electric fields, high-quality materials with dense microstructures must be fabricated (a larger $E_{BD}$).

Capacitor dielectric failure is thought to be caused by several causes, including inherent electric strength, thermal runaway, partial discharge, and electrostatic compressive pressures [14–17]. The intrinsic electric breakdown is electronic and is a material feature. Electron production causes a well-defined current to flow when a gradually rising voltage is applied to a dielectric [16]. After achieving saturation, the current will rapidly rise ($10^{-8}$ sec), culminating in an electron avalanche and collapse. Increase in current may result due to field-assisted emission from metal electrodes, emission from bulk impurity centers, and ionization caused by collisions of conduction electron. When the heat created by Joule heating and other dielectric losses cannot be dispersed rapidly enough by conduction or convection to the surroundings, thermal runaway occurs [17].

$$C_V \frac{\partial T}{\partial t} - \nabla(K.\nabla T) = \sigma E^2 \tag{9.9}$$

where $C_V$ is for specific heat per unit volume, $T$ stands for dielectric temperature, $K$ stands for thermal conductivity, and $E$ denotes electric field, while $\sigma$ is electrical conductivity. Formula (9.9) shows that the breakdown field shrinks exponentially manner as the temperature rises.

Partial discharges produce discharge breakdown occuring at microstructural level voids, fractures, holes, interfaces, and sharp electrode sites within a dielectric [17]. At these places, the average level of electric stress rises dramatically causing partial breakdown due to stress. Dielectric discharge causes erosion, tree removal, tracking, local melting, and pitting changes in electrochemistry. The accumulation of space charges and long-term dielectric aging failure is caused by dielectric losses. Electrostatic compressive forces caused by the attraction between opposing charges in the dielectric medium are responsible for the electromechanical break-down [17]. Due to compressive stress, the thickness of the dielectric decreases with an applied voltage $V$, which is normally balanced by elastic dielectric deformation. The reduction in thickness, on the other hand, results in a greater field ($E = V/d$) and more charged ions on the electrodes ($Q = CV$, $C = \epsilon_0 \epsilon_r A/d$). This causes a higher compressive force, which reduces the thickness even more, and so on until shear stresses inside the dielectric induce plastic deformation and finally breakage. The electromechanical breakdown ($E_{BD}$) is represented by the equations below [14].

$$\frac{\epsilon_0 \epsilon_r}{2d^2} = Y \ln\left(\frac{d_0}{d\tau}\right) \tag{9.10}$$

$$E_{BD} = 0.6\left(\frac{Y}{\epsilon}\right)^{\frac{1}{2}} \tag{9.11}$$

where $Y$ denotes Young's modulus, $d_0$ and $d_r$ are the original and reduced thicknesses of the specimen, and ($=\epsilon_0 \epsilon_r$) is the permittivity of the insulating medium. Even if several capacitors are made of the same material, their $E_{BD}$ values may vary due to changes in the weak links, such as contaminated particles, internal microstructures, associated faults, etc. As a result, statistical analysis of electric field breakdown data is recommended, and the Weibull distribution may usually provide a valid estimate of dielectric failure [18, 19]. The Weibull distribution of $E_{BD}$ for dielectric capacitors can be calculated using the equations below [20].

$$X_i = \ln(E_i) \tag{9.12}$$

$$Y_i = \ln\left\{\ln\left[1/\left(1 - \frac{i}{n+1}\right)\right]\right\} \tag{9.13}$$

where $n$ is the total number of samples examined, and $E_i$ is the breakdown electric field of the $i^{th}$ sample, in ascending order. To find the most likely breakdown field, apply the linear connection between $X_i$ and $Y_i$, where the fitting line intersects the horizontal axis at $Y_i = 0$. Using the two-parameter Weibull distribution, the cumulative probability of dielectric breakdown can be expressed as [18]

$$P(E) = 1 - \exp\left[1 - \left(\frac{E}{\alpha}\right)^\beta\right] \tag{9.14}$$

The probability of a dielectric breakdown in an electric field is denoted by $P(E)$. $W$ represents the Weibull modulus, which denotes the data dispersion, and $E$ is the scale parameter calculated from the experimental breakdown electric field with a cumulative failure probability of 63.2%. A higher score suggests that the data is more evenly distributed.

## 9.2.4 Discharge Time

Because $I = dQ/dt$ and power $= I \times V$, dielectric capacitors have a unique attribute in that they discharge energy relatively rapidly. This rapid discharge provides a large current, which results in high power in the dielectric capacitor. The discharge speed is expressed in discharge time, which is denoted by $\tau_{0.9}$. The experimental discharge time can be estimated directly from discharge characteristics and is defined as the time it takes for the discharged energy in the load resistor to reach 90% of its final value over an indefinite duration. The time that it takes for a capacitor to discharge its energy is determined by permittivity and thickness of dielectric material, as well as operating circumstances, i.e., load resistance and applied voltage. The discharge period should be as short as feasible for pulsed power applications.

## 9.2.5 Fatigue Endurance and Thermal Stability

Large fatigue durability is essential for long-term physical safety and efficient working of dielectric capacitors throughout the charging-discharging process. The $P–E$ loops, which are determined at a specific electric field cycling frequency, are commonly used to calculate the capacitor's endurance. At any observed level, the fatigue effects of paraelectric are absent [21]. A variety of structural changes associated with strain energy relaxation in lattices, diffusion and aggregation of defects, pinning of domain walls and low permittivity dead layers at the electrode/dielectric interface are likely to contribute to fatigue behavior in dielectric materials with high permittivities such as ferroelectrics, relaxor ferroelectrics, and antiferroelectrics [22]. The storage capacity of dielectric capacitors degrades as the capacitance (polarization) of the dielectric material decreases as the number of charge-discharge cycles increases. Under high electric field conditions, the logarithm of time to breakdown is proportional to the applied field for ferroelectrics, but not for paraelectric [21]. Capacitor fatigue life can range from a few hundred to tens of billions of charge-discharge cycles. The dielectric material's temperature stability is just as important as its fatigue resistance in energy storage capacitors. Dielectric capacitors with a high maximum working temperature are required for good thermal control in power electronics and other device systems. In this case, capacitors must be able to withstand significant temperature cycling over a long period. Furthermore, heat build-up in the capacitor material due to energy loss due to leakage currents and dielectric loss reduces its energy storage capability. Depending on the application and operating conditions, capacitors can be exposed to temperatures ranging from 90 to 250 °C. Additionally, heat conditions may differ based on where the capacitor electronics are installed within the final product. In extreme weather conditions, dielectric materials with low leakage current densities, low dielectric losses, and thermally stable permittivities (polarization) are potential energy storage capacitors.

## 9.3 DIELECTRIC MATERIALS FOR HIGH ENERGY STORAGE APPLICATION

Different kinds of materials have been investigated for energy storage capacitor applications, i.e., linear dielectrics, paraelectrics, ferroelectrics, relaxor ferroelectrics, and antiferroelectrics. Typical dipolar and ferroelectric domain designs, as well as electric field-dependent variations in permittivity and polarization behavior of linear and nonlinear dielectric materials, are shown schematically in Figure 9.4. Linear dielectrics exhibit essentially constant permittivity and a linear proportionality change in polarization when the $e_d$ field is applied due to the lack of permanent dipoles (Figure 9.4a). $P–E$ and $r–E$ characteristics are mildly non-linear in paraelectric materials with persistent dipoles but no ferroelectric domains (Figure 9.4b). When exposed to an electric field, they become polarized, but when the field is removed, they revert to a non-polar condition. Ferroelectrics, on the other hand, have a net electric dipole moment in the absence of an external electric field and so exhibit spontaneous polarization with extensive hysteresis loops (Figure 9.4c). Relaxor ferroelectrics have polar

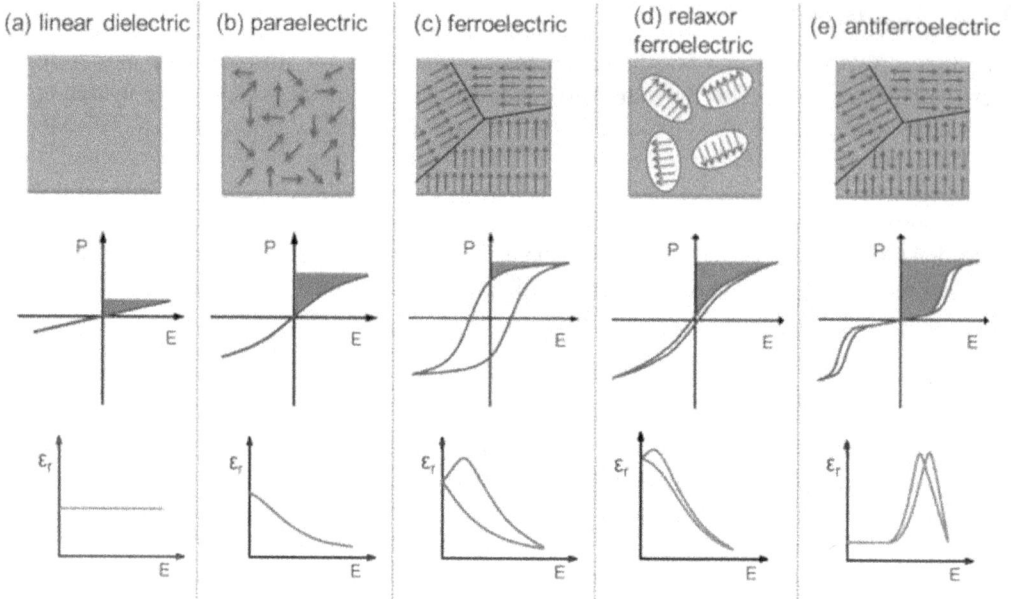

**FIGURE 9.4** Top panel: structures of ferroelectric domain and dipole. Middle panel: dependence of polarization. Bottom panel: permittivity vs electric field for (a) linear dielectric; (b) paraelectric; (c) normal ferroelectric; (d) relaxor ferroelectric; and (e) antiferroelectric materials. The shaded area above the polarization hysteresis loop represents recoverable energy density and the area inside the loop represents energy loss from the corresponding material.Top/middle panel: reproduced with permission [20]. Copyright 2013, Elsevier. Bottom panel: reprinted (adapted) with permission from [21]. Copyright 2013 American Chemical Society. Reproduced with permission. Copyright 2016, AIP Publishing [22].

nanodomains that minimize cooperative coupling across ferroelectric domains, resulting in lower spontaneous polarization and smaller hysteresis loops (Figure 9.4d). In antiferroelectrics, opposing polarizations align neighboring crystalline lattices, resulting in zero net polarization in the virgin state. They reveal multiple hysteresis loops generated by reversible AFE-FE phase transitions under high electric fields (Figure 9.4e) [7, 23, 24]. Although each of these dielectric materials has its own set of benefits, they all have limitations in terms of energy storage capacity.

Linear dielectrics have a low dielectric constant and a high dielectric breakdown strength, which are two frequent properties. In comparison, ferroelectric materials have a high dielectric constant but a low dielectric breakdown strength, making them the second most popular capacitor material. Ferroelectric materials also suffer dielectric losses during charging and discharging cycles due to ferroelectric hysteresis. Despite this, contemporary improvements in materials chemistry and manufacturing processes have made it feasible to adjust the dielectric response of such materials [25, 26]. Furthermore, energy densities have been obtained in comparison to high voltage linear dielectrics. Dipolar glasses with no long-range order in ferroelectric domains are known as relaxor ferroelectrics. As a result, they have very little residual polarization and a coercive electric field, and no ferroelectric hysteresis. As a result, they are gaining traction as a viable choice for capacitor applications over traditional ferroelectric materials. However, relaxor-like material systems are uncommon [18, 2] and typically consist of lead-based materials. As a result, a more appropriate material is being sought to reduce the usage of lead in ferroelectric materials, which is becoming a rising environmental problem.

A dielectric material with high permittivity or polarizability, low dielectric/hysteresis loss, low electronic/ionic conductivity, and a large breakdown field are required for a capacitor with high storage performance, i.e., high recoverable energy density and energy storage efficiency as discussed. All of these characteristics, on the other hand, are difficult to obtain in a single dielectric material. As a result, the majority of research into high energy storage dielectric materials has focused on integrating the aforementioned features. Dielectric materials based on polymers, glasses, and ceramics have been studied for application in energy storage capacitors [2, 7, 8, 19].

## 9.3.1 CERAMIC-BASED DIELECTRICS

Linear dielectric ceramics have a slightly higher permittivity than polymers but have a much lower breakdown field. Nonlinear dielectric ceramics, on the other hand, have far higher permittivities than linear dielectric ceramics, allowing them to store large amounts of energy with much less applied fields. Ceramic materials also have higher thermal stability, making them ideal for high-temperature (up to 250 °C) capacitor applications. The energy storage properties of various dielectric ceramics have been investigated in bulk as well as in film form. Different solid solutions of $Pb_{0.97}Y_{0.02}[(Zr_{0.6}Sn_{0.4})_{0.925}Ti_{0.075}]O_3$ (PYZST), $PbZr_{0.4}Ti_{0.6}O_3$, $(Pb_{1-x}La_x)(Zr_{0.90}Ti_{0.10})_{1-x/4}O_3$(PLZTx), $Pb_{0.89}La_{0.06}Sr_{0.05}(Zr_{0.95}Ti_{0.05})O_3$ have been studied [27–32] in addition to undoped and doped dielectric ceramic compounds. Ceramic bulks provide limited energy storage because their energy density rapidly decreases. Glass phase addition, special sintering technology, and refining grain size can both improve $E_b$ of ceramic bulks. Researchers have studied heterostructures such as superlattices, multilayer structures, and core-shell structures to improve the energy storage capability of dielectric ceramics. Despite a large amount of work done so far, most bulk ceramics (10 J cm$^{-3}$) have insufficient energy densities due to their low breakdown fields (1 MV cm$^{-1}$) and low energy densities [33]. Numerous tests on dielectric ceramic films have shown ultrahigh-energy storage qualities with energy densities >100 J cm$^{-3}$ and efficiency >80%, as well as the ability to endure huge applied fields >3 MV cm$^{-1}$, similar to those conducted on bulk ceramics [27, 32, 34]. Table 9.1 shows the summary of energy storage performance of some ceramics-based dielectrics.

Recently, $Ca_{0.5}Sr_{0.5}Ti_{0.97}Sn_{0.03}O_3$ ceramics have been synthesized possessing a high energy storage density of 2.06 J/cm$^3$ and 95% efficiency under 330 kV/cm breakdown strength. This system also achieves outstanding stability with minimum fluctuation (15%) throughout a frequency range of 1–1000 Hz and a temperature range of 20–120 °C [27].

$NaNbO_3$-based ceramics [37] (0.80$NaNbO_3$-0.20$SrTiO_3$) with a high storage density (3.02 J/cm$^3$) and storage efficiency of 80.7% were devised and prepared by Zhou and his workers. As demonstrated in Figures 9.5 and 9.6, the energy storage capability of as-prepared ceramics improves as frequency and temperature rise. After 105 unipolar fatigue cycles, the energy storage capacity of $NaNbO_3$-based ceramics remained unchanged. Quick discharge periods (0.220 sec), a high current density, and a high power density are other features of the as-prepared ceramics. Charge-discharge performance further demonstrates the thermal stability of the 0.80NN-0.20ST ceramics.

## TABLE 9.1
### Ceramics-Based Dielectrics and Their Energy Storage Performance

| Materials | $W_{rec}$ (J cm$^{-1}$) | $\eta$ (%) | $E_b$ (kV cm$^{-1}$) | Ref. |
|---|---|---|---|---|
| $K_{0.48}Nb_{0.52})_{0.88}Bi_{0.04}NbO_3$ | 1.04 | – | 189 | [23] |
| $Na_{0.7}Bi_{0.1}NbO_3$ | 3.44 | 85.4 | 250 | [24] |
| $Ba_{0.3}Sr_{0.7}TiO_3$ | 0.23 | 95.7 | ~90 | [35] |
| (0.96)$NaNbO_3$-(0.04)$CaZrO_3$ | 0.55 | 63 | 130 | [36] |
| (0.8)$NaNbO_3$-(0.2)$SrTiO_3$ | 3.02 | 80.7 | 323 | [37] |
| $Ba_{0.3}Sr_{0.7}TiO_3$(SPS) | 1.13 | 86.8 | 230 | [38] |
| $Ba_{0.4}Sr_{0.6}TiO_3$ | 1.28 | - | 243 | [39] |

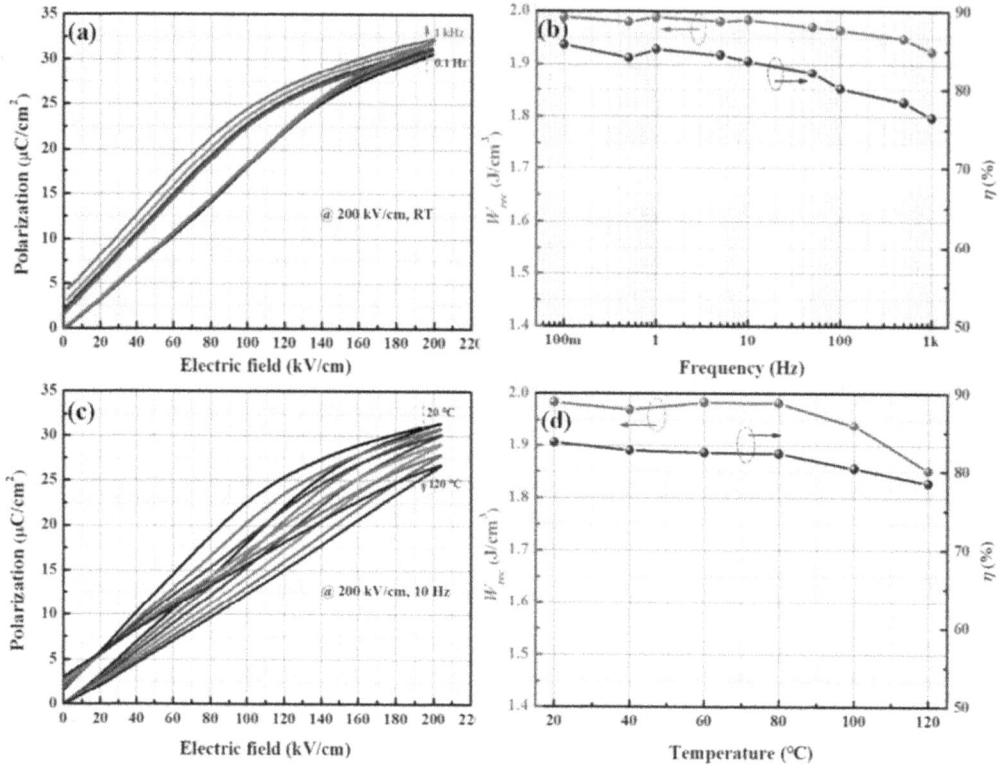

**FIGURE 9.5**  Frequency and temperature dependence of $NaNbO_3$-based energy storage properties of the $NaNbO_3$-based ceramics. Reprinted (adapted) with permission from [37]. Copyright [2018] American Chemical Society.

## 9.3.2  GLASS-BASED DIELECTRICS

Another dielectric material alternative for energy storage has been glass-based oxide ceramics. A mixture of an amorphous form with one or more crystalline forms is known as glass-ceramic. This type of material generally has qualities that are similar to both glass and ceramics. Glass ceramic is often made in one of two ways: body crystallization technique and composite method [28, 29]. The relevant glass and ceramic powders are first made in the former; they are then blended in a predetermined ratio and pressed into specific geometries; and lastly, sintered at a predetermined temperature to make the result. The compositions of glass ceramics synthesized in this technique are normally perfect, however, pores are frequently created. In another, an amorphous phase containing the appropriate chemical constituents is first created using a standard glass-making method; then, after cooling and shaping, the amorphous is partially crystallized at a certain temperature to produce the final glass-ceramic. Glass-ceramics may readily achieve a pore-free microstructure in this instance, but their composition is difficult to determine. High electrical breakdown field and increased permittivity may be attained concurrently in a glass-ceramic system by appropriately regulating the chemical composition of the amorphous and crystalline phases, indicating that these materials could have a high storage density. However, it should be noted that there is frequently a large interface between the amorphous and crystalline phases, which hurts the energy discharge. In comparison to polymer and ceramic dielectrics, fewer investigations on the storage performance of glass-ceramics have been done. The existence of crystallites binded by an amorphous glass matrix characterizes the core-shell structure of a glass-ceramics. When compared to polymer dielectrics, these glass- ceramics,

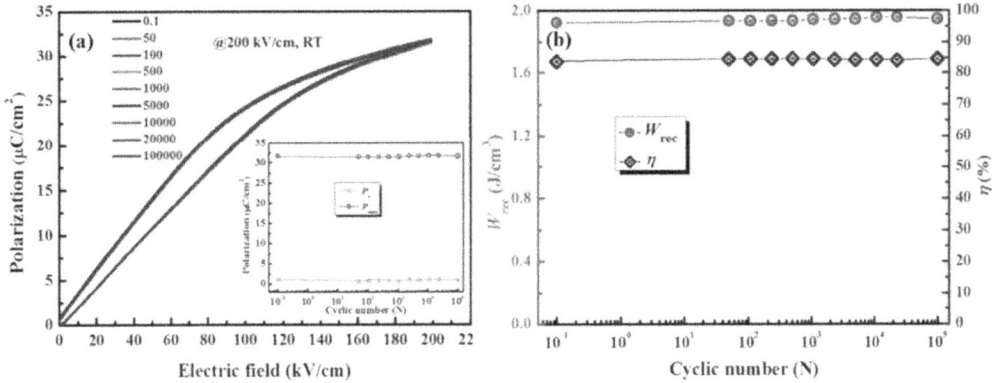

**FIGURE 9.6**   Energy storage behavior of the NaNbO$_3$-based ceramics as functions of cycle number. Reprinted (adapted) with permission from [37]. Copyright [2018] American Chemical Society.

which are created by controlled crystallization of glass, have higher permittivities and substantially larger breakdown fields.

A variety of glass-ceramic systems based on barium titanate (BaO–TiO$_2$–SiO$_2$–Al$_2$O$_3$), barium strontium titanate, alkali niobates (Na$_2$O–BaO–Nb$_2$O$_5$–SiO$_2$, Na$_2$O–Nb$_2$O$_5$–SiO$_2$–B$_2$O$_3$) and other glasses oxide have been investigated for energy storage applications [7, 8, 38, 39]. However, the need for extremely high electric field intensities in dielectric glasses to achieve large energy densities poses various safety issues regarding their usage as a dielectric medium in capacitors. Leakage, electrical fatigue, catastrophic dielectric breakdown, and insulation failure are all risks associated with the high applied voltage [32]. Furthermore, because of the enormous volume of the interface between fewer amorphous and crystalline phases, interfacial polar glass-ceramics allow more charge to be stored and increase the total (stored) energy characterize by glass-ceramic. The less charge is trapped in the discharge cycle owing to the extended relaxation process of interfacial polarization, which reduces the discharge energy density and energy storage efficiency of the glass-ceramic [33].

Zhang and co-authors synthesized Ba$_{0.4}$Sr$_{0.6}$TiO$_3$ ceramics with the addition of 520 vol. percent BaO–SiO$_2$–B$_2$O$_3$ glass in a typical sample [29]. The grain size, permittivity, and polarization values of the glass-ceramics steadily decreased as the glass content increased, resulting in linear $P$–$E$ loops, although their electric breakdown field was increased owing to the reduction of porosity and pore size. In 20 vol. percent glass added ceramics, the greatest breakdown field of 240 kV/cm was achieved. Overall, the samples containing 5% glass exhibited the maximum recoverable energy-storage density of 0.89 J/cm$^3$, which is 2.4 times greater than that of pure Ba$_{0.4}$Sr$_{0.6}$TiO$_3$ ceramics (0.37 J/cm$^3$).

The effects of SrO–B$_2$O$_3$–SiO$_2$ glass addition on the energy storage capabilities of Ba$_{0.4}$Sr$_{0.6}$TiO$_3$ ceramics were then investigated by Chen and co-authors [34]. This study came up with similar conclusions. In comparison to pure ceramics, the dielectric breakdown strength of glass-ceramics (approximately 150 kV/cm) was increased by about twice by adding 10% glass. In 2 wt. percent glass added ceramics, the highest discharged energy storage density was 0.44 J/cm$^3$. The glass additives can increase the breakdown strength, but they also cause a declination of the maximum polarization, according to the aforementioned findings. As a result, the glass composition must be carefully managed to get the highest energy storage density.

The body crystallization technique, on the other hand, is a more extensively utilized method of synthesizing glass-ceramics for high energy storage. Glass-ceramics based on (Ba, Sr)TiO$_3$ and NaNbO$_3$ are now the most commonly investigated systems. The energy storage characteristics of BaO–SrO–TiO$_2$–Al$_2$O$_3$–SiO$_2$ glass-ceramics with (Ba, Sr) TiO$_3$ phase were investigated by

Gorzkowski and co-authors [40–43]. The relative permittivity of 1000 and a breakdown field of 800 kV/cm were attained; however, because of the significant hysteresis loss, the recoverable energy storage density from P–E findings was only 0.3–0.9 J/cm³.

### 9.3.3 POLYMER-BASED DIELECTRICS

Polymer dielectrics are the preferred material for room-temperature capacitors due to their inherent advantages of a lightweight, ease of production, scalability, high $E_{BD}$, and great fatigue resistance. Low-cost fabrication, simplicity of synthesis into desired configurations, and strong mechanical flexibility are all advantages of polymer dielectrics. Polymers have high breakdown fields (>7 MV cm⁻¹) and failure at high electric fields, in addition to their high electrical resistance and low dielectric loss. A polymer dielectric can keep insulating qualities and may still be useful due to evaporation of the surrounding electrode material near the breakdown field. However, polymer's low dielectric permittivity and poor heat stability (maximum working temperature 100 °C) are significant stumbling blocks to their employment in energy storage applications. Several dielectric polymers have been studied for use in capacitors, including polyvinylidene fluoride (PVDF) and its copolymers, biaxially oriented polypropylene, polyethylene terephthalate (PET), polycarbonate (PC), polyimide (PI), and others. Polyimides, polythioureas, [44] polyurethanes, and polyureas [45–47] have all been investigated for dielectric applications, however, they suffer from production difficulties, heat constraints, and large dielectric loss. Ramprasad et al. look at the most prevalent dielectric materials in use today and try to find new polymers that may be employed in energy storage applications [48]. Alternative polymeric materials with increased energy density, low loss, excellent performance at extreme temperatures, and ease of processing are critical in reducing the volume and weight of pulsed power capacitor systems.

Various temperature-stable polymer dielectrics have been developed with much effort. The various polymer based dielectrics and their properties are described in Table 9.2. Zhang et al. [6] examined the dielectric characteristics of high-temperature resistant polymer dielectrics (HTPDs) and their applications in electrical/electronic insulation in a systematic manner. Polymers like BOPP, poly(ethylene terephthalate) (PET), polycarbonate (PC), polyetheretherketone (PEEK) and imide polymers including polyimide (PI), PEI, polyamide-imide (PAI), etc., exhibit low loss factor, high volume resistivity, high electromechanical breakdown ($E_{BD}$), as well as good temperature stability. Thus, enormous attempts have been made to validate these polymers as dielectrics. Because of the nature of covalent molecule bonding, the dielectric constant ($\varepsilon_r$) for organic polymers is substantially lower than for inorganic polymers [49]. Indeed, a density functional theory investigation of hundreds of hydrocarbon-based polymers confirms the dielectric constant of polymer limitation

## TABLE 9.2
## Summary of Polymer-Based Dielectrics

| Polymer | Dielectric constant (kHz) | Loss tangent (kHz) | Dielectric strength (kV/cm) | Energy density (J/cm³) | Ref. |
|---|---|---|---|---|---|
| Polystyrene (PS) | 2.4–2.7 | 0.008 | 2000 | 0.55(at 2000 kV/cm) | [61][62] [51][63] |
| PVDF/BT@ BN | 11.6 | 0.027 | 580 | 17.6 | [64] |
| PEI/Al$_2$O$_3$ | 3.0 | 0.003 | 582 500 (100 °C) 500 (150 °C) | 5 3.8(100 °C) 3.7(150 °C) | [65] |
| PVDF/TiO$_2$ | 1.2 | 0.02 | 90.3 | 0.32 | [66] |
| PVDF/Ba$_{0.6}$ Sr$_{0.4}$TiO$_3$ | 13.5 | 0.04 | 575 | 19.4 | [67] |
| Polycarbonate (PC) | 2.55 | - | 300 | 0.8 | [68] |
| Polyimide (PI) | 3.33 | - | 250 | 0.82 | [69] |

[50, 51]. Recent research has focused on addressing orientational polarization to synthesis polymers with a high dielectric constant and minimal dielectric loss (dipolar glass and paraelectric polymers, as well as relaxor ferroelectric) [49].

Despite extensive attempts to improve the $\varepsilon_r$ of polymer film capacitors, little attention has been paid to the polymer films' $E_{BD}$. Because the energy density has a quadratic reliance on $E_{BD}$, upgrading $E_{BD}$ provides a greater potential for boosting the current energy densities of polymer film capacitors than just increasing $\varepsilon_r$. The development and spread of electrical trees that resemble fractal conducting channels within polymer dielectrics are typically thought to occur via an electron avalanche mechanism [52–54].

Electrical trees are started by structural faults in the material and grow between the electrodes, eventually leading to a breakdown. The breakdown process can be delayed if the propagation of these electrical trees is hampered by a barrier, improving total breakdown strength. The "barrier effect" is a well-known concept used to raise the breakdown voltage in high-voltage engineering. In the application of an external field, it is postulated that the existence of a barrier, i.e., a material with a different permittivity or conductivity than the matrix, induces the production of space charge near the interface, causing the field to redistribute across the barrier-matrix interface [55, 56].

The creation of polymer-based composites including dielectric ceramic fillers and/or conductive organic dopants spread in a polymer matrix has recently gotten a lot of attention [57–60]. The composites should have a high energy density due to the combination of ceramics with high permittivities and polymers with strong breakdown fields. When a polymer is mixed with modest amounts of conducting organic dopants like carbon black, graphene, and carbon nanotubes, the composites permittivities increase. Nonetheless, significant practical issues connected with interfacial effects from agglomeration and inhomogeneity of the fillers on the polymer matrix might impact polymer-based composites high energy storage capacity. Furthermore, the substantial electrical differences between the fillers and the matrix may cause local electric field concentration and interfacial polarization, both of which are unfavorable to the polymers' breakdown field.

## 9.4 DIELECTRIC CERAMIC FILMS FOR ENERGY STORAGE CAPACITORS

Ceramic film capacitors with small footprints are ideal for microelectronic systems, mobile platforms, and miniature power devices. Ceramic films with dense and pore-free microstructures with fewer flaws and lattice faults may endure higher electric fields and so have larger energy storage densities than their bulk equivalents. In general, the dielectric strength of dielectric samples rises as the thickness of the samples decreases. When the thickness approaches a certain threshold, the dielectric strength becomes dimensionless and achieves the inherent value. This explains why dielectric films have more dielectric strength and energy storage density than their bulk counterparts. It should be noted, however, that the total amount of energy that a film capacitor can store is typically restricted by its thickness/volume. Many investigations on dielectric ceramic films have been undertaken in order to prove energy storage performance. High dielectric constant and low dielectric loss, high polarization and low hysteresis loss, and broad breakdown field, as well as high fatigue endurance and thermal stability, are among the optimum features required for high energy storage.

For the development of dielectric ceramic films for energy storage capacitors, a variety of film processing technologies have been used, including screen printing, tape casting, sol–gel method, magnetron sputtering, pulsed laser deposition (PLD), chemical solution deposition (CSD), atomic layer deposition (ALD), metal-organic deposition (MOD), and aerosol deposition (AD). On a range of substrates, including glass, metals, and alloys, $Al_2O_3$, MgO, Si, $SrTiO_3$ (STO), $LaAlO_3$ (LAO), and $(La,Sr)(Al,Ta)O_3$, several types of ceramic films with polycrystalline, textured, and epitaxial structure have been developed. Pt, indium tin oxide (ITO), $LaNiO_3$ (LNO), $La_{0.7}Sr_{0.3}MnO_3$ (LSMO), $SrRuO_3$ (SRO), and $Ca_2Nb_3O_{10}$ (CNO) layers were most often used to buffer these substrates. Composition, grain size, film thickness, substrate clamping, stress condition between the film and

**TABLE 9.3**
**Several Approaches Adopted for Enhancing the Energy Storage Performance of Dielectric Ceramic Films**

| Physical modification | Chemical modification | Microstructural modification |
|---|---|---|
| Film thickness control | Composition optimization | Phase structure tuning |
| Compositional gradient sequence of layers | Elemental doping and chemical substitution | Crystallinity, grain size, defects, and impurities |
| Substrate material and its orientation | Low melting point glass additives | Crystallographic orientation |
| Top and bottom electrodes | Alloying/composite/solid solution | Domain engineering |
| Buffer layer, seed layer, insert layer, capping layer | Humidity/moisture control | Interface engineering – epitaxial or misfit strain, charge transport, and |

substrate, and other factors may all influence the energy storage ability of ceramic films. To adapt and improve the energy storage capabilities of dielectric ceramic films, a variety of ways, including modification or control of these parameters have been used. Some of these techniques, which may be divided into three groups, are explored in Table 9.3.

### 9.4.1 Linear Dielectric Ceramic Film

Linear dielectric materials are suitable for energy storage capacitors due to their high $E_{BD}$ and low dielectric loss, allowing for high $U_{rec}$ and $\eta$. Pan et al. [70] used pulsed laser deposition (PLD) to create a 5 mol percent Mn-doped $(SrTiO_3)_{0.6}$–$(BiFeO_3)_{0.4}$ $(_{0.6}ST-_{0.4}BF)$ thin film over an Nb-doped $SrTiO_3$ single crystal substrate, and found that $U_{rec}$ and were 51 J/cm³ and 64%, respectively. Furthermore, the energy storage performances showed good temperature stability over a temperature range of (40)–140 °C, as well as fatigue endurance after $2 \times 10^7$ cycles. The modification of macro/microarchitectures to improve the energy storage capabilities of $SrTiO_3$-based ceramic films has yielded several useful and fascinating results. The amorphous phase is well recognized having a greater $E_b$ but a lower $\in_r$ than their crystalline counterparts. Gao et al. [71] investigated the energy storage properties of amorphous $SrTiO_3$ thin films with various top electrodes and proposed a "self-healing" mechanism based on the anodic oxidation reaction in aluminium electrolytic capacitors. Amorphous $SrTiO_3$ films with an Al top electrode obtain a $U_{rec}$ of 15.7 J/cm³ at 3500 kV/cm at a relative humidity of 60%, which is nearly eight times that of samples with an Au electrode. The energy storage of certain linear dielectric ceramic films are summarized and listed in Table 9.4.

Spahr et al. used an ALD approach to create thin film capacitors made of pure $Al_2O_3$ (3 nm thick) and nanolaminates made of $Al_2O_3/ZrO_2$ and $Al_2O_3/TiO_2$ (Figure 9.7a,b) with a variety of dielectric layer thicknesses up to 300 nm [78]. The relative permittivity and capacitance density of nanolaminates made of high-dielectric materials ($ZrO_2$, $TiO_2$) were greater than those of the $Al_2O_3$ film. Furthermore, increasing the layer thickness caused the capacitance density to drop while the breakdown voltage increased (Figure 9.7c). As a result, as the layer thickness increased, the volume energy density of the film capacitors fell. The optimum capacitance density (1643 nF cm⁻²) and $U_{rec}$ (60 J cm⁻³) were found in the $Al_2O_3/TiO_2$ nanolaminate with a layer thickness of 7.6 nm.

Lee et al. also construct single layer dielectric capacitors of $Ca(Zr_{1-x}Ti_x)O_3$ (CZT) solid solution using tape casting to with varied Zr/Ti ratios. These capacitors are used to store energy, which displayed exceptional thermal stability (up to 250 °C) and a high degree of reproducibility dependable and steady functioning over a longer period of time (12000 h) without any failure (at 150 °C) [79].

**TABLE 9.4**
**Linear Dielectric Ceramic Films**

| Material | Film thickness | Substrate | Fabrication method | $U_{rec}$ [J cm$^{-3}$] | $\eta$ [%] | $E_{BD}$ [MV cm$^{-1}$] | Ref. |
|---|---|---|---|---|---|---|---|
| $Al_2O_3$ | 5 nm | ITO/glass | ALD | 50 | - | 20 | [72] |
| $Ba[(Ni_{1/2}W_{1/2})_{0.1}Ti_{0.9}]O_3$ | 290 nm | Pt/Si | CSD | 34 | - | 3 | [73] |
| $Bi_{1.5}Zn_{0.9}Nb_{1.5}O_{6.9}$ | 150 nm | Pt/Si | CSD | 60.8 | - | 5.1 | [74] |
| $HfO_2$ | 63 nm | LNO/Pt/Si | ALD | 21.3 | 75 | 4.25 | [75] |
| $\delta\text{-}Bi_2O_3@Bi_{1.5}Zn_{0.9}Nb_{1.35}Ta_{0.15}O_6.$ | 110 nm | Pt/Si | CSD | 40.2 | 82 | 3.8 | [76] |
| $Bi_3NbO_7$ | 140 nm | Pt/Si | CSD | 39 | 72 | 3.6 | [76] |
| $PbO@Pb_{1.1}TiO_{3.1}$ | 400 nm | Pt/Si | CSD | 28 | - | 5 | [77] |

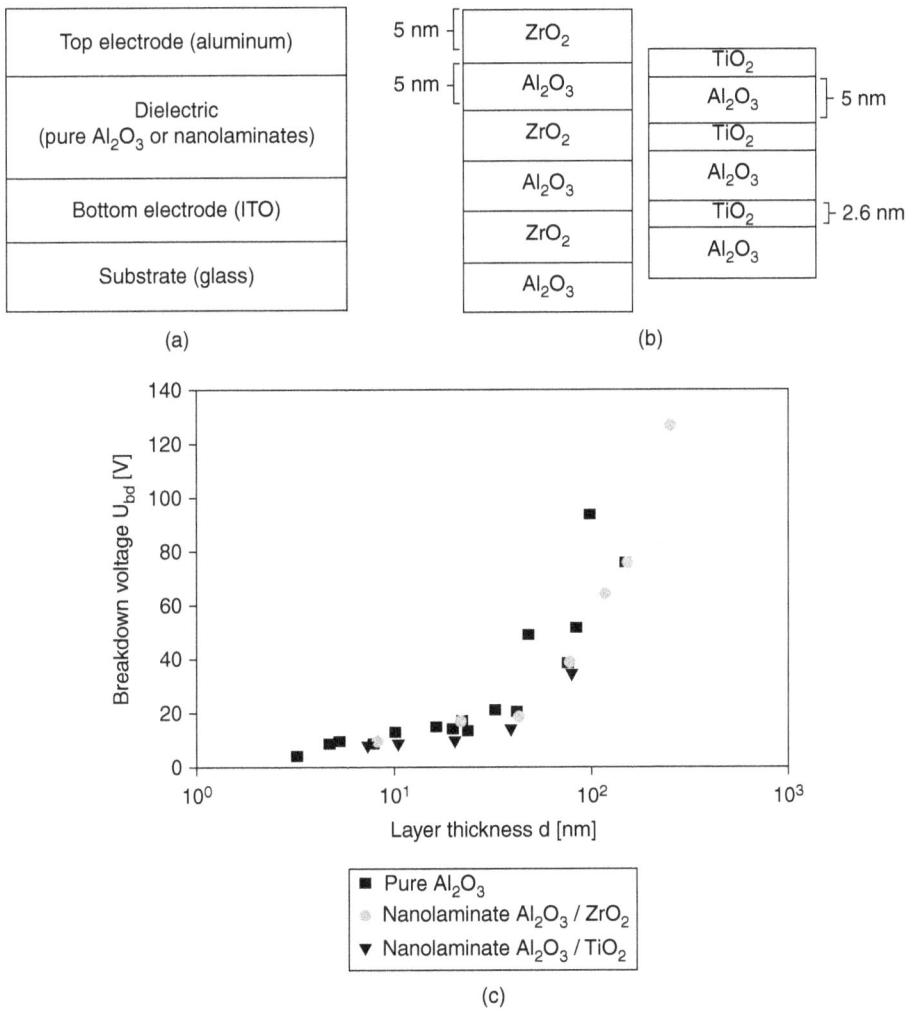

**FIGURE 9.7** (a, b) The layer configuration of $Al_2O_3$ and its nanolaminate-based thin film capacitor. (c) Variation of breakdown voltages of $Al_2O_3$ and its nanolaminates with increasing layer thicknesses. Reproduced with permission [78]. Copyright 2013, AIP Publishing.

### 9.4.2 PARAELECTRIC CERAMIC FILMS

When compared to linear dielectrics, paraelectric materials with somewhat nonlinear polarization responses have much greater dielectric constants but lower $E_{BD}$ values. Paraelectric materials also have low dielectric and hysteresis losses, making them ideal for energy storage applications. Table 9.5 lists the energy storage characteristics of certain paraelectric ceramic films that have been described in the literature. Xie et al. investigated the energy storage capabilities of Fe-doped BST thin films (100 nm grain size) with low concentrations of Fe, namely $Ba_{0.7}Sr_{0.3}Fe_xTi_{1-x}O_3$ with $x = 0.004, 0.008, 0.01$, and $0.015$ [80]. When $x = 0.008$, grain size and dielectric constant increased, resulting in enhanced polarization, $U_{rec}$, and $E_{BD}$. Excessive Fe doping resulted in smaller grains and less efficient energy storage. The electrical and energy storage capabilities of $Ba_{0.4}Sr_{0.6}TiO_3$ films were also affected by Mn doping (1, 3, and 5 mol%) [80]. The $U_{rec}$ of the 1 mol% Mn-doped BST films with lower leakage current was higher, while the $U_{rec}$ of the 3 mol% Mn-doped BST films with a bigger difference between saturation and remnant polarization was higher.

Peng et al. developed Al electroded STO thin films that were doped with $Al_2O_3$ nanoparticles [63]. Under an applied electric field, the $Al_2O_3$ nanoparticles created ion transport channels, which stimulated anodic oxidation processes, generating an $Al_2O_3$ layer at the Al/STO interface (Figure 9.8), which was not the case with pure STO films (Figure 9.8a). From an initial dielectric strength of 23.3 MV cm$^1$, STO films now have a dielectric strength of 50.7 MV cm$^{-1}$. As a result, the $U_{rec}$ of the $Al_2O_3$-doped STO films was significantly higher (19.3 J cm$^3$) than that of the STO films (3.2 J cm$^3$) [81].

The use of interfacial modification to improve the $E_{BD}$ of dielectric films has been suggested. The effect of the interface and film thickness on the energy storage capabilities of STO films grown on (La,Sr)MnO$_3$ bottom electrodes was investigated by Hou et al. [12]. With increasing layer thickness (410–710 nm), the dielectric permittivity of the STO capacitor increased. The film's polarization and $E_{BD}$ showed opposing trends as a function of thickness (Figure 9.8a, b). At an applied field of >6 MV cm$^1$, the STO films achieved an ultrahigh $U_{rec}$ of >300 J cm$^3$, as shown in Figure 9.8c. STO/LSMO capacitors' compact hysteresis loops resulted in higher energy storage efficiency (85–90%) (Figure 9.8d). The creation of an ion interdiffusion layer composed of (Sr, La) TiO$_3$ and (La, Sr)(Mn,Ti)O$_3$ was responsible for the very high $E_{BD}$ (5.5–6.8 MV cm$^1$) of STO/ LSMO capacitors. The interface composition gradient (Figure 9.8e) ensured gradual permittivity fluctuation and smoothed the distribution of the interfacial electric field, resulting in high electrical tolerances for STO/LSMO.

### 9.4.3 FERROELECTRIC CERAMIC FILMS

High permittivities and extremely nonlinear hysteresis curves with substantial polarization are common characteristics of ferroelectric materials. Due to the presence of internal stresses, flaws, and electrostriction effects, they exhibit higher dielectric losses and lower $E_{BD}$ values when

**TABLE 9.5**
**Paraelectric Thin-Film-Based Energy Storage Devices**

| Material | Film thickness | Substrate | Fabrication method | $U_{rec}$ [J cm$^{-3}$] | $\eta$ [%] | $E_{BD}$ [MV cm$^{-1}$] | Ref. |
|---|---|---|---|---|---|---|---|
| $Ba_{0.7}Sr_{0.3}Fe_xTi_{1-x}O_3$ | 100nm | Pt/Si | Sol–gel | 7.6 | 65 | 0.4 | [83] |
| $Ba_{0.7}Sr_{0.3}Mn_xTi_{1-x}O_3$ | 100nm | Pt/Si | Sol–gel | 8.48 | 42.4 | 1.63 | [84] |
| $SrTiO_3$ | 300 nm | Pt/Si | Sol–gel | 15.7 | - | 3.5 | [82] |
| $SrTiO_3$ | 610 nm | LSMO/STO | Magnetron | 307 | 89 | 6.6 | [85] |
|  | 643 nm | Pt/STO | sputtering | 150 | - | 4.26 | |

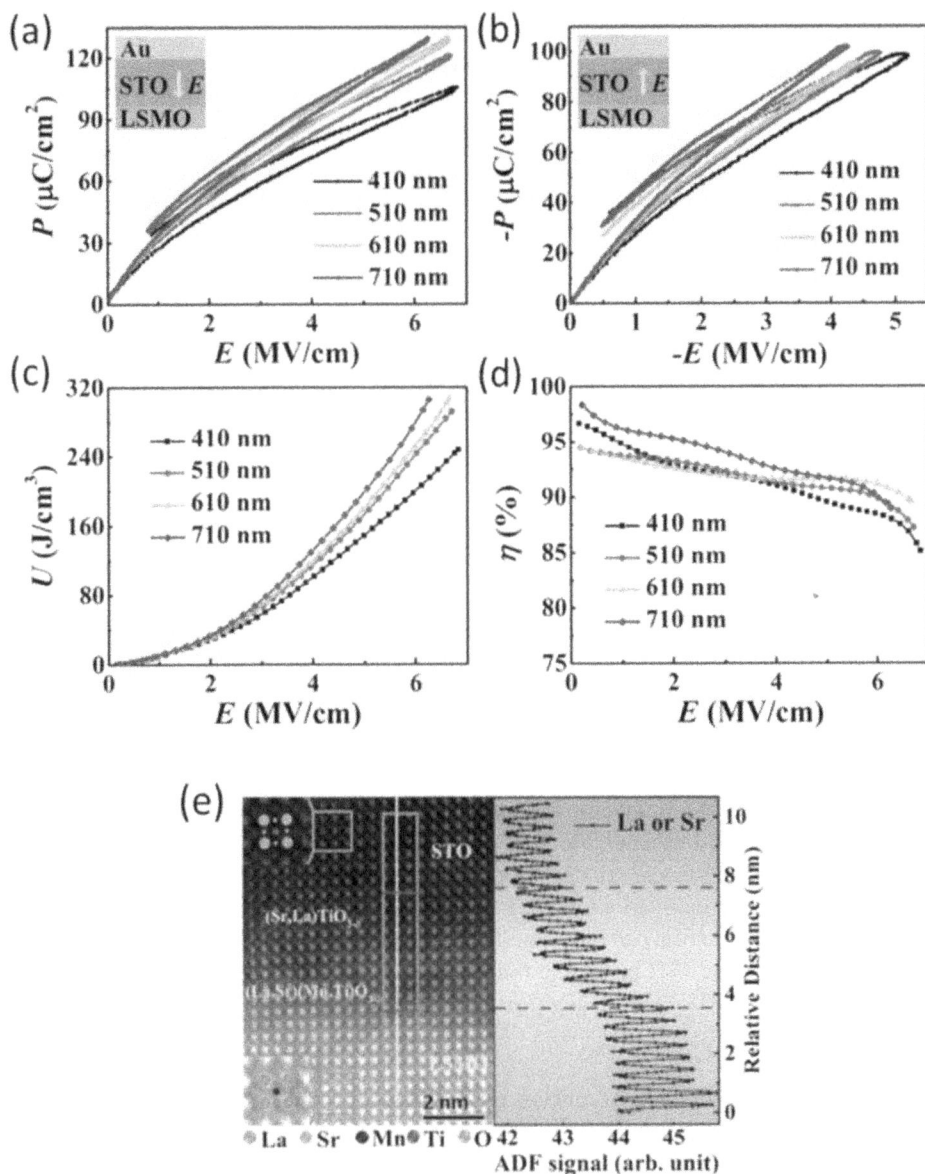

**FIGURE 9.8** Polarization hysteresis loops of Au/STO/LSMO capacitors under (a) positive fields and (b) negative fields, and the insets show the orientation of $E$. (c) Variations of recoverable energy density and (d) variations of energy storage efficiency under positive electric fields for Au/STO/LSMO capacitors. (e) High angle angular dark-field (HAADF) scanning transmission electron microscopic (STEM) cross-sectional image of STO/LSMO interface, and the insets show the magnifying images of STO and LSMO. Reproduced with permission [82]. Copyright 2017, ACS Publishing.

compared to paraelectrics and linear dielectrics [83]. The reported energy storage capabilities of several ferroelectric ceramic films are detailed in Table 9.6. Although there have been few studies on films of typical perovskite ferroelectrics such as $BiFeO_3$ (BFO)[86, 87], $BaTiO_3$ [88–90], $PbTiO_3$ [91], and $Pb(Zr,Ti)O_3$ [92, 93], solid solutions based on these ceramics have been extensively researched.

**TABLE 9.6**
**Ferroelectric Ceramics-Based Energy Storage Devices**

| Material/composition | Film thickness | Substrate | Fabrication method | $U_{rec}$ [J cm$^{-3}$] | $\eta$ [%] | $E_{BD}$ [MV cm$^{-1}$] | Ref. |
|---|---|---|---|---|---|---|---|
| BiFeO$^3$ | 40 nm | Pt/Si | CSD | 3.2 | – | 0.5 | [95] |
| (Ti, Zn) co-doped Bi$_{0.97}$Nd$_{0.03}$FeO$_3$ | 320 nm | ITO/glass | MOD | 10.43 | 75 | 0.35 | [86] |
| (La,Ca) co-doped PbTiO$_3$ | 300 nm | Pt/Si | Sol–gel | 14.9 | 77.6 | 0.4 | [91] |
| PbZrO$_3$/Pb(Zr$_{0.52}$Ti$_{0.48}$)O$_3$ | 60 nm/ 350 nm | Pt/Si | Sol–gel | 28.2 | 50 | 2.61 | [96] |
| Bi$_{3.25}$La$_{0.75}$Ti$_3$O$_{12}$ | 455 nm | Pt/Si | CSD | 44.7 | 78.4 | 2.04 | [87] |
| 0.5Bi(Ni$_{1/2}$Ti$_{1/2}$)O$_3$–0.5PbTiO$_3$ | 400 nm | Pt/Si | CSD | 45.1 | 43.4 | 2.25 | [97] |
| Fe-doped (Na$_{0.5}$Bi$_{0.5}$)TiO$_3$ | 300 nm | ITO/glass | MOD | 30.15 | 61.05 | 1 | [98] |
| Fe-doped (Na$_{0.85}$K$_{0.15}$)$_{0.5}$Bi$_{0.5}$TiO$_3$ | 1.15 μm | LNO/Pt/Si | Sol–gel | 33.3 | 51.3 | 2.3 | [99] |
| CaBi$_2$Nb$_2$O$_9$–Bi$_4$Ti$_3$O$_{12}$ | 300 nm | (100) MgO | Magnetron sputtering | 76 | 66.3 | 2.5 | [100] |
| PZT/Al$_2$O$_3$/PZT | 330 nm | Pt/Si | Sol–gel | 63.7 | 81.3 | 5.7 | [96] |

The energy storage capability of BFO can be enhanced by lowering its leakage current through doping, according to Zhang et al. [86] the oxygen vacancies emerging from a Bi deficiency due to its high volatility were projected to be minimized in (Ti, Zn) codoped Bi$_{0.97}$Nd$_{0.03}$FeO$_3$ (BNFTZ) thin films by Ti$^{4+}$ and Zn$^{2+}$ atoms substituting for Fe$^{3+}$, resulting in a lower leakage current than pure BFO. However, the BNFTZ films stated polarization hysteresis loops were unsaturated with some leakage, resulting in reduced $U_{rec}$ and $E_{BD}$ values. Gao et al. recently presented a method for increasing energy density in a Sn-doped BTO ferroelectric system at low electric fields by compositionally inducing a tricritical point [94]. At 10 kV cm$^{-1}$, the ideal dielectric permittivity in Ba(Ti$_{0.895}$Sn$_{0.105}$)O$_3$ reached 5.4 104, and the corresponding energy density exceeded 30 mJcm$^3$, which is higher than most ferroelectric materials at the same field strength. Furthermore, Zhang et al. demonstrated that a space charge dominated BTO thin film can have significantly better energy storage ($U_{rec}$ = 35.4 J cm$^3$) than a conventional ferroelectric BTO insulating film [94].

### 9.4.4 RELAXOR FERROELECTRIC FILMS

Relaxor ferroelectrics or relaxors as a subclass of disordered crystals have piqued interest since their discovery by Smolensky and co-authors in 1954 because of their unique properties, such as large permittivity, high piezoelectric coefficient $d_{33}$, and large field-induced strain, which make them a candidate for use in advanced microelectronic devices [101].

Normal ferroelectrics retain their high permittivities and saturation polarization, whereas relaxor ferroelectrics have a lower residual polarization and a thinner hysteresis. Circuits allow them to provide substantially greater energy densities as well as storage efficiencies.

Several relaxors, such as Pb(Mg$_{1/3}$Nb$_{2/3}$O$_3$PbTiO$_3$, (Pb,La)(Zr,Ti)O$_3$, and Ba(Ti,Sn)O$_3$, have been proven in the last 60 years [102]. Relaxor's characteristics are thought to be derived from their nanometer-sized polar regions and their reactivity to external stimuli in general [103]. These polarization-related areas arise at the "Burns" temperature TD, which is often much higher than the permittivity maximum point $T_m$. The presence of a large peak in the temperature-dependent permittivity, in which $T_m$ is pushed to higher temperature as the measured frequency increases, is a common property of relaxors. Additionally, the polarization curves often display a slender trend with a bigger difference between saturation polarization and residual polarization as a function of external electric field. Because of their dispersed phase switching, the P–E loops of relaxors also

**TABLE 9.7**
**Relaxor Ferroelectric Film-Based Energy Storage Devices**

| Material/composition | Film thickness | Substrate | Fabrication method | $U_{rec}$ [J cm$^{-3}$] | $\eta$ [%] | $E_{BD}$ [MV cm$^{-1}$] | Ref. |
|---|---|---|---|---|---|---|---|
| Ba(Zr$_{0.2}$Ti$_{0.8}$)O$_3$ | 350 nm | LSAT LAO STO | Magnetron sputtering | 115 | 69 | 5.7 | [57] |
| 0.4BiFeO$_3$–0.6SrTiO$_3$ | 500 nm | Nb:STO | PLD | 70.3 | 70 | 3.85 | [13] |
| 0.25BiFeO$_3$–0.75SrTiO$_3$ | | | | 70 | 68 | 4.46 | |
| Na$_{0.5}$K$_{0.5}$NbO$_3$–BiMnO$_3$ | 700 nm | Pt/Si | Sol–gel | 14.8 | 79 | 0.985 | [105] |
| Mn-doped Pb$_{0.91}$La$_{0.09}$(Zr$_{0.65}$Ti$_{0.35}$)O$_3$ | 1.5 $\mu$m | LNO/Si | Sol–gel | 30.8 | 68.4 | 1.7 | [106] |
| Mn-doped 0.4(Na$_{0.5}$Bi$_{0.5}$)TiO$_3$–0.6SrTiO$_3$ | 190 nm | Pt/Si | Sol–gel | 33.58 | 64 | 3.13 | [107] |
| 0.8Pb(Mg$_{1/3}$Nb$_{2/3}$)O$_3$–0.2PbTiO$_3$ | 400 nm | LNO/Pt/Si | Magnetron sputtering | 31 | 64 | 2 | [108] |

display polarization loss with temperature. Relaxors have the potential to be used in high-energy storage because of these feature [104]. An overview of energy storage characteristics of several relaxor ferroelectric ceramics have been documented. Table 9.7 shows a list of relaxor ferroelectric film-based energy storage devices.

The energy storage behavior of relaxor thin films has recently awakened interest, owing to the growing need for thin film capacitors. Yao and co-authors used a PEG-assisted chemical solution method to effectively manufacture 0.462Pb (Zn$_{1/3}$Nb$_{2/3}$) O$_3$–0.308Pb(Mg$_{1/3}$Nb$_{2/3}$) O$_3$–0.23PbTiO$_3$ relaxor thin films in 2011 [109]. At an electric field of 700 kV/cm, the films displayed a higher maximum polarization of 108 C/cm$^2$ and a modest remnant polarization of 20 C/cm$^2$. As a result, this film has a higher recoverable energy storage density of 15.8 J/cm$^3$. The energy storage in 600-nm-thick BaTiO3/Ba$_{0.3}$Sr$_{0.7}$TiO$_3$ relaxor superlattices produced on (100) MgO single crystal substrates was reported by Ortega and co-authors [18]. The current field measurements showed that the thin films had a very high breakdown field of 5.86 MV/cm and a low leakage current density of 1035 mA/cm$^2$, implying that a greater recovered energy storage density of 46 J/cm$^3$ may be obtained.

The typical Pb$_{0.91}$La$_{0.09}$(Zr$_{0.65}$Ti$_{0.35}$)O$_3$ relaxor ferroelectric in thin-film form, which were formed on platinum-buffered silicon substrates through a sol–gel process, had dielectric characteristics and energy-storage performance shown in Figure 9.9 [110]. The random orientation of the crystalline films with a thickness of roughly 1000 nm exhibited a homogeneous and dense macrostructure. In thin film, a capacitance density of 925 nF/cm$^2$ at 1 MHz was achieved, as well as a higher critical breakdown field of 2177 kV/cm. This relaxor film likewise had a narrow P–E loop with a higher difference between the saturated and residual polarization, which was 49.1 C/cm$^2$ at 1580 kV/cm, as seen in Figure 9.9. The recovered energy storage density at 1 kHz rose from 1.7 to 28.7 J/cm$^3$ as the external field increased from 200 to 2040 kV/cm. However, the energy efficiency remained consistent across the measuring range, ranging from 52.4 percent to 59.4 percent. The relaxor thin films were also shown to have good temperature and frequency-dependent stability. All of these findings suggest that relaxor ferroelectrics might be a good choice for high energy storage applications.

### 9.4.5 ANTIFERROELECTRIC CERAMIC FILMS

There are various subclasses of antiferroelectric materials, such as the perovskite group, the pyrochlore group, the liquid crystal group, and so on. The most significant antiferroelectric materials are those having a perovskite structure, which is generally abbreviated as ABO$_3$. At ambient temperature,

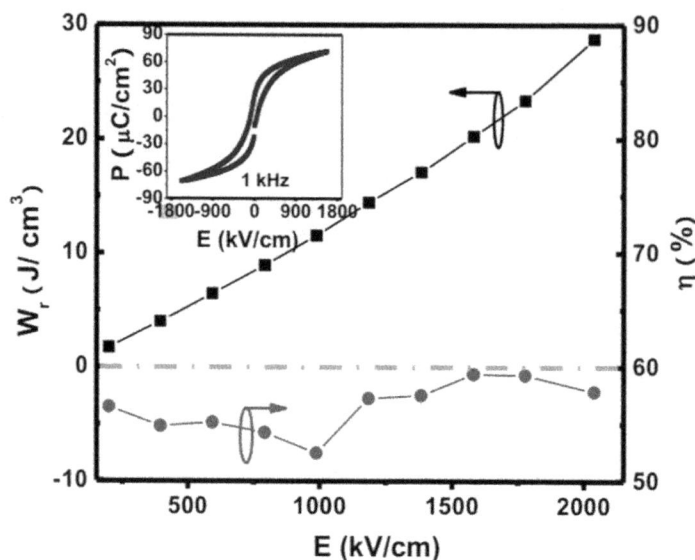

**FIGURE 9.9**  The electric field dependence of energy-storage density and energy-storage efficiency of $Pb_{0.91}$ $La_{0.09}(Zr_{0.65}Ti_{0.35})O_3$ thin film measured at room temperature and 1 kHz. The insert shows $P–E$ loop of the thin film measured at room temperature and at 1 kHz [104] copyright 2013 [Open Access].

$PbZrO_3$, $PbHfO_3$, $NaNbO_3$ and their combinations have been proven to be typical antiferroelectrics with perovskite structure [111–115].

Due to the absence of ferroelectric domains, antiferroelectric materials have minimal dielectric loss, a small coercive field, and no hysteresis at low electric fields. They go through a high-field-driven AFE–FE phase transition and the realignment of antiparallel dipoles results in a twofold hysteresis polarization response with a large $U_{rec}$ [2]. In antiferroelectric materials, reduced hysteresis losses and reversible phase shift have been linked to high power and energy density [116–118]. Furthermore, a quick discharge rate and high discharge efficiency in these materials are due to the loss of ferroelectric properties at low electric fields. Energy-storage characteristics, charging and discharging process for antiferroelectrics are depicted in Figure 9.10. As a result, antiferroelectric materials have the potential to be high-performing rivals in commercial capacitor applications. Excellent reviews on the development and uses of a wide range of antiferroelectric materials may be found in the literature [119, 120]. As a result, these materials can store a lot of energy with improved material efficiencies and faster discharge rates. As a result of weighing the facts connected with each material family, it is clear that antiferroelectric materials are a logical option as viable candidates for electrical energy storage capacitors. Xu et al. projected that rare-earth (Dy, Sm, Nd, La, and Gd) ions replaced in the $BiFeO_3$ system will show tremendous potential for achieving a large energy density using first-principles-based atomistic simulations [121]. Under acceptable electric fields (2–3 MV $cm^{-1}$), a typical compound $Bi_{1-x}Nd_xFeO_3$ was calculated to have significant energy densities (100–150 J $cm^{-3}$) and efficiencies (80–88%). The energy storage characteristics of several antiferroelectric ceramic films are listed in Table 9.8. $PbZrO_3$, (Pb, La)(Zr,Ti)$O_3$, and $HfO_2$ have been the primary materials studied for antiferroelectric ceramic films for energy storage applications.

It is generally recognized that the microstructure of ceramics with certain chemical compositions, such as density, homogeneity, and grain size, has a significant impact on their ultimate electrical characteristics. Meanwhile, lead loss during preparation methods has a significant impact on the electrical properties of lead-containing materials. As a result, choosing the right production procedure is a good strategy to improve the qualities of lead-containing ceramics. Other innovative sintering

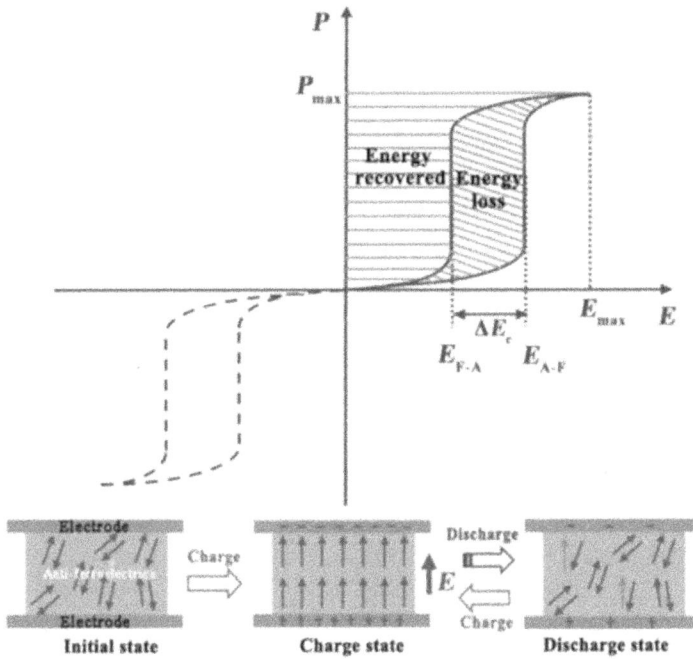

**FIGURE 9.10** Energy-storage characteristics, charging and discharging process for antiferroelectrics. Reproduced with permission [78] Copyright 2022, Elsevier Publishing.

**TABLE 9.8**
**Antiferroelectric Ceramic Film-Based Energy Storage Devices**

| Material | Thickness of film | Substrate | Fabrication method | $U_{rec}$ [J cm$^{-3}$] | $\eta$[%] | $E_{BD}$ [MV cm$^{-1}$] | Ref. |
|---|---|---|---|---|---|---|---|
| La-doped PbZrO$_3$ | 1 $\propto$m | LNO/Pt/Si | CSD | 17.3 | 80.8 | 1 | [127] |
| Hf$_{0.3}$Zr$_{0.7}$O$_2$ | 9.2 nm | TiN/SiO$_2$/Si | ALD | 45 | 50 | 4.35 | [128] |
| Si-doped Hf$_{0.5}$Zr$_{0.5}$O$_2$ | 10 nm | TiN/Si | ALD | 50 | >80 | 4.5 | [129] |
| Al-doped Hf$_{0.5}$Zr$_{0.5}$O$_2$ | | | | 35 | | | |
| (Pb$_{0.97}$La$_{0.02}$)(Zr$_{0.98}$Ti$_{0.02}$)O$_3$ | 3.3 $\propto$m | LNO/Si | Sol-gel | 58.1 | 37.3 | 2.8 | [130] |
| Pb$_{0.79}$Ba$_{0.11}$La$_{0.1}$(Zr$_{0.9}$Ti$_{0.1}$)O$_3$ | 1.5 $\propto$m | LNO/Si | Sol-gel | 42.3 | 68 | 2.3 | [131] |
| Pb$_{0.97}$Y$_{0.02}$(Zr$_{0.6}$Sn$_{0.4}$)$_{0.925}$ Ti$_{0.075}$)]O$_3$ | 500 nm | Pt/Si | ALD | 21 | 91.9 | 1.3 | [132] |
| Pb$_{1-3x/2}$La$_x$Zr$_{0.85}$Ti$_{0.15}$O$_3$ | - | - | PLD | 38.0 | 65 | 2.1 | [133] |
| 4 mol.%Nb-doped PbZr$_{0.4}$Ti$_{0.6}$O$_3$ | - | - | PLD | 20 | 70 | 1.87 | [134] |

procedures (such as hot press sintering and spark plasma sintering) have been employed to generate dense AFE ceramics at relatively low sintering temperatures, in addition to the conventional synthesis methods such as solid-state reaction method, wet-chemistry methods, high-energy ball milling method, molten salt method. The process of hot pressing sintering (HPS) is widely used to increase the densification of ceramics and hence their electrical characteristics. Markowski et al. used this approach to make AFE ceramics for the first time in 1996. Hot isostatic pressing at 20 MPa for 2 hours produced PLZST ceramics with a relative density of 98 percent. The dielectric properties of the AFE ceramics made in this manner were excellent. From the perspective of practical use, AFE thin/thick films are more appealing. To begin with, AFE films typically develop at temperatures below

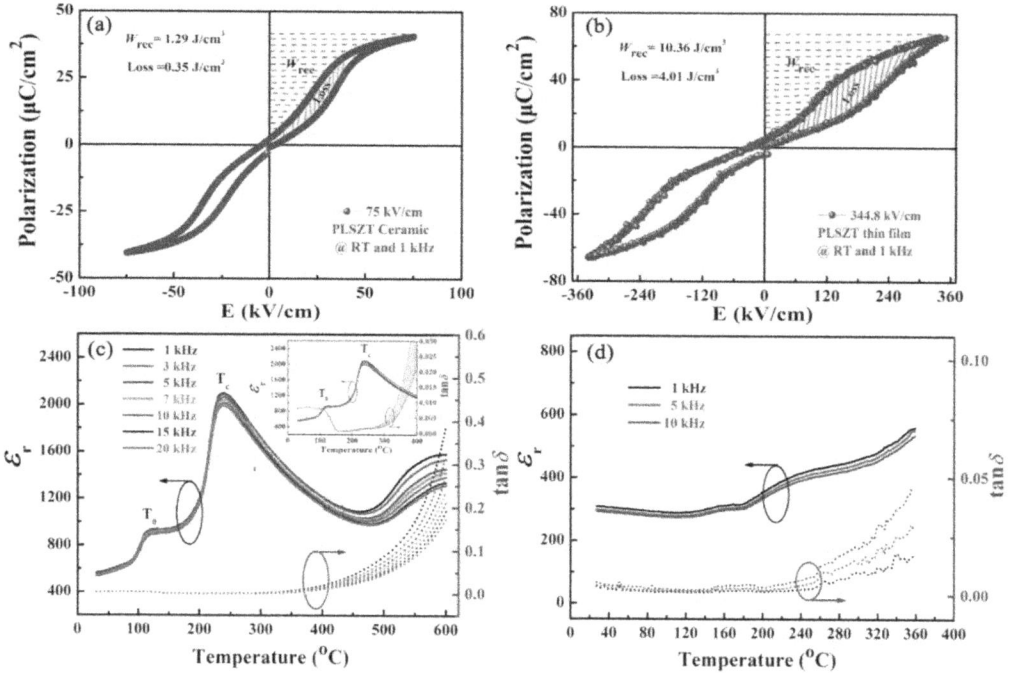

**FIGURE 9.11** The *P–E* hysteresis loops as well as storage characteristics of (a) PLSZT ceramic (b) thin film; temperature dependence of dielectric constant as well as loss at different frequencies: (c) PLSZT ceramic and the enlarged area from 0 to 400 °C in the inset; (d) PLSZT thin film. Reproduced with permission [78]. Copyright 2022, Elsevier Publishing.

800 °C, making them ideal for integration with silicon technology. Second, films often have larger densities, allowing them to withstand higher external electric fields. Finally, because to their thinness, film operating voltages are substantially lower than those needed by their ceramic counterparts.

Various approaches for depositing thin (1000 nm) and thick (>1000 nm) films have been devised. Two major approaches used for thin-film depositions are chemical routes, such as sol–gel, chemical solution deposition (CSD), chemical gas evaporation, and hydrothermal oxidation (HTO), and physical routes, such as physical gas evaporation, sputtering, pulsed laser deposition (PLD), and ion beam assisted deposition [122, 123] (IBAD). In comparison, only a few technologies for fabricating thick films have been accessible. Modified chemical solution deposition, spraying methods, screen printing, and tap casting are some of the thick film production processes [124–126]. Chemical solution deposition, PLD, and sputtering are the most often utilized procedures for fabricating lead-based AFE films, particularly chemical solution deposition.

The ferroelectric, dielectric, and energy storage characteristics of antiferroelectric ceramics $Pb_{0.89}La_{0.06}Sr_{0.05}(Zr_{0.95}Ti_{0.05})O_3$ and film capacitor were effectively constructed [135] as shown in Figure 9.11. When compared to PLSZT ceramic, PLSZT thin-film capacitors had a higher BDS of 2068.9 kV/cm, a higher storage density of 52.9 J/cm³ with 67.7%, temperature stability, quick discharge time, and high fatigue endurance, which could be attributed to the size effect and oxygen defects' mechanism. Furthermore, the charging and discharging characteristics of the thin-film capacitor demonstrated its faster charging and discharging time and large dielectric strength, suggesting that antiferroelectric PLSZT could be a promising material for applications in high-reliability energy-storage devices.

Figure 9.11(a) and (b) show the *P–E* hysteresis loops of the PLSZT ceramic and thin film measured at 1 kHz, respectively. The two AFE hysteresis loops with high $P_{max}$ and low $P_r$ clearly reveal the AFE nature of the PLSZT ceramic and thin film. At applied electric fields of 75 and 344.8 kV/cm, the $P_{max}$ and $P_r$ for the PLSZT bulk ceramic and thin film are 40.58 and 66.11 $\mu C/cm^2$, 2.25 and 4.85 $\mu C/cm^2$, respectively. The computed $W_{rec}$ and $\eta$ for the PLSZT ceramic and thin film, respectively, are 1.29 J/cm$^3$ with a $h$ of 78.7% and 10.36 J/cm$^3$ with a $h$ of 72.1 percent. The temperature dependences of dielectric permittivity ($\epsilon_r$) and dielectric loss (tan$\theta$) for the antiferroelectric PLSZT ceramic and film at various frequencies from 1 to 20 kHz are shown in Figure 9.11(c) and (d), respectively. In the case of the antiferroelectric PLSZT ceramic with multiphase regions, the results show parallel dielectric constant curves with two peaks, and heating from RT to T0 (~120 °C) most likely indicates the antiferroelectric-to-ferroelectric transition with the ferroelectric relaxation process, and an additional increase to $T_c$ of 232°C denotes the temperature of ferroelectric to paraelectric phase transition [136, 137]. In the frequency range, PLSZT ceramics could sustain a maximum dielectric constant of 2069.2 and a low dielectric loss of less than 0.4 [Figure 9.11(c)]. The results of temperature-dependent dielectric constant ($\epsilon_r$-$T$) for a PLZST thinfilm show a continuous increase of $\epsilon_r$ during the heating process, indicating that no phase transition occurs at the stable antiferroelectric phase [Figure 9.11(d)], which could be due to size effect, defects, and damping effect.

The thermal stability of the energy storage performance of the antiferroelectric PLSZT ceramic and thin film was also examined at various temperatures, and the $P$–$E$ loops for the antiferroelectric PLSZT ceramic and thin film, respectively, are shown in Figure 9.12(a) and (b). Furthermore, the $P$–$E$ hysteresis loops obtained at different temperatures under an electric field of 75 and 344.8

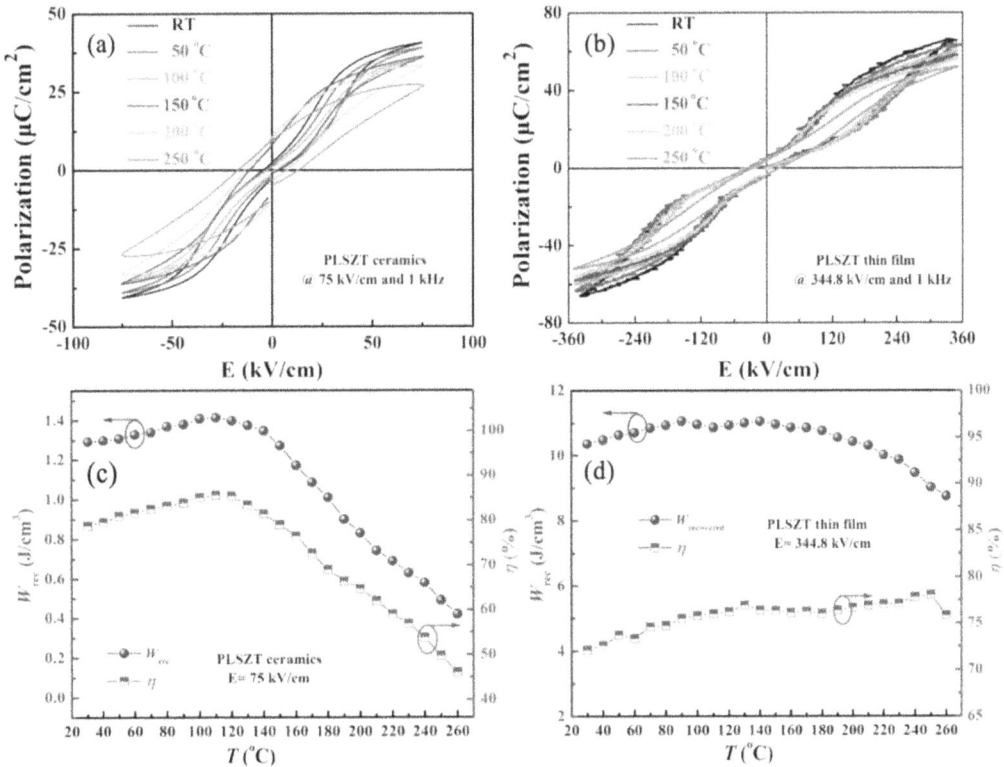

**FIGURE 9.12** (a) $P$-$E$ hysteresis loops measured at different temperatures for PLSZT ceramic; (b) $P$–$E$ hysteresis loops measured at different temperatures for PLSZT thin film; (c) calculated energy densities as well as conversion efficiencies measured at different temperatures for (PLSZT ceramic at 75 kV/cm; and (d) thin film at 344.8 kV/cm. Reproduced with permission [78]. Copyright 2022, Elsevier Publishing.

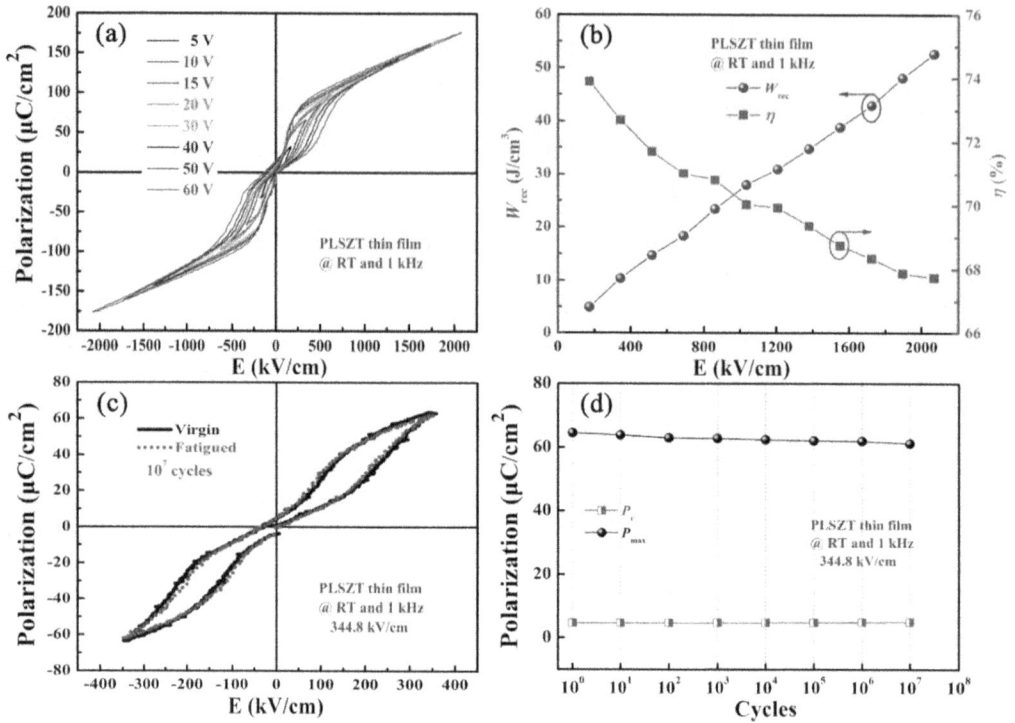

**FIGURE 9.13** (a) *P–E* hysteresis loops for PLSZT thin film at the different electric fields. (b) The calculated energy densities and conversion efficiencies for PLSZT thin film. (c) Virgin and fatigued *P–E* hysteresis loops for PLSZT thin film. (d) $P_r$ and $P_{max}$ for PLSZT thin film after fatigued cycles. Reproduced with permission [78]. Copyright 2022, Elsevier Publishing.

kV/cm were used to compute the recoverable energy storage density and efficiency for the PLSZT ceramic and thin film, as shown in Figure 9.12(c) and (d), respectively. The antiferroelectric PLSZT thin film exhibits improved temperature stabilities of energy storage capabilities when compared to the ceramic form, which may be related to the size effect and flaws [138–140].

Furthermore, due to the demand of strong electric fields, various safety issues about dielectric capacitors arose for the implementation of high energy density. The greater the BDS of a capacitor in film form, the higher the $W_{rec}$. The maximum electric field applied to the PLSZT ceramic is 75 kV/cm (7.5 MV/m), which is significantly less than the dielectric breakdown strength of the PLSZT thin film. As a result, the antiferroelectric PLSZT thin-film capacitor's energy storage ability at high electric fields, fatigue behaviors, and charging and discharging properties are explored in more depth [140, 141].

The *P–E* loops for the antiferroelectric PLSZT thin film at various electric fields are shown in Figure 9.13(a), and the results show that the antiferroelectric PLSZT film can withstand an electric field of up to 2068.9 kV/cm, a $P_{max}$ of up to 176.5 μC/cm², and a stable small $P_r$ of 13.2 μC/cm², indicating an ultra-high recoverable energy. Under a high electric field of 2068.9 kV/cm [see Figure 9.13(b)], an excellent recoverable energy storage density of 52.9 J/cm³ with a corresponding efficiency of 67.7% can be obtained, and the energy efficiency also shows high values over 65.0 percent, which is a comparable result in earlier reports [138, 136]. In general, films have a significant loss tangent, which results in inferior energy efficiency. Furthermore, because any change in $P_{max}$ and $P_r$ during cycling would basically decrease energy storage density, the polarization fatigue resistance of ferroelectric film must be mentioned as an important criteria in energy storage applications. As a result, the antiferroelectric PLSZT thin-film's reliability and fatigue behaviors were depicted as a function of polarization switching cycles at a frequency of 1 kHz with an external electric field of 344.8 kV/cm in a shorter time (24 h) to avoid possible dielectric failure caused by extrinsic measurement factors [Figure 9.13(c)]. The results showed that the antiferroelectric PLSZT thin film had good wear durability after 107 cycles, with a polarization loss of roughly 8.0 percent as shown in Figure 9.13(d)[142–144].

## 9.5 FUTURE OUTLOOK AND PERSPECTIVES

Dielectric capacitors with excellent storage capacity can continue to be developed as the demand for pulsed power electronics grows for a range of applications. Energy storage materials should have high permittivity, large electric displacement, low dielectric and hysteresis losses, low electrical conductivities, large breakdown region, and high fatigue endurance and excellent temperature stability. Nonetheless, meeting these stringent parameters in a single material is quite difficult. To identify a suitable material for high energy density capacitors, several dielectric based on ceramics (bulk, film) polymers, and glasses have been explored. Although each of these material systems has its own set of advantages and disadvantages in terms of energy storage capacity, it is heartening to note that a great trend has been made in the creation of dielectrics with enhanced energy storage capabilities. The storage capabilities of epitaxial and textured/oriented ceramic films were likewise found to be more superior than polycrystalline films. Dielectric ceramic films have energy densities in range of 50–300 J cm$^{-3}$, having storage efficiencies of 60–90%, breakdown regions of 1–6 MV cm$^{-1}$, a fatigue period of 106–109 cycles, and excellent temperature stability of –50 to 200 °C, in particular, offers a lot of promise for use in practical device applications. These features may be found in a variety of dielectric films, including linear dielectric, paraelectric, ferroelectric, relaxor ferroelectric, and antiferroelectric ceramics.

Synthesis of solid ceramic solutions, film fabrication, synthesis of solid ceramic solutions, film designation with coexistence of RFE–AFE phases or amorphous and crystalline nanocrystals, control of crystallite graphic alignment of the films, lattice mismatch, and misfit strain between the film and the substrate are all effective ways to improve the energy storage properties of dielectric ceramic films. Despite the progress made through these methods, there is still potential for growth because a number of difficulties need to be addressed. How to overcome dielectric material's inherent limitations, such as low permittivity of linear dielectrics and paraelectric, high hysteresis loss and low $E_{BD}$ of ferroelectrics, and piezoelectric noise/mechanical vibration associated with field-induced phase transitions in antiferroelectrics is still a question. Furthermore, strong electric fields and high temperatures cause conduction losses and polarization fatigue in energy storage dielectrics. Similarly, the frequency dependency of dielectric energy storage characteristics has received little attention and should be investigated further. It is vital to examine features of device design and operating circumstances while implementing dielectric ceramic films in pulsed power capacitors. Despite this thought, ceramic films have considerably high volume-specific energy volume-specificities, the absolute energy stored in them is quite small due to their thickness/volume restrictions. As a result, the matching devices must be developed with series and parallel connection arrangements.

Alternatively, dielectric ceramics may be used to create thick films and multi-layered films. The fundamental energy storage qualities of dielectric capacitors must be kept once the capacitor is integrated with other electronic components for its final use to ensure long-term performance. The capacitor must be dependable under all operational situations, including voltages, temperatures, and a wide range of operating settings. To expedite the implementation of energy storage dielectric capacitors in practical applications, future research and development activities should focus on (i) enhancing the basic knowledge of various framework relationships; (ii) using high-throughput computational screening combined with experimental synthesis and testing; and (iii) simple and scalable synthesis and processing approach for fabrication of large-area (wafer-scale).

## REFERENCES

[1]  Z.-M. Dang *et al.*, "Flexible nanodielectric materials with high permittivity for power energy storage," *Adv. Mater.*, vol. 25, no. 44, pp. 6334–6365, Nov. 2013, doi: 10.1002/ADMA.201301752

[2]  A. Chauhan, S. Patel, R. Vaish, and C. R. Bowen, "Anti-ferroelectric ceramics for high energy density capacitors," *Mater.*, vol. 8, no. 12, pp. 8009–8031, Nov. 2015, doi: 10.3390/MA8125439

[3]  Q. Li *et al.*, "High energy and power density capacitors from solution-processed ternary ferroelectric polymer nanocomposites," *Adv. Mater.*, vol. 26, no. 36, pp. 6244–6249, Sep. 2014, doi: 10.1002/ADMA.201402106

[4]  R. W. Johnson, J. L. Evans, P. Jacobsen, J. R. Thompson, and M. Christopher, "The changing automotive environment: High-temperature electronics," *IEEE Trans. Electron. Packag. Manuf.*, vol. 27, no. 3, pp. 164–176, Jul. 2004, doi: 10.1109/TEPM.2004.843109

[5]  "High-temperature electronics pose design and reliability challenges | analog devices." www.ana log.com/en/analog-dialogue/articles/high-temperature-electronic-pose-design-challenges.html (accessed Mar. 21, 2022).

[6]  X.-M. Zhang, J.-G. Liu, S.-Y. Yang, X.-M. Zhang, J.-G. Liu, and S.-Y. Yang, "A review on recent progress of R&D for high-temperature resistant polymer dielectrics and their applications in electrical and electronic insulation." *Rev.Adv. Mater. Sci.*, vol. 46, p. 22-38, 2016.

[7]  Z. Yao *et al.*, "A review on the dielectric materials for high energy-storage application," *Adv. Mater.*, vol. 29, no. 20, p. 1601727, 2017, doi: 10.1002/ADMA.201601727

[8]  X. Hao, "A review on the dielectric materials for high energy-storage application," *J. Adv. Dielectrics,*vol. 03, no. 01, p. 1330001, May 2013, doi: 10.1142/S2010135X13300016

[9]  X. Hu, K. Yi, J. Liu, and B. Chu, "High energy density dielectrics based on PVDF-based polymers," *Energy Technol.*, vol. 6, no. 5, pp. 849–864, May 2018, doi: 10.1002/ENTE.201700901

[10]  R. Wen, J. Guo, C. Zhao, and Y. Liu, "Nanocomposite capacitors with significantly enhanced energy density and breakdown strength utilizing a small loading of monolayer titania," *Adv. Mater. Interfaces*, vol. 5, no. 3, p. 1701088, Feb. 2018, doi: 10.1002/ADMI.201701088

[11]  K. Zou *et al.*, "Recent advances in lead-free dielectric materials for energy storage," *Mater. Res. Bull.*, vol. 113, pp. 190–201, May 2019, doi: 10.1016/J.MATERRESBULL.2019.02.002

[12]  Y. Dan *et al.*, "Energy storage characteristics of (Pb,La)(Zr,Sn,Ti)O3 antiferroelectric ceramics with high Sn content," *Appl. Phys. Lett.*, vol. 113, no. 6, p. 063902, Aug. 2018, doi: 10.1063/1.5044712

[13]  J. Ma *et al.*, "Giant energy density and high efficiency achieved in bismuth ferrite-based film capacitors via domain engineering," *Nat. Commun.*, vol. 9, no. 1, Dec. 2018, doi: 10.1038/S41467-018-04189-6

[14]  J. R. Laghari and W. J. Sarjeant, "Energy-storage pulsed-power capacitor technology," *IEEE Trans. Power Electron.*, vol. 7, no. 1, pp. 251–257, Jan. 1992, doi: 10.1109/63.124597

[15]  T. Zhang *et al.*, "A highly scalable dielectric metamaterial with superior capacitor performance over a broad temperature," *Sci. Adv.*, vol. 6, no. 4, Jan. 2020, doi: 10.1126/SCIADV.AAX6622/SUPPL_FILE/AAX6622_SM.PDF

[16]  M. Stewart, M. G. Cain, and P. Weaver, "Electrical measurement of ferroelectric properties," *springer*, pp. 1–14, 2014, doi: 10.1007/978-1-4020-9311-1_1

[17]  "S. O. Kasap - Principles of electronic materials and devices McGraw-Hill (2006) | PDF." www.scr ibd.com/doc/311514185/S-O-Kasap-Principles-of-Electronic-Materials-and-Devices-McGraw-Hill-2006-pdf (accessed Mar. 24, 2022).

[18]  N. Ortega *et al.*, "Relaxor-ferroelectric superlattices: High energy density capacitors," *J. Phys. Condens. Matter*, vol. 24, no. 44, p. 445901, Oct. 2012, doi: 10.1088/0953-8984/24/44/445901

[19]  Q. Li, F. Z. Yao, Y. Liu, G. Zhang, H. Wang, and Q. Wang, "High-temperature dielectric materials for electrical energy storage," *Ann. Rev.*, vol. 48, pp. 219–243, Jul. 2018, doi: 10.1146/ANNUREV-MATSCI-070317-124435

[20]  L. Yang, X. Li, E. Allahyarov, P. L. Taylor, Q. M. Zhang, and L. Zhu, "Novel polymer ferroelectric behavior via crystal isomorphism and the nanoconfinement effect," *Polymer (Guildf).*, vol. 54, no. 7, pp. 1709–1728, Mar. 2013, doi: 10.1016/J.POLYMER.2013.01.035

[21]  S. Tong *et al.*, "Lead lanthanum zirconate titanate ceramic thin films for energy storage," *ACS Appl. Mater. Interfaces*, vol. 5, no. 4, pp. 1474–1480, Feb. 2013,doi:10.1021/AM302985U/ASSET/IMAGES/AM302985U.SOCIAL.JPEG_V03

[22] H. Pan, Y. Zeng, Y. Shen, Y. H. Lin, and C. W. Nan, "Thickness-dependent dielectric and energy storage properties of (Pb0.96La0.04)(Zr0.98Ti0.02)O3 antiferroelectric thin films," *J. Appl. Phys.*, vol. 119, no. 12, p. 124106, Mar. 2016, doi: 10.1063/1.4944802

[23] H. Tao and J. Wu, "Optimization of energy storage density in relaxor (K, Na, Bi)NbO$_3$ ceramics," *J. Mater. Sci. Mater. Electron.*, vol. 28, no. 21, pp. 16199–16204, Nov. 2017, doi: 10.1007/S10854-017-7521-2/FIGURES/8

[24] M. Zhou, R. Liang, Z. Zhou, and X. Dong, "Superior energy storage properties and excellent stability of novel NaNbO$_3$-based lead-free ceramics with A-site vacancy obtained via a Bi2O$_3$ substitution strategy," *J. Mater. Chem. A*, vol. 6, no. 37, pp. 17896–17904, Sep. 2018, doi: 10.1039/C8TA07303A

[25] N. Setter *et al.*, "Ferroelectric thin films: Review of materials, properties, and applications," *J. Appl. Phys.*, vol. 100, no. 5, p. 051606, Sep. 2006, doi: 10.1063/1.2336999

[26] R. Ramesh and N. A. Spaldin, "Multiferroics: Progress and prospects in thin films," *Nat. Mater.*, vol. 6, no. 1. pp. 21–29, Jan. 10, 2007, doi: 10.1038/nmat1805

[27] W. Wang, Y. Pu, X. Guo, R. Shi, M. Yang, and J. Li, "Combining high energy efficiency and fast charge-discharge capability in calcium strontium titanate-based linear dielectric ceramic for energy-storage," *Ceram. Int.*, vol. 46, no. 8, pp. 11484–11491, Jun. 2020, doi: 10.1016/J.CERAMINT.2020.01.174

[28] J. Du, B. Jones, and M. Lanagan, "Preparation and characterization of dielectric glass-ceramics in Na$_2$O-PbO-Nb$_2$O5-SiO$_2$ system," *Materials Letters*, vol. 59, no 22, pp. 2821-2826, 2005, doi: 10.1016/j.matlet.2005.02.090

[29] Q. Zhang, L. Wang, J. Luo, Q. Tang, and J. Du, "improved energy storage density in barium strontium titanate by addition of BaO–SiO$_2$–B$_2$O$_3$ glass," *J. Am. Ceram. Soc.*, vol. 92, no. 8, pp. 1871–1873, Aug. 2009, doi: 10.1111/J.1551-2916.2009.03109.X

[30] Y. Pu, X. Liu, Z. Dong, P. Wang, Y. Hu, and Z. Sun, "Influence of crystallization temperature on ferroelectric properties of Na0.9K0.1NbO3 glass-ceramics," *J. Am. Ceram. Soc.*, vol. 98, no. 9, pp. 2789–2795, Sep. 2015, doi: 10.1111/JACE.13671

[31] W. Huang *et al.*, "Ultrahigh recoverable energy storage density and efficiency in barium strontium titanate-based lead-free relaxor ferroelectric ceramics," *Appl. Phys. Lett.*, vol. 113, no. 20, p. 203902, Nov. 2018, doi: 10.1063/1.5054000

[32] A. Chauhan, S. Patel, R. Vaish, and C. R. Bowen, "Anti-ferroelectric ceramics for high energy density capacitors," *Mater. (Basel, Switzerland)*, vol. 8, no. 12, pp. 8009–8031, 2015, doi: 10.3390/MA8125439

[33] S. Zhang, B. Neese, K. Ren, B. Chu, and Q. M. Zhang, "Anti-ferroelectric ceramics for high energy density capacitors," *J. Appl. Phys.*, vol. 100, no. 4, 2006, doi: 10.1063/1.2335778

[34] K. Chen, Y. Pu, N. Xu, and X. Luo, "Effects of SrO-B$_2$O$_3$-SiO$_2$ glass additive on densification and energy storage properties of Ba$_{0.4}$Sr$_{0.6}$TiO$_3$ ceramics," *J. Mater. Sci. Mater. Electron.*, vol. 23, no. 8, pp. 1599–1603, Aug. 2012, doi: 10.1007/S10854-012-0635-7

[35] Y. Wang, Z. Y. Shen, Y. M. Li, Z. M. Wang, W. Q. Luo, and Y. Hong, "Short communication," *Ceram. Int.*, vol. 6, no. 41, pp. 8252–8256, Jul. 2015, doi: 10.1016/J.CERAMINT.2015.02.156

[36] Z. Liu, J. Lu, Y. Mao, P. Ren, and H. Fan, "Energy storage properties of NaNbO$_3$-CaZrO$_3$ ceramics with coexistence of ferroelectric and antiferroelectric phases," *J. Eur. Ceram. Soc.*, vol. 38, no. 15, pp. 4939–4945, Dec. 2018, doi: 10.1016/J.JEURCERAMSOC.2018.07.029

[37] M. Zhou, R. Liang, Z. Zhou, S. Yan, and X. Dong, "novel sodium niobate-based lead-free ceramics as new environment-friendly energy storage materials with high energy density, high power density, and excellent stability," *ACS Sustain. Chem. Eng.*, vol. 6, no. 10, pp. 12755–12765, Oct. 2018, doi: 10.1021/ACSSUSCHEMENG.8B01926/SUPPL_FILE/SC8B01926_SI_001.PDF

[38] Y. J. Wu, Y. H. Huang, N. Wang, J. Li, M. S. Fu, and X. M. Chen, "Effects of phase constitution and microstructure on energy storage properties of barium strontium titanate ceramics," *J. Eur. Ceram. Soc.*, vol. 5, no. 37, pp. 2099–2104, May 2017, doi: 10.1016/J.JEURCERAMSOC.2016.12.052

[39] Z. Song *et al.*, "Effect of grain size on the energy storage properties of (Ba0.4Sr0.6)TiO3 paraelectric ceramics," *J. Eur. Ceram. Soc.*, vol. 34, no. 5, pp. 1209–1217, May 2014, doi: 10.1016/J.JEURCERAMSOC.2013.11.039

[40] J. Luo, Q. Tang, Q. Zhang, J. Zhu, D. Han, and J. Du, "Preparation of bulk Na$_2$O-BaO-PbO-Nb$_2$O$_5$-SiO$_2$ glass-ceramic dielectrics for energy storage sources," *IOP Conf. Ser. Mater. Sci. Eng.*, vol. 18, no. SYMPOSIUM 14, 2011, doi: 10.1088/1757-899X/18/20/202024

[41] J. Luo, J. Du, Q. Tang, and C. Mao, "Lead sodium niobate glass-ceramic dielectrics and internal electrode structure for high energy storage density capacitors," *IEEE Trans. Electron Devices*, vol. 55, no. 12, pp. 3549–3554, 2008, doi: 10.1109/TED.2008.2006541

[42] Y. Zhang, J. Huang, T. Ma, X. Wang, C. Deng, and X. Dai, "Sintering temperature dependence of energy-storage properties in (Ba,Sr)TiO$_3$ glass-ceramics," *J. Am. Ceram. Soc.*, vol. 94, no. 6, pp. 1805–1810, Jun. 2011, doi: 10.1111/J.1551-2916.2010.04301.X

[43] J. Huang, Y. Zhang, T. Ma, H. Li, and L. Zhang, "Correlation between dielectric breakdown strength and interface polarization in barium strontium titanate glass ceramics," *Appl. Phys. Lett.*, vol. 96, no. 4, p. 042902, Jan. 2010, doi: 10.1063/1.3293456

[44] L. Jiang, W. Li, J. Zhu, X. Huo, L. Luo, and Y. Zhu, "Great reduction of loss at high electric field in the polyvinylidene fluoride/aromatic polythiourea blend films along with an irreversible phase transition," *Appl. Phys. Lett.*, vol. 106, no. 5, p. 052901, Feb. 2015, doi: 10.1063/1.4907549

[45] S. Wu, M. Lin, Q. Burlingame, and Q. M. Zhang, "Meta-aromatic polyurea with high dipole moment and dipole density for energy storage capacitors," *Appl. Phys. Lett.*, vol. 104, no. 7, p. 072903, Feb. 2014, doi: 10.1063/1.4865931

[46] Y. Thakur, M. Lin, S. Wu, and Q. M. Zhang, "Aromatic polyurea possessing high electrical energy density and low loss," *J. Electron. Mater. 2016 4510*, vol. 45, no. 10, pp. 4721–4725, Jul. 2016, doi: 10.1007/S11664-016-4759-Z

[47] Z. Cheng *et al.*, "Aromatic poly(arylene ether urea) with high dipole moment for high thermal stability and high energy density capacitors," *Appl. Phys. Lett.*, vol. 106, no. 20, p. 202902, May 2015, doi: 10.1063/1.4921485

[48] Y. Hambal *et al.*, "Effect of composition on polarization hysteresis and energy storage ability of P(VDF-TrFE-CFE) relaxor terpolymers," *Polym.* , vol. 13, no. 8, p. 1343, Apr. 2021, doi: 10.3390/POLYM13081343

[49] L. Zhu, "Exploring strategies for high dielectric constant and low loss polymer dielectrics," *J. Phys. Chem. Lett.*, vol. 5, no. 21, pp. 3677–3687, Nov. 2014,doi:10.1021/JZ501831Q/ASSET/IMAGES/JZ501831Q.SOCIAL.JPEG_V03

[50] C. C. Wang, G. Pilania, S. A. Boggs, S. Kumar, C. Breneman, and R. Ramprasad, "Computational strategies for polymer dielectrics design," *Polymer (Guildf).*, vol. 55, no. 4, pp. 979–988, Feb. 2014, doi: 10.1016/J.POLYMER.2013.12.069

[51] V. Sharma *et al.*, "Rational design of all organic polymer dielectrics," *Nat. Commun.*, vol. 5, no. 1, pp. 1–8, Sep. 2014, doi: 10.1038/ncomms5845

[52] M. Ieda, M. Nagao, and M. Hikita, "High-field conduction and breakdown in insulating polymers present situation and future prospects," *IEEE Trans. Dielectr. Electr. Insul.*, vol. 1, no. 5, pp. 934–945, 1994, doi: 10.1109/94.326660

[53] M. Ieda, "Dielectric breakdown process of polymers," *IEEE Trans. Electr. Insul.*, vol. EI-15, no. 3, pp. 206–224, 1980, doi: 10.1109/TEI.1980.298314

[54] J. Artbauer, "Electric strength of polymers," *J. Phys. D. Appl. Phys.*, vol. 29, no. 2, pp. 446–456, Feb. 1996, doi: 10.1088/0022-3727/29/2/024

[55] Y. Abdullahi Hassan and H. Hu, "Current status of polymer nanocomposite dielectrics for high-temperature applications," *Compos. Part A Appl. Sci. Manuf.*, vol. 138, p. 106064, Nov. 2020, doi: 10.1016/J.COMPOSITESA.2020.106064

[56] V. A. Kioussis, M. Danikas, D. D. Christantoni, G. E. Vardakis, and A. Bairaktari, "Electrical trees in a composite insulating system consisted of epoxy resin and mica: The case of multiple mica sheets for machine insulation," *Eng. Technol. Appl. Sci. Res.*, vol. 4, no. 4, pp. 662–668, Aug. 2014, doi: 10.48084/ETASR.468

[57] H. Cheng *et al.*, "Demonstration of ultra-high recyclable energy densities in domain-engineered ferroelectric films," *Nat. Commun. 2017 81*, vol. 8, no. 1, pp. 1–7, Dec. 2017, doi: 10.1038/s41467-017-02040-y

[58] T. Zhang *et al.*, "A highly scalable dielectric metamaterial with superior capacitor performance over a broad temperature," *Sci. Adv.*, vol. 6, no. 4, Jan. 2020, doi: 10.1126/SCIADV.AAX6622/SUPPL_FILE/AAX6622_SM.PDF

[59] G. Wang, X. Huang, and P. Jiang, "Tailoring dielectric properties and energy density of ferroelectric polymer nanocomposites by high-k nanowires," *ACS Appl. Mater. Interfaces*, vol. 7, no. 32, pp. 18017–27, Aug. 2015, doi: 10.1021/acsami.5b06480

[60] Prateek, V. K. Thakur, and R. K. Gupta, "Recent progress on ferroelectric polymer-based nanocomposites for high energy density capacitors: synthesis, dielectric properties, and future aspects," *Chem. Rev.,* vol. 116, no. 7, pp. 4260–4317, Apr. 2016, doi: 10.1021/ACS. CHEMREV.5B00495

[61] W. D. Callister and D. G. Rethwisch, "Materials science and engineering: an introduction." *Hoboken, NJ : Wiley*: ISBN-13: 9781119321590 edn. 10th , 2018.

[62] C. G. Hardy *et al.*, "Converting an electrical insulator into a dielectric capacitor: End-capping polystyrene with oligoaniline," *Chem. Mater.*, vol. 25, no. 5, pp. 799–807, Mar. 2013, doi: 10.1021/ CM304057F/ASSET/IMAGES/CM304057F.SOCIAL.JPEG_V03

[63] G. G. Raju, "Dielectrics in electric fields," *Dielectr. Electr. Fields,* vol. 5, no. 4 p. 792 Jan. 2003, doi: 10.1201/9780203912270

[64] S. Luo *et al.*, "Significantly enhanced electrostatic energy storage performance of flexible polymer composites by introducing highly insulating-ferroelectric microhybrids as fillers," *Adv. Energy Mater.*, vol. 9, no. 5, p. 1803204, Feb. 2019, doi: 10.1002/AENM.201803204

[65] M. Fan, P. Hu, Z. Dan, J. Jiang, B. Sun, and Y. Shen, "Significantly increased energy density and discharge efficiency at high temperature in polyetherimide nanocomposites by a small amount of $Al_2O_3$ nanoparticles," *J. Mater. Chem. A*, vol. 8, no. 46, pp. 24536–24542, Dec. 2020, doi: 10.1039/ D0TA08908G

[66] C. Zhu *et al.*, "Enhanced energy storage of polyvinylidene fluoride-based nanocomposites induced by high aspect ratio titania nanosheets," *J. Appl. Polym. Sci.*, vol. 138, no. 16, p. 50244, Apr. 2021, doi: 10.1002/APP.50244

[67] X. Xiong, D. Shen, Q. Zhang, H. Yang, J. Wen, and Z. Zhou, "Achieving high discharged energy density in PVDF-based nanocomposites loaded with fine $Ba_{0.6}Sr_{0.4}TiO_3$ nanofibers," *Compos. Commun.*, vol. 25, p. 100682, Jun. 2021, doi: 10.1016/J.COCO.2021.100682

[68] D. Ai *et al.*, "Tuning nanofillers in in situ prepared polyimide nanocomposites for high-temperature capacitive energy storage," *Adv. Energy Mater.*, vol. 10, no. 16, p. 1903881, Apr. 2020, doi: 10.1002/ AENM.201903881

[69] Q. Li *et al.*, "Flexible high-temperature dielectric materials from polymer nanocomposites," *Nat.*, vol. 523, no. 7562, pp. 576–579, Jul. 2015, doi: 10.1038/nature14647

[70] H. Pan *et al.*, "$BiFeO_3$–$SrTiO_3$ thin film as a new lead-free relaxor-ferroelectric capacitor with ultrahigh energy storage performance," *J. Mater. Chem. A*, vol. 5, no. 12, pp. 5920–5926, Mar. 2017, doi: 10.1039/C7TA00665A

[71] W. Gao, M. Yao, and X. Yao, "Improvement of energy density in $SrTiO3$ film capacitor via self-repairing behavior," *Ceram. Int.*, vol. 43, no. 16, pp. 13069–13074, Nov. 2017, doi: 10.1016/ J.CERAMINT.2017.06.162

[72] H. Yang, F. Yan, Y. Lin, and T. Wang, "High-performance dielectric ceramic films for energy storage capacitors: progress and outlook," *Appl. Phys. Lett.*, vol. 111, no. 25, p. 253903, Dec. 2017, doi: 10.1063/1.5000980

[73] A. A. Instan, S. P. Pavunny, M. K. Bhattarai, and R. S. Katiyar, "Ultrahigh capacitive energy storage in highly oriented $Ba(Zr_xTi_{1-x})O_3$ thin films prepared by pulsed laser deposition," *Appl. Phys. Lett.*, vol. 111, no. 14, p. 142903, Oct. 2017, doi: 10.1063/1.4986238

[74] D. M. Marincel *et al.*, "A-site stoichiometry and piezoelectric response in thin film $PbZr_{1-x}Ti_xO_3$," *J. Appl. Phys.*, vol. 117, no. 20, p. 204104, May 2015, doi: 10.1063/1.4921869

[75] L. Zhang *et al.*, "ALD preparation of high-k $HfO_2$ thin films with enhanced energy density and efficient electrostatic energy storage," *RSC Adv.*, vol. 7, no. 14, pp. 8388–8393, Jan. 2017, doi: 10.1039/ C6RA27847G

[76] X. Chen, B. Huang, Y. Liu, W. Wang, and P. Yu, "High energy density and high efficiency achieved in the $Ca_{0.74}Sr_{0.26}Zr_{0.7}Ti_{0.3}O_3$ linear dielectric thin films on the silicon substrates," *Appl. Phys. Lett.*, vol. 117, no. 11, p. 112902, Sep. 2020, doi: 10.1063/5.0024307

[77] E. K. Michael and S. Trolier-Mckinstry, "Amorphous-nanocrystalline lead titanate thin films for dielectric energy storage," *J. of The Ceramic Society OF JAPAN*, Vol. 122, no. 1424, pp. 250-255, Apr. 2014, doi: 10.2109/jcersj2.122.250

[78] H. Spahr *et al.*, "Enhancement of the maximum energy density in atomic layer deposited oxide based thin film" *Appl. Phys. Lett.*, vol. 103, no. 4, p. 042907, Jul. 2013, doi: 10.1063/1.4816339

[79] H. Lee, J. R. Kim, M. J. Lanagan, S. Trolier-Mckinstry, and C. A. Randall, "High-energy density dielectrics and capacitors for elevated temperatures: Ca(Zr,Ti)O3," *J. Am. Ceram. Soc.*, vol. 96, no. 4, pp. 1209–1213, Apr. 2013, doi: 10.1111/JACE.12184

[80] J. Xie *et al.*, "Energy storage properties of low concentration Fe-doped barium strontium titanate thin films," *Ceram. Int.*, vol. 44, no. 6, pp. 5867–5873, Apr. 2018, doi: 10.1016/J.CERAMINT.2017.11.218

[81] Y. Peng, M. Yao, and X. Yao, "Interfacial origin of enhanced energy density in SrTiO3-based nanocomposite films," *Ceram. Int.*, vol. 44, no. 3, pp. 3032–3039, Feb. 2018, doi: 10.1016/J.CERAMINT.2017.11.060

[82] W. Gao, M. Yao, and X. Yao, "Ultrahigh energy density in $SrTiO_3$ film capacitors," *Ceram. Int.*, vol. 43, no. 16, pp. 13069–13074, Nov. 2017, doi: 10.1016/J.CERAMINT.2017.06.162

[83] J. Xie *et al.*, "Homogeneous/inhomogeneous-structured dielectrics and their energy-storage performances," *Ceram. Int.*, vol. 44, no. 6, pp. 5867–5873, Apr. 2018, doi: 10.1016/J.CERAMINT.2017.11.218

[84] C. Diao, H. Liu, H. Hao, M. Cao, and Z. Yao, "Enhanced recoverable energy storage density of Mn-doped $Ba_{0.4}Sr_{0.6}TiO_3$ thin films prepared by spin-coating technique," *J. Mater. Sci. Mater. Electron.*, vol. 29, no. 7, pp. 5814–5819, Apr. 2018, doi: 10.1007/S10854-018-8553-Y

[85] C. Hou, W. Huang, W. Zhao, D. Zhang, Y. Yin, and X. Li, "Ultrahigh energy density in $SrTiO_3$ film capacitors" *ACS Appl. Mater. Interfaces*, vol. 9, no. 24, pp. 20484–20490, Jun. 2017, doi: 10.1021/ACSAMI.7B02225/SUPPL_FILE/AM7B02225_SI_001.PDF

[86] Y. X. Zhang, C. H. Yang, Y. C. Guo, J. H. Song, Q. Yao, and Z. D. Yi, "The microstructure, energy storage and dielectric behaviours of (Ti,Zn)-doped $Bi_{0.97}Nd_{0.03}FeO_3$ thin films," *Mater. Technol.*, vol. 33, no. 1, pp. 10–15, Jan. 2017, doi: 10.1080/10667857.2017.1370816

[87] B. B. Yang *et al.*, "Energy storage properties in $BaTiO_3$-$Bi_{3.25}La_{0.75}Ti_3O_{12}$ thin films," *Appl. Phys. Lett.*, vol. 113, no. 18, p. 183902, Oct. 2018, doi: 10.1063/1.5053446

[88] S. Cho *et al.*, "Strongly enhanced dielectric and energy storage properties in lead-free perovskite titanate thin films by alloying," *Nano Energy*, vol. 45, pp. 398–406, Mar. 2018, doi: 10.1016/J.NANOEN.2018.01.003

[89] M. Liu, H. Zhu, Y. Zhang, C. Xue, and J. Ouyang, "Energy storage characteristics of $BiFeO_3$/$BaTiO_3$ bi-layers integrated on Si," *Mater.*, vol. 9, no. 11, p. 935, Nov. 2016, doi: 10.3390/MA9110935

[90] J. Gao *et al.*, "Enhancing dielectric permittivity for energy-storage devices through tricritical phenomenon," *Sci. Rep.*, vol. 7, no. 1, pp. 1–10, Jan. 2017, doi: 10.1038/srep40916

[91] H. Zhu, H. Ma, and Y. Zhao, "Enhanced energy storage and pyroelectric properties of highly (100)-oriented $(Pb_{1-x-y}La_xCa_y)Ti_{1-x/4}O_3$ thin films derived at low temperature," *Phys. Lett. A*, vol. 382, no. 21, pp. 1409–1412, May 2018, doi: 10.1016/J.PHYSLETA.2018.03.047

[92] J. K. Faherty *et al.*, "Energy density and storage capacity of $La^{3+}$ and $Sc^{3+}$ co-substituted $Pb(Zr_{0.53}Ti_{0.47})O_3$ thin films," *Nano Express*, vol. 2, no. 2, p. 020007, Apr. 2021, doi: 10.1088/2632-959X/ABF58F

[93] A. P. Sharma, D. K. Pradhan, B. Xiao, S. K. Pradhan, and M. Bahoura, "Lead-free epitaxial ferroelectric heterostructures for energy storage applications Introduction." *AIP Advances*, vol. 8, p. 125112, 2018, doi: 10.1063/1.5046089

[94] J. Gao *et al.*, "Effects of substrate-controlled-orientation on the electrical performance of sputtered BaTiO3 thin films," *Sci. Rep.*, vol. 7, no. 1, pp. 1–10, Jan. 2017, doi: 10.1038/srep40916

[95] B. B. Yang *et al.*, "Ultrahigh energy storage in lead-free $BiFeO_3$/$Bi_{3.25}La_{0.75}Ti_3O_{12}$ thin film capacitors by solution processing," *Appl. Phys. Lett.*, vol. 112, no. 3, p. 033904, Jan. 2018, doi: 10.1063/1.5002143

[96] T. Zhang, W. Li, Y. Zhao, Y. Yu, and W. Fei, "High energy storage performance of opposite double-heterojunction ferroelectricity–insulators," *Adv. Funct. Mater.*, vol. 28, no. 10, Mar. 2018, doi: 10.1002/ADFM.201706211

[97] Z. Xie, B. Peng, S. Meng, Y. Zhou, and Z. Yue, "High–energy-storage density capacitors of $Bi(Ni_{1/2}Ti_{1/2})O_3$-$PbTiO_3$ thin films with good temperature stability," *J. Am. Ceram. Soc.*, vol. 96, no. 7, pp. 2061–2064, Jul. 2013, doi: 10.1111/JACE.12443

[98] J. Suchanicz *et al.*, "Electric properties of Fe-doped $Na_{0.5}Bi_{0.5}TiO_3$ ceramics in unpoled and poled state," *Phase. Trans.*, vol. 93, no. 9, pp. 877–882, Sep. 2020, doi: 10.1080/01411594.2020.1804902

[99] J. Wang *et al.*, "Effects of $Fe^{3+}$ doping on electrical properties and energy-storage performances of the $(Na_{0.85}K_{0.15})_{0.5}Bi_{0.5}TiO_3$ thick films prepared by sol-gel method," *J. Alloys Compd.*, vol. 727, pp. 596–602, Dec. 2017, doi: 10.1016/J.JALLCOM.2017.08.169

[100] C. Xue *et al.*, "Natural-superlattice structured $CaBi_2Nb_2O_9$-$Bi_4Ti_3O_{12}$ ferroelectric thin films," *Ceram. Int.*, vol. 43, no. 11, pp. 8459–8465, Aug. 2017, doi: 10.1016/J.CERAMINT.2017.03.197

[101] C. Chen, Y. Wang, J. Li, C. Wu, and G. Yang, "Piezoelectric, ferroelectric and pyroelectric properties of $(100 - x)Pb(Mg_{1/3}Nb_{2/3})O_3 -xPbTiO_3$ ceramics," *J. Adv. Dielectrics,* vol. 12, no. 3, pp. 2250002, Mar. 2022, doi: 10.1142/S2010135X22500023

[102] A. A. Bokov and Z. G. Ye, "Recent progress in relaxor ferroelectrics with perovskite structure," *J. Mater. Sci.*, vol. 41, no. 1, pp. 31–52, Jan. 2006, doi: 10.1007/S10853-005-5915-7

[103] L. E. Cross, "Relaxorferroelectrics: An overview," *Ferroelectrics*, vol. 151, no. 1, pp. 305–320, 2011, doi: 10.1080/00150199408244755

[104] X. Hao, "A review on the dielectric materials for high energy-storage application," *J. Adv. Dielectr.*, vol. 03, no. 01, p. 1330001, Jan. 2013, doi: 10.1142/S2010135X13300016

[105] Y. Sun *et al.*, "High energy storage efficiency with fatigue resistance and thermal stability in lead-free $Na_{0.5}K_{0.5}NbO_3$/$BiMnO_3$ solid-solution films ferroelectrics www.pss-rapid.com," *Phys. Status Solidi RRL*, vol. 12, p. 1700364, 2018, doi: 10.1002/pssr.201700364

[106] Y. Liu, X. Hao, and S. An, "Significant enhancement of energy-storage performance of $(Pb_{0.91}La_{0.09})$ $(Zr_{0.65}Ti_{0.35})O_3$ relaxor ferroelectric thin films by Mn doping," *J. Appl. Phys.*, vol. 114, no. 17, p. 174102, Nov. 2013, doi: 10.1063/1.4829029

[107] J. Wang and H. Fan, "Enhanced energy storage performance and fatigue resistance of Mn-doped $0.7Na_{0.5}Bi_{0.5}TiO_3$–$0.3Sr_{0.7}Bi_{0.2}TiO_3$ lead-free ferroelectric ceramics," *J. Mater. Res.*, pp. 1–10, 6374, doi: 10.1557/JMR.2020.291

[108] B. Fang, K. Qian, N. Yuan, J. Ding, X. Zhao, and H. Luo, "Large pyroelectric response of $0.8Pb(Mg_{1/3}Nb_{2/3})O_3$-$0.2PbTiO_3$ ceramics prepared by reaction-sintering method," *Mater. Lett.*, vol. 84, pp. 91–93, Oct. 2012, doi: 10.1016/J.MATLET.2012.06.061

[109] K. Yao *et al.*, "Nonlinear dielectric thin films for high-power electric storage with energy density comparable with electrochemical supercapacitors," *IEEE Trans. Ultrason. Ferroelectr. Freq. Control*, vol. 58, no. 9, pp. 1968–1974, Sep. 2011, doi: 10.1109/TUFFC.2011.2039

[110] X. Hao, Y. Wang, J. Yang, S. An, and J. Xu, "High energy-storage performance in $Pb_{0.91}La_{0.09}(Ti_{0.65}Zr_{0.35})O3$ relaxor ferroelectric thin films," *J. Appl. Phys.*, vol. 112, no. 11, p. 114111, Dec. 2012, doi: 10.1063/1.4768461

[111] E. C. Subbarao, "Ferroelectric and antiferroelectric materials," *Ferroelectrics*, vol. 5, no. 1, pp. 267–280, Jan. 2011, doi: 10.1080/00150197308243957

[112] D. Bernard, J. Pannetier, and J. Lucas, "Ferroelectric and antiferroelectric materials with pyrochlore structure," *Ferroelectrics*, vol. 21, no. 1, pp. 429–431, Jan. 2011, doi: 10.1080/00150197808237288

[113] "Willson Research Group–Liquid Crystals." https://willson.cm.utexas.edu/Research/Sub_Files/Liquid_Crystals/index.htm (accessed Apr. 06, 2022).

[114] G. A. Samara, "Pressure and temperature dependence of the dielectric properties and phase transitions of the antiferroelectric perovskites: $PbZrO_3$ and $PbHfO_3$," *Phys. Rev. B,* vol. 1, no. 9, p. 3777, May 1970, doi: 10.1103/PhysRevB.1.3777

[115] O. A. Zhelnova and O. E. Fesenko, "Phase transitions and twinning in $NaNbO_3$ crystals," *Ferroelectrics*, vol. 75, no. 1, pp. 469–475, 2011, doi: 10.1080/00150198708215068

[116] F. Gao *et al.*, "Energy-storage properties of $0.89Bi_{0.5}Na_{0.5}TiO_3$–$0.06BaTiO_3$–$0.05K_{0.5}Na_{0.5}NbO_3$ lead-free anti-ferroelectric ceramics," *J. Am. Ceram. Soc.*, vol. 94, no. 12, pp. 4382–4386, Dec. 2011, doi: 10.1111/J.1551-2916.2011.04731.X

[117] S. E. Young, J. Y. Zhang, W. Hong, and X. Tan, "Mechanical self-confinement to enhance energy storage density of antiferroelectric capacitors," *J. Appl. Phys.*, vol. 113, no. 5, p. 054101, Feb. 2013, doi: 10.1063/1.4790135

[118] A. Chauhan, S. Patel, and R. Vaish, "Mechanical confinement for improved energy storage density in BNT-BT-KNN lead-free ceramic capacitors," *AIP Adv.*, vol. 4, no. 8, p. 087106, Aug. 2014, doi: 10.1063/1.4892608

[119] X. Hao, J. Zhai, L. B. Kong, and Z. Xu, "A comprehensive review on the progress of lead zirconate-based antiferroelectric materials," *Prog. Mater. Sci.*, vol. 63, pp. 1–57, Jun. 2014, doi: 10.1016/J.PMATSCI.2014.01.002

[120] G. H. Haertling, "PLZT electrooptic materials and applications – a review," *Ferroelectrics,* vol. 75, no. 1, pp. 25–55, 2011, doi: 10.1080/00150198708008208

[121] B. Xu, J. I. Niguez, and L. Bellaiche, "Designing lead-free antiferroelectrics for energy storage," *Nat. Commnication.*, vol. 8, 2017, doi: 10.1038/ncomms15682

[122] R. W. Schwartz, "Chemical solution deposition of perovskite thin films," *Chem. Mater.*, vol. 9, no. 11, pp. 2325–2340, 1997, doi: 10.1021/CM970286F

[123] C. Adamo *et al.*, "Effect of biaxial strain on the electrical and magnetic properties of (001) $La_{0.7}Sr_{0.3}MnO_3$ thin films," *Appl. Phys. Lett.*, vol. 95, no. 11, 2009, doi: 10.1063/1.3213346

[124] J. Cinert, P. Ctibor, and J. Sedláček, "Barium titanate dielectric ceramics fired by spark plasma sintering with and without annealing," 2019, Accessed: Apr. 06, 2022. [Online]. Available: www.pccc.icrc.ac.ir

[125] L. Wang, K. Yao, and W. Ren, "Piezoelectric $K_{0.5}Na_{0.5}NbO_3$ thick films derived from polyvinylpyrrolidone-modified chemical solution deposition," *Appl. Phys. Lett.*, vol. 93, no. 9, p. 092903, Sep. 2008, doi: 10.1063/1.2978160

[126] D. A. Barrow, T. E. Petroff, R. P. Tandon, and M. Sayer, "Characterization of thick lead zirconate titanate films fabricated using a new sol gel based process," *J. Appl. Phys.*, vol. 81, no. 2, p. 876, Aug. 1998, doi: 10.1063/1.364172

[127] H. J. Lee *et al.*, "Flexible high energy density capacitors using La-doped $PbZrO_3$ anti-ferroelectric thin films," *Appl. Phys. Lett.*, vol. 112, no. 9, p. 092901, Feb. 2018, doi: 10.1063/1.5018003

[128] M. H. Park *et al.*, "Study on the size effect in $Hf_{0.5}Zr_{0.5}O_2$ films thinner than 8 nm before and after wake-up field cycling," *Appl. Phys. Lett.*, vol. 107, no. 19, p. 192907, Nov. 2015, doi: 10.1063/1.4935588

[129] P. D. Lomenzo, C. C. Chung, C. Zhou, J. L. Jones, and T. Nishida, "Doped $Hf_{0.5}Zr_{0.5}O_2$ for high efficiency integrated supercapacitors," *Appl. Phys. Lett.*, vol. 110, no. 23, p. 232904, Jun. 2017, doi: 10.1063/1.4985297

[130] X. Hao, Y. Wang, J. Yang, S. An, and J. Xu, "High energy-storage performance in $Pb_{0.91}La_{0.09}(Ti_{0.65}Zr_{0.35})O_3$ relaxor ferroelectric thin films," *J. Appl. Phys.*, vol. 112, no. 11, p. 114111, Dec. 2012, doi: 10.1063/1.4768461

[131] H. Gao *et al.*, "Enhanced electrocaloric effect and energy-storage performance in PBLZT films with various $Ba^{2+}$ content," *Ceram. Int.*, vol. 42, no. 15, pp. 16439–16447, Nov. 2016, doi: 10.1016/J.CERAMINT.2016.08.054

[132] C. W. Ahn, G. Amarsanaa, S. S. Won, S. A. Chae, D. S. Lee, and I. W. Kim, "Antiferroelectric thin-film capacitors with high energy-storage densities, low energy losses, and fast discharge times," *ACS Appl. Mater. Interfaces*, vol. 7, no. 48, pp. 26381–26386, Dec. 2015, doi: 10.1021/ACSAMI.5B08786/SUPPL_FILE/AM5B08786_SI_001.PDF

[133] B. Ma, Z. Hu, R. E. Koritala, T. H. Lee, S. E. Dorris, and U. Balachandran, "PLZT film capacitors for power electronics and energy storage applications," *J. Mater. Sci. Mater. Electron.*, vol. 26, no. 12, pp. 9279–9287, Dec. 2015, doi: 10.1007/S10854-015-3025-0/FIGURES/8

[134] B. Peng, Z. Xie, Z. Yue, and L. Li, "Improvement of the recoverable energy storage density and efficiency by utilizing the linear dielectric response in ferroelectric capacitors," *Appl. Phys. Lett.*, vol. 105, no. 5, p. 052904, Aug. 2014, doi: 10.1063/1.4892454

[135] Z. Tang *et al.*, "Enhanced energy-storage density and temperature stability of $Pb_{0.89}La_{0.06}Sr_{0.05}(Zr_{0.95}Ti_{0.05})O_3$ anti-ferroelectric thin film capacitor," *J. Mater.*, vol. 8, no. 1, pp. 239–246, Jan. 2022, doi: 10.1016/J.JMAT.2020.12.012

[136] B. Lu *et al.*, "Large electrocaloric effect in relaxor ferroelectric and antiferroelectric lanthanum doped lead zirconate titanate ceramics," *Sci. Rep.*, vol. 7, Mar. 2017, doi: 10.1038/SREP45335

[137] T. Zhang *et al.*, "High-energy storage density and excellent temperature stability in antiferroelectric/ferroelectric bilayer thin films," *J. Am. Ceram. Soc.*, vol. 100, no. 7, pp. 3080–3087, Jul. 2017, doi: 10.1111/JACE.14876

[138] Z. H. Niu, Y. P. Jiang, X. G. Tang, Q. X. Liu, and W. H. Li, "B-site non-stoichiometric $(Pb_{0.97}La_{0.02})(Zr_{0.95}Ti_{0.05})O_3$ antiferroelectric ceramics for energy storage," *J. Asian Ceram. Soc.*, vol. 6, no. 3, pp. 240–246, Jul. 2018, doi: 10.1080/21870764.2018.1501126

[139] M. Sharifzadeh Mirshekarloo, K. Yao, and T. Sritharan, "Large strain and high energy storage density in orthorhombic perovskite $(Pb_{0.97}La_{0.02})(Zr_{1-x-y}Sn_xTi_y)O_3$ antiferroelectric thin films," *Appl. Phys. Lett.*, vol. 97, no. 14, p. 142902, Oct. 2010, doi: 10.1063/1.3497193

[140] H. He and X. Tan, "Electric-field-induced transformation of incommensurate modulations in antiferroelectric $Pb_{0.99}Nb_{0.02}[(Zr_{1-x}Sn_x)_{1-y}Ti_y]_{0.98}O_3$," Physical Review B, vol.72, p. 024102 2005, doi: 10.1103/PhysRevB.72.024102

[141] J. Parui and S. B. Krupanidhi, "Enhancement of charge and energy storage in sol-gel derived pure and La-modified $PbZrO_3$ thin films," Appl. Phys. Lett., vol. 92, no. 19, p. 192901, May 2008, doi: 10.1063/1.2928230

[142] B. Peng et al., "Giant Electric Energy Density in Epitaxial Lead-Free Thin Films with Coexistence of Ferroelectrics and Antiferroelectrics," Adv. Electron. Mater., vol. 1, no. 5, p. 1500052, May 2015, doi: 10.1002/AELM.201500052

[143] Z. G. Liu et al., "Excellent energy storage density and efficiency in lead-free Sm-doped $BaTiO_3$–$Bi(Mg_{0.5}Ti_{0.5})O_3$ ceramics," J. Mater. Chem. C, vol. 8, no. 38, pp. 13405–13414, Oct. 2020, doi: 10.1039/D0TC03035J

[144] J. Frederick, X. Tan, and W. Jo, "Strains and polarization during antiferroelectric-ferroelectric phase switching in $Pb_{0.99}Nb_{0.02}[(Zr_{0.57}Sn_{0.43})_{1-y}Ti_y]_{0.98}O_3$ ceramics," J. Am. Ceram. Soc., vol. 94, pp. 1149–1155, 2011, doi: 10.1111/j.1551-2916.2010.04194.x

# 10 Ferrofluid-Based Nanogenerators for Self-Power Generation in Electronic Devices

*Balwinder Kaur, Ajay Singh, Manju Arora, Archna Sharma, Meenakshi Dhiman, and Baljinder Kaur*

## 10.1 INTRODUCTION

Scientists always have an urge to develop power sources that are permanent and portable by collecting the excess of unutilized energy from the environment. A lot of research work is carried out for developing efficient and economical energy harvesting devices based on the piezoelectric, triboelectric, pyroelectric, thermoelectric, or photovoltaic effects [1–5]. The working of these devices to harvest energy effectively mainly depends on the design of the transducer.

The main reason behind the energy harvesting is the need of wireless and small-scale devices to get converted into the global market. The utilization of portable sources is necessary because of the advancement in technology such as wireless sensor networks (WSNs), micro-electromechanical systems (MEMS), and the Internet of Things (IoT) [6].

These smart devices should be simple, durable, wearable, wireless, portable, self-charging, cost-efficient, and even implantable electronic equipment to overcome conventional and unsustainable power supply limitations [7]. The devices used to harvest small energies with nanomaterials or nanostructures are nowadays called nanogenerators (which convert thermal or mechanical change into electricity) and the first nanogenerator was reported in 2006 with zinc oxide (ZnO) nanowires [1].

These smart small devices used for various purposes such as health monitoring, biochemical detection, environmental protection, remote controls, wireless transmission, and security require power in the range of microwatt (μW) to milliwatt (mW). For powering these devices, batteries are used as they have the characteristics of mobility, sustainability, and availability. But the management and recycling of these batteries is another challenging and difficult task because of the leftover chemicals in the exhausted batteries [8].

The power generation by electromagnetic generators (EMGs), invented in 1831, can only harvest high-frequency mechanical energies [9] but the low-frequency mechanical energies, which is required for these small-scale devices can be harvested by NGs based on the effects of piezoelectric [1], triboelectric [10], pyroelectric [11], thermoelectric [12], ion streams [13], etc., and also by using batteries and supercapacitors.

Nanogenerators can be classified into different categories like piezoelectric (PENG), triboelectric (TENG), pyroelectric nanogenerators, electromagnetic generators (EMG), solar cells, and electrochemical cells capable of converting mechanical, thermal, magnetic, solar, and chemical energy into electricity as shown in Figure 10.1. In particular, mechanical energy found in the surroundings is exploited by means of electrostatic [14], electromagnetic [15], piezoelectric [16], and triboelectric effects [17] to generate electric energy. The potential applications of PENGs and TENGs bring new

DOI: 10.1201/9781003340539-10

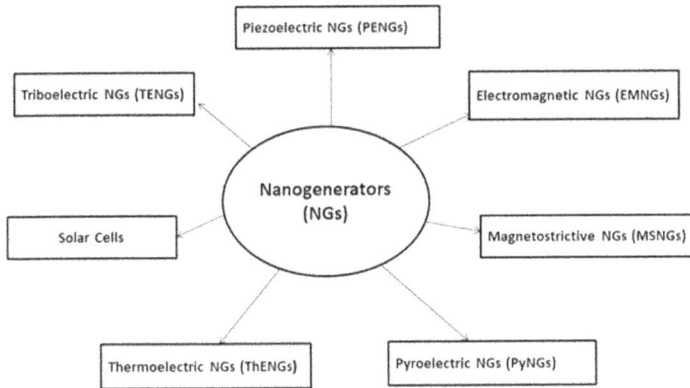

FIGURE 10.1 Classification of nanogenerators.

challenges for nanogenerators as different environmental conditions may affect the working of these devices and it may vary with time span. So, it is important to have a clear idea from their working mechanisms and construction materials to know about the environmental effects and upto what limit they can work properly [18].

The electric power generated by these nanogenerators is still limited in terms of sustainability and availability and it was thought to be overcome by coordinating two or more types of nanogenerators largely known as hybrid nanogenerators, which increase the total electric power. These hybridized NGs provide a new way to maximize the scavenged multiple energies from ambient environment [19–23].

Different types of NGs including hybridized NGs are discussed in detail in this chapter and the schematic diagram of these different types are shown in Figure 10.2.

(A) Triboelectric nanogenerator (TENG) working modes.
(B) Electromagnetic generator (EMG) working modes.
(C) Piezoelectric nanogenerator (PENG) working modes.
(D) Principle diagram of electrochemical cell.
(E) Principle diagram of solar cell.
(F) Principle diagram of pyroelectric nanogenerator.

Reproduced with permission, from Chen et al. [24], Copyright 2020, Elsevier and Zhang et al [25] 2020 The Author(s). This is an open access article under the CC BY-NC-ND license http://crea tivecommons.org/licenses/by-nc-nd/4.0/)

Nanomagnetic particles offer immense potential applications and advancements in electronics, optoelectronics, energy devices, magnetic storage, MEMS, EMI shielding, and biomedical applications. Nanoscale magnetic particles prepared with surface stabilizers constitute magnetic fluids, also known as ferrofluids, which display superparamagnetic properties. The significant property of these materials is that they are single domain particles having discrete randomly oriented magnetic moments. When placed in an external magnetic field their moments rapidly rotate in the direction of applied magnetic field and enhance the magnetic flux. When the external magnetic field is removed from these particles the Brownian motion is sufficient to randomize their magnetic moments and thus having no magnetic remanence. The benefit of such materials is that they can be designed as per requirement, e.g., biocompatible surface stabilizers for new biomedical applications, which include retinal detachment therapy, cell separation methods, tumor hyperthermia, improved

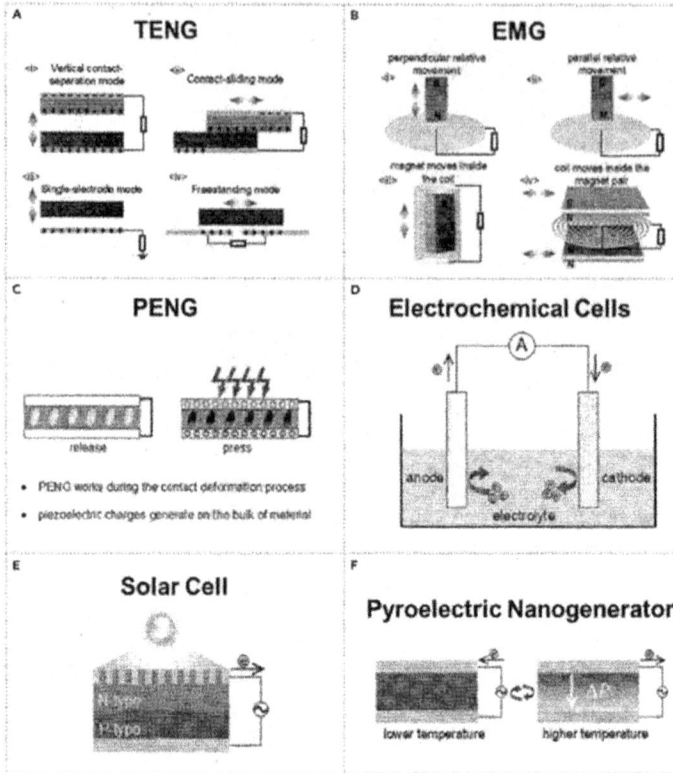

**FIGURE 10.2**   Schematic of different types of nanogenerators.

MRI diagnostic contrast agents, and as magnetic field-guided carriers for localizing drugs or radio-active therapies.

Ferrofluids have a very good property of low friction shape adaptability that has been utilized in the design and development of new generation devices like electric power conversion devices converting wind energy or mechanical energy into electrical power and sensors like vibration and temperature sensors, etc. The magnetic properties of ferrofluids make them attractive for varieties of applications where the combination of liquid state and magnetic properties is essential [26]. An important property of concentrated ferrofluids is that they are strongly attracted by permanent magnets, while their liquid character is preserved. The attraction can be strong enough to overcome the force of gravity. Many applications of ferrofluids are based on this property. Moving a magnet back and forth through a conductive coil induces an electromotive force that generates an AC current in the coil. Ferrofluid bearings levitate the magnets and make their motion easy, thus making electrical power generation devices more efficient.

## 10.2   PRINCIPLE OF NANOGENERATORS

The mechanism of PENG is a transductive mechanism usually based on the piezopotential created in the piezoelectric materials when these materials are subjected to external mechanical stress thus modifying the central symmetry of the crystalline structure. This mechanical deformation affects the overall orientation of the electric dipole and oppositely charged species are induced on the surface of the material. An electric potential difference is generated across an external load due to these induced charges [27]. This output is a promising power source for sustainable operation of micro nano systems. The efficiency of the PENGs depends on the type of material (ceramic, single crystals, and polymers) and the direction of applying external force, and thus the piezoelectric-dependent current is given by Equation (10.1), as follows [28]:

$$I_{piezo} = d_{33} A \frac{d\sigma}{dt} \tag{10.1}$$

where $d_{33}$ is the piezoelectric coefficient, $A$ is the effective area of the device, and $\frac{d\sigma}{dt}$ is the rate of mechanical stress change.

Triboelectrification is a well-known phenomenon. The basic principle of triboelectric nanogenerators [29, 30] is the combination of triboelectrification and electrostatic induction between the surfaces of two different friction materials during periodic contact/separation. It is a kind of developing charge on the two different materials when they are separated from each other after being in contact with one another [31]. It is generally believed that a chemical bond is formed when the two materials are in contact with each other and the charge's movement balanced the electrochemical potential difference. Triboelectric charges are induced on the surface of the material due to extra electrons retained by the bonding atoms. These triboelectric charges on the surface of the dielectric could be a driving force for driving the flow of electrons in the electrode to balance the resulting potential drop [32]. These transferred charges may be electrons, ions, or molecules.

This is also known as contact electrification as it occurs only when two materials are brought into contact with each other and then separate. According to the "surface state model," electrons get transferred from higher Fermi level surface to lower Fermi level when the two surfaces are in contact with each other [33]. Based on above principle, four different TENG modes have been invented, viz, vertical contact–separation (CS) mode, lateral sliding mode (LS), single-electrode (se) mode, freestanding triboelectric-layer (FT) as shown in Figure 10.3. For all these basic modes of TENGs, there are at least one pair of triboelectric surfaces that contact together initially (for FT mode, there are usually two pairs of triboelectric surfaces). And there are also at least two electrodes, which are carefully insulated from each other.

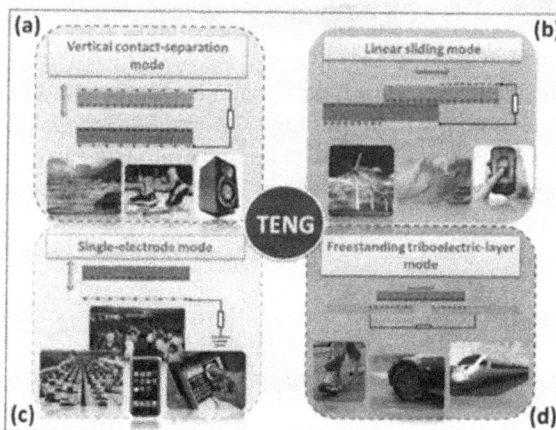

**FIGURE 10.3** Four basic operation modes for TENG. (a) Vertical contact-separation (CS) mode; (b) in-plane lateral sliding (LS) mode; (c) single electrode (SE) mode; (d) freestanding triboelectric-layer (FT) mode. Reprinted from Z. L. Wang et al. [32] with standard acknowledgement. Copyright 2015 The Royal Society of Chemistry.

The vertical CS mode is the earliest working mode and it uses relative motion perpendicular to the interface of two dielectric materials with a metal layer deposition thus serving as an electrode. The potential arise between the electrodes is governed by the gap distance between material surfaces. The lateral–sliding mode causes high triboelectrification on the surface as it involves the relative displacement in a direction parallel to the interface between two dielectric materials. The single-electrode mode takes the ground as the reference electrode that comes into contact with a single dielectric material and hence static or grounded. The freestanding triboelectric-layer mode is similar to the single-electrode mode using a pair of symmetric electrodes and electrical output is induced from asymmetric charge distribution caused by the changing position of the freely moving object. One thing worth noting is that practical application of TENG is not limited to one single mode, but relies more on conjunction or hybridization of different modes to harness their full advantages [34, 35].

## 10.3   TYPES OF NANOGENERATORS

As stated above, different types of NGs available are piezoelectric, triboelectric, and pyroelectric nanogenerators, etc.

### 10.3.1   PIEZOELECTRIC NGS (PENGS)

PENGs are used to convert mechanical energy into electrical energy and the deformation in the structures occurs in this process [36]. Piezoelectric effect can be understood in two ways. In one way, electric field is generated on application of mechanical stress and this process is applicable in the sensors. And the second is the reverse effect in which the electric field is applied and deformation in the material arises and these materials acts as an actuator [37, 38]. PENGs are built from one-dimensional piezoelectric nanostructures, such as wurtzite ZnO [1], CdS [39], ZnS [40], and GaN [41], $\beta$ phase PVDF polymer [42], perovskite PZT [43], $BaTiO_3$ [44], $NaNbO_3$ [45], trigonal Te [46], etc. The mechanical deformation in the materials can take place because of body motion, acoustic wave, wind, and machine vibration and the separation of charge builds up a piezoelectric potential, which induces a current in the external circuit and drives devices. Utilizing the strain-induced piezo-electric polarization in materials, PENG has been developed for advanced materials, and varieties of demonstrated applications [8]. Wang's group invented the piezoelectric nanogenerator (PENG) in 2006 [1]. PENG is used to harvest small-scale and low-frequency mechanical energy efficiently [47] and also they are small in size and easy to carry.

Piezoelectric materials can be single crystals, ceramics, polymers, and thin films. The piezo-electric ceramic element can be used for both low- and high-power applications. Earlier the most studied piezoelectric material was ZnO because of its simple structure and found suitable for making nanogenerators. The drawback of ZnO material was that the potential developed was not sufficient to meet the requirements of electronic devices. Lead zirconium Tttanate (PZT) was found suitable having high piezoelectric coefficient.

### 10.3.2   TRIBOELECTRIC NGS (TENGS)

TENGs can convert mechanical energy in the surroundings into electrical energy and become the center of attraction in the field of energy due to its variety of energy sources, high efficiency with a wide application range. Triboelectricity arises when two different substances come into contact with each other and charges move from one side to the other. A potential difference can be developed if the two substances are separated or some sort of unbalance is created in the charge densities. This potential difference helps to drive the electrons through the load connected across the electrode.

The potential difference vanishes if the two materials are brought into contact once again and the electrons flow back to the electrodes completing one cycle. These electric discharges produced in this way proved to be dangerous in many ways causing unpredictable losses in many industries as this can generate very high voltages of the order of 35,000 V [48]. But in many ways, it has also produced many meaningful applications, such as electrospinning, electrostatic printing, and TENGs [49–52].

After the PENGs, Wang's group in 2012 [10], first invented the TENGs, and it has been quickly developed to be a revolutionary breakthrough in the self-powered [53, 54] and energy harvesting systems [55–57]. The novel TENG makes use of Maxwell's displacement current to power portable electronic devices due to electrostatic induction and triboelectrification [58]. TENGs can harvest the mechanical energy present in the surroundings and also the ability to convert low-frequency mechanical energy arising from walking, waving, and eye-blinking into electricity. The working of TENGs can be understood in four basic operation modes [32], including vertical contact–separation mode [59], lateral-sliding mode [60], single-electrode mode [61], and freestanding triboelectric-layer mode [62]

TENGs have proven their potential in wearable devices, to provide high output power to the Internet of Things and also as a human machine interface device.

## 10.3.3  Electromagnetic NGs (EMNGs)

The power generation by conventional approaches are carried by EM (electromagnetic) generators, which are based on Faraday's law of electromagnetic induction. In EMNGs, electrical energy is generated whenever there is relative motion between the magnetized body and conductive coil. Permanent magnets or electromagnets are more suitable to magnetize the magnetic circuit or configuration used for power generation, for low-power devices. Ferromagnetic materials are commonly used as they have high electrical resistance resulting in less eddy current effects. Generally, four types of magnets are present for the application: ceramic, neodymium iron boron, alnico, and samarium cobalt [63]. The researchers have worked on decreasing the resonant frequency of the system or configuration used for power generation and to widen the frequency bandwidth only to enhance the effectiveness and efficiency of the EMNGs. The low-frequency vibrations present in the surroundings is very small [64] and to harvest energy from this requires a generator, which functions at low frequency [65]. The first micro-scaled EMNG was designed by Shearwood and Yates in 1997 [66, 67] to generate power from low frequencies. This device was an inertial generator used to generate electrical energy from mechanical vibrations in its environment. In this device, an inertial mass "$m$" is attached with a spring of spring-constant, $k$. When the generator is vibrated, the mass moves out of phase with the generator housing, resulting in a net movement between the mass and the housing. This relative motion is converted to electrical energy by an electromechanical transducer. The topology of the electromagnetic transducer considered here consists of a moving magnet that induces a voltage on a coil attached to the housing. This moving magnet configuration maximizes the inertial mass and avoids the problems of attaching electrical conductors to the mass.

The response of the configuration used for EMNG decreases significantly due to the occurrence of frequency mismatch between environment and system due to different physical parameters. Due to this, it was thought to develop multi-frequency EMNG by widening bandwidth. In 2014, a novel EMNG was designed by Ooi and Gilbert [68] that implements a dual resonator technique to improve the operating range of frequency. This technique consists of two different resonator systems. Because of this multi-vibration mode, corresponding to a particular range, several frequencies of different modes are adjusted resulting in broader bandwidth [6] and exhibits a wide operating range of frequencies.

### 10.3.4 MAGNETOSTRICTIVE NGs (MSNGs)

Magnetostrictive materials (MSM) consist of small ferromagnetic materials like iron, cobalt, nickel and thus have a small magnetic moment. These materials either shrink or stretch on applying the magnetic field [69]. These materials produce kinetic energy from magnetic energy. It is a two-step process in which NGs make use of the MS effect to generate electricity from mechanical vibration. First mechanical energy is converted to magnetic energy through the magneto-mechanical coupling and then using electromagnetic coupling, this magnetic energy is converted to electrical energy [70]. It follows the Villari/Magnetoelastic effect. The Villari effect is the inverse magnetostrictive effect, which gives us the change in magnetic susceptibility of a material when subjected to a mechanical stress.

Numerical modeling and experimentation have already been done by different researchers to know the MS response of the cobalt ferrite nanomaterials [71, 72]. Polymer-based composites are also used as MSM. The dimension as well as the elastic and magnetic properties of these materials are changed simultaneously when it comes to the contact of the magnetic field. Different MS bulk materials are terfenol-D, galfenol, alfenol, cobalt ferrite, metglas 2605SC amorphous alloy, etc., [70, 73] and various MS polymer composites are nickel/vinyl ester, galfenol flakes/epoxy, CoO–$Fe_2O_3$/phenol, galfenol particles/epoxy, silicon steel/epoxy, carbonyl iron/silicon, etc. MS polymer composites are used to sense ultrasonic wave in actuators, biomedical applications, sensors, vibration isolation and control, and health monitoring.

### 10.3.5 PYROELECTRIC NGs (PYNGs)

A pyroelectric nanogenerator is used to harvest energy by converting the external thermal energy into electrical energy with the help of nano-structured pyroelectric materials. Its working depends upon the Seebeck effect in which the diffusion of charge carriers take place due to temperature difference maintained at the two ends of the device [74]. The pyroelectric effect can be a better choice if there is no temperature gradient and there exist time-dependent temperature fluctuation as in the case of certain anisotropic solids [75]. The first pyroelectric nanogenerator was introduced by Prof. Zhong Lin Wang at Georgia Institute of Technology in 2012 [76]. By harvesting the waste heat energy, this new type of nanogenerator has potential applications, such as wireless sensors, temperature imaging, medical diagnostics, and personal electronics.

The pyroelectric materials have the advantage of being used to harvest energy from temperature variations as compared to the materials used for thermoelectric energy harvesters as they are generally expensive [77]. The pyroelectric coefficient is the key factor to know the ability of a material for harvesting energy and this coefficient is related to the changing rate of the polarization density with respect to the temperature. The energy harvested by the PyENG has been demonstrated for powering small electronics like charging of Li ion battery, to drive a liquid-crystal display (LCD), a light-emitting diode (LED), and a wireless sensor. The first PyENG is fabricated based on ZnO nanowire arrays [78]. Another type of PyENG reported is based on the $KNbO_3$ nanowire and polydimethylsiloxane (PDMS) polymer composite, with the pyroelectric coefficient demonstrated as around 0.8 nC/(cm2 K) [79]. In general, the pyroelectric nanogenerator gives a high output voltage, but the output current is small. It not only can be used as a potential power source, but also as an active sensor for measuring temperature variation [80].

### 10.3.6 THERMOELECTRIC NGs (THENG)

Thermal energy can also be harvested based on the Seebeck effect when there is temperature gradient existing in the environment. Such a gradient possibly exists between the human body and ambient environment. Therefore, the thermoelectric nanogenerator (ThENG) can be fabricated as a wearable energy harvester to power devices for human-machine interface [8].

Thermoelectric nanogenerators (ThENGs) gives an ideal way of harvesting thermal energy from the ambient environment to overcome the energy crisis and environmental degradation. It is considered as a significant technology to exhibit excellent thermoelectric performance and thus gradually becoming one of the trending research fields. The optimization about materials and flexible structural designs for potential applications in wearable electronics are systematically discussed and described by Zhang et al. [80, 81]. With the development of flexible and wearable electronic equipment, flexible ThENGs show increasingly great application prospects in artificial intelligence, self-powered sensing systems, and other fields in the future.

### 10.3.7  HYBRIDIZED NANOGENERATORS

Although there were significant advancements in the output power density and conversion efficiency, but still the total electric power of these types of nanogenerators is limited in the aspect of sustainability and availability. This was solved by blending two or more kinds of NGs known as a hybridized nanogenerator, which opens a new way of harvesting multiple energies thus increasing the total electric power [19, 21–23, 81]. The main purpose of the hybridized nanogenerators [82, 83] is to combine several diverse nanogenerators into a single unit thus enabling the unit to provide any type of energy as per the requirement at any time. These types of devices can be used in a number of situations, such as wearable products [84], self-powered sensors [85], charging Li-ion batteries [86, 87], and charging supercapacitors [88].

The idea of producing hybrid nanogenerators for wearables assure to compensate shortcomings from each type of nanogenerator in a more efficient energy-capture process from multiple sources [27]. The different possibilities of combination of hybrid devices are triboelectric–piezoelec7tric generators [89–92], triboelectric–thermoelectric generators [93], photovoltaic–triboelectric generators [94], UV photovoltaic–piezoelectric generators [95], electromagnetic–triboelectric generators [96], electromagnetic–triboelectric–piezoelectric generators [97], photovoltaic–piezoelectric generators [98], and piezoelectric–pyroelectric generators [28].

## 10.4  APPLICATIONS OF NANOGENERATORS

The technology of nanogenerators finds applications in plenty of new devices such as biosensors, health-monitoring human–computer interfaces [99] (like the improvement of the performance of and in the detection of vital signals – specific analytes in sensors and biosensors) in addition to the improvement in the performance of electrochemical devices (energy storage systems such as supercapacitors and batteries). The sensing mechanism can be explained with the example of the creation of gate potential as reported by Selvarajan et al. [100] creating a direct relationship between the generated voltage and the amount of analyte (glucose). This process is understood by the changes in the piezoelectric components with enzymes that interact with the analyte and modulate the generated power [101]. Nanogenerators also replace the conventional sensors as they can be used for monitoring physiological signals for diaphragmatic breathing monitoring. The relation between output voltage from the nanogenerator and the respiration signal represents an important application for this self-powered prototype of the sensor [102].

A remarkable progress has been done in the field of raindrop, ocean wave, ultrasonic, and wind energy harvesting utilizing TENG technology. These energy sources are renewable and address the environmental issues caused by fossil fuel consumption. It was a major challenging job to manufacture the generators for harvesting energy from tidal and ocean wave as they are expensive and bulky. But TENG technology proved to be a promising alternative for its efficiency and cost-effective water wave energy harvesting [17].

Biomechanical energy includes daily wasted mechanical energy in the form of human body motion, like walking, arm and hand stretching, breathing, etc. biocompatible TENGs were used to harvest biomechanical energy wasted through walking. This was done by putting the first synthetic-based TENGs in common shoes, which are activated by the press and release motion while walking [103, 104]. The other energy harvesting from human motion can be done by the wearable and stretchable fabrics that induce triboelectric energy [105, 106]. The other diverse potential applications of synthetic polymer based TENGs in the medical field are cardiac pacemakers powered by animal breathing [107], electrical stimulation for cell modulation [108], nerve stimulation [109], implantable drug delivery system [110], healthcare monitoring [111, 112] among many other appliances. TENGs have also become a major driver of the Internet of Things (IoT) as they are versatile and have practical applications in the paramount networks and interconnected machines and devices [113].

## 10.5   MATERIALS USED IN FABRICATION OF NGS

The three pillars that are important in the technology for the development of society are materials, energy, and information. Amongst the three, the high-performance materials play an important role for device formation and make energy harvest or storage possible in hybrid nanogenerators. Nanomaterials owing to their large specific surface area and nanostructures show great prospect in energy devices.

The material type and the direction of external applied force decide the efficiency of a piezo-electric generator. Generally, organic materials like PVDF (polyvinylidene fluoride) and inorganic materials such as PZT (lead zirconate titanate), ZnO, $BaTiO_3$ are important piezoelectric materials. Nanocomposites based on polymer–piezoelectric and organic–inorganic piezoelectric are also used as piezoelectric materials. Xue et al. [114] fabricated a self-charging power cell (SCPC) based on Li-ion battery and PENG. They used $\beta$ phase of organic piezoelectric PVDF film as a separator, which can generate strong piezoelectric property. By optimizing the structure of piezoelectric PVDF film, a hybrid PENG-based energy storage device is developed [115]. Wang et al. [116] prepared a flexible and transparent composite film based on PDVFNWs–PDMS (polyvinylidene fluoride nanowires-polydimethylsiloxane). This film shows excellent triboelectric, piezoelectric, and pyroelectric effects. A 4 × 4 sensing array system was developed using perovskite $BaTiO_3$ (BTO) nanoparticles for sensing the change of strain and temperature [117]. A Li-ion battery was designed by using piezoelectric PVDF–PZT nanocomposite as a piezo-separator with self-charging function [118]. The selection of material plays an important role for the fabrication of TENGs as energy storage devices. A hybrid TENG and solid Li-ion battery was fabricated by Liu et al. [119]. They used $TiO_2$ nanotubes as both triboelectric layer and anode of Li-ion battery and LATP (lithium aluminium titanium phosphate) nanoparticles as the solid electrolyte for mechanical energy harvest and storage. To overcome the disadvantage of high voltage and low current [120], high-performance TENGs based on Ag nanoparticles were developed [121]. They show high output performance because of excellent electric conductivity of Ag nanoparticles. A hybrid nanogenerator was designed by Quan et al. [122] consisting of six electromagnetics and a TENG. It converts wrist motion energy into electricity for driving a watch [122]. To obtain excellent flexibility, researchers have fabricated TENGs by using flexible composite or textile such as (polyvinyl butyral PVB) NWs and PDMS and designing other flexible structures.

Various factors such as stretchability, breathability, washability, and efficiency are the basic requisite for the development of wearable nanogenerators (tribo and piezo) devices. A hybrid generator in wearables is based on successive deposition of layers of active materials and electrodes excited by different sources (electromagnetic, heat, wind, pressure, or contact). A nanogenerator for wearables uses the combination of triboelectric and piezoelectric devices [8].

## 10.6   USE OF FERROFLUIDS IN NGS

Although smart and deformable electronics have gained remarkable attention but still inclusion of safety hazardous stretchable multifunctional sensors into wearables is a challenging job as per the personal safety and healthcare are concerned. Apart from many other risk factors, it includes a high level of noise or high value of magnetic field, which are risky for human health [123]. A TENG, based on triboelectrification and electrostatic induction is a promising type of sensor because of their fabrication simplicity, small size, low cost, and light weight. These devices limited sensing capabilities, durability, and deformable ability. To overcome this, a ferrofluid-based triboelectric nanogenerator (FO-TENG) that significantly enables multifunctional sensing ability with extremely confirmability is developed [123]. The FO-TENG function as a wearable multifunctional self-powered sensor by integrating with the human body in three different arrangements, i.e., textile-like, ring-like, and bracelet-like. The development of this sensor is a good effort to avoid hazards in wearable electronics and it serves as a self-powered safety sensor.

Although the energy efficiency of TENG has rapidly improved, certain issues like wear and tear against abrasion need to be addressed. This can be overcome if the solid-solid contact electrification is replaced by the liquid-solid contact electrification [124, 125]. In a ferrofluid-based triboelectric-electromagnetic (FFTEEM) generator [126], the TENG and the EMG components are packaged into a single-body structure, which are simultaneously activated when the ferrofluid comes under the effect of an external mechanical excitation. The proposed FFTEEM generator is composed of a hollow cylindrical tube, half filled with a ferrofluid (Figure 10.4). The inner sidewall is made of polytetrafluoroethylene (PTFE), which has the most negative order of electrification [127]. The outer sidewall is surrounded by two separate top and bottom electrodes, which consist of Al. A wire coil is wound around the outer wall for 200 turns. A 2230 Gauss permanent magnet (NdFeB) is positioned at the bottom end of the device to polarize the magnetic nanoparticles inside the ferrofluid. No magnet is placed at the top end in order to concentrate all of the ferrofluid in the bottom half in a neutral state.

Both the electrostatic properties of the water solvent, and the magnetic properties of the ferrofluid containing magnetic particles, are used independently for the TENG and the EMG components, respectively [126]. The hybrid design offers a complementary synergy in output performance as well as the practical benefit of mass-normalized output energy [128] with mechanical endurance. In addition, the packaged structure guarantees high immunity against environmental effects such as humidity and airborne debris [129]. FFTEEM has the unique advantage of providing amplitude and frequency performance capabilities. FFTEEM was fabricated to overcome the friction level faced

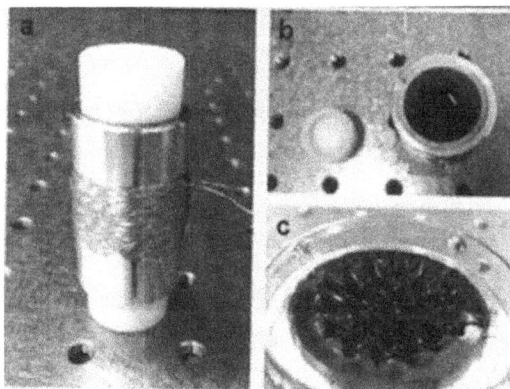

**FIGURE 10.4** Structure of the FF-TEEM generator. (a) An image of the outer body. (b) An image of inner body when the upper cover is open. (c) Ferrofluid pattern when the permanent magnet is positioned in the lower section. A stronger magnet was used to create this image. There is no clear ferrofluid pattern in the actual experimental device because it uses a lower intensity magnet. (Reprinted from Seol et al. [126] with permission from Elsevier under licence no. 5351371260574 dated 17 July 2022.)

in general solid-solid contact-based generators. But this FFTEEM generator can be operated using extremely weak vibrations having amplitude of 1 mm also thus making it an effective system for harvesting subtle and irregular vibrations also.

Ferrofluids are colloidal suspensions of magnetically soft, multidomain NPs (Fe nanoparticles) in a carrier liquid, which can respond immediately when an external magnetic field is applied showing a dramatic change from liquid to semisolid [130, 131]. The ferrofluid behavior is similar to metal particles in suspension when magnetic field is removed. These magnetic NPs are very well suspended and over the time maintain the same shape as long as a magnetic field is present. This smart ferrofluid enables the emergence of smart functionalities such as multimodal energy harvesting, and multifunctional sensing capabilities [123].

A flow nanogenerator was developed by using a mixture of magnetic nanofluid (MNF) and bubbles in a fluid circulating system, and all the phenomena related to the power generation properties of the nanogenerator have been explored. This system could be used as a device to reproduce as energy from the wasted heat discarded in the surroundings from the industries and proved as a promising energy reproduction technology [132]. The proposed technique works on the principle to obtain induced electromotive force (EMF) based on Faraday's law due to the flow of MNF in a closed-circulating system. The magnetic nanoparticles (MNPs) should pass through the induction coil with a perpendicular magnetization direction in accordance with Faraday's law to get maximum induced EMF. The magnetization direction of the MNPs can be controlled by employing a permanent magnet to produce an external magnetic field considering both the Brownian and Néel motions. A circulation system was implemented ensuring the flow of the MNF in the closed cycle to get a continuous induced voltage. Further, power generation properties were investigated considering electric, magnetic, and fluidic effects.

A portable electric power generator with improved output has been developed using ferrofluid by Pant et al. Patent No. **1364/DEL/2008**. The novel generator consists of a non-obvious combination of a plurality of permanent magnets, ferrofluid bearings, and solenoid coils, which is capable of converting with increased efficiency the prime mover input of movement to generate improved quantum of electric power. The plurality of permanent magnets being configured in combination with the said ferrofluid bearings at the top and bottom poles of the said permanent magnets in such a manner that the adjacent magnets have N-S polarity and are placed in a vertical fashion and the said combination of a plurality of permanent magnets and ferrofluid bearings being provided with plurality of separate solenoid coils on both the upper and lower ends. An alternating current has been produced and rectified by using bridge circuit for providing continuous power source. The device of the present invention is able to start functioning at very low wind speeds of even less than 2 m/s as compared to normally available wind turbines, which starts working at nearly 4 m/s wind motion. Further, the portable ferrofluid-based electric power generator of the present invention is capable of generating minimum of 1.2 Watt electric power as driven by wind energy with the efficiency of 24% at wind speed of 4 m/s and 410 mW electric power as driven by mechanical motion using 6 to 7 Hz frequency. Adding more similar modules/devices can increase the capacity of this power generator.

In other reported works, US Patent No. 6812598 Jeffrey et al. have used the wave motion of sea or wind as a source of input power but their method of using these sources is a very crude way as they are moving the whole structure. Inventors in Patent No. US 6,809,427 B2 have generated 0.4 Watt power by shaking of the device whereas others have not mentioned the quantum of power generation in their invention. In US Patent 7,095,143 the inventor has utilized the motion of nanoparticles of ferrofluid by convection method but the power generated by this technique is very small as the magnetic flux density of ferrofluid particles is very small as compared to magnets. In US Patent No. 7,095,143 the inventor uses only ferrofluid motion within the coils for generation of electric power but the magnetization of ferrofluid particles is very small as compared to permanent magnets of rare earth magnets so, power generated by them should be very small, the same has not been mentioned in the patent specifications.

An electromagnetic wearable 3-degree-of-freedom (3-DoF) resonance human body motion energy harvester using ferrofluid as a lubricant have been investigated [133]. It consists of a permanent magnet supported by two elastic strings within a rectangular box, which form a 3-DoF vibrator. The Copper windings are attached to the outer surface of the box to generate electrical energy when the permanent magnet is forced to vibrate. Ferrofluid is used to minimize the frictional losses and kept in such a way that the poles of permanent magnet are cushioned by the ferrofluid, to the effect that the permanent magnet will not touch the inner of the box. Simulation results show that the ferrofluid can keep the permanent magnet "contactless'" from the box even subject to 10 times gravity acceleration. A prototype is built and tested under different loading conditions. Resistance load experimental results indicate the proposed harvester can generate 1.1 mW in walking condition and 2.28 mW in running condition. An energy storage circuit is employed and the energy storage experimental results show that the average storage power during walking and running conditions are 0.014 and 0.149 mW, respectively. It is shown that the developed harvester can be readily attached on a shoe to offer continuous power supply for wearable sensors and devices. The main breakthrough out of this work was that the ferrofluid improves 40% harvesting efficiency by reducing friction.

## 10.7   FUTURE OF FERROFLUID-BASED NGS

As explained above, a stretchable, multimodal, ferrofluid-based triboelectric nanogenerator (FO-TENG) is developed, which has multifunctional sensing ability. There are other ferrofluid-based nanogenerators also which can be used to harvest energy in one form or other. Despite many exciting and compelling achievements, there are several important issues to be addressed to make ferrofluids more applicable in thrust areas. Future endeavors could be focused on the applications of ferrofluid-based nanogenerators, FO-TENG, which could be reconfigured and becomes adaptive providing innovative solution to many future applications such as space exploration in extreme or otherwise challenging environments, compliant wearable devices, and even in the medical field for *in vivo* applications. We can also move further in fabricating the simple and flexible devices and their combination with other platforms, which can offer a generic platform broadly applicable to a multitude of intelligent wearable electronics.

On the other side, future work could include the development of more such devices for better sensor performance by including signal processing unit [134], artificial intelligence [135], and wireless technology [136]. All this can be achieved by knowing more about the ferrofluid topography and manipulation [137], optimizing the structure of device to be more skin-like [138]. By knowing this, the sensor information can be interpreted, identified thus enabling fully automatic hazard monitoring system towards internet of prevention and wearable hazard avoidance electronics.

## CONFLICT OF INTEREST

The authors declare no conflict of interest.

## REFERENCES

[1]   Wang, Z. L., Song, J. H. Piezoelectric nanogenerators based on zinc oxide nanowire arrays *Science*, 2006, *312(5771)*, 242–246.

[2]   Wang, Z. L., Chen, J., Lin, L. Progress in triboelectric nanogenerators as a new energy technology and self-powered sensors, *Energy Environ. Sci.,* 2015, *8*, 2250.

[3]   Wang, Z., Yu, R., Pan, C., Li, Z., Yang, J., Yi, F., Wang, Z. L. Light-induced pyroelectric effect as an effective approach for ultrafast ultraviolet nanosensing, *Nat. Commun.,* 2015, *6*, 8401.

[4]   Bell, L. E. Cooling, heating, generating power, and recovering waste heat with thermoelectric systems, *Science,* 2008, *321*, 1457.

[5] Spanier, J. E., Fridkin, V. M., Rappe, A. M., Akbashev, A. R., Polemi, A., Qi, Y., Gu, Z., Young, S. M., Hawley, C. J., Imbrenda, D., Xiao, G., Bennett-Jackson, A. L., Johnson, C. L. Power conversion efficiency exceeding the Shockley-Queisser limit in a ferroelectric insulator, *Nat. Photonics,* 2016, *10,* 611.

[6] Mohanty, A., Parida, S., Behera, R.K., Roy, T. Vibration energy harvesting: A review, *J. Adv. Dielectrics,* 2019, *9,* 1930001.

[7] Zhu, G., Peng, B., Chen, J., Jing, Q., Wang, Z. L. Triboelectric nanogenerators as a new energy technology: From fundamentals, devices, to applications. *Nano Energy,* 2014, *14,* 126–138.

[8] Zil, Y., and Wang, Z. L. Nanogenerators: An emerging technology towards nano energy, *APL Mater.,* 2017, *5,* 074103.

[9] Zi, Y., Guo, H., Wen, Z., Yeh, M.-H., Hu, C., Wang, Z. L. Harvesting low-frequency (<5 Hz) irregular mechanical energy: A possible killer application of triboelectric nanogenerator, *ACS Nano,* 2016, *10*(4), 4797.

[10] Fan, F.-R., Tian, Z.-Q., Wang, Z. L. Flexible triboelectric generator, *Nano Energy,* 2012, *1*(2), 328.

[11] Yang, Y., Guo, W., Pradel, K. C., Zhu, G., Zhou, Y., Zhang, Y., Hu, Y., Lin, L., Wang, Z. L. Pyroelectric nanogenerators for harvesting thermoelectric energy, *Nano Lett.,* 2012, *12*(6), 2833.

[12] Yang, Y., Pradel, K. C., Jing, Q., Wu, J. M., Zhang, F., Zhou, Y., Zhang, Y., Wang, Z. L. Thermoelectric nanogenerators based on single Sb-doped ZnO micro/nanobelts, *ACS Nano,* 2012, *6*(8), 6984.

[13] Zhang, R., Wang, S., Yeh, M.-H., Pan, C., Lin, L., Yu, R., Zhang, Y., Zheng, L., Jiao, Z., Wang, Z. L. A streaming potential/current-based microfluidic direct current generator for self-powered nanosystems, *Adv. Mater.,* 2015, *27*(41), 6482.

[14] Boisseau, S., Despesse, G., Ahmed, B. Chapter 5: Electrostatic conversion for vibration energy harvesting. Intechopen Book: *Small-scale energy harvesting,* Edited by M. Lallart, London, England, 2012, 91–134.

[15] Yang, B., Lee, C., Xiang, W., Xie, J., Han He, J., Kotlanka, R. K., Low, S. P., Feng, H). Electromagnetic energy harvesting from vibrations of multiple frequencies. *J. Micromechan. Microeng.,* 2009, *19*(3), 35001.

[16] Narita, F., Fox, M. A review on piezoelectric, magnetostrictive, and magnetoelectric materials and device technologies for energy harvesting applications. *Adv. Eng. Mater.,* 2018, *20*(5), 1700743.

[17] Tian, J., Chen, X., Wang, Z. L. Environmental energy harvesting based on triboelectric nanogenerators. *Nanotechnology,* 2020, *31*(24), 242001.

[18] Nguyen, V., Zhu, R., Yang, R. Nano energy, *Environ. Eff. Nanogener.,* 2015, *14,* 49–61.

[19] Yang, Y., Wang, Z. L. Hybrid energy cells for simultaneously harvesting multi-types of energies, *Nano Energy,* 2015, *14,* 245–256.

[20] Weinstein, L. A., McEnaney, K., Strobach, E., Yang, S., Bhatia, B., Zhao, L., Huang, Y., Loomis, J., Cao, F., Boriskina, S. V., Ren, Z., Wang, E. N., Chen, G. A hybrid electric and thermal solar receiver *Joule,* 2018, *2,* 962.

[21] Yang, X., Daoud, W. A. Synergetic effects in composite-based flexible hybrid mechanical energy harvesting generator, *J. Mater. Chem. A,* 2017, *5,* 9113.

[22] Lee, J.-H., Kim, J., Kim, T. Y., Al Hossain, M. S., Kim, S.-W., Kim, J. H. All-in-one energy harvesting and storage devices, *J. Mater. Chem. A,* 2016, *4,* 7983.

[23] Yang, Y., Zhang, H., Zhu, G., Lee, S., Lin, Z.-H., Wang, Z. L. Flexible hybrid energy cell for simultaneously harvesting thermal, mechanical, and solar energies, *ACS Nano,* 2013, *7,* 785.

[24] Chen, X., Ren, Z., Han, M., Wan, J., and Zhang, H. Hybrid energy cells based on triboelectric nanogenerator: From principle to system. *Nano Energy,* 2020, *75,* 104980.

[25] Zhang, T., Yang, T., Zhang, M., Bowen, Chris R., Yang, Ya. Recent progress in hybridized nanogenerators for energy scavenging, *iScience,* 2020, *23,* 101689.

[26] Raj, K. "Ferrofluids: Applications," *Encyclopedia of Materials: Science and Technology,* Oxford: Elsevier Science Ltd, 2003, pp. 3083–3087.

[27] Oliveira, H. P. D. Wearable nanogenerators: working principle and self-powered biosensors applications. *Electrochem.* 2021, *2,* 118–134.

[28] You, M. H., Wang, X. X., Yan, X., Zhang, J., Song, W. Z., Yu, M., Fan, Z. Y., Ramakrishna, S., Long, Y. Z. A self-powered flexible hybrid piezoelectric-pyroelectric nanogenerator based on non-woven nanofiber membranes. *J. Mater. Chem. A,* 2018, *6,* 3500–3509.

[29] Gao, L. X., Hu, D. L., Qi, M. K., Gong, J., Zhou, H., Chen, X., Chen, J. F., Cai, J., Wu, L. K., Hu, N., et al. A double-helix-structured triboelectric nanogenerator enhanced with positive charge traps for self-powered temperature sensing and smart-home control systems. *Nanoscale,* 2018, *10,* 19781–19790.

[30] Liu, X., Zhao, K., and Yang, Y. (Effective polarization of ferroelectric materials by using a triboelectric nanogenerator to scavenge wind energy. *Nano Energy,* 2018, *53,* 622–629.

[31] Ghomian, T., Mehraeen, S. Survey of energy scavenging for wearable and implantable devices. *Energy,* 2019, *178,* 33–49.

[32] Wang, Z. L. Triboelectric nanogenerators as new energy technology and self powered sensors – Principles, problems and perspectives. *Faraday Discuss.* 2014, *176,* 447–458.

[33] Harper, W. R. The generation of static charge. *Adv. Phys.,* 1957, *6*(24), 365–417.

[34] Fernando, G. T., Gabriel, E. De-la-Torre, Polysaccharide-based triboelectric nanogenerators: A review, *Carbohyd. Polym.,* 2021, *251,* 117055.

[35] Wu, C., Wang, A. C., Ding, W., Guo, H., Wang, Z. L. Triboelectric nanogenerator: A foundation of the energy for the new era, *Adv. Energy Mater.,* 2019, *9,* 1802906.

[36] Ramírez, J. M., Gatti, C. D., Machado, S. P., Febbo, M. An experimentally validated finite element formulation for modelling 3D rotational energy harvesters, *Eng. Struct.,* 2017, *153,* 136.

[37] Fang, L. H., Hassan, S. I. S., Rahim, R. B. A., Malek, M. F. A. A study of vibration energy harvester, *ARPN J. Eng. Appl. Sci.,* 2016, *11,* 5028.

[38] Srikanth, K. V. State of art: Piezoelectric vibration energy harvesters, *Mater. Today Proc.,* 2017, *4,* 1091.

[39] Lin, Y.-F., Song, J., Ding, Y., Lu, S.-Y., Wang, Z.L. Piezoelectric nanogenerator using CdS nanowires, *Appl. Phys. Lett.,* 2008, *92,*022105.

[40] Lu, M.-Y., Song, J., Lu, M.-P., Lee, C.-Y., Chen, L.-J., Wang, Z.L. ZnO– ZnS heterojunction and ZnS nanowire arrays for electricity generation, *ACS Nano,* 2009, *3*(2), 357–362.

[41] Huang, C.-T., Song, J., Lee, W.-F., Ding, Y., Gao, Z., Hao, Y., Chen, L.-J., Wang, Z. L. GaN nanowire arrays for high-output nanogenerators, *J. Am. Chem. Soc.,* 2010, *132,* 4766–4771.

[42] Chang, C., Tran, V.H., Wang, J., Fuh, Y.-K., Lin, L. Direct-write piezoelectric polymeric nanogenerator with high energy conversion efficiency, *Nano Lett.,* 2010, *10,* 726–731.

[43] Chen, X., Xu, S., Yao, N., Shi, Y. 1.6 V nanogenerator for mechanical energy harvesting using PZT nanofibers, *Nano Lett.,* 2010, *10,* 2133–2137.

[44] Park, K.-I., Xu, S., Liu, Y., Hwang, G.-T., Kang, S.-J.L., Wang, Z.L., Lee, K. J. Piezoelectric BaTiO$_3$ thin film nanogenerator on plastic substrates, *Nano Lett.,* 2010, *10,*4939–4943.

[45] Jung, J. H., Lee, M., Hong, J.-I., Ding, Y., Chen, C.-Y., Chou, L.-J., Wang, Z. L. Lead-free NaNbO$_3$ nanowires for a high output piezoelectric nanogenerator, *ACS Nano,* 2011, *5,*10041–10046.

[46] Lee, T. I., Lee, S., Lee, E., Sohn, S., Lee, Y., Lee, S., Moon, G., Kim, D., Kim, Y. S., Myoung, J. M., Wang, Z. L. High-power density piezoelectric energy harvesting using radially strained ultrathin trigonal tellurium nanowire assembly, *Adv. Mater.,* 2013, *25,* 2920–2925.

[47] Chen, J., Yang, J., Li, Z. et al. Networks of triboelectric nanogenerators for harvesting water wave energy: A potential approach toward blue energy. *ACS Nano,* 2015, 9, 3324.

[48] Wang, N., Liu, Y., Ye, E., Li, Z., Wang, D. Control methods and applications of interface contact electrification of triboelectric nanogenerators: A review, *Mater. Res. Lett.,* 2022, *10*(3), 97–123,

[49] Wang, W., Yu, A., Liu, X., et al. Large-scale fabrication of robust textile triboelectric nanogenerators. *Nano Energy.* 2020, *71,*104605.

[50] Deng, W., Zhou, Y., Zhao, X., et al. Ternary electrification layered architecture for high-performance triboelectric nanogenerators. *ACS Nano.* 2020, *14,* 9050–9058.

[51] Liu, Z., Li, H., Shi, B., et al. Wearable and implantable triboelectric nanogenerators. *Adv Funct Mater.,* 2019, 29, 180 8820.

[52] Zou, Y., Raveendran, V., Chen, J. Wearable triboelectric nanogenerators for biomechanical energy harvesting. *Nano Energy.* 2020, 77, 105303.

[53] Meng, K., Zhao, S., Zhou, Y., Wu, Y., Zhang, S., He, Q., Wang, X., Zhou, Z., Fan, W., Tan, X., Yang, J., Chen, A wireless textile-based sensor system for self-powered personalized health care, *J. Matter* 2020, *2,* 896–907.

[54] Tat, T., Libanori, A., Au, C., Yau, A., Chen, J. Advances in triboelectric nanogenerators for biomedical sensing, *Biosens. Bioelectron.* 2021, *171*, 112714.

[55] Deng, W., Zhou, Y., Zhao, X., Zhang, S., Zou, Y., Xu, J., Yeh, M.-H., Guo, H., Chen, J. Ternary electrification layered architecture for high-performance triboelectric nanogenerators, *ACS Nano* 2020, *14*, 9050–9058.

[56] Jin, L., Xiao, X., Deng, W., Nashalian, A., He, D., Raveendran, V., Yan, C., Su, H., Chu, X., Yang, T., Li, W., Yang, W., Chen, J. Manipulating relative permittivity for high-performance wearable triboelectric nanogenerators, *Nano Lett.* 2020, *20*, 6404–6411.

[57] Zou, Y., Raveendran, V., Chen, J. Wearable triboelectric nanogenerators for biomechanical energy harvesting, *Nano Energy* 2020, *77*, 105303.

[58] Wang, Z. L. On Maxwell's displacement current for energy and sensors: the origin of nanogenerators, *Mater. Today* 2017, *20*, 74–82.

[59] Zhang, B., Tang, Y., Dai, R., Wang, H., Sun, X., Qin, C., Pan, Z., Liang, E., Mao, Y. Breath-based human–machine interaction system using triboelectric nanogenerator, *Nano Energy* 2019, *64*, 103953.

[60] Zhu, G., Zhou, Y. S., Bai, P., Meng, X. S., Jing, Q., Chen, J., Wang, Z. L. A shape-adaptive thin-film-based approach for 50% high-efficiency energy generation through micro-grating sliding electrification, *Adv. Mater. (Weinheim, Ger.)* 2014, *26*, 3788–3796.

[61] Wang, M., Zhang, N., Tang, Y., Zhang, H., Ning, C., Tian, L., Li, W., Zhang, J., Mao, Y., Liang, E. Single-electrode triboelectric nanogenerators based on sponge-like porous PTFE thin films for mechanical energy harvesting and self-powered electronics, *J. Mater. Chem. A* 2017, *5*, 12252–12257.

[62] Wang, S., Xie, Y., Niu, S., Lin, L., Wang, Z. L. Freestanding triboelectric-layer-based nanogenerators for harvesting energy from a moving object or human motion in contact and non-contact modes, *Adv. Mater. (Weinheim, Ger.)* 2014, *26*, 2818–2824.

[63] Priya, S., Inman, D. J. *Energy Harvesting Technologies* (Springer, New York, 2009).

[64] Matiko, J. W., Grabham N. J., Beeby S. P., Tudor, M. J. Review of the application of energy harvesting in buildings, *Meas. Sci. Technol.*, 2014, 25, 012002.

[65] Najafi, K., Galchev T., Aktakka, E. E. Peterson, R. L., McCullagh, J. Microsystems for energy harvesting, *2011 16th Int. Solid-State Sensors, Actuators and Microsystems Conf. TRANSDUCERS'11* (IEEE, New York, 2011), pp. 1845–1850.

[66] William, C. B. et al., Development of an electromagnetic microgenerator, *IEE Proc. Circuits Dev. Syst.*, 2001, *148*, 337.

[67] Shearwood, C., Yates, R. B. Development of an electromagnetic micro-generator, *Electron. Lett.*, 1997, *33*, 1883.

[68] Ooi, B. L., Gilbert, J. M. Design of wideband vibration-based electromagnetic generator by means of dual-resonator, *Sens. Actuators A Phys.*, 2014, *213*, 9.

[69] Lee, E. W. Magnetostriction and magnetomechanical effects, *Rep. Prog. Phys.*, 1955, *18*, 305.

[70] Deng, Z., Dapino, M. J. Review of magnetostrictive vibration energy harvesters, *Smart Mater. Struct.*, 2017, *26*, 103001.

[71] Kholmetska, I., Chleboun, J., Krejčí, P. Numerical modeling of Galfenol magnetostrictive response, *Appl. Math. Comput.*, 2018, *319*, 527.

[72] Ajroudi L. et al., Magnetic, electric and thermal properties of cobalt ferrite nanoparticles, *Mater. Res. Bull.*, 2014, *59*, 49.

[73] Yan, B., Zhang, C., Li, L. Magnetostrictive energy generator for harvesting the rotation of human knee joint, *AIP Adv.*, 2018, *8*, 056730.

[74] Yang, Y., Pradel, K. C., Jing, Q., Wu, J. M., Zhang, F., Zhou, Y., Zhang, Y., Wang, Z. L. Thermoelectric nanogenerators based on single Sb-doped ZnO micro/nanobelts, *ACSNano.*, 2012, *6* (8), 6984–6989.

[75] Zook, J. D., Liu, S. T. (1978). Pyroelectric effects in thin film, *J. Appl. Phys.*, 1978, *49*(8), 4604.

[76] Yang, Y., Guo, W., Pradel, K. C., Zhu, G., Zhou, Y., Zhang, Y., Hu, Y., Lin, L., Wang, Z. L. Pyroelectric nanogenerators for harvesting thermoelectric energy, *Nano Lett.*, 2012, *12* (6), 2833–2838.

[77] Androja, N., Mehta, S. B., Shah, P. J. Emerging technology *Innovative Res.*, 2015, *2*(3), 847–850.

[78] Yang, Y., Guo, W., Pradel, K. C., Zhu, G., Zhou, Y., Zhang, Y., Hu, Y., Lin, L., Wang, Z. L. Pyroelectric nanogenerators for harvesting thermoelectric energy, *Nano Lett.*, 2012, *12*(6), 2833.

[79] Yang, Y., Jung, J. H., Yun, B. K., Zhang, F., Pradel, K. C., Guo, W., Wang, Z. L. Flexible pyroelectric nanogenerators using a composite structure of lead-free $KNbO_3$ nanowires, *Adv. Mater.*, 2012, *24*(39), 5357.

[80] Zhang, D., Wang, Y., Yang, Y. Design, Performance, and application of thermoelectric nanogenerators, *Small*, 2019, *15*, 1805241.

[81] Weinstein, L. A., McEnaney, K., Strobach, E., Yang, S., Bhatia, B., Zhao, L., Huang, Y., Loomis, J., Cao, F., Boriskina, S. V., Ren, Z., Wang, E. N., Chen, G. *Joule*, 2018, *2*, 962.

[82] Zhang, K. W., Wang, S. H., and Yang, Y. A one-structure-based piezo-tribo-pyrophotoelectric effects coupled nanogenerator for simultaneously scavenging mechanical, thermal, and solar energies. *Adv. Energy Mater.*, 2017, *7*, 1601852.

[83] Zhang, K. W., Wang, Y. O. H., and Yang, Y. Structure design and performance of hybridized nanogenerators. *Adv. Funct. Mater.*, 2019, *29*, 1806435.

[84] Lee, S., Hinchet, R., Lee, Y., Yang, Y., Lin, Z. H., Ardila, G., Montes, L., Mouis, M., Wang, Z. L. Ultrathin nanogenerators as selfpowered/active skin sensors for tracking eye ball motion. *Adv. Funct. Mater.*, 2014, *24*, 1163–1168.

[85] Zhang, H. L., Yang, Y., Su, Y. J., Chen, J., Adams, K., Lee, S., Hu, C. G., Wang, Z. L. Triboelectric nanogenerator for harvesting vibration energy in full space and as selfpowered acceleration sensor. *Adv. Funct. Mater.*, 2014, *24*, 1401–1407.

[86] Gao, T. T., Zhao, K., Liu, X., and Yang, Y. Implanting a solid Li-ion battery into a triboelectric nanogenerator for simultaneously scavenging and storing wind energy. *Nano Energy*, 2017, *41*, 210–216.

[87] Zhao, K., Yang, Y., Liu, X., Wang, Z. L. Triboelectrification-enabled self-charging lithium-ion batteries. *Adv. Energy Mater.*, 2017, *7*, 1700103.

[88] Zhao, K., Qin, Q. Q., Wang, H. F., Yang, Y., Yan, J. A., Jiang, X. M. Antibacterial triboelectric membrane-based highly-efficient self-charging supercapacitors. *Nano Energy*, 2017, *36*, 30–37.

[89] Zhu, Q., Dong, L., Zhang, J., Xu, K., Zhang, Y., Shi, H., Lu, H., Wu, Y., Zheng, H., Wang, Z. All-in-one hybrid tribo/piezoelectric nanogenerator with the point contact and its adjustable charge transfer by ferroelectric polarization. *Ceram. Int.* 2020, *46*, 28277–28284.

[90] Guo, Y., Zhang, X. S., Wang, Y., Gong, W., Zhang, Q., Wang, H., Brugger, J. All-fiber hybrid piezoelectric-enhanced triboelectric nanogenerator for wearable gesture monitoring. *Nano Energy*, 2018, *48*, 152–160.

[91] Lee, D. W., Jeong, D. G., Kim, J. H., Kim, H. S., Murillo, G., Lee, G. H., Song, H. C., Jung, J. H. Polarization-controlled PVDF-based hybrid nanogenerator for an effective vibrational energy harvesting from human foot. *Nano Energy*, 2020, *76*, 105066.

[92] Wang, W., Zhang, J., Zhang, Y., Chen, F., Wang, H., Wu, M., Li, H., Zhu, Q., Zheng, H., Zhang, R. Remarkably enhanced hybrid piezo/triboelectric nanogenerator via rational modulation of piezoelectric and dielectric properties for self-powered electronics. *Appl. Phys. Lett.*, 2020, *116*, 023901.

[93] Jo, S., Kim, I., Byun, J., Jayababu, N., Kim, D. Boosting a power performance of a hybrid nanogenerator via frictional heat combining a triboelectricity and thermoelectricity toward advanced smart sensors. *Adv. Mater. Technol.*, 2020, *6*, 2000752.

[94] Jung, S., Oh, J., Yang, U. J., Lee, S. M., Lee, J., Jeong, M., Cho, Y., Kim, S., Baik, J. M., Yang, C. 3D Cu ball-based hybrid triboelectric nanogenerator with non-fullerene organic photovoltaic cells for self-powering indoor electronics. *Nano Energy*, 2020, *77*, 105271.

[95] Liu, X., Li, J., Fang, Z., Wang, C., Shu, L., Han, J. Ultraviolet-protecting, flexible and stable photovoltaic-assisted piezoelectric hybrid unit nanogenerator for simultaneously harvesting ultraviolet light and mechanical energies. *J. Mater. Sci.*, 2020, *55*, 15222–15237.

[96] Zhong, Y., Zhao, H., Guo, Y., Rui, P., Shi, S., Zhang, W., Liao, Y., Wang, P., Wang, Z. L. An easily assembled electromagnetic- triboelectric hybrid nanogenerator driven by magnetic coupling for fluid energy harvesting and self-powered flow monitoring in a smart home/city. *Adv. Mater. Technol.*, 2019, *4*, 1900741(9 pp.).

[97] He, J., Wen, T., Qian, S., Zhang, Z., Tian, Z., Zhu, J., Mu, J., Hou, X., Geng, W., Cho, J. et al. Triboelectric-piezoelectric-electromagnetic hybrid nanogenerator for high-efficient vibration energy harvesting and self-poweredwireless monitoring system. *Nano Energy*, 2018, *43*, 326–339.

[98] Ahmed, R., Kim, Y., Zeeshan, Chun, W. Development of a tree-shaped hybrid nanogenerator using flexible sheets of photovoltaic and piezoelectric films. *Energies*, 2019, *12*, 229.

[99] Li, Z., Zheng, Q., Wang, Z. L., Li, Z. Nanogenerator-based self-powered sensors for wearable and implantable electronics. *Research,* 2020, *2020,* 1–25.

[100] Selvarajan, S., Alluri, N. R., Chandrasekhar, A., Kim, S. J. Unconventional active biosensor made of piezoelectric batio$_3$ nanoparticles for biomolecule detection. *Sens. Actuators B Chem.,* 2017, *253,* 1180–1187.

[101] Mao, Y., Yue, W., Zhao, T., Shen, M. L., Liu, B., Chen, S. A self-powered biosensor for monitoring maximal lactate steady state in sport training. *Biosensors,* 2020, *10,* 75.

[102] Liu, Z., Zhang, S., Jin, Y. M., Ouyang, H., Zou, Y., Wang, X. X., Xie, L. X., Li, Z. Flexible piezoelectric nanogenerator in wearable self-powered active sensor for respiration and healthcare monitoring. *Semicond. Sci. Technol.,* 2017, *32,* 064004.

[103] Hou, T. C., Yang, Y., Zhang, H., Chen, J., Chen, L. J., Wang, Z. L. (2013). Triboelectric nanogenerator built inside shoe insole for harvesting walking energy. *Nano Energy,* 2013, *2*(5), 856–862.

[104] Huang, T., Wang, C., Yu, H., Wang, H., Zhang, Q., Zhu, M. Human walking-driven wearable all-fiber triboelectric nanogenerator containing electrospun polyvinylidene fluoride piezoelectric nanofibers. *Nano Energy,* 2014, *14,* 226–235.

[105] Kim, K. N., Chun, J., Kim, J. W., Lee, K. Y., Park, J. U., Kim, S. W., Wang, Z.L., Baik, J. M. Highly stretchable 2D fabrics for wearable triboelectric nanogenerator under harsh environments. *ACS Nano,* 2015, *9*(6), 6394–6400.

[106] Zhou, T., Zhang, C., Han, C. B., Fan, F. R., Tang, W., Wang, Z. L.. Woven structured triboelectric nanogenerator for wearable devices. *ACS Appl. Mater. Interfaces,* 2014, *6*(16), 14695–14701.

[107] Zheng, Q., Shi, B., Fan, F., Wang, X., Yan, L., Yuan, W., Wang, S., Liu, H., Li, Z., Wang, Z. L. In vivo powering of pacemaker by breathing-driven implanted triboelectric nanogenerator. *Adv. Mater.,* 2014, *26*(33), 5851–5856.

[108] Tian, J., Shi, R., Liu, Z., Ouyang, H., Yu, M., Zhao, C., Zou, Y., Jiang, D., Zhang, J., Li, Z. Self-powered implantable electrical stimulator for osteoblasts' proliferation and differentiation. *Nano Energy,* 2019, *59,* 705–714.

[109] Lee, S., Wang, H., Shi, Q., Dhakar, L., Wang, J., Thakor, N. V., Yen, S.C., Lee, C. Development of battery-free neural interface and modulated control of tibialis anterior muscle via common peroneal nerve based on triboelectric nanogenerators (TENGs). *Nano Energy,* 2017, *33,* 1–11.

[110] Song, P., Kuang, S., Panwar, N., Yang, G., Tng, D. J. H., Tjin, S. C., Ng, W. J., Abdul Majid, M. B., Zhu, G., Yong, K. T., Wang, Z. L. A self-powered implantable drug-delivery system using biokinetic energy. *Adv. Mater.,* 2017, *29*(11), 1605668.

[111] Bai, P., Zhu, G., Jing, Q., Yang, J., Chen, J., Su, Y., Ma, J., Zhang, G., Wang, Z. L. (2014). Membrane-based self-powered triboelectric sensors for pressure change detection and its uses in security surveillance and healthcare monitoring. *Adv. Funct. Mater.,* 2014, *24* (37), 5807–5813.

[112] Zhao, Z., Yan, C., Liu, Z., Fu, X., Peng, L. M., Hu, Y., Zheng, Z.. Machine-washable textile triboelectric nanogenerators for effective human respiratory monitoring through loom weaving of metallic yarns. *Adv. Mater.,* 2016, *28*(46), 10267–10274.

[113] Lee, I., Lee, K. (2015). The Internet of Things (IoT): Applications, investments, and challenges for enterprises. *Bus. Horizons,* 2015, *58*(4), 431–440.

[114] Xue, X., Wang, S., Guo, W. et al. (2012). Hybridizing energy conversion and storage in a mechanical-to-electrochemical process for self-charging power cell. *Nano Lett.,* 2012, 12, 5048.

[115] Xue, X., Deng, P., Yuan, S. et al. (2013). CuO/PVDF nanocomposite anode for a piezo-driven self-charging lithium battery. *Energ. Environ. Sci.,* 2013, *6*: 2615.

[116] Wang, S., Wang, Z.L., Yang, Y. A one-structure-based hybridized nanogenerator for scavenging mechanical and thermal energies by triboelectric–piezoelectric–pyroelectric effects. *Adv. Mater.,* 2016, *28,* 2881.

[117] Song, K., Zhao, R., Wang, Z. L., Yang, Y. Conjucted pyro-piezoelectric effect for self-powered simultaneous temperature and pressure sensing. *Adv. Mater.,* 2019, *31,* e1902831.

[118] Zhang, Y., Zhang, Y., Xue, X. et al. (2014). PVDF-PZT nanocomposite film based self-charging power cell. *Nanotechnology,* 2014, *25,* 105401.

[119] Liu, X., Zhao, K., Wang, Z. L., Yang, Y. (2017). Unity convoluted design of solid li-ion battery and triboelectric nanogenerator for self-powered wearable electronics. *Adv. Energy Mater.,* 2017, *7,* 1701629.

[120] Cao, X., Jie, Y., Wang, N., Wang, Z. L. Triboelectric nanogenerators driven self-powered electrochemical processes for energy and environmental science. *Adv. Energy Mater.*, 2016, *6*, 1600665.

[121] Jiang, Q., Chen, B., Zhang, K., Yang, Y Ag nanoparticle-based triboelectric nanogenerator to scavenge wind energy for a self-charging power unit. *ACS Appl. Mater. Inter.*, 2017, *9*, 43716.

[122] Quan, T., Wang, X., Wang, Z.L., Yang, Y. Hybridized electromagnetic-triboelectric nanogenerator for a self-powered electronic watch. *ACS Nano*, 2015, *9*, 12301.

[123] Ahmed, A., Hassan, I., Mosa, I. M., Elsanadidy, E., Sharafeldin, M., Rusling, J. F., Ren, S. An ultra-shapeable, smart sensing platform based on a multimodal ferrofluid-infused surface, *Adv. Mater.* 2019, *31*, 1807201.

[124] Lin, Z. H., Cheng, G., Lin, L., Lee, S., Wang, Z. L. Water-solid surface contact electrification and its use for harvesting liquid-wave energy, *Angew. Chemie–Int. Ed.*, 2013, *52*, 12545–12549.

[125] Jeon, S.-B., Kim, D., Seol, M.-L., Park, S.-J., Choi, Y.-K. 3-dimensional broadband energy harvester based on internal hydrodynamic oscillation with a package structure, *Nano Energy,* 2015, *17*, 82–90.

[126] Seol, M.-Lok., Jeon, S.-Bae, Han, J.-Woo, Choi, Y.-Kyu, Ferrofluid-based triboelectric-electromagnetic hybrid generator for sensitive and sustainable vibration energy harvesting, *Nano Energy*, 2017, *31*, 233–238.

[127] Zi, Y., Niu, S., Wang, J., Wen, Z., Tang, W., Wang, Z. L. Standards and figure-of-merits for quantifying the performance of triboelectric nanogenerators, *Nat. Commun.*, 2015, *6*, 8376.

[128] Seol, M.-L., Han, J.-W., Park, S.-J., Jeon, S.-B., Choi, Y.-K. Hybrid energy harvester with simultaneous triboelectric and electromagnetic generation from an embedded floating oscillator in a single package, *Nano Energy,* 2016, *23*, 50–59.

[129] Seol, M., Han, J., Jeon, S., Meyyappam. M., Choi, Y. Floating oscillator-embedded triboelectric generator for versatile mechanical energy harvesting, *Sci. Rep.*, 2015, *5*, 16409.

[130] Genc, S., Derin, B. Synthesis and rheology of ferrofluids: a review, *Curr. Opin. Chem. Eng.* 2014, *3*, 118.

[131] Chiolerio, A., Quadrelli, M. B. Smart fluid systems: the advent of autonomous liquid robotics, *Adv. Sci.* 2017, *4*, 1700036.

[132] Kim, S.-H., Park, J.-H., Choi, H.-S., Lee, S.-H. Power generation properties of flow nanogenerator with mixture of magnetic nanofluid and bubbles in circulating system, *IEEE Trans. Magnet.,* 2017, *53*(11), 4600904

[133] Wu, S., Luk, P.C.K., Li, C., Zhao, X., Jiao, Z., Shang, Y. An electromagnetic wearable 3-DoF resonance human body motion energy harvester using ferrofluid as a lubricant, *Applied Energy*, 2017, *197*, 364–374,

[134] Servati, A., Zou, L., Wang, Z. J., Ko, F., Servati, P. Novel flexible wearable sensor materials and signal processing for vital sign and human activity monitoring, *Sensors* 2017, *17*, 1622.

[135] Wan, C., Chen, G., Fu, Y., Wang, M., Matsuhisa, N., Pan, S., Pan, L., Yang, H., Wan, Q., Zhu, L. an artificial sensory neuron with tactile perceptual learning, *Adv. Mater.* 2018, *30*, 1801291.

[136] Araki, H., Kim, J., Zhang, S., Banks, A., Crawford, K. E., Sheng, X., Gutruf, P., Shi, Y., Pielak, R. M., Rogers, J. A., Materials and device designs for an epidermal UV colorimetric dosimeter with near field communication capabilities, *Adv. Funct. Mater.*, 2017, *27*, 1604465.

[137] Wang, W., Timonen, J. V., Carlson, A., Drotlef, D.-M., Zhang, C. T., Kolle, S., Grinthal, A., Wong, T.-S., Hatton, B., Kang, S. H. et al., Multifunctional ferrofluid-infused surfaces with reconfigurable multiscale topography, *Nature,* 2018, *559*, 77.

[138] Wang, S., Xu, J., Wang, W., Wang, G.-J. N., Rastak, R., Molina-Lopez, F., Chung, J. W., Niu, S., Feig, V. R., Lopez, J. et al., Skin electronics from scalable fabrication of an intrinsically stretchable transistor array, *Nature* 2018, *555*, 83.

# 11 Wind Energy Harvesting and Involved Power Electronics Conversion Systems for Smart Grid Interfacing

*Sachidananda Sen and Maneesh Kumar*

## 11.1 INTRODUCTION: RENEWABLE ENERGY SOURCES AND WIND ENERGY SYSTEMS

The climate change and adverse environmental impact of using fossil fuels like coal and petroleum for the generation of electrical energy have raised an alarm to switch towards greener alternatives. As the global population is increasing at a steady pace, and so is the aspiration of people, a surge in energy demand year after year has been observed in the last three decades, which puts pressure on the energy generation sector in meeting this demand. Thermal (coal fired) power plants have been the largest contributor of the electricity needs for mankind. The consumption of electric energy by a country signifies the living standard of its citizens and human development index. Electricity is used to provide better livelihood opportunities by establishing different industries and improving the lifestyle by better residence condition with various comforting appliances like heating, cooling, ventilation, etc. In the late 20th century, the unawareness regarding the harmful environmental impact and concerns of global warming, lack of any alternative or cost-effective energy source have been some of the major reasons for setting up of these thermal power plants by both government and private parties. Therefore, people were not concerned about the different forms of pollution, such as air, water, and land, that it caused over the decades [1].

Since the beginning of the 21st century, after fulfilling the basic energy requirements of the masses, all the world leaders and international bodies have identified the harmful consequences of fossil-fuel-based energy production on the planet, and are gradually taking initiatives by signing pacts towards reducing the carbon footprint. In this regard, the use of renewable energy sources (RESs), viz., wind turbines, solar photovoltaic (PV) panels, fuel cells, etc., have been promoted and incentivized for increasing their implementation for the generation of greener energy. These RESs can be interfaced to the low voltage (LV) distribution network, i.e., less than 1 kV, near the demand locations like within the city outskirts. By doing so, the need for transmission lines are eliminated that were used to bring the power from the far off thermal stations to the city that ultimately reduces the transmission losses that approximately accounts for 20% total transmitted power. Furthermore, integration of the RESs empowers consumers, such as individual households and industries, to generate their own power that introduces the term called prosumers, i.e., producer and consumer simultaneously [2, 3].

The concepts of microgrids (MGs), i.e., a small-scale grid comprising its own loads and energy sources, have evolved gradually with the increasing number of integration of RESs on the distribution network that is also called an active distribution network as its characteristics changes from passive to active. Such interconnection of multiple MGs are versioned to establishing a smart grid with RESs, monitoring units and cyber security arrangements as an integrated part of the overall

DOI: 10.1201/9781003340539-11

power grid. Therefore, MGs are also considered as the building blocks of a smart grid. Various goals within the smart grid are identified, viz., smart load scheduling, increased penetration of RESs, energy management systems for uninterrupted power supply to critical loads [4].

Now, coming to the increasing usage of the RESs, like solar PV and wind turbines (WTs) that, respectively, rely on the solar irradiance and speed of the blowing wind. Both these sources are of intermittent nature, i.e., they are not able to generate electric power continuously. As solar PV arrays produce energy only during the day time when sun is available, and for WTs, the wind does not always blow. Therefore, to have a buffer that nullifies the effect of such generation intermittency, battery-based storage devices are also integrated as an important component of the microgrids. Moreover, as both alternating current (AC) and direct current (DC) are involved, the integration of RESs and storage units needs proper power electronic conversion systems. Among the different RESs the solar PV and WTs are the most preferred ones. They have their own advantages and disadvantages. In present work, our studies are confined to the wind energy harvesting, various aspects, and its implications.

### 11.1.1 Wind Energy Systems (WESs)

The wind flows because of the air pressure difference mostly created by the sun. The coastal receives maximum wind beyond a certain speed throughout the year. This is a renewable energy source that is available free of cost, emission less (clean and green), and can decrease the dependency upon the non-renewables. Because of these features and recent developments in power electronics' domain the WESs have gained attention worldwide.

As per the latest statistics, in 2018, about 5% of the global energy consumption is extracted from the wind that amounts to 750 GW. From 2010 to 2020 there has been an annual increase at 14% for wind energy harvesting capability. Furthermore, it is expected to grow at a rate of 15% here onward and capture approximately (20% and 35%) of the gross energy requirements of the globe by 2030 and 2050, respectively.

The energy present in the wind can be utilized to rotate the blades of the wind turbine (WT) that is subsequently connected to the generator via the gearbox arrangement. In other words, WTs are applied to transform the kinetic energy available in the wind into mechanical energy, which is later fed to the electrical generators for obtaining the electrical energy. Based on their rotational axis, the WTs can be categorized into two types:

a. Horizontal axis wind turbine (HAWT).
b. Vertical axis wind turbine (VAWT).

Out of these two, HAWTs are mostly used worldwide because of their multiple advantages over VAWTs, like simple configuration (especially during high wind speed), higher efficiency, and cost-effectiveness [5].

Wind generation systems have seen tremendous growth due to fast expansion of the alternate energy industry and amalgamation of knowledge from various fields, viz., power electronics converters, turbine blades design, mechanical gearbox interfacing, special electrical machines, and advanced control system schemes have pushed forward the development of the wind energy conversion system (WECS). The basic functionality of the WECS remains to be able to convert or adapt the kinetic energy of wind first into mechanical energy (via turbine blades and gearbox arrangement), which is subsequently transformed to electricity or electrical energy using an electrical machine. As the generated electricity is at different voltage and frequency levels, it is not suitably equipped for supplying into the utility or for grid integration and calls for parameter adjustments or grid synchronism. For this purpose various power electronics' converter topologies are utilized for controlling the overall machine-side converter (MSC) and the grid-side converter (GSC). Hence, the

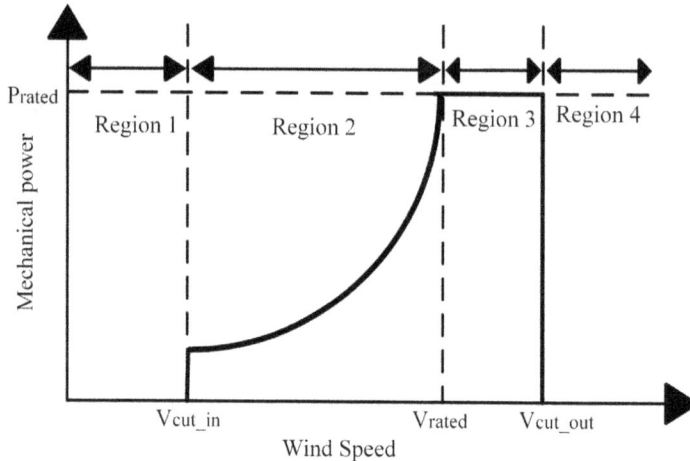

**FIGURE 11.1**  Operating regions of the wind turbine system.

WECS is usually controlled to operate these power electronics' interfacings (PEI) to achieve the requirements of the utility power grid under time-varying wind speeds. The reliable operation and output of the wind energy system highly rely on the time-varying weather and unpredictable/inter-mittent behavior of the wind speed. Therefore, it is inevitably necessary to mandatorily analyze and identify the characteristics of range-bound variations of wind speed and to set or specify some of its operating regions for integrating the WECS with the upstream grid based upon the wind speed measured. Therefore, by carefully operating the WTs within the pre-defined range of wind speeds, one can successfully achieve the goal of extracting the power available in wind velocity [6]. The lower and upper limitations of wind velocity are known as the cut-in ($V_{cut\_in}$) and cut-out ($V_{cut\_out}$) speeds as illustrated in Figure 11.1.

The operational area or regions of WT can be elaborated as given in the following:

- **Regions 1 and 4:** Here, the wind speed is less than the lower threshold $V_{cut\_in}$ or more than the upper limit $V_{cut\_out}$, respectively. This would result in the rotational speed of the turbine blades being either too slow or too high. Therefore, for safety concerns and as a precautionary measure, the WT has to be halted/stopped and isolated from the main upstream power grid at the point of common coupling (PCC).
- **Region 2:** Within the mid region of operation, the maximum power point tracking (MPPT) algorithms or methodologies are implemented for extracting or harvesting the maximum possible power from the wind.
- **Region 3:** This region represents the speed above the middle area and near to the upper limit of the wind speed. The output power generated by the WECS should be restrained/bounded to the maximum power rating of the WT by applying the controller mechanism for regulating the pitch angle of the turbine blades so that the mechanical stresses on the blades are decreased within the acceptable range under the high wind speed scenarios.

## 11.1.2 BENEFITS OF WIND ENERGY SYSTEMS

By using wind as a source of energy, there are multiple advantages that make it one of the fastest-growing energy sources around the globe. Due to these benefits, researchers are currently focused on addressing the challenges for enhanced utilization of wind energy. Some of the advantages of wind energy sources are as follows:

- **Wind energy is less expensive:** After installation of WTs, the additional cost involved in maintenance is very low. The wind mills comprising multiple WTs is considered to be one of the cheapest energy sources in present times. As the power produced by the wind farms is sold at a fixed price for a longer time period say 15–20 years with free fuel, hence, it lowers the uncertainty in pricing of fossil fuels' existing conventional energy sources.
- **Wind energy systems creates jobs and livelihood:** The wind energy sector needs several area experts and WT technicians for manufacturing of various components, installation and assembling of WTs, maintenance as well as supporting services from both electrical and mechanical background.
- **Wind energy is emission free and a clean source:** Wind energy does not involve any harmful emissions, viz., nitrogen oxides, and sulfur dioxide like fossil fuels that can lead to air pollution or causing acid rain, smog, or accumulation of greenhouse gases that could lead to health problems in humans and damage the economy.
- **Wind is a sustainable and domestic source of energy:** Wind energy is actually a derived form of solar energy. It blows because of the heating of the atmosphere by the sun, the rotational motion of the earth, and its surface irregularities. As long as the sun shines and the wind energy can be harnessed making it a sustainable one. Also, a nation's wind supply is domestic in nature, i.e., it is not required to be imported from other countries and is abundantly available and of inexhaustible nature.
- **WTs can be placed on the sea shore, existing farming lands or ranches:** The empty lands in coastal areas and rural areas can be used to establish wind mills, where most of the high speed blowing wind sites are available. It only needs a fraction of the land and can increase the earnings of farmers and ranchers. The wind power plant companies can pay rent to the farmer for using their land.
- **Wind energy mills enables industrial growth and competitiveness of a country:** The establishment of new wind projects brings investments into a country's economy. Also, it helps to harness the existing vast domestic resources of a country and create a workforce of high-skilled personnel, and boosts the competitiveness of the economy of clean and sustainable energy at the world forum.

This book chapter provides an extensive state-of-the-art wind energy harvesting techniques' literature on the smart buildings as important "building blocks" for realizing the vision of smart cities. It explains various components of the wind energy systems (WES), different types of WES, their power electronics interfacing (PEI), and available storage arrangement, the role of WES in realizing the vision of a smart grid through MGs, implementation challenges, and future trends.

## 11.2 COMPONENTS OF WES

Figure 11.2 shows the various components or parts that constitutes a typical variable-speed type, standalone wind energy system, along with its involved power electronics converters. The major parts are as follows: (1) wind turbine blades attached to the rotor; (2) mechanical gearbox arrangement for maintaining required speed ratio; (3) electrical machine or generator with its stator and rotor; (4) generator-side or source-side power electronic interfacing/converter; (5) load-side power electronic interface; (6) storage-side power electronic interface; (7) DC-bus link interfacing along with energy storage unit/device; (8) low-pass filter for attenuation of harmonics; (9) power transformer; (10) energy storage components like battery and/or supercapacitor; (11) dumping load for absorbing surplus power under full charge condition of the storage device; (12) control system mechanism for regulating the overall WES.

The controller design for WES includes maximum power point tracking (MPPT) control on the generator-side converter that helps to extract maximum power by aligning turbine blade angle with

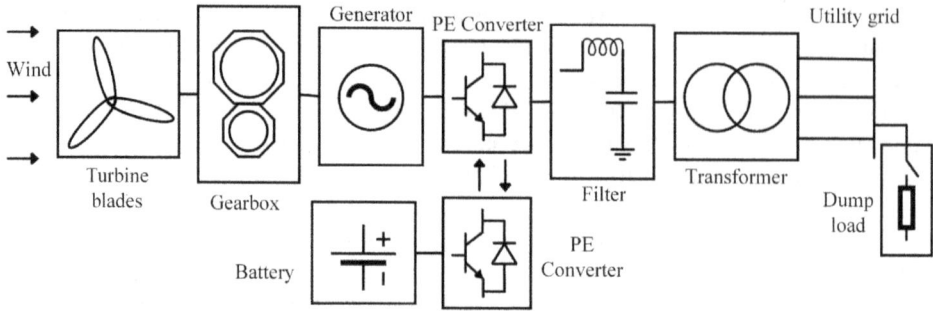

**FIGURE 11.2**   Different components of the wind energy system.

respect to striking wind. After extracting maximum possible power, the next control objective is to maintain the rated voltage-frequency and power factor of the injected power at the grid-interfacing junction or bus by the assistance of load-side power electronics converter or inverter. Lastly, another important function of the control scheme for WES is power/energy management for both storage device converter and dump load-side converter. These two components are used as energy buffers that store excess power generated (because of high speed wind) by charging the storage battery or dissipating it on the dumping load if storage battery is full capacity. The storage devices also supply power to the dc-bus link to maintain its rated voltage under deficient generation condition due to low wind speed [7].

## 11.3 TYPES OF WES

There can be two broad perspective under which the WTs can be categorized into (1) depending upon the rotational axis, and (2) based on the electrical machine used.

### 11.3.1 CATEGORIZATION BASED UPON ORIENTATION OF ROTATIONAL AXIS

Based upon the orientation of rotation of turbine blades, WTs can be divided into two categories, viz., horizontal-axis WT (HAWT) and vertical-axis WT (VAWT).

#### 11.3.1.1 Horizontal Axis Wind Turbine (HAWT)

The HAWTs are suitable even for slower wind speed as it uses gearbox with multiple gear ratios to compensate for the slower rotating blades of the WT. In this type of WT, the electrical generator is placed at the top of the tower. Because of HAWT's tall tower base structure, it is able to generate maximum possible energy in comparison to other existing WESs. The relationship between the height of the tower and energy produced is also important. It suggests that with increase in height by 10 m, wind speed enhances by 20%, which leads to 34% increased generation of the output power. Moreover, it has the features of variable blade pitch that are capable of adjusting the blades of the WT in an optimal manner to achieve is maximum wind strike resulting in heightened energy production. All these advantages lead to relatively increased efficiency of HAWTs that is further aided by the perpendicularly rotating turbine blades with respect to the blowing wind.

With the above said merits there are some demerits of HAWT configuration as well. Firstly, the construction and erection of the tower for holding the electrical machines, mechanical gear box system, and the heavy turbine blades is highly costly. Secondly, the tall tower of HAWTs affects the radar installation for a mobile communication tower that creates distortions in the signal. Lastly, this configuration of WTs having adjustable turbine blades needs an additional mechanism or controller strategy for properly adjusting their direction to get exposure that creates maximum thrust by the wind. Because of enhanced performance of the aerodynamics, cost-effectiveness, and balanced loading of the drive-train in the mechanical gearbox, the HAWT configuration that is also fitted with three blades is highly prevalent and a globally accepted arrangement in the modern industrial grade WT systems [8].

Furthermore, depending upon the behavior or time-varying speed of the wind, the HAWTs can is designed to efficiently work as the fixed-speed WT (FSWT) and the variable-speed WT (VSWT) by integrating different types of existing electrical machines/generators being assembled with or without a mechanical gearbox for speed ratio control of the blades and rotor. However, there are some critical issues that are observed in FSWTs like the limited/restricted speed range of operation, huge amount of mechanical stress, and need for multi-level gearbox systems. While the VSWTs are successfully able to overcome the aforesaid demerits. As per the variation in wind speed, the VSWT can be tasked into controlled operation for harvesting the maximum possible power at each variable wind speed position, such that the mechanical stress on the WT is mitigated with allowable power variations within a limited range. Moreover, the speed rotor can be continuously adjusted according to instantaneous fluctuations in speed of the wind speed so that the rotor speed to the wind speed ratio is preserved at a constant value. If the WT's gearbox fails to maintain the constant ratio, there will be a reduction in the extracted wind power. As mentioned earlier, the VSWT configurations of WESs can be discussed by utilizing various available electrical generators. More about such machines will be discussed subsequently.

### 11.3.1.2  Vertical Axis Wind Turbine (VAWT)

In vertical-axis wind turbines (VAWTs), the blades rotate around a short tower (known as shaft) that is configured in a vertical manner with respect to the ground. In other words, the rotational axis of the VAWT is perpendicular with respect to the ground. All the equipment are attached and assembled near the foundation surface or base. The fact that the wind always strikes perpendicular to the blades is the main advantage of this WT as this eliminates the use of any additional control system design as required in the case of HAWT-type WES. Furthermore, the configuration of the VAWT has its mechanical gear box setup and the electrical machine attached at the base level of the ground. Hence, these types of WTs are easier for maintenance purposes. Due to the lower speed of operation, the VAWTs makes less noise in comparison to its other counterpart. The VAWT-based WESs can be placed at several locations such as the roofs of the buildings, in both sides of long roads or highways, etc. Because of the short height and lower operational speed of the VAWTs, its efficiency is relatively lower than that of the HAWTs. Moreover, as high wind speeds could not accommodate short towers, the power output produced by the VAWT type is much less in comparison to the HAWT type. Although the VAWT have been widespread in many locations and research facilities, the limitations faced by its various configurations for commercial expansion at large-scale are related to the different shortcomings in arrangements like blade lift forces and lower rated capacity of power generation. Table 11.1 presents the detailed comparative features of the HAWT- and VAWT-based WESs.

### 11.3.2  Categorization Based upon the Type of Electrical Generator Used

### 11.3.2.1  Asynchronous or Induction Generators

The technology of the induction or asynchronous generators are considered to be very mature and needs quite low maintenance, is cost-effective, along with providing good transient or dynamic response, simplicity of operational and controller design mechanism. In addition to that induction generators (IGs) are highly robust in construction and inherently gives protection against faults arising due to the short circuit. Its main disadvantage is the fact that it requires continuous excitation from a reactive power ($Q$) source for setting up the magnetic field for generating voltage and supplying the active power. In single or standalone type WTs, the needed reactive power support is usually provided with the help of an external VAR source. i.e., switched capacitors, or using a proper power electronic interfacing (PEI) arrangement and such types of WTs are known

**TABLE 11.1**

**Comparison of the Two Types of Rotational Axis Wind Turbines: HAWTs and VAWTs**

| Sl. No. | Performance features | HAWTs | VAWTs |
|---|---|---|---|
| 1 | Starting/base wind speed | High (3 to 5 m/sec) | Low (1.5 to 3 m/sec) |
| 2 | Speed of rotation | High (5 to 12 m/sec) | Low (3 to 7 m/sec) |
| 3 | Efficiency | High (more than 70%) | Low (less than 60%) |
| 4 | Maintenance process | Complicated and costly | Simpler and cheaper |
| 5 | Height of the tower | Large (30 to 100 m) | Small (less than 10 m) |
| 6 | Rotational area covered by the blades | Larger | Smaller |
| 7 | Need of control mechanism for blade adjustment | Yes | No |
| 8 | Location of generator | Top of tower | Ground level |
| 9 | Noise level | High (5 to 60 dB) | Low (0 to 10 dB) |
| 10 | Power output | More | Less |
| 11 | Deployment location | On shore and off shore | On shore |
| 12 | Birds hit cases | High | Low |

as self-excited induction generators (SEIGs). Moreover, a mechanical gearbox system is used to increase the low turbine/rotor blade speed so as to match the high speed requirement of the electric generator. Depending upon the type of construction of the rotor of the electrical machine, the IGs are categorized into two varieties, viz., wound-rotor induction generator (WRIG) and squirrel-cage induction generator (SCIG) [9, 10].

In WRIG, a three-phase winding that is similar to the stator is applied over the rotor, while SCIG-type WTs consist of rotors with embedded short-circuited conducting/metallic bars, which is shaped as a squirrel cage having some skewed position. Further, the WRIG-type WTs can be reconfigured or converted into two forms: (1) doubly-fed induction generator (DFIG) and (2) brushless doubly-fed induction generator (BDFIG). The applications and utility of the above said different types of WTs, i.e., WRIG, DFIG, BDFIG, and SCIG as tools for wind energy harvesting are described as follows.

**(a)    Wound-rotor induction generator (WRIG)**

A standalone, variable-speed WRIG-type WES can be represented by a simplified configuration. The stator winding of the WRIG is tied with the point of common coupling (PCC) and an external resistance ($R_{ext}$) with variable features is attached to the rotor, which is regulated by a power electronic interfacing. This generator can be operated at different ranges by varying the magnitude of $R_{ext}$. This type of IG has starting issues, and hence, needs a soft starter that also assists in reducing the inrush current during start-up. The issues of the inrush effect during the starting can also be mitigated by using a proper sized transformer and VAR compensator. The WES made up of standalone-type WRIG mainly produces the stable rated voltage having a pre-decided constant magnitude and frequency by controlling the speed of the rotor that is varying in the range of a certain percentage due to time-varying intermittent wind.

The main disadvantages of WRIG-based wind energy harvesting system are its requirement for a soft starter facility, only operating within some range-bound speed limits and the need for external resistance that causes power loss due to heat dissipation and ultimately reduces the energy efficiency of the whole WES. Furthermore, inclusion of other accessories like the slip ring and brushes, which needs proper maintenance works as well as needing replacement at a regular time-interval that is not cost-effective and quite tedious too. Because of these reasons, the WRIG types are not preferred for applications in remote places/locations. Also, some of the issues can be prevented by implementing a variable resistance setup connected to the rotor winding in series configuration placed on the generator shaft using a fixed resistor and controlled PEI; but, this arrangement raises the dissipated heat inside the generator that results into decreasing or limiting the range of speed variation only within 10% (approx.) of the set rated speed.

### (b) Doubly-fed induction generator (DFIG)

A DFIG-type standalone WES can be built by directly attaching its stator on to grid PCC, whereas a power electronic converter is used for connecting its rotor terminal to the bus. The DFIG components and construction are represented in Figure 11.3. There is a unidirectional power flow via the stator windings, and based upon the operating scenario or mode of the IG, the power flow direction through the rotor windings are determined. There can be three different modes of operation for an asynchronous/induction generator, viz., sub-synchronous or below synchronous speed, synchronous speed, and super-synchronous or above the synchronous speed. Now, if the DFIG is operating under sub-synchronous mode then the rotor draws power from the distribution grid to catch up to the synchronous speed by decreasing the slip speed. If the generator is operating at super-synchronous speed then power is delivered/fed into the grid by the rotor converter. At synchronous speed, no power is transferred to or from the rotor circuit. The DFIG can be functioned up to 30% variation (i.e., having a slip speed of 0.3 pu) above or below the synchronous speed; therefore, the ratings of the power electronic converters of the rotor should be close to 30% of the stator power. In addition to that, the stator does not have any PEI, hence, DFIG-type WES drastically decreases the required power converters' rating and harmonic filter ratings that reduces overall cost by a significant amount in comparison to other WES implementing power electronic converters at full-scale.

This cost or economic advantage of DFIG because of its reduced size of PEI makes it a highly preferable selection for high-power grid-tied operation of the WES. Although this cost-effectiveness is not of large value in the case of standalone DFIG-type WT systems, having power output only from a few kWs to a few 100 kWs, which is relatively quite low. Another merit or feature of having DFIG-based WES is for attaining excitation in stator winding without the requirement of an external VAR compensation arrangement as the power converters interfacing available in the rotor circuit is used for exciting the stator circuit. The reactive power needed in the stator of the electrical machine is created by using the load-side converter (LSC), as well as with the assistance from the energy storage unit and its PEI. A block diagram representation of this feature is presented in Figure 11.3. The rotor-side converter (RSC) is responsible for controlling the generator's speed,

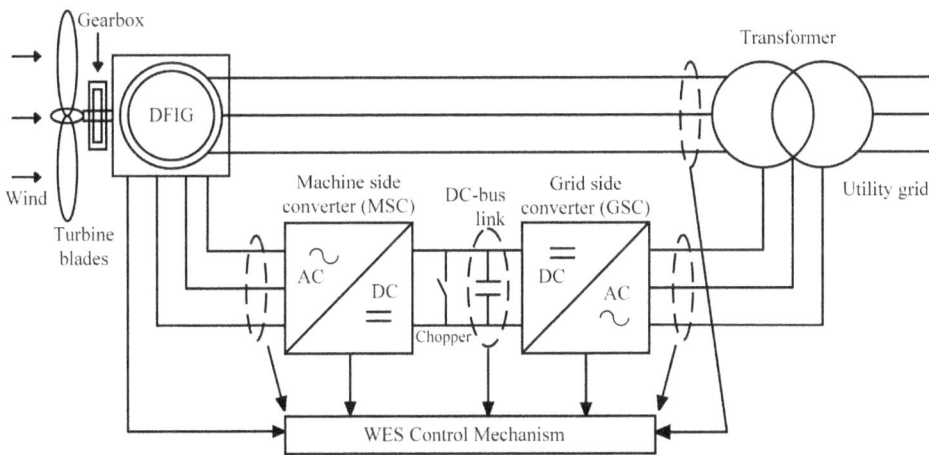

**FIGURE 11.3** DFIG-type wind energy system.

and therefore, the reactive power flow direction. The DC-bus link voltage is regulated by the load-side converter.

Several controller design mechanisms have been built and analyzed and show that regulating the stator voltage output of a standalone DFIG is quite simple for its operation as WES. However, careful attention into details are needed while controller for regulating the voltage and current of the rotor mainly during the transient stage at the starting, because the power electronics converters used in the rotor circuits are of reduced size and the dynamic/transient magnitude of the starting current can be too high to handle for the PEI. Similar to the WRIG-based WTs, the DFIG WES too has the demerit of using the slip ring and brush arrangement inevitably, which decreases the reliability and regularly calls for maintenance work that is difficult and costly. Therefore, to avoid these issues, the recent advances in the new technologies for brushless operation of electrical machines have compelled the application of brushless DFIG (BDFIG)-type WTs that are being configured, developed, and adopted at a large-scale in the WESs.

### (c)  Brushless DFIG (BDFIG)

The configuration of BDFIG is slightly more complicated than its other counterparts as it consists of two wound-rotor induction machines connected in a cascade manner. Among the two electrical machines one is used for generation purposes while the other one is applied as a controller of the first one. Alternatively, a single stator circuit along with two three-phase windings, in combination with a special cage rotor structure can also be considered to construct the BDFIG-type WES. In both these processes for making a BDFIG would have two groups of stator windings, which are called the power winding (PW) and control winding (CW). Here, the power winding is attached to the PCC as a direct connection, while the controller winding is connected to the distribution network at PCC via two power electronic converters connected in back-to-back fashion or configuration, viz., machine-side converter (MSC) and load-side converter (LSC), both of which are of reduced size. Another important point here is that the number of poles for the CW and PW should not be the same, so that direct coupling between these is prevented. However, there is a need to achieve cross-coupling through the rotor circuit by choosing the number of poles equal to the aggregation of the number of pole pairs of the power winding and control winding.

The main advantages of BDFIG-type WES are more or less similar to those of the DFIG types. Unfortunately, it is usually larger in size, and because of its double-stator winding cascade connected structure, BDFIG needs a higher level of complex controller design and assembling procedure. Even though having these demerits, BDFIG-based WESs are highly preferred for large-scale wind mills operated in grid-connected mode, due to its nearly maintenance-free working and higher reliability that are required especially for off-shore applications. Furthermore, it also demonstrates quite robust performance under unbalanced grid conditions. For the standalone application of WES of the BDFIG especially for small WTs, it is not considered to be attractive enough because of the large-sized construction and complex control of the machinery, which outweighs the merits of the reduced ratings of the power electronic converter.

Also, it is worth mentioning another existing brushless design of the wind harvesting system, i.e., brushless doubly-fed reluctance generator (BDFRG), developed using the reluctance-type rotor instead of the conventional BDFIG made up of the wound-type rotor. Despite having more efficiency and reliable than BDFIG, the BDFRG-type WES are not in many applications in the field mainly because of the fact that its rotor design is quite complex, and results in a large-sized construction due to a low torque-to-volume ratio.

### (d)  Squirrel-cage induction generator (SCIG)

In the earlier discussion on WRIG, where the slip rings and brushes are used to connect with insulated windings of the rotors. In the case of the SCIG-type design of WTs, the rotor is built by attaching the

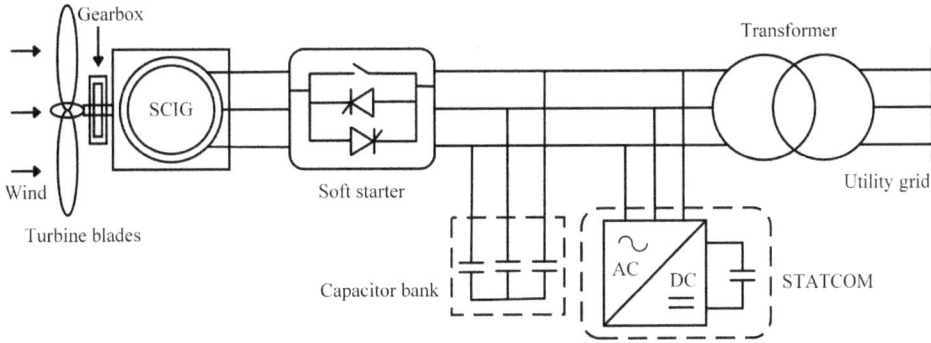

**FIGURE 11.4** SCIG-type wind energy system.

longitudinal conducting bars that are set/fixed into grooves and their terminals are short circuited by considering the shorting rings. Therefore, most of the issues and problems highlighted due to the use of brushes and slip rings in WRIG- and DFIG-based WESs, as well as the involved complex design in developing and controlling BDFIG, have been eliminated in the SCIG-type WESs. Hence, among the four configurations of IGs, the SCIG is the smallest in size, having the most robust constructional structure, and is available at a cost that is lowest of all. However, the main disadvantage in a SCIG-based WES is that it needs the power electronic converters at full or rated capacity so as maximum power from the striking wind can be harvested, along with both active and reactive power can be delivered to the PCC in a stable and complete controlled manner. As a mature electrical machine in the SCIG-based WTs are highly researched for the field of WES and smart grid applications. The research projects include design and development of wind turbine simulator and emulator hardware setup, various possible PEI and their control strategies, and fault ride through, voltage build up mechanisms, self-excitation techniques and suitability for functionality as standalone entity as well as in tandem with other types of RESs within the hybrid microgrids' environment. The structure of SCIG-type WES is given in Figure 11.4.

### 11.3.2.2 Synchronous Generators

Synchronous generators (SGs) are the electrical machines whose rotor revolves at a rated or synchronous speed only. Stator of the SG is designed in the same way as that of an IG. Rotor of SG can be developed into two types of shapes, i.e., either cylindrical type having distributed winding and uniform air gap, or salient-pole type with concentrated windings on the poles and the air gap is not uniform. Out of the two types of SGs, the salient-pole rotor is mostly preferred for low-speed applications because it has a shorter axial length and larger diameter, along with the number of poles being comparatively more than the other counterpart. The main benefit of SG-type WES over IG type is that it eliminates the requirement of including the gearbox for maintaining a speed ratio, and therefore, reduces the need of regular maintenance and saves the cost. Moreover, it also makes it possible to attain an enhanced efficiency and increasing the overall reliability of the whole system. For this reason, the SG-based WTs are also known as gearless or direct-drive WTs. Elimination of gearbox can help to reduce some spending, however, the gearless (or direct-drive) WT generators are usually slightly larger in size, heavier, and more costly because of the requirement of a greater number of rotor poles so that the speed of the generator gets reduced and matches the speed of the revolving turbine fitted with no gearbox. Thus, indirect-drive SG having a gearbox with a lesser number of poles are still widely accepted WTs in the wind energy systems. The SG in the application of WESs can be further classified into two varieties, viz., wound-rotor synchronous generator (WRSG) and permanent-magnet synchronous generator (PMSG), which are described in the following.

### (a)   Wound-rotor synchronous generator (WRSG)

In WRSG-type machines, an additional arrangement for excitation of the field winding of the rotor is needed with the DC power that can be supplied using either an external DC source via path of slip rings and brushes or applying a brushless-type exciter. The first technique is although very simple and easy to implement, however, it asks for maintaining procedure as well as replacement of the slip rings and brushes regular interval. At the other end, the second method does not require much maintenance, but it has a higher level of complexity and is more expensive because of the involvement PEI and the use of an AC generator as an auxiliary source. Due to the involved DC power-source-based excitation in the WRSG, this structure is also called the electrically excited synchronous generator (EESG). Similar to the earlier WESs, the MPPT algorithm is implemented by the generator-side converter (GSC) for extracting maximum power from wind, whereas the voltage and frequency at the intersecting junction of PCC is controlled by applying the load-side converter (LSC) under different working scenarios like the unbalanced loads and continuous load variation in the active distribution network. A power management strategy (PMS) is also realized to regulate the DC-bus link voltage with the help of controller design under time-varying load conditions and wind speed variation with the assistance of the power buffering from the storage battery and dumping load power electronic converters. The field winding of the rotor is generally provided with the excitation from the DC powered auxiliary source through a DC/DC PEI converter that is controlled for maintaining a fixed voltage at the WRSG's stator terminals. This task can also be accomplished by using an automatic voltage regulator (AVR) along with one of the classical or modern control mechanism for regulating the stator winding's voltage of the standalone WRSG-based wind energy system.

Although the WRSG-based standalone WT systems can be considered a suitable choice for remote location applications with isolated loads, this has some disadvantages, which reduces its chances of becoming a promising alternative to serve load demands that are remotely located. The main disadvantage being the requirement of an external DC power source for exciting the windings of the rotor through the available options of slip rings and brushes' integration, or an arrangement of a brushless excitation system that needs complex PEI and is costly, acting as a hindrance in the path to adopt an EESG alternative for the off-grid operation of WRSG.

### (b)   Permanent-magnet synchronous generator (PMSG)

The PMSG-type electrical machines are inherently having the advantage of being the brushless self-excited synchronous systems unlike the WRSG counterpart that needs external excitation system. The permanent magnets used in PMSG are responsible for creating magnetic flux in the air-gap for the working of the rotor circuit. Now, in the absence of rotor copper losses, the effective stress on the rotor due to thermal/heating is decreased significantly and attainment of higher power density is guaranteed. Since the last two decades, PMSG has been considered to be an attractive alternative in comparison to the WRSG-type WESs because of the fact that there has been a substantial deduction in the cost of power electronic converters and permanent magnet materials. Unfortunately, there is a lot of uncertainties regarding the ease of extraction and availability of the permanent magnet materials and their expensiveness in the future, which acts as a demerit for its large-scale use in WESs. This mostly happens because of the limited number of miners and suppliers of permanent magnets in the market worldwide as well as its dependency on global geopolitical stability. A general configuration of the standalone PMSG-type WES is shown in Figure 11.5.

Now, as per the constructional difference in the manner that the permanent magnets are attached or mounted on the rotor, the PMSGs can be classified into one of these varieties: (a) surface mounted PMSG and (b) inset-magnet PMSG. The surface-mounted type PMSG, the permanent magnets are fixed on the surface of the rotor that compromises the robustness of the mechanical structure, which

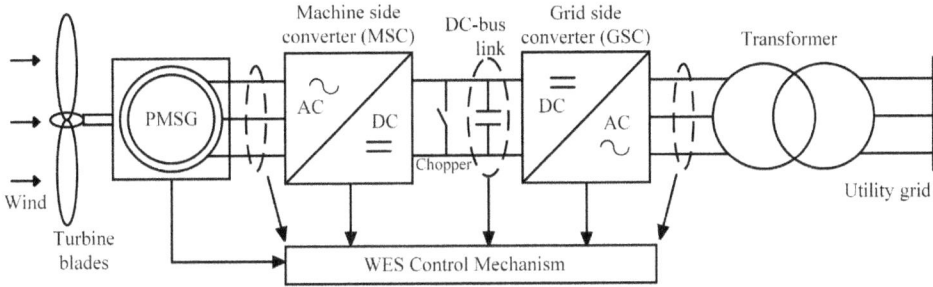

**FIGURE 11.5** PMSG-type wind energy system.

is risking the detachment of the permanent magnet during the high-speed operation. Because of this main reason and others like regarding the overall efficiency of the electrical machine and effective power density, the surface-mounted based PMSGs are restricted only to the low-speed applications of the WTs. In the second variety of this risk is eliminated as the permanent magnets are inserted into the body of the rotor, which makes the structural design appropriately suitable for WT applications under the high-speed operating conditions. Moreover, the inserted permanent magnets in the construction of PMSG gives the feature of utilizing the magnetic torque as well as the torque offered due to reluctance ultimately results in achieving a higher efficiency drive that are the need of the hour in the WESs.

For the above discussed reasons, since the early years of the 2000s, researchers and industrial persons acknowledged the importance of PMSG as a robust solution for both isolated and grid-attached operation of small-scale WTs as direct-drive machines. In the case of large-scale WTs, the PMSG structure having inserted permanent magnets around the rotor, i.e., a salient-pole rotor design gives the highest overall efficiency among the various possible rotor design structures. But, due to the utilization of large-sized and heavy permanent magnets the PMSGs might not be considered for designing large-scale WTs. This issue can be resolved by using the permanent magnets' structure of lighter weight for PMSG construction that could become a solution for manufacturing large-scale direct drive WTs.

Several PMSG configurations can be used in WESs by implementing different controller design techniques for the various PEIs involved. Depending upon the proposed control scheme and realized construction, the performance and effectiveness of the PMSG-type WESs can be appropriately used in standalone applications and integrated within the hybrid ac/dc MGs in the presence of other RESs like solar PV modules. Applications pertaining to the usage of WESs in combination with other intermittent sources shall be discussed in the subsequent section.

From the above discussion, it can be said that the most simple, reliable, and cost-effective solutions for the WT is SCIGs that are implemented as both fixed and variable-speed WES functionalities. Unfortunately, these types of WTs get affected from restricted fault ride-through (FRT) capability, huge mechanical stress, and grid's reactive power loading. Alternatively, the DFIGs and the SG-based WTs are preferably employed as the WESs. The DFIGs type that needs partial or fractional scale power electronic converters are considered to be an economic option, even though these call for multi-level mechanical gearbox arrangement and externally supplied excitation current. Hence, in recent times the researchers in power and energy fraternity are more attractive towards utilizing the PMSGs owing to its highly advantageous features like improved efficiency, more reliability, good enough FRT capability, and adequate power density that supports overall WT performance. Furthermore, it removes the requirement of the external DC excitation system and mechanical gearboxes, and helps to expand the operational range of wind speed. Now, the multi-phase PMSG implementations can be extended further into reduced phase current, attaining lower ripple torque with greater power density, as well as the approved fault-tolerant capability (FTC).

There can be multi-phase connections of such PMSG-based WTs like for instance the dual 3-phase, and the 6-phase, which contains a complex configuration of the controllers along with the need of costly PEI of frequency converters. Therefore, the PMSGs having 5-phases are extremely applied in both small-scale and large-scale WESs. Moreover, several such multi-phase PMSG-type WTs with different benefits and features can be built/developed to be used for WESs. From the aforementioned points, it is observed that the DFIG, SCIG and PMSG types seem to be the appropriate choices of generator for WES for standalone applications.

## 11.4 EXTRACTION OF MAXIMUM POWER FROM WIND ENERGY USING MPPT

To tackle the stochastic behavior of wind, it is of utmost importance to harvest or extract the maximum power possible by the WESs at each instant of time. To accomplish the task of extracting maximum power from the wind energy, controllers with maximum power point tracking (MPPT) algorithms are implemented in an optimal manner. In this section, a short discussion on the state-of-the-art MPPT algorithms as well as their critical studies to improve the decision making quality of the controller working appropriately within the operating range of the algorithm as per their specific application scenarios [6, 11].

The MPPT algorithms are inevitably applied in RESs such as solar PV systems as well as in the WES. Implementation of MPPT techniques involves estimation of multiple system parameters and state/control variables, viz., optimal power, dc-bus link voltage, PCC voltage values, or duty cycle of the various power electronic converters that are quite troublesome and challenging in real-time applications. Therefore, the purpose of extracting maximum power at every instant, estimating these control variables becomes the optimal necessities and making the MPPT algorithms an essential part while installing WTs. These MPPT algorithms can be usually divided into four major categories: the indirect power controller (IPC), the direct power controller (DPC), hybrid or mixed type, and smart/evolutionary algorithms. Each of these categories have their own advantages and disadvantages. Here, the first method or IPC handles the output electrical power $(P_e)$ by controlling the mechanical input power $(P_m)$ as per the pre-recorded information in the power-speed characteristics/curves of the WTs. While in the other end, the DPC-type strategy analyzes the deviation of electric power output for tracking the maximum power point (MPP) for generating the highest electric power output on an instantaneous basis. There exists several other MPPT algorithms that applies the recently developed artificial intelligence (AI), machine learning (ML)-based controllers that are extensively used in various industrial applications.

Among the existing several types of MPPT algorithms, one of these that is highly preferred in the industry for mitigating the fluctuations of wind speed is known as the perturb and observe (P&O) algorithms, also called the name hill climbing search (HCS), by readjusting the state/control variables. Another advantage of these P&O algorithms is that they do not need any sensor or device for measurements and are consideredan efficient sensor-less mathematics-based optimization technique to determine or search for the MPP, which provides a pre-defined effective procedure for tracking of various essential variables needed for functioning of the controller. In recent years, many trending research opportunities have evolved around the application of the P&O algorithm, such as the comparison of the different available MPPT algorithms, to increase the capturing of wind power in an optimal process with the assistance of WESs.

These WESs implementing P&O algorithms must have some essential features as mentioned:

- It is an algorithm that does not need any sensing element, i.e., sensor-less.
- The algorithm is less complex to implement and cost-effective.
- There is no requirement of knowing the system parameters and variables, needs less memory allocation and computational resources.

- It is independent of parameter variation, i.e., sensitive to any parameter change.
- Operates in real-time and performs online updates.

Some of the important details and various features of the P&O algorithm pertaining to its wide application for various configurations of WESs are presented here. A classification of the P&O algorithms can be done with regards to the generated step-sizes and tracking methodologies. The first category that is based upon generated step-sizes, which can be further segregated into various types of step-sizes viz., fixed, variable, adaptive, and hybrid. In this type of categorization fails to highlight the operating policy of the P&O algorithms and tools or mechanism for their respective performance assessment. The second type of classification is based upon their tracking method that is further divided as the following two types: conventional and modified algorithms. Moreover, the conventional P&O-type algorithms are subdivided into algorithms with fixed and adjustable step-sizes. Whereas the modified P&O algorithms can be divided into subcategories of dividing power curve, using a generic objective function, integration of the various mathematical optimization methodologies and hybrid/mixed techniques. From the presented extensive analysis, on the use of P&O algorithms it is suggested that multiple future research paths can be approached and envisioned in accomplishing the MPPT algorithms with better robustness, achieving efficient, and effective dynamic performance of the overall WT systems, and decreasing the number of components and arrangements in WESs.

## 11.5  TYPES OF POWER ELECTRONIC CONVERTERS FOR GRID INTERFACING OF WES

As mentioned in previous sections, the role of power electronic converters are to provide an energy buffer between the intermittent generation and grid with stringent parameter conditions. Over the past two decades, several topologies of PEI or converters were applied and investigated for the WESs to achieve proper power conditioning under different operating scenarios. All the designed power electronic converters have their own merits and demerits. The different types are discussed here. Out of all these categories, the commercial WESs usually applies the topologies of unidirectional diode rectifier type converter and bidirectional back to back converter [12].

### 11.5.1  Unidirectional Diode Rectifier and Inverter-Based Converter

The harvested electric AC power from wind turbine generator (WTG) has the characteristics of time-varying magnitude and frequency. In unidirectional diode rectifier and inverter-based converter topology firstly converts the output of WTG to a DC power (at a constant voltage) using the diode rectifier circuit (called machine side converter) and then at the second stage, this topology converts the DC power back to the AC power at the desired voltage and frequency level to be matched with the upstream grid using the circuit of controlled inverter (called grid side converter). The uncontrolled diode rectifier-type converter topology only transfers the extracted electric power in one direction, i.e., from WTG to the interfacing grid. This category of PEI is commonly implemented in the synchronous generator, viz., the permanent magnet synchronous generator (PMSG) and the wound rotor synchronous generator (WRSG)-based WTG systems and not for the induction-type WTGs. In addition to that, for achieving the variable speed operation, the WRSG system also considers using an extra circuit for providing the excitation to the rotor excitation winding of the WRSG system. Similarly, the PMSG-type WTG systems makes use of a step-up chopper circuit that works as adapting/modifying the rectifier output voltage to the level of DC-bus link voltage at the input of the inverter. Here, in the PMSG-based system, if the inductor current of the step-up chopper is controlled then it can be used to regulate the torque and speed of the WTG. Also, in the present type of power electronic converter mechanism, the grid side inverter

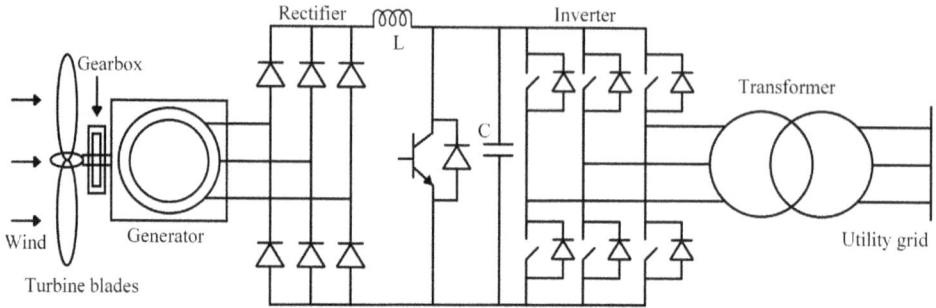

**FIGURE 11.6**   Diode rectifier topology for WECS.

is able to regulate the active and reactive power ($P$ and $Q$) that is being delivered to the utility grid at the intersection point. The range of voltage magnitude output from this PEI is 380–690 V. Hence, a step-up transformer is usually considered for injecting the harvested energy into the medium voltage (MV) distribution network. A general configuration of the diode rectifier topology for WECS is shown in Figure 11.6.

Advantages:

- As the production cost of this PEI system less, it is highly economical.
- The PEI architecture is quite simple and easy to implement.

Disadvantages:

- Diode rectifier existing in this converter produces a large amount of harmonics and ripples in the input current, hence, affecting the performance of the overall WECS.
- The PEI structure results in high harmonic losses in the obtained output voltage.
- The power handling capability of this converter setup is only limited to one direction.

### 11.5.2   BACK TO BACK BIDIRECTIONAL CONVERTER

The back to back bidirectional PEI too contains a rectifier and an inverter pair of converters connected in a back to back manner that are fully controlled type as it consists of two traditional pulse width modulated (PWM) voltage source inverters (VSIs). It is different from the earlier discussed diode rectifier type PEI converter in the rectification stage only, where the diode rectifier along with the step-up chopper circuit has been replaced by a fully controlled type rectifier. The advantage of this controlled rectifier is that it is capable of allowing power flow in two directions, i.e., from MSC to GSC and vice-versa that is not possible to realize in the diode rectifier type PEI. The advantage of using a controlled rectifier significantly mitigates the harmonics present in the input current and power losses cause due to harmonics. The GSC allows controlling of the $P$ and $Q$ power flow to the upstream grid as well as maintains the voltage of the DC-bus link at a constant level, which helps to improve the power quality of the output by decreasing the total harmonic distortion (THD). At the other end, the GSC functions as a driver that controls the demand for magnetic field inside the rotor and the WTG's speed. Furthermore, the decoupling capacitor or the DC-bus link between MSC and GSC gives controlling capabilities independently for these two back to back tied converters.

This back to back connected power electronic converter is mostly used for PMSG- and SCIG-type wind energy systems. Multi-national companies like Siemens implement back to back PEI for

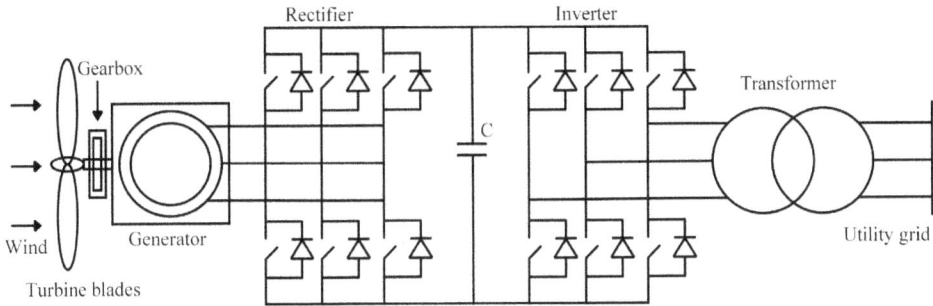

**FIGURE 11.7** Back to back converter topology for WECS.

the SCIG-based WESs. The rated voltage level/magnitude of the above WT generators mostly lie within the 380–690 V range. Hence, the PEI too have the same voltage ratings that uses of two-level converter topology. For connecting the WTs with MV distribution network, a step-up transformer is also used. In the past 20 years, a lot of research and development works were conducted on this converter and a lot of improvement on the overall performance has been achieved. Several novel controller design strategies have been developed mainly for the inverter section to improve the power quality of the supplied power.

In recent times, since the early 2000s this back to back bidirectional power electronic converter has also been considered for applications requiring partial rated converters such as the WECSs using the DFIG-type electrical machines. Figure 11.7 shows the schematic diagram representation of a DFIG-type WTG connected with a back to back bidirectional power electronic converter. There has been a gradual shift in manufacturing of WTs in recent times. Because of the lower rating of PEI most of the manufacturers like ABB, GE Wind, Ecotecnia, Repower, and Vestas presently install the DFIG machines for modern WTs. Such generators operate at both sub-synchronous and super-synchronous speed and the power electronic converter attached to the rotor should be capable of managing the flow of slip power under both the operational directions. When the speed of the WTG is lower than its rated synchronous speed, the rotor draws power as input from the grid and helps to speed up the rotor speed such that a balance can be maintained. Alternatively, when the shaft speed goes beyond the synchronous speed, the surplus power is fed into the grid through the rotor circuit. Several control topologies have been implemented to enhance the performance of the back to back bidirectional converter-type WECSs. Some of the important controller design architectures used in rotor circuits with this PEI are the vector control scheme, as well as recently included features of fault ride-through capabilities (also known as low voltage ride through, i.e., preventing any tripping during small duration of voltage sags) and crowbar dynamics of the system have also been implemented.

Advantages:

- The back to back architecture of this converter gives it the ability of the bidirectional power flow.
- The voltage magnitude at the DC-bus link can be boosted or stepped-up to a level greater than the grid line to line voltage ratings for attaining full control of the injected current into the distribution grid.
- The capacitor attached at the DC-bus link lies between the rectifier and inverter topologies allows one for controlling the two converters in a decoupled manner, which allows to compensate for the asymmetries existing on both the MSC and the GSC.
- As only partially rated PEI components are attached, the total cost is considerably less and is also available in modular form for commercial purposes.

Disadvantages:

- Because of use of large-sized bulky DC-bus link capacitor, it enhances the overall costs and decreases the lifespan of the PEI.
- Large switching losses as each commutation in both pair of converters involves steps of a hard switching and a natural commutation.
- To achieve high switching frequency, the back to back converter might need additional filter circuits.
- The complex controller design strategy are needed for combined operation of the fully controlled rectifier and inverter circuits.

### 11.5.3  Matrix Converter

The earlier PEI involved conversion of AC to DC to AC that needed bulky intermediate capacitor and filtering operation. This issue was eliminated by applying the matrix converter (MC) that is having some unique topology for converting AC to AC power without the need for intermediate DC conversion, i.e., it basically has a direct/single stage AC to AC conversion architecture. The matrix converter comprises of several bidirectional switches placed at the intersection junction of the input and output phases. The desired AC output can be obtained from the (wind energy harvested) AC input by properly choosing the switches to open and close. By attaching a filter at the input terminal of the MC helps to accomplish smooth commutation and avoids propagation of the switching harmonics to the output side of the MC. The converters provide the scope to reduce the size and weight significantly because of the absence of any large storage or filter component. Its size is much smaller when compared to a traditional-type back to back bidirectional converter. There, in recent times the MC has gained the attention of power electronic fraternity for its various merits and different applications, and since the later years of the last decade only, the matrix converters have attracted the interest of researchers for applications in WECSs. A schematic diagram depicting the MC used for the WTG system is shown in Figure 11.8.

Advantages:

- It gives an AC to AC power conversion architecture without using bulky any intermediate DC power components like filters and capacitors, hence, significantly reducing the size and weight of PEI.

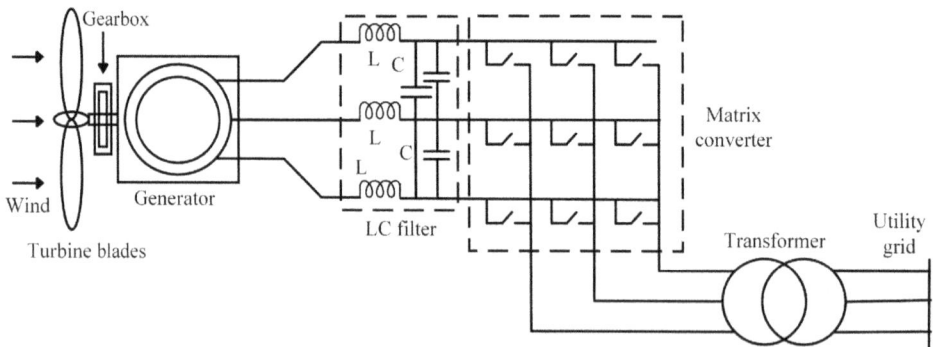

FIGURE 11.8   Matrix converter topology for WECS.

- The architecture of MC requires six extra power switches in comparison with the back to back bidirectional-type converter, even then, the MC-based PEI have better efficiency and greater lifetime as it lacks the bulky capacitor at the DC-bus link.
- By proper arrangement of these bidirectional switches, half of the switches in MC can be commutated naturally that would lead to a decrease in switching losses.
- Gives high power quality and output current having sinusoidal waveforms containing only a small amount of lower order harmonics.
- The power electronic semiconductor components in MC are exposed to a decreased level of thermal stresses than in other traditional converters.

Disadvantages:

- The voltage level at the DC-bus link cannot be stepped-up like in the case of back to back-type converter. In fact, the MC can only provide a maximum output voltage equal to 0.866 times the input voltage, if the modulation index is not violated.
- To obtain the same level of output voltage (or power) for the matrix converter to be at a par with the back to back converter, its output current should be 1.15 times more, which shall lead to increased conducting losses.
- As the DC-bus link is non-existing in the matrix converter, a decoupling is not available between the input side and output terminal. Under ideal operating conditions, this does not create any issue, however, during an unbalanced load condition or unbalanced/distorted input voltages, results in distortion of the output voltage and input current.
- In MC, the controller design for commutation as well as modulation schemes are highly complex as compared to the traditional PWM-type converters.
- The protection arrangement of the MC are not at a par with the bidirectional-type back to back converter in case of any fault condition arises.

### 11.5.4   Z-Source Converter

Z-source converters are novel types of PEI that are based upon the impedance source or impedance fed. This can be considered to realize the conversion of power for all the possible combination, viz., AC to AC, DC to AC, DC to DC, and AC to DC. It is able to eliminate some of the issues and limitations of the conventional voltage source and current source-type converters with its new concept of converting one form of power to another. In the Z-source converter, a two-port network consisting of split type inductors L1 and L2 and capacitors C1 and C2 tied in X shape are implemented to obtain an impedance source coupling for the PEI to the DC source, load, or other converters. Here, the DC source (or load) can be any component like either a current or may be a voltage source or even the load. Hence, active devices like generator or battery, power electronic converters like the diode rectifier, thyristor converter, or passive elements like an inductor, a capacitor, and any combination of these can be considered as the DC source. In recent times, a lot of research has been done to achieve proper steady state and transient characteristics of Z-source converter by working upon several some topologies having their own improvements in the certain performance. Furthermore, because of cost-effective and high efficiency it is quite suitable for the applications of WTGs. In recent times, multiple research works have applied the Z-source converter in WECSs. It can be designed for both unidirectional- and bidirectional-type wind energy harvesting systems. Figure 11.9 represents the schematic diagram Z-source converter used upon a WTG system.

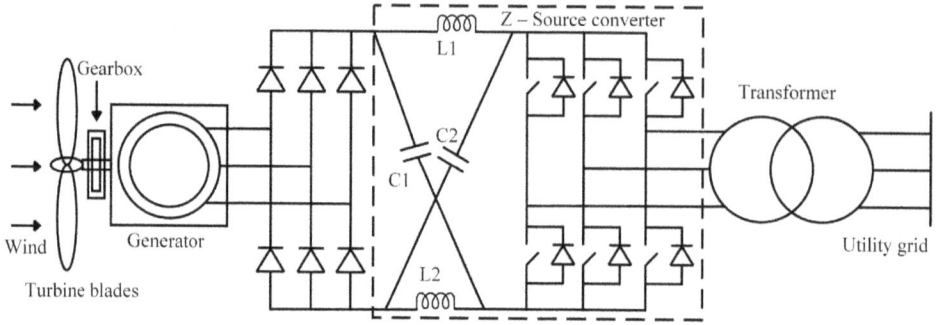

**FIGURE 11.9** Z-source converter topology for WECS.

Advantages:

- For any input voltage magnitude, the output voltage can be obtained of any magnitude between lying between zero and infinity, i.e., it is capable of performing both buck and boost-type converter operation.
- The Z-source converter needs fewer components than other existing traditional converters.
- It is more efficient and because of its ability to tackle the shoot-through or faulty conditions it is considered to be more reliable.
- As the Z-source converter allows to choose optimal values of inductors and capacitors, it is smaller in size and cost-effective in comparison to other traditional power electronics converter.

Disadvantages:

- The size and cost might increase due to use of Z-capacitors with high voltage rating for performing the voltage boost operation at the level of Z-source stage with lesser input voltage.
- During the start-up state, it fails to mitigate the rush current and is not able to prevent the resonance condition for using the Z-capacitors and Z-inductors, ultimately might lead to surge in voltage and current magnitude resulting into damaging the unit.
- Due to inherent nature Z-source converters are unidirectional type.

## 11.5.5 Improved Z-Source Converter

About a decade ago, an improved Z-source converter keeping intact all the characteristics of the Z-source converter and removing its demerits was developed. In this improved design, the same components of the Z-source converter are implemented, whereas exchanging the positions of the converter-bridge and diode as well as reversing their direction on the connecting switches. After this modification, the voltage polarity of the Z-capacitors remains the same as the polarity of the input voltage, which leads to achieving the same magnitude (or ratio) of step-up/boosting across the whole converter bridge also at reduced voltage stress on the Z-capacitor. This design does not provide any current path during the start-up, therefore, it inherently has the capability for limiting the inrush current. The block diagram layout of the WTG system fitted with an improved Z-source converter is represented in Figure 11.10.

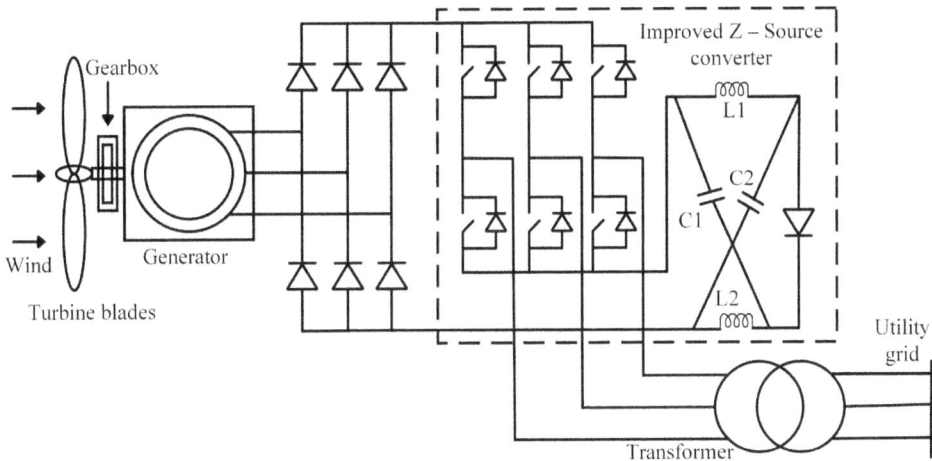

**FIGURE 11.10**   Improved Z-source converter topology for WECS.

## 11.5.6   CYCLOCONVERTER

Similar to the operation of matrix converter, the cycloconverters are the power electronic devices that convert the AC to AC directly without the requirement of intermediate DC conversion. The voltage level at the input and the output of this PEI usually remains the same and only the frequency is changed, hence, it is also called frequency converter. The high voltage AC (HVAC) and high voltage DC (HVDC) are considered for power transmission systems to decrease the losses. If the distance of power transmission is < 80 km than the HVAC-based transmission lines operating at 50 or 60 Hz are considered to be suitable. Whereas for the HVDC, because of the involvement of costly converter stations at both ends of the transmission line, it is selected for longer distance power transmission only. To decrease the electrical length of the AC transmission line and enhance the capacity of power transmission, some novel idea of lower-frequency (50/3 Hz) transmission line was investigated and suggested, where the system frequency was decreased by using the cycloconverters.

On a similar note, from the last decade onwards the concept of using a cycloconverter for offshore wind mills is being worked on. It has been observed that the establishment cost of the transmission system for extraction of power from remote wind farms using the low-frequency technology is less expensive in comparison to the HVDC-type transmission lines as well as lower expenses for the maintenance purpose also. Many works in the literature show the design, simulation, and performance assessment of lower-frequency transmission lines using the cycloconverter coastal wind mills. It is important to note that when the WTGs are extracting power at lower frequencies then the PEI needs one step-up cycloconverter to bring it up to the used frequency at the end of the transmission line. Whereas if the WTGs are operating at 50 or 60 Hz, then both the terminals of the transmission system needs frequency converters.

Advantages:

- As lower frequency is used, hence, the commutation and conduction losses are reduced.
- The cycloconverters have the inherent ability of allowing bidirectional power flow.
- The design of overall power circuit is quite compact in nature.

Disadvantages:

- There exists a limit to upper cutoff output frequency of the cycloconverter.
- The power factor is quite poor at the input side/terminal of the attached PEI.
- It has a lower voltage transfer ratio and needs complicated controller design circuitry.

### 11.5.7 MULTI-LEVEL CONVERTERS

The multi-level converters are the power electronic devices/circuits having the two operational mode as an inverter or a rectifier. In the multi-level converter design, several switching units having LV DC sources are connected in series to obtain an increased/higher required voltage level. The RESs like WTGs, solar PV panels, and fuel cells can be considered as the multiple sources of dc voltages. To obtain higher output voltage, a stringent controller mechanism is needed for these many DC sources such that the switching units operate in a superimposed staircase manner. In present times, the inter-facing of multiple sources to the grids have drastically heightened the focus of the researchers towards multi-level power converters giving special importance for applications in medium voltage lines. In WTGs, the multilevel converter gives a mechanism to directly inject the harvested energy from the source into the upstream 3-phase utility network. For this purpose there are three important topologies of multilevel converters, viz., neutral point clamped (NPC), flying capacitor (FC), and series connected H-bridge (SCHB) that are generally used in the case of medium voltage and high power applications.

Advantages:

- It has the most important benefit of being able to tackle high voltage even with the low rating sources or devices.
- The multi-level converter has high overall efficiency and less losses due to switching.
- Uses reduced sized filter elements that improves the transient response of the converter.

Disadvantages:

- Needs multiple semiconductor devices that leads to increased complexity of the developed controlling mechanism for the circuit.
- There exists imbalance voltage within the upper and the lower DC-bus link capacitor.
- Results in applied current stress on the semiconductor devices in an unequal manner.

## 11.6  DIFFERENT APPLICATIONS OF WES

With the technological developments in various aspects like materials to construct the various components of wind turbines, PEI, usage of the extracted or harvested energy are now becoming more and more economical and environmental friendly. This source of electricity can be used in several different types of applications depending upon the country of location of the wind farms, and new use of WESs are also going to be developed in the coming 10 to 20 years.

In India, where there is a variety of terrains, large coastal or offshore, hilly areas, and desert having different behavior of the blowing wind. The harvested wind energy is highly significant as the generated electricity has to be carefully exploited or consumed for fulfilling basic necessities like bringing water out of underground through running a pumping system. As wind energy systems can be located in any part of the country, like both urban and rural places and the usage can be different as per the requirements of the peoples. Furthermore, the extracted energy from the wind can be converted into different forms like mechanical (to run a motor) and electrical (in a storage battery) energies. Hence, wind energy in a diversified country like India are usually implemented to operate wind-energy-based water pumps, battery-based storage using proper charging system, wind electricity generators, etc. Presently, in India, about 20,000 MW of electricity is being generated from the blowing wind by several wind farms or mills established in different states of the country. It is also considered as one of the world's fastest growing sources of energy just after the solar PV systems [13].

Different applications of WESs can be summarized as follows:

1. The energy in high-speed wind is considered for propelling of the sailboats across the rivers for transporting peoples and goods from one side to other.
2. Wind energy is also suitable for running water pumping systems to bring out the underground water via wind mills for both irrigation and drinking purposes.
3. Another utilization of the wind energy system has been to run flourmills for grinding the harvested grains/crops like wheat as well as corn to prepare the flour.
4. The best application of wind energy is in generating electricity that can subsequently be converted into any desired form as discussed in the above.

## 11.7   FUTURE TRENDS AND RESEARCH CHALLENGES IN WES

From the aforesaid discussion on various aspects of wind energy conversion systems, there are various future research directions/trends and challenging topics that are needed to be explored and implemented. Some of these valuable trending points in WECS are represented here.

- There is a need for accurate estimation of the mechanical stress falling upon the turbine blades and gearboxes during its operation at higher speed limits of the wind that can help in preventing any fatigue or breakdown of the WES.
- Use of small-scale WTs over a floating base on the water surface as well as tulip inspired vertical axis type rooftop WTs are also considered for establishing WESs at community level MGs nearby the demand premise [14].
- More work needs to be done in the design and development of portable or small-scale WTs that are easy to transport to remote locations and easy to assemble/install and can act as a substantial energy source if the area falls under high wind speed zone.
- For maintaining uninterrupted electricity to be fed into MGs having critical loads, the WTs should be protected from short circuit faults, and voltage surges. To accomplish this task, fast-acting relays like field programmable grid array (FPGA)-based digital relays can be explored for switching it from the grid-connected mode to the isolating and reconnecting, if the fault is mitigated.
- Improving the efficiency of various power electronics converters used in WESs, and the storage device that act as buffer between the intermittent energy production as well as time-varying load [15].
- Application of artificial intelligence (AI) and machine learning (ML)-based forecasting techniques to predict the renewable generation from WTs and the load demand profile within an MG entity are required to be researched for achieving accurate matching of the energy demand and supply.

From the above discussion on the future trends and research gaps for wind energy conversion systems, the technical challenges for further improvement can be mentioned as follows:

- Wind energy must be affordable or cost-effective at a par with traditional energy generation sources. Although the wind energy cost has been in declining mode in recent times, wind mill projects should be established after full techno-economic analysis at any particular location (must be windy enough) that must be competitive enough to provide electricity at lowest possible price.

- As wind sites with large land base are often situated in remote locations, which are usually at far off distance from the cities or load centers where the electricity is consumed. There is a need to bring more connectivity of newly setup transmission lines with the existing ones for expanding the use of wind energy.

- Improving the understanding of atmospheric phenomena that determines the flow of wind throughout the year during different seasons and physics of Earth's landscape according to the location to effectively design the WT components that increases the efficiency of the wind energy system [16].

- The rotating turbines blades might make too much of noise and aesthetic pollution. Proper greasing of the rotating components must be done from time to time to prevent it. Although WTs have considerably less environmental impact in comparison to the existing traditional sources, concerns like noise, regular maintenance, and visual impacts to the landscape are still a challenge to address.

- There is a harmful effect of wind mills on the habitat of local wildlife. Mostly the flying birds that enter into the spinning/rotating turbine blades are getting killed frequently. Also, to clear the ground or land and make space for blades' rotation, a lot of trees might be need to be cut down, which further downgrades the environment.

- Understanding the aerodynamics, structural dynamics, especially of the large-sized blade WTs located on offshore sites. The ratings of such WTs goes beyond 1 MW and installing such a giant machine needs amalgamation of civil, mechanical, and electrical engineering to handle forces and the moment of inertia from all directions, work efficiently, and last over its complete operational lifespan extending up to or more than 20 years.

- The large-scale land that is used to develop a wind energy plant is not considered to be the most profitable way to make use of the available land. Therefore, a land suitable for installing WTs must be capable of generating electricity whose value can be compared with other uses for the land.

- Controller design mechanism to tackle various state variables associated while integrating of wind energy systems into the smart electric grid. As these state variables are operating on different time scales, providing instantaneous power, maintaining the rate values of voltage and current, meeting the demand throughout the day are come important objectives of control strategies implemented on WESs [17].

- Depending upon the control objective needs like stability and reliability of overall grid, its operational dynamics, as well as long-term planning are of varying time periods/scales from the interval of sub-seconds to decades. Hence, keeping both short-term and long-term goals in mind while installing new wind energy projects is of high priority [18].

## BIBLIOGRAPHY

1.  S. Sen, and V. Kumar, "Assessment of various MOR techniques on an inverter-based microgrid model," 14th IEEE India Council International Conference, pp. 1–6, 2017.
2.  S. Sen, and V. Kumar, "Robust decentralized output-feedback type LQR voltage controller for inverter-based microgrid," *10th Int. Conf. on Computing, Communication and Networking Technologies (ICCCNT)*, pp. 1–6, July 2019.
3.  M. Kumar, S. Sen, and S. Kumar, "A robust performance analysis of a solar PV-battery based islanded microgrid inverter output voltage control using dual-loop PID controller," *IEEE IAS Global Conf. on Emerging Technologies*, May 2022. (Accepted)
4.  S. Sen, and V. Kumar, "Microgrid control: a comprehensive survey," *Annual Reviews in Control*, vol. 45, pp. 118–151, 2018.

5.  K. B. Tawfiq, A. S. Mansour, H. S. Ramadan, M. Becherif, and E. E. El-kholy, "Wind Energy Conversion System Topologies and Converters: Comparative Review," *Energy Procedia*, vol. 162, pp. 38–47, April 2019.

6.  H. H. H. Mousa, A. R. Youssef, and E. E. M. Mohamed, "State of the art perturb and observe MPPT algorithms based wind energy conversion systems: A technology review," *International Journal of Electrical Power and Energy Systems*, vol. 26, p. 106598, 2020.

7.  S. Sen, and V. Kumar, "Microgrid modelling: A comprehensive survey," *Annual Reviews in Control*, vol. 46, pp. 216–250, 2018.

8.  Z. Alnasir, and M. Kazerani, "An analytical literature review of stand-alone wind energy conversion systems from generator viewpoint," *Renewable and Sustainable Energy Reviews*, vol. 28, pp. 597–615, Sept. 2013.

9.  A. M. Howlader, and T. Senjyu, "A comprehensive review of low voltage ride through capability strategies for the wind energy conversion systems," *Renewable and Sustainable Energy Reviews*, vol, 56, pp. 643–658, Nov. 2015.

10. D. Rajababu, and K. Raghu Ram, "Performance analysis of isolated wind energy conversion system connected to unbalanced loads," Materials Today: Proceedings, March 2021.

11. S. Sen and V. Kumar, "Decentralized output-feedback based robust LQR V-f controller for PV-battery microgrid including generation uncertainties," *IEEE Systems Journal*, vol. 14, no. 3, pp. 4418–4429, Sept. 2020.

12. M. R. Islam, Y. Guo, and J. Zhu "Power converters for wind turbines: Current and future development," Materials and processes for energy: communicating current research and technological developments. pp. 559–571, Aug 2013.

13. M. Kumar, and B. Tyagi, "Multi-variable constrained nonlinear optimal planning and operation problem for isolated microgrids with stochasticity in wind, solar, and load demand data," *IET Generation, Transmission and Distribution*, vol. 14, no. 11, pp. 2181–2190, 2020.

14. Z. Tang, Y. Yang, and F. Blaabjerg, "Power electronics: The enabling technology for renewable energy integration," *CSEE Journal of Power and Energy Systems*, vol. 8, no. 1, pp. 39–52, Jan. 2022.

15. S. Sen, and V. Kumar, "Simplified modeling and HIL validation of solar PVs and storage based islanded microgrid with generation uncertainties," *IEEE Systems Journal*, vol. 14, no. 2, pp. 2653–2664, June 2020.

16. M. Kumar, and B. Tyagi, "An optimal multivariable constrained nonlinear (MVCNL) stochastic microgrid planning and operation problem with renewable penetration," *IEEE Systems Journal*, vol. 14, no. 3, pp. 4143–4154, 2020.

17. S. Sen, and M. Kumar, "MPC based energy management system for grid-connected smart buildings with EVs," IEEE IAS Global Conference on Emerging Technologies, May 2022. (Accepted)

18. T. J. Ding, C. C. W. Chang, M. A. S. Bhuiyan, K. N. Minhad, and K. Ali, "Advancements of wind energy conversion systems for low-wind urban environments: A review," *Energy Reports*, vol. 8, pp. 3406–3414, Feb. 2022.

# 12 Trends in Energy Storage Devices Incorporating Lead-Acid Batteries and Supercapacitors for Smart Grid Applications

*Maneesh Kumar and Sachidananda Sen*

## 12.1 INTRODUCTION

A reliable electricity supply is one of the main catalysts to the smooth running of a civilization. The demand for energy is continuously rising day by day, and conventional energy sources, such as coal and petroleum, have their own limitations in terms of availability and environmental impact. Currently, the utilization of renewable energy in the system is in full swing, and researchers are finding the best ways to integrate these sources with the existing grid. The storage system provides support against the variability of the RESs and also supports the grid as a dispatchable unit. It also provides virtual inertia to the low inertia microgrid systems. Energy can be stored in these units in thermal form, chemical form, mechanical form, or electrical form. Chemical energy storage involves the use of batteries and fuel cells. Batteries and supercapacitors are electrochemical energy storage devices.

In batteries, the electrochemical reaction occurs between electrodes (anode and cathode) and the electrolyte. In comparison, in supercapacitors the energy is stored and released by reversible adsorption and desorption of ions at the interfaces between electrode materials and electrolytes. Based upon the application, there are various types of batteries available, such as nickel-cadmium (Ni-Cd), nickel-metal hydride (Ni-MH), lithium-ion (Li-ion), and lead acid (PbA) batteries. Out of the various available storage batteries, Li-ion and lead acid are more prominently used with renewable energy sources because of their high energy densities. The use of supercapacitors also increases with the photovoltaic system nowadays because of their strong energy storage capacity and very strong pulse power. In this chapter, a comprehensive study of the recent trends in the storage system is discussed more specifically with lead acid and supercapacitors in relation to the smart grid applications. The classifications of various storage energy technologies are provided in Figure 12.1. This figure shows the classification of various available storage devices based on their operating cycle, characteristics, power, and energy densities, etc. The flywheel, compressed air, and hydro-pumped storage belong to electromechanical energy storage. The superconducting material storage device comes under electromagnetic energy storage. The storage batteries come under electrochemical storage, and the fuel cells are based on the chemical reaction. The supercapacitors, inductors, and capacitors belong to the electrostatic storage principle. Further, the operating time also classifies these storage devices as given in the figure. The energy storage devices can be of different types: spinning and non-spinning type, or it could be based upon their constituent material or functional characteristics like electromechanical, electromagnetic, electrochemical, chemical,

DOI: 10.1201/9781003340539-12

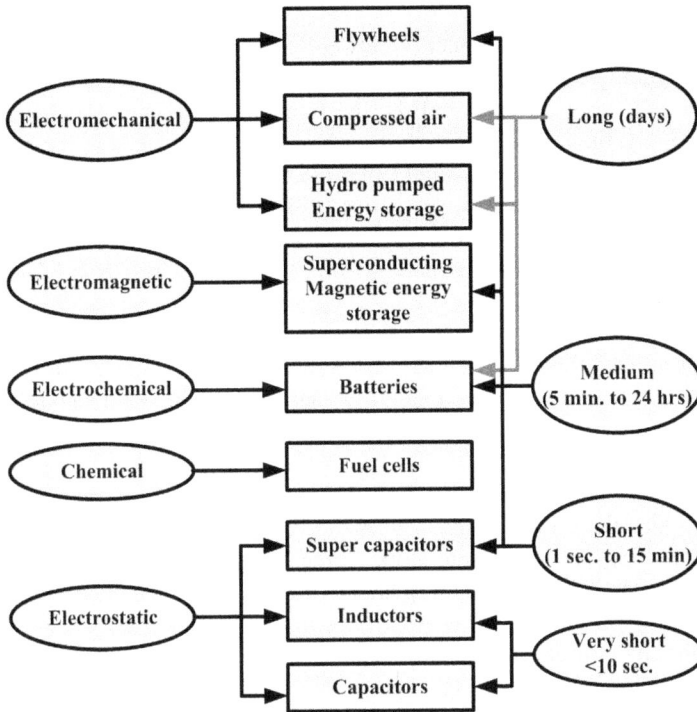

FIGURE 12.1   Flowchart enumerating the classification of various ESSs.

and electrostatic type. Under each category some energy storage devices are being used. Among all these devices lead-acid batteries fall under the electrochemical category having the support duration ranging from minutes to days as per their connected capability. Similarly, the supercapacitor falls under the electrostatic-type energy storage device that is applicable for low-duration energy support applications that ranges from seconds to minutes.

## 12.2   APPLICATIONS/BENEFITS OF USING ENERGY STORAGE DEVICES

In energy harvesting, the energy storage devices play a very crucial role for storing and releasing the stored energy as per the need of the overall smart grid. Furthermore, the storage devices are also useful to accommodate the generation intermittency due to integrated renewable energy sources (RESs) [1]. Hence, because of the use of energy storage devices both grid operators and the customers benefit.

- **It gives flexibility during peak load demand**
  As peak demand of the load occurs only for a few hours of the day, the energy stored in the storage devices like the lead-acid batteries and supercapacitors can be used to supply the deficient in generated power during the peak load demand.
- **Energy arbitrage**
  The distribution system operator (DSO) can use the stored energy for arbitrage purposes, i.e., benefiting from the price difference of electricity in locational marginal price (LMP) at different time instants like peak and off-peak load situation. It can be used to offset the cost incurred for installation and maintenance of storage devices.

- **Frequency regulation**

  The dynamic balancing of the energy between the effective power generation/supply and load demand dictates the frequency deviation from the rated value. More the gap between the demand and supply, greater will be the frequency deviation. Therefore, if demand is more than supply, then frequency decreases (this scenario is taken as the positive frequency deviation). Similarly, when the supply is more than the demand, then the grid frequency increases (frequency deviation is negative). By using the energy storage devices, i.e., absorbing surplus generation and supplying the same during energy-deficient condition will help to regulate or control the frequency. Here, the active power is managed that is related with the frequency.

- **Voltage support**

  On a similar ground, the reactive power difference can also be managed by using the energy storage devices. By providing the support for reactive power, the grid voltage can be regulated/supported.

- **Black start**

  Under the power-outage condition, where the whole grid is shut down. Black start is the process by which the grid is established by turning on each generator as per their ascending order of their ratings. Here, the role of the storage device is to provide the excitation and initial support for starting the initial few generators after these can handle the overall load of the network.

- **Reserve capacity (spin/non-spin)**

  A certain level of energy generation is kept under reserve to tackle the sudden surge in the load demand. This surplus energy capacity is known as the reserve capacity. This can be divided under two categories: spin and non-spin. By keeping the generators' running leads to inefficiencies, puts additional cost burden and waste of energy. Therefore, by using proper energy storage devices having the characteristics of being fast-acting like the supercapacitors, lead-acid batteries (non-spinning types) and flywheels (spinning type) can be applied as a suitable replacement.

- **Increasing self-consumption capacity using storage and solar photovoltaic**

  In the present active distribution network, the consumers are preferring to install the solar photovoltaic (PV)-based energy sources to reduce the dependency on the grid as well as reduce the electricity bill. The load, installed solar PV and grid forms a microgrid (MG) in the demand premise of the consumer. This MG can be either operated as the grid-connected mode as well as the off-grid isolated mode. By using the energy storage devices the uncertainties in the renewable energy usage can be reduced up to a certain extent [2].

## 12.3   PARAMETERS PERTAINING TO ENERGY STORAGE DEVICES

**Charge (or Ahr) capacity**: the electric charge that an electrochemical or electrostatic storage device can provide is one of the crucial parameters. Its SI unit is expressed in Coulomb, which signifies that the amount of charge when one Ampere of current flows for one second. Suppose the capacity of an energy storage unit is represented to be 50 Amp hours. This simply reflects the fact that it is capable of supplying 5 Amp for 10 hours.

**Terminal voltage magnitude:** all electrical energy storage units possess a nominal or rated voltage magnitude, which represents the approximate voltage level of electrical power that is being delivered. Multiple cells or storage units that are tied in series to achieve the desired level of aggregate voltage. Most of such storage components like the battery can be expressed as having a fixed voltage potential $E$, however, it has a different voltage level $V$, due to some voltage drop occurring because of its internal resistance $R$, i.e., $V = E - \mathrm{IR}$.

**Specific energy**: it is defined as the amount of energy stored (mostly electrical energy) per kilogram of mass of the storage device (mostly batteries). Its units is expressed as Wh/kg.

**Energy density**: it is the amount of electrical energy stored per unit volume (cubic meter) of the energy storage unit. It normally has units of Wh/m$^3$.

**Specific power**: it is the amount of power that can be extracted or obtained per kilogram of the storage device. This parameter can be highly variable rather than being an anomalous or constant value or quantity. As the power supplied by the any storage unit (especially the lead-acid and lithium-ion type batteries) mainly depends more on the attached load at its terminal than the battery or the storage device itself.

Suppose a storage device having a high specific energy, and low specific power characteristics, would imply that the device is capable to store a lot of energy, however, exergy can be extracted at a much slower pace.

**Ahr (or charge) efficiency**: ideally in the case of a battery or any storage device, it is supposed to return or give back the complete electrical charge stored or put into it. Under this scenario, the ampere hour or Ahr efficiency of the storage unit is 100%. Unfortunately, no storage device or any battery is capable to do so, hence, their charging efficiencies are observed to be always less than 100%. The exact value would change or vary depending on the type of the storage battery, ramping rate of charging and discharging, surface temperature, and state of charge.

**State of charge (SOC)**: an important parameter in the storage devices especially in the case of batteries is the SOC. It is defined as the measure or amount of the residual/remaining capacity of a battery. It is expressed as the percentage of remaining capacity with respect to the full capacity that is represent as 100%. It gives an idea of how much energy is still left in the battery and can be used to estimate how long it can further support a load.

**Depth of discharge (DOD)**: another complementary parameter that is similar to the SOC of the batteries is the depth of discharge (DOD). It is defined as the percentage of battery capacity that has been used up or in other words up to what level the battery is discharged. The allowable maximum DOD of batteries lies between the 10–20% range and going below this range frequently shall decrease the lifespan of the battery due to high deterioration effect.

**Energy efficiency**: the efficiency of any device is an effective way to show the working relationship between output and input. For storage devices, the energy efficiency is represented as the ratio of electrical energy provided by the storage device to the amount of electrical energy needed to get back to the state before discharging was initiated. This ratio mainly depends upon the internal resistance of the storage device preferably a battery. It can be calculated as the ratio of electrical energy provided by a battery to the amount of energy given to the battery by charging so that it returns back to its earlier SOC status just before the discharging happened. Presently, the range of 55–95% is considered to be suitably existing energy efficiency of a battery. The more the energy efficiency, the greater the cost of the battery. Hence, it is selected based upon the application and cost analysis of the project.

**Ramping rate**: it is defined as the rate at which discharging (or charging) is possible for a battery. Rate of change in energy delivered or received dictates the operational speed of the battery to tackle the load demand.

**Self-discharge rates:** most of the storage devices especially the batteries automatically discharge when these are left unused or in the idle state for some time. This phenomenon of losing energy by themselves is known as self-discharging of the batteries. This parameter is highly important as it implies that some of the batteries or storage units cannot be left idle for long periods without recharging to avoid complete draining that could damage the battery. The slower the self-discharging rate, the better the quality of the battery. This self-discharging rate varies or depends upon the type of the battery being considered, and the operating or surrounding temperature, which greatly increases the self-discharge.

**Temperature, heating, and cooling needs**: the working environment also dictates the operational efficiency of the storage devices. Even though most of the storage units especially batteries run at ambient temperature, a few of them operate at higher temperatures and require initial heating from the surroundings during the starting phase with and then further need a cooling system during their running. Therefore, the performance of the batteries decreases at lower temperatures, which is considered to be not appropriate and it could reduce the overall performance. However, this issue can be compensated by providing some initial heat to the battery. This point needs to be considered while selecting a battery as an engineer must have the knowledge of the various requirements, viz., temperature, heating, and cooling of the battery during the design process itself.

**Battery life and number of deep cycles**: the life expectancy of the battery depends upon its operating conditions and how much sudden or sharp charging and discharging conditions are varied. The lead-acid or lithium-ion batteries can perform or sustain a few hundred deep discharging cycles, i.e., going below 20% of the battery SOC. However, the precise number of deep discharging cycles can be decided based on the type of the battery, its design details, and on how well the battery has been used by the operator. This is considered to be a crucial factor or parameter in a battery specification, as it gives an idea about the lifespan of a battery, which suggests its replacement duration and overall cost of the storage device.

## 12.4   LEAD-ACID BATTERIES

The most common and cheapest vehicle batteries are the flooded PbA batteries. These are generally of two types:

- Deep cycle batteries. These types of PbA batteries are utilized to run forklifts, etc. These are utilized as auxiliary batteries in recreational vehicles, although they require multi-stage charging. With more than 50% discharge, these batteries face the issue of low battery life.
- PbA as starter batteries: these batteries are used to start the automobile with a low percentage of their capacity. It provides a high charging rate, which helps in the start of the vehicle's engine.

Flooded types of PbA batteries require a frequent inspection of the electrolyte levels and also the occasional replacement of water. The reason behind this is the vaporization during the charging cycle. The PbA batteries were used in most electric vehicles (EVs) because of their availability, well-developed technology, and low cost. Similar to other batteries, the PbA also has a lower specific energy than petroleum fuel. Therefore, it did not get much encouragement for EVs further. Although, the operation and charging of these batteries produce gases such as hydrogen, oxygen, and sulfur, which are normally harmless if these are vented properly and also occur naturally.

## 12.4.1  FACTORS AFFECTING LIFE CYCLE OF PbA BATTERIES

### 12.4.1.1  Over and Undercharging

The overcharging of a battery beyond its capacity has some harmful impacts, such as the following:

- Reduction in water level: the water level reduces because of its breakdown in the electrolyte, and a gas is formed. This gas tends to scrub the active material of the electrode.
- Electrode plate deformation: overcharging of PbA tries to deform the plates, which in turn damage the separators.

The insufficient charge for a long duration also creates issues in PbA as given:

- Plate sulphation: this results in the deformation of the plates and a reduction in the specific gravity of the plates.
- With low gravity, the chances of freezing are greater.
- An insufficiently charged battery also fails to provide the required power to the system.

### 12.4.1.2  Loss of Active Material

The active material loses when a successive expansion, as well as the contraction, occurs during the charging and discharging of the battery. This action increases with increased use of the cell and imposes constraints on the battery life. One of the indications of active material loss is the formation of sediments at the bottom of the cell.

### 12.4.1.3  Local Aalvanic Action

This is referred to as the electrolytic action that occurs between the grid and the active material of the plate. This in turn causes the production of hydrogen gas form the negative plate and reduce the voltage output of the battery. In this case, the galvanic action occurs uniformly inside the battery, no major issues will take place in the performance as sheen experimentally. On the other hand, if this confined nonuniformly to only a few cells of the battery, the floating voltage gets reduced, which indicates that the PbA is not getting charged.

### 12.4.1.4  Electrolytic Action

The primary reason for this is the direct contact between the electrolyte used and the battery's positive plate. This issue is less serious with a battery in use, provided it is not localized. Local confinement of this effect may increase the plate material degradation rapidly.

### 12.4.1.5  Excessive Rate of Charge/Discharge

This can cause damage to the battery by increasing the temperatures to a very high limit. This also deteriorates the battery life.

### 12.4.1.6  Impurities Insertion

It occurs with the addition of water to the electrolyte of the battery. This causes the local galvanic action inside the battery and hence affects the battery life.

### 12.4.1.7  Low Water Levels

The low water level increases the acid concentration in the cell, which in turn affects the wood separator. This affects the operation of the plates and impairs them permanently. As discussed earlier, the local galvanic action also increases because of low water levels, which reduces the cell voltage.

Therefore, before charging a battery, the appropriate amount of water should always be present for the mixing with acid.

## 12.4.2  FACTORS AFFECTING BATTERY SOC

### a. Charging/discharging rate

With the change in rate of charging and discharging, battery SOC and the battery capacity also changes. The relationship between battery capacity and charging/discharging rate can be obtained by Peukert's equation:

$$C = Ki_c^{1-n}$$

where $C$ is the battery capacity, $n$, $K$ are the battery constants and $i_c$ is charging current of battery [3, 4].

### b. Temperature

A common model used to provide the relationship between temperature and SOC is shown below.

$$C = C_{std}\left[1-\alpha\left(25-T\right)\right]$$

where $C_{std}$ is the standard battery operating temperature, $\alpha$ is the temperature coefficient, and $T$ is the battery temperature.

### c. Battery aging

As the internal chemical structure of the battery will change with calendar aging and cycle aging, its performance will decrease. The effect of the battery aging can be seen with the help of battery present and initial capacity.

$$\text{Battery state of health} = \frac{C_p}{C_{in}} * 100\left(\%\right)$$

where $C_p$ is the present capacity of the battery. $C_{in}$ is the battery initial capacity.

### d. Self-discharging

Generally provided by the battery manufacturer. This phenomenon is related to internal chemical properties of the battery.

### e. Internal resistance of the battery.

The internal resistance of the battery decreases with the aging of battery. It influences the SOC of the battery and can be demonstrated through the battery equivalent circuits.

## 12.4.3  BATTERY MODELING

The battery models can be categorized into two types: traditional and improved models as discussed in the following.

### 12.4.3.1 Traditional Models

There are three basic traditional models of equivalent circuits.

 a. Thevenin's model.
 b. Internal resistance ($R_{in}$) model.
 c. Partnership with a new generation of vehicle (PNGV) model.

These are discussed in the following:

 a. Thevenin's model: It is the simplest model as shown in Figure 12.2.

Where $V_{oc}$ is the open circuit voltage, $R_s$ is the series resistance, $C_p$ represents the polarization capacitor, $R_p$ denotes the polarization resistance. $U_t$ is the load voltage.

 b. $R_{in}$ model

This is the simplest battery model with series resistance $R_s$, open circuit voltage $V_{oc}$ and the load voltage $U_t$ as shown in Figure 12.3.

 c. PNGV model

In this model an additional capacitor is added into the circuit to make the $V_{oc}$ constant. Figure 12.4 shows the PNGV model of storage battery.

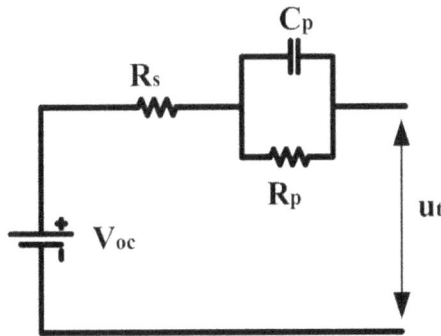

**FIGURE 12.2** Thevenin's equivalent model of storage battery.

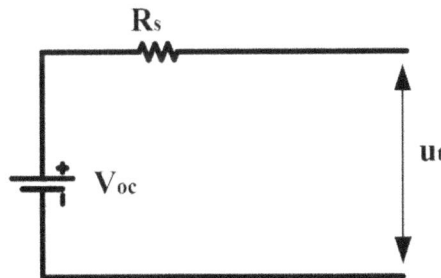

**FIGURE 12.3** internal resistance model of storage battery.

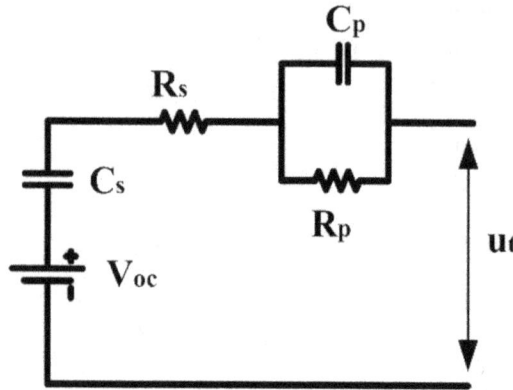

**FIGURE 12.4**  PNGV model of storage battery.

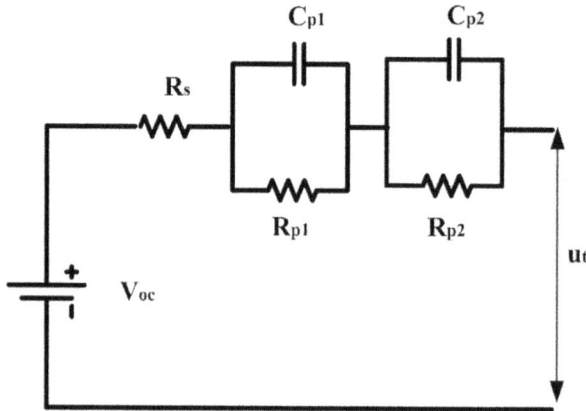

**FIGURE 12.5**  DP model of storage battery.

## 12.4.4  IMPROVED BATTERY MODELS

### a. DP model

In this model an addition RC circuit is added. To increase the model accuracy, the polarization characteristics has also been considered. $R_{p1}$ and $R_{p2}$ are the electrochemical polarization resistance, and the concentration polarization resistance, respectively. The equivalent capacitances, which include $C_{p1}$ and $C_{p2}$ (polarization capacitors), are used to describe the transient response of the battery during charging/discharging processes as shown in Figure 12.5.

### b. Extended Thevenin's model

In the extended Thevenin model, as shown in Figure 12.6, more RC circuits are added to increase the overall accuracy of the model.

Comprehensive model
The model is divided into two parts. The left part simulates the lifetime of the battery and the right part shows the transient response of the battery as given in Figure 12.7. There exist several factors important affecting the battery model as summarized in Table 12.1.

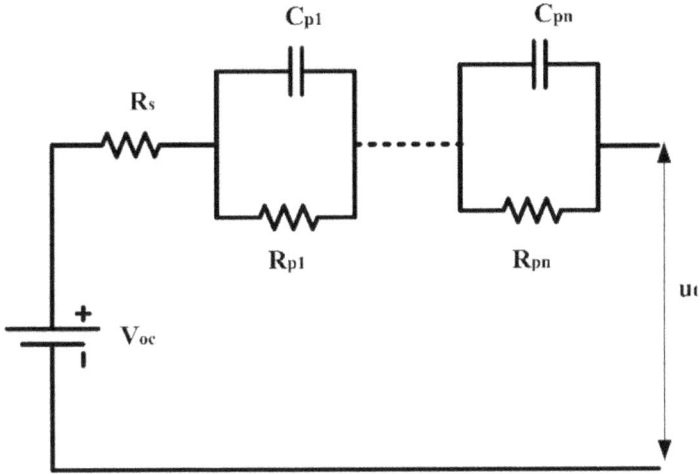

**FIGURE 12.6**   Extended Thevenin's model of storage battery.

**TABLE 12.1**
**Key Factors Affecting the Battery Model**

| Model | Polarization | Self-discharging | Aging |
|---|---|---|---|
| Thevenin | YES | NO | NO |
| $R_{in}$ | NO | NO | NO |
| PNGV model | YES | NO | NO |
| DP model | YES | NO | NO |
| Extended thevenin model | YES | NO | NO |
| Comprehensive battery model 1 | YES | YES | NO |

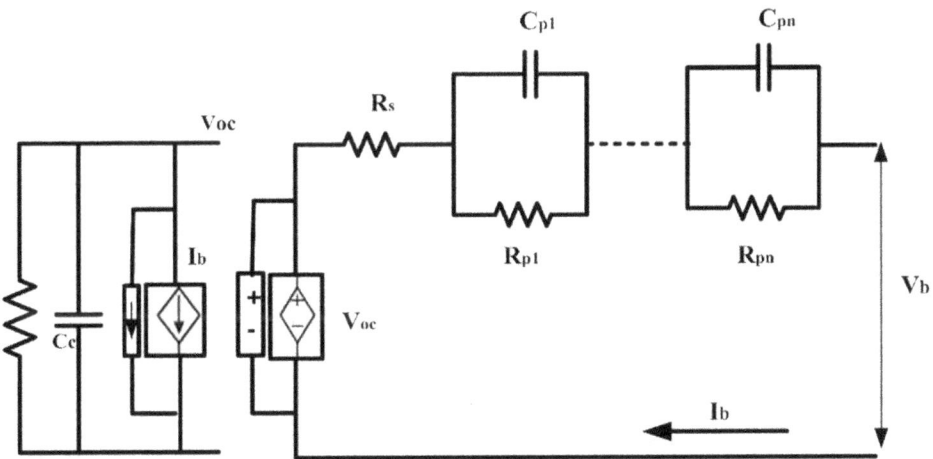

**FIGURE 12.7**   Comprehensive model of storage battery.

Battery life
Factors affecting battery life can be mentioned to be as follows:

a. High voltage
b. High current rate
c. Low voltage
d. High pressure
e. Temperature

Out of the five factors, the temperature plays a vital role in the degradation of battery life. Therefore, a proper thermal management is required for the battery to operate more accurately [5–7].

The heating issues related to a battery can be incorporated with the help of a proper battery management system.

### 12.4.5 ADVANCED LEAD-ACID BATTERIES FOR GRID APPLICATIONS

In this section, we will discuss the development of advanced lead-acid battery technology for grid applications. The first lead-acid battery was assembled by Gaston Plante more than 150 years ago. After that, a comprehensive refinement was observed in the lead-acid-based technology. A consortium was formed in the year of 1992, which has been the primary sponsor of the advancement of a new generation of lead-acid battery design over the past 25 years. The consortium was formed to investigate the performance improvements and limitations of the lead-acid batteries. The valve-regulated lead-acid (VRLA) batteries are the further evolution of the lead-acid batteries. The VRLA technology solved a significant problem of frequent maintenance. These types of cells require minimal maintenance and also much less electrolytic content. Integrating the carbon in the negative electrode of the cell has allowed the VRLA cells to enter the new application areas and extend the cycling [8–14]. The life span of the VRLA cells depends upon many factors, such as the number of failure mechanisms, corrosion, breakdown of positive plate material, negative material sulfation, etc. [15]. The battery cycling is more associated with the positive plate material breakdown and the sulfation of the negative plate [16]. As discussed earlier, the use of carbon opens a number of improvement paths, initially as a conductive material for electrodes and recently as a capacitive component [17]. A carbon lead-acid (PbC) battery is developed by Axion, which is different from the conventional VRLA battery. This developed carbon lead-acid battery is claimed to accept high charge and a comprehensive improvement in the life cycle, approximately four times as compared to VRLA.

However, discharging of PbC is associated with a wide voltage range, unlike VRLA and, in turn, increases the costs associated with the AC/DC conversion. A USA-based firm named Firefly Int. Co. introduced a design of a cell that incorporates carbon foam as a substrate for the active material of the electrode. An Australian Firm Common-Wealth Scientific and Industrial Research has found that instead of mixing carbon with negative active material (NAM) of PbA battery, the carbon could be placed in VRAL to differentiate it from the NAM [18–20]. This allows to form an ultra-battery configuration by utilizing both. This allowed the formation of an ultra-battery by utilizing an asymmetric supercapacitor construction, which is to be merged with VRLA cell construction. The configuration is shown in Figure 12.8.

The ultra-batteries have low-power handling capacities while discharging as compared to conventional VRAL, but at the same time, they have higher-power charging capacity. Many competing technologies have constraints over charging/discharging, while SoC lies within certain limits. This makes the grid operation more complex while utilizing the energy storage in the system. The power engineers have started discussing mileage terms over the working period. This is nothing but the

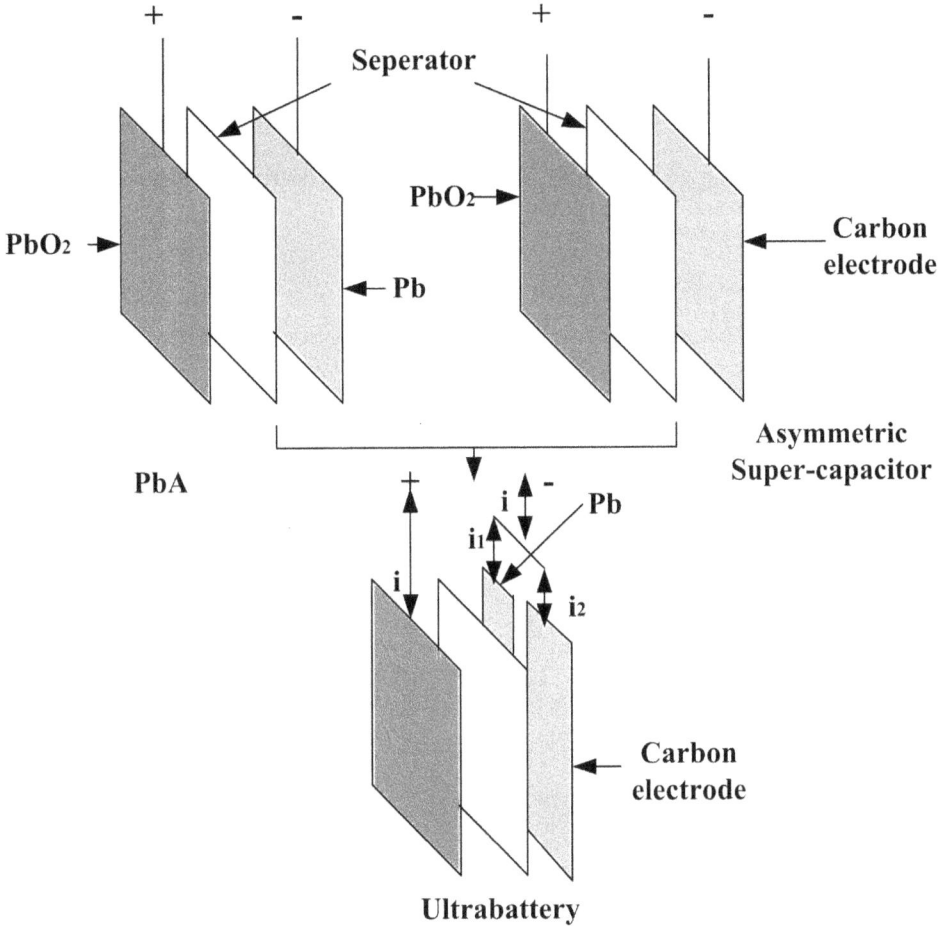

**FIGURE 12.8**  Ultra-battery formation using lead-acid battery and a suppercapacitor.

summation of absolute values of charge and discharge (in watt-hours) over that particular period. Some battery manufacturers also use the term throughput, which is the sum of all discharge watt-hours. Storage devices have a wide variety of applications in the grid system. These devices provide dispatchable support to the supply system for reliability purposes where the penetration of renewable energy is very high [21, 22]. It also helps the grid system in the case of intentional or unintentional islanding events. It also supports the external grid by adding some power to the system while producing it at the distribution level by utilizing renewable energy [23]. The inclusion of storage devices into the existing grid system also reduces the use of fossil fuels in the system, which in turn reduces land, soil, and air pollution [24].

## 12.5   LEAD-ACID BATTERY AGING MODEL

For a comprehensive energy system planning and operation utilizing a storage system, it is evident to estimate the life cycle of the battery system. Various models and approaches can be seen in the available literature to estimate the battery life cycle [25]. Although, [26] provides a more detailed and comprehensive lifetime model of a PbA incorporating the actual operating conditions. The capacity left of the storage battery (Cleft.bat) was modeled as given by equation (12.1)

$$C_{\text{left.bat}} = C_{d0} - C_{cor}(t) - C_{deg}(t) \tag{12.1}$$

where $C_{d0}$ is the normalized discharge capacity of the battery, $C_{cor}(t)$ is the decrement in the capacity of storage due to corrosion and $C_{deg}(t)$ is the decrement in the capacity of storage due to degradation.

It is also worthwhile to note that the corrosion and the battery degradation also affects the overall battery life and hence should be discussed.

### 12.5.1 Battery Corrosion

The total internal resistance of the cell increases with a new corrosion coating with a weaker conductivity. Therefore, corrosion directly impacts the storage cell resistance.

Also, as discussed earlier, the loss of active material of the cell proportional to the thickness of corrosion layer $\Delta W_c(t)$. Therefore, the loss in storage cell capacity can be correlated with the corrosion layer as given by equation (12.2).

$$C_{cor} = C_{cor,max} * \frac{\Delta W_c(t)}{\Delta W_{c,max}} \tag{12.2}$$

where $\Delta W_{c,max}$ represents the thickness of the corrosion layers when the storage cell acquires its maximum life time. It is also worthwhile to note that the corrosion layer $\Delta W_c(t)$ is a function of temperature and the voltage due to corrosion [26].

### 12.5.2 Storage Battery Degradation

This occurs due to continuous charging and discharging of the battery or in other terms, it relates to the loss in the storage system because of the number of cycles a battery completes during its lifetime. The standard operating conditions are used to obtain the nominal number of cycles of a storage battery. The storage cell life reduces with the real operating conditions such as actual SOC and the discharge current rates. Hence, the term $W_z(t)$, which represents the weighted cycles is used to obtain the storage loss because of degradation as given by equation (12.3):

$$C_{deg}(t) = C_{deg,max} * \exp\left(-C_z\left(1 - \left(W_z(t)/C_0\right)\right)\right) \tag{12.3}$$

Here $C_0$ represents the operational cycles given in the data sheet of a storage system, $C_z$ is a constant, $C_{deg,max}$ represents the storage system capacity before its replacement.

## 12.6 TYPES OF SUPERCAPACITORS

The supercapacitors are classified as per their charge storing mechanism. They are electrochemical double-layer capacitors (EDLCs), hybrid supercapacitors (HSCs) and pseudocapacitors (PCs). In Figure 12.9, the basic structure of the supercapacitor is shown, highlighting the basic components of a SC. An EDLC stores the energy in electrostatic form. It is generally of non-dielectric nature. As compared to EDLCs, the PCs has large current density and the capacitance. HSC is classified as an asymmetric class supercapacitor. It utilizes battery as an anode and supercapacitive material as a cathode.

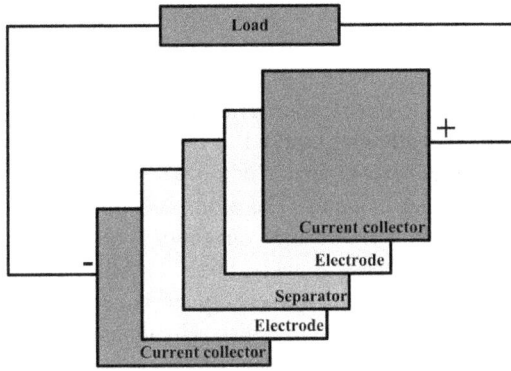

**FIGURE 12.9** Schematic structures of the supercapacitors.

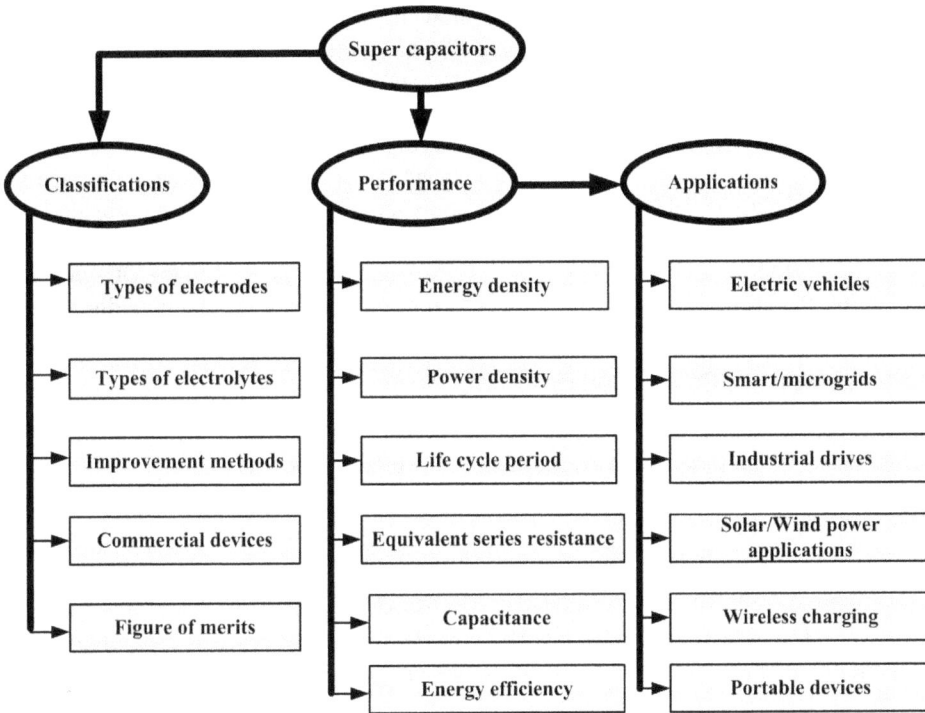

**FIGURE 12.10** Various important features of the supercapacitors.

### 12.6.1 SUPERCAPACITORS FOR SMART GRID APPLICATION

Figure 12.10 shows the various important features of supercapacitors incorporating their classification, performance, and applications. The classifications of the supercapacitors are based upon the type of electrode used, type of electrolyte, various methods of improvements, commercialization, and the figure of merit. Similarly, the performance of the supercapacitor is based upon its power and energy densities, life cycle period, equivalent series resistance, capacitance, and energy efficiency. The supercapacitor has various applications in the area of EVs, smart grid/microgrids, industrial drives, and renewable energy systems, such as solar and wind. It is also being utilized in wireless charging and as a portable storage device.

In a smart grid environment, the use of storage systems is versatile, which provides a non-variable as well as dispatchable power supply to the load under the penetration of renewable energy sources (RESs), such as solar PV or wind. These storage units, either as a battery system or as a supercapacitor, reduce the power fluctuations. Although, because of the low-power density of storage cells, a large number of cells are required to overcome the issues of power fluctuations. Overheating can also damage batteries because of continuous charging and discharging cycles with pulsed load and high peak load requirements. Therefore, in many applications, the supercapacitor finds a more suitable place as compared to the conventional faradic battery system. The supercapacitor has the capability to charge and discharge quickly, has a long life cycle, and has a swift dynamic response time. But at the same time, it has a low-energy storing capacity.

In many applications, a hybrid configuration is also reliably used, which consists of both PbA or other battery systems and a supercapacitor. It combines the advantages of both batteries as well as the supercapacitor. Some smart techniques are used to realize the power-sharing between the storage battery and the supercapacitor, such as the adaptive intelligence technique given in [27], a fuzzy and ANN-based technique presented in [28] and [29], respectively. The hybrid configuration can be realized by formulating an optimization problem while maximizing the storage system life. This can comprehensively increase the storage battery lifetime and also improves the dynamic response. However, a good power and energy management strategy is required for this hybrid system.

The power quality application can be significantly improved by replacing the battery storage system with a supercapacitor while providing the charging/ discharging in a very short time duration. Several strategies are provided to control and manage the supercapacitor energy storage system (SCESS) in hybrid electric vehicles (HEVs). The application of SCESS has been provided with the grid system incorporating SPV and wind energy sources. The existing SCESS uses the bidirectional DC-DC converter for the exchange of power between the supercapacitor and the DC-link bus. This system controls the DC-link voltage and the supercapacitor current while charging/discharging.

### 12.6.2  PERFORMANCE COMPARISON OF VARIOUS SUPERCAPACITORS

Various types of supercapacitors can be classified as per their performance, as summarized in Table 12.2. This comparison is based upon the mechanism of storage, types of electrolytes, and electrodes, specific power, operating temperature, specific energy, etc.

## 12.7  ROLE OF ENERGY STORAGE DEVICES IN SMART GRID TECHNOLOGIES

The vision of smart grid technologies involves several aspects of the energy supplying network. Some of the important functionalities of the smart grid are increasing the penetration of RESs

**TABLE 12.2**
**Performance Comparison of Supercapacitors**

| | Parameter | EDLCs | HSCs | PCs |
|---|---|---|---|---|
| 1 | Storage mechanism | Electrostatic | Faradic/non faradic | Faradic, reversible redox reaction |
| 2 | Electrolyte type | Protic or Aprotic | Aprotic | Protic |
| 3 | Electrode material | Carbon-based | Metal oxide/carbon-based | Conducting polymer/ metal oxide |
| 4 | Specific power (maximum) (W/Kg) | 1e4 | 5e3 | 4e3 |
| 5 | Operating temp. (Deg. C) | -40 to 65 | -40 to 65 | -40 to 65 |
| 6 | Cycle life | 1e6 | 5e5 | 1e5 |
| 7 | Specific energy (Wh/Kg) | 3–5 | 180 | 10 |

beyond 70% of the total generation, smart metering and realizing the concept of prosumers, i.e., both producer and consumer of energy, using green technology for both energy generation and storage, including the two tier energy pricing for base load and peak load, making each prosumer to participate in the energy market, etc. In all these features of smart grid technologies, there is a significant role of the energy storage devices to be played. Some of these roles are discussed as follows:

- **Maintaining the power balance using green technologies**: The energy storage devices like lithium-ion batteries, supercapacitors, and fuel cells are emission free and these help to absorb the surplus power output from the generation units and supply the same under power-deficient condition. Hence, helps to manage the power balance of the smart distribution grid. This shock absorbing characteristics of storage devices increases the flexibility and reliability of the grid [30].
- **Improves the system reliability and resilience**: Due to dynamic nature of the active distribution network, and add on effect of uncertainties of RESs, various short duration and long duration faults frequent disruption occurs. The purpose of keeping the energy storage devices is to assist as a backup power and help to restart the system. The restarting method is to use the storage device to provide uninterrupted supply to some of the critical loads and further use it to start one of the smallest generators and keep on doing that for higher rated generators. This technique can be initiated in a smart building and can be scaled up to the distribution grid at large. Hence, storage devices ensure uninterrupted power to consumers that increases both reliability and resilience [31].
- **Allows integration of diverse distributed energy resources**: In general, different types of RESs like solar PV, wind energy system, fuel cells, etc., which have different power generation characteristics throughout the day. Like the PV panels generates only during the day, and wind turbines produce power output whenever there is sufficient wind speed. Under such circumstances it is necessary to include the storage devices that act as a buffer between these different types of RESs and the active distribution network. For this purpose, several power electronics converters are used for interfacing the storage devices as well as various RESs. Hence, overall diverse characteristics of the smart grid is also fulfilled by storage units.
- **Making the residential and industrial users to be market participants by giving incentives**: The government is encouraging the consumers to become energy producers. For this purpose, some incentives are provided during installation of RESs and storage devices. Furthermore, by storing energy during off-peak load hours and supplying the same during peak load demand can help the prosumers to earn some monthly incentives as per their contribution. All these things leads to empowerment of the end residential and industrial participants. This will improve the service of the energy provider as a monopoly would not be encouraged.

## 12.8 FUTURE TRENDS AND RESEARCH CHALLENGES

The future trends in using the energy storage devices from the batteries and supercapacitor perspective can be mentioned as follows:

- **Future devices for energy storage are worked upon reducing the cost**: The cost of energy storage devices can be reduced by using low-cost membranes for batteries, innovative compounds and materials that are cheaper. All these areas need to be explored for obtaining new storage devices.
- **Grid interfacing as a standalone storage unit as well as a support for RESs**: The recent increased integration of solar PV and wind energy systems have increased the intermittency of the active distribution grid. Hence, more and more energy storage devices are needed to be installed to introduce increased flexibility to the grid, decrease the uncertainty and improve the reliability.

- **Increasing the capacity or ratings of the energy storage units for supporting the existing resources**: As the costs of energy storing components are coming down, their energy storage capacity or the ratings are increasing. The future trends is to further increase the capacity within a compact size of the device. All these researches leads to reduced cost and volume, as well as increased capacity.
- **Grid forming control technologies using proper power electronic converters**: Different storage units can be integrated to form and support a local grid microgrid with own loads, generation units (mostly RESs), and storage devices. Such control techniques are needed to be developed for improving the performance of the microgrid using the interfacing power electronic converters [32].
- **Hybridization or use of two or more storage devices in parallel**: Some applications need both high specific power and specific energy, hence, by combining two or more energy storages together. The merits of each one can be brought in together as well as compensating one another's demerits. For example, the hybridization of lead-acid batteries with supercapacitors can compensate for the issue of low specific power in battery and low specific energy in supercapacitors, hence, attaining both the characteristics of high specific power and high specific energy. Such combinations can be found in hybrid electric vehicles.
- **Energy management system of microgrids**: The storage devices can also be used to design the energy management system (EMS) that helps to maintain uninterrupted power supply for the critical loads under low generation or during off-grid mode of operation. Here, the storage devices are acting as a secondary energy source during power supplying mode and it also acts as a load while absorbing excess generated energy from the primary sources [33].

From the above discussion on the future trends in the research of energy storage devices, the following challenges can be identified for the same.

- **High cost of implementation in energy-related projects**: In the last one decade, there has been a tremendous reduction in the cost of storage elements like lead-acid and lithium-ion batteries, as well as supercapacitors. However, for rapid integration of RESs into the active distribution network, a faster acceptance of storage devices that are portable in nature are needed, which is possible if the consumers find the needed initial investment within their budget.
- **Lack of standardization in energy storage systems**: As proper standardization is not available, each manufacturer develops their own storage systems that ultimately creates compatibility issues if a project requires multiple energy sources and storage systems, which can cause replacement cost, scalability, and expansion issues. Also, if these issues are not addressed then use of different standard energy storage devices can lead to inefficient systems degrading the reliability of the overall system.
- **Outdated national level regulatory policy and market design**: The energy sector is observing tremendous change due to frequent up gradation in emerging technology. However, there has been very less development in the regulation and policy making that causes concerns among the end user. There has to have involvement of different players like the residential, industrial, and energy market participants and if proper standard procedures are not dictated by policy makers then it would be highly complex to operate the energy sector with multiple energy storage devices.
- **Modernization of the smart grid**: The increasing use of both batteries and supercapacitors towards developing smart grid features. These energy storage devices can be used to integrate intermittent RESs, and electric vehicles into the existing active distribution grid and implement various smart technologies to improve the reliability by using the storage devices as energy buffers.

# REFERENCES

1. S. Sen and V. Kumar, "Robust decentralized output-feedback type LQR voltage controller for inverter-based microgrid," 10th Int. Conf. on Computing, Communication and Networking Technologies (ICCCNT), pp. 1–6, July 2019.

2. S. Sen, and V. Kumar, "Microgrid control: a comprehensive survey," *Annual Reviews in Control*, vol. 45, pp. 118–151, 2018.

3. A. Azzollini *et al.*, "Lead-acid battery modeling over full state of charge and discharge range," *IEEE Transactions on Power Systems*, vol. 33, no. 6, pp. 6422–6429, Nov. 2018.

4. M. Greenleaf, O. Dalchand, H. Li, and J. P. Zheng, "A temperature-dependent study of sealed lead-acid batteries using physical equivalent circuit modeling with impedance spectra derived high current/power correction," *IEEE Transactions on Sustainable Energy*, vol. 6, no. 2, pp. 380–387, April 2015.

5. X. Han, L. Lu, Y. Zheng, X. Feng, Z. Li, J. Li, and M. Ouyang "A review on the key issues of the lithium ion battery degradation among the whole life cycle" *eTransportation*, vol 1, p. 100005, 2019.

6. X. Chen, A. Chu, and D. Li. Development of the cycling life model of Ni-MH power batteries for hybrid electric vehicles based on real-world operating conditions, *Journal of Energy Storage, vol. 34*, p. 101999, 2021.

7. M. Shiomi, T. Funato, K. Nakamura, and T. Takahashi, "Effects of carbon in negative plates on cycle-life performance ofvalve-regulated lead/acid batteries," *Journal of Power Sources*, vol. 64, no. 1–2, pp. 147–152, 1997.

8. B. B. McKeon, "Advanced lead–acid batteries and the development of grid-scale energy storage systems," *Proceedings of the IEEE,* vol. 102, no. 6, June 2014.

9. T. Hund, N. Clark, and W. Baca, "Ultrabattery test results for utility cycling applications," Presented at the 18th Int. *Seminar Double Layer Capacitors Hybrid Energy Storage Devices*, Deerfield Beach, FL, USA, Dec. 8–10, 2008.

10. E. Dickinson, "Axion PbC lead-carbon hybrid battery/supercapacitor," ELBC, Paris, France, Operational Stability of PbC Batteries and Battery Systems, 2012.

11. K. Kelley, "Lead-acid vs. lithium for the smart-grid: Balancing the Illinois Battery Research Facility efforts," Firefly International Energy Co., Peoria, IL, USA, 2012. [Online]. Available: http://firefly internationalenergy. com/wp-content/uploads/2013/07/Lead-Acid-vs-Lithium.pdf.

12. M. Terada, and H. Takabayas, "Industrial storage device for low-carbon society," *Hitachi Reviews*, vol. 60, no. 1, pp. 22–27, 2011. [Online]. Available: www.hitachi.com/rev/pdf/2011/r2011_01_104. pdf.

13. H. Yoshida, T. Mangahara, H. Noguchi, D. Kikuchi, W. Tezuka, M. Miura, and J. Furukawa, "Development of ultra long-life (6000 cycles) VRLA for deep-cycle service," Proceedings of the 13th Asian Battery Conference, Macau, China, 2009. [Online]. Available: www.conferenceworks. net.au/archive/13abc/extended-bstractsdownload.php.html.

14. D. A. J. Rand, P. T. Moseley, J. Garche, and C. D. Parker, Eds., Valve-Regulated Lead-Acid Batteries. Amsterdam: Elsevier, 2004, pp. 3–7.

15. T. Tsujikawa, T. Matsushima, K. Yabuta, and T. Matsushita, "Estimation of the lifetimes of valve-regulated lead-acid batteries," *Journal of Power Sources*, vol. 187, pp. 613–619, 2009.

16. ALABC, "Technology development. Duration: 15 months, continuation from Program 2007–2009," Status: Completed Rep. 0709 C1.2A, 2012.

17. L. T. Lam and R. Louey, "Development of ultraBattery for hybrid electric vehicle applications," *Journal of Power Sources*, vol. 158, pp. 1140–1148, 2006.

18. L. T. Lam, R. Louey, N. P. Haigh, O. V. Lim, D. G. Vella, C. G. Phyland, and L. H. Vu, "ALABC Project DP 1.1. Production and test of hybrid VRLA ultrabatteryTM designed specifically for high-rate partial-state-of-charge operation," *CSIRO Energy Technology, Investigation Rep. ET/IR967R, Apr.* 2007.

19. L. T. Lam, R. Louey, N. P. Haigh, O. V. Lim, D. G. Vella, C. G. Phyland, L. H. Vu, J. Furukawa, T. Takada, D. Monma, and T. Kano, "VRLA ultraBattery for high-rate partial-state-of-charge operation," *Journal of Power Sources*, vol. 174, pp. 16–29, 2007.

20. B. J. Kirby, "Frequency regulation basics and trends," U.S. Dept. Energy, ORNL/TM-2004/291, Dec. 2004.

21. M. Kumar and B. Tyagi, "An optimal multivariable constrained nonlinear (MVCNL) stochastic microgrid planning and operation problem with renewable penetration," *IEEE Systems Journal*, vol. 14, no. 3, pp. 4143–4154, 2020.
22. M. Kumar and B. Tyagi, "Multi-variable constrained nonlinear optimal planning and operation problem for isolated microgrids with stochasticity in wind, solar, and load demand data," *IET Generation, Transmission and Distribution*, vol. 14, no. 11, pp. 2181–2190, 2020.
23. S. Sen and V. Kumar, "Decentralized output-feedback based robust LQR V-f controller for PV-battery microgrid including generation uncertainties," *IEEE Systems Journal*, vol. 14, no. 3, pp. 4418–4429, Sept. 2020.
24. S. Sen, and V. Kumar, "Assessment of various MOR techniques on an inverter-based microgrid model," 14th IEEE India Council International Conference, pp. 1–6, 2017.
25. T. Hund and W. Baca, "Accelerated cycle-life testing on the cyclon lead-acid battery," Presented at the Electrical *Energy Storage Application and Technology Conference, San Francisco, CA, USA*, Oct. 17–19, 2005.
26. X. Li, D. Hui, and X. Lai, "Battery energy storage station (BESS)-based smoothing control of photo-voltaic (PV) and wind power generation fluctuations," *IEEE Transactions on Sustainable Energy*, vol. 4, no. 2, pp. 464–473, Apr. 2013.
27. H. Yin, W. Zhou, M. Li, C. Ma, and C. Zhao, "An adaptive fuzzy logic-based energy management strategy on battery/ultracapacitor hybrid electric vehicles," *IEEE Transactions on Transportation Electronics*, vol. 2, no. 3, pp. 300–311, Sept. 2016.
28. J. Shen and A. Khaligh, "A supervisory energy management control strategy in a battery/ultracapacitor hybrid energy storage system," *IEEE Transactions on Transportation Electrification*, vol. 1, no. 3, pp. 223–231, Oct. 2015.
29. J. Shen and A. Khaligh, "Optimization of sizing and battery cycle life in battery/ultracapacitor hybrid energy storage systems for electric vehicle applications," *IEEE Transactions on Industrial Informatics*, vol. 10, no. 4, pp. 2112–2121, Nov. 2014.
30. S. Sen, and V. Kumar, "Microgrid modelling: A comprehensive survey," *Annual Reviews in Control*, vol. 46, pp. 216–250, 2018.
31. S. Sen and V. Kumar, "Simplified modeling and HIL validation of solar PVs and storage based islanded microgrid with generation uncertainties," *IEEE Systems Journal*, vol. 14, no. 2, pp. 2653–2664, June 2020.
32. S. Sen an M. Kumar, "MPC based energy management system for grid-connected smart buildings with EVs," IEEE IAS Global Conference on Emerging Technologies, May 2022. (Accepted)
33. S. Sen and V. Kumar, "Distributed adaptive-MPC type optimal PMS for PV-battery based isolated microgrid," *IEEE Systems Journal*, April 2022.

# Index

*Note*: Page numbers in **bold** refer to tables and those in *italic* refer to figures.

For Product Safety Concerns and Information please contact our EU
representative GPSR@taylorandfrancis.com
Taylor & Francis Verlag GmbH, Kaufingerstraße 24, 80331 München, Germany

www.ingramcontent.com/pod-product-compliance
Lightning Source LLC
Chambersburg PA
CBHW080927220326
41598CB00034B/5704

* 9 7 8 1 0 3 2 3 7 5 0 9 0 *